Environmental Science and Engineering

Series Editors

The ultimate goal of this series is to contribute to the protection of our environment, which calls for both profound research and the ongoing development of solutions and measurements by experts in the field. Accordingly, the series promotes not only a deeper understanding of environmental processes and the evaluation of management strategies, but also design and technology aimed at improving environmental quality. Books focusing on the former are published in the subseries Environmental Science, those focusing on the latter in the subseries Environmental Engineering.

Dhani Arya · Naveen Chandra · Ravindra Kumar ·
Manju Lata Upadhayay · Arun Pratap Mishra
Editors

Climate Change Impact on Himalayan Biodiversity

Springer

Editors
Dhani Arya
Department of Botany
Soban Singh Jeena University
Almora, Uttarakhand, India

Naveen Chandra
Department of Botany
Soban Singh Jeena University
Almora, Uttarakhand, India

Ravindra Kumar
Department of Botany
Soban Singh Jeena University
Almora, Uttarakhand, India

Manju Lata Upadhayay
Department of Botany
Soban Singh Jeena University
Almora, Uttarakhand, India

Arun Pratap Mishra
Department of Forestry and Remote Sensing
Earthtree Enviro Pvt. Ltd.
Shillong, Meghalaya, India

ISSN 1863-5520 ISSN 1863-5539 (electronic)
Environmental Science and Engineering
ISBN 978-3-031-77148-4 ISBN 978-3-031-77149-1 (eBook)
https://doi.org/10.1007/978-3-031-77149-1

This Springer imprint is published by the registered company Springer Nature Switzerland AG
The registered company address is: Gewerbestrasse 11, 6330 Cham, Switzerland

About This Book

Climate Change Impact on Himalayan Biodiversity is an informative and profound book that explores the impact of climate change on the unique and delicate environment of the Himalayan region. The book meticulously examines the different plants and animals that live there and how they are affected by changing temperatures and weather patterns. The book showcases how species are changing their habits and distribution patterns, and what this means for the future of the Himalayan environment. The book also provides practical suggestions for how local communities and policymakers can work together to protect the region's biodiversity for the future.

Overall, Climate Change Impact on Himalayan Biodiversity is an excellent resource for researchers and academics who are interested in learning more about the environmental impacts of climate change on the Himalayan region. The book is written in a clear and accessible manner, making it easy for anyone to understand the profound implications of climate change on this unique and sensitive environment.

Climate change in the Himalayas manifests through various phenomena, including rising temperatures, altered precipitation patterns, melting glaciers, and shifting phenology. These changes disrupt the delicate balance of ecosystems, leading to habitat loss, species range shifts, altered species interactions, and increased susceptibility to invasive species and diseases. The book meticulously examines these impacts across different taxonomic groups and ecosystems, shedding light on the cascading effects on biodiversity and ecosystem services. Drawing on a wealth of scientific research and field observations, the book presents case studies highlighting the specific vulnerabilities of key species and habitats in the Himalayan region. From the impact of retreating glaciers on freshwater biodiversity to changes in flowering and fruiting patterns affecting pollinators and frugivores, each case study offers valuable insights into the complex dynamics of climate change and biodiversity loss.

In the face of these challenges, the book emphasizes the urgent need for robust mitigation and adaptation strategies to safeguard Himalayan biodiversity. It discusses ecosystem-based approaches, such as habitat restoration, protected area management, and sustainable land use practices, as well as policy interventions aimed at reducing greenhouse gas emissions and promoting climate resilience. Furthermore, it underscores the importance of community engagement and interdisciplinary collaboration in implementing effective conservation measures.

Climate Change Impact on Himalayan Biodiversity concludes by outlining future research directions and policy implications to address the looming biodiversity crisis in the Himalayas. It calls for enhanced monitoring and modeling efforts to better understand the complex interactions between climate, biodiversity, and human activities. Moreover, it advocates for transboundary cooperation and knowledge sharing to foster regional resilience and ensure the long-term sustainability of this critical ecosystem.

The book opens with an in-depth survey of medicinal plants in the Himalaya: Current knowledge and future directions. The analysis of scientific literature spanning from 1959 to 2022 reveals a significant increase in research output concerning Himalayan medicinal plants, particularly in ethnobotany, phytochemistry, and ecology. India leads in research contributions and funding, with a notable focus on the western Himalayan states and Nepal, while the eastern Himalayan regions remain comparatively understudied.

Moving forward, the book narrates the status of prehistoric forts (Garh/Kot) and explores possibilities for historical tourism in the Uttarakhand state, western Himalaya.

Shifting focus to Arunachal Pradesh addresses the comprehensive exploration of traditional medicinal plant knowledge in Arunachal Pradesh, Eastern Himalaya. It underscores the value of traditional ecological knowledge and the potential for sustainable healthcare practices rooted in local biodiversity.

The study investigates the influence of climate variability on threatened medicinal plants in the Niti Valley within the Nanda Devi Biosphere Reserve, employing Landsat-8 for NDVI analysis. Population assessment of medicinal plant species, namely *Allium stracheyi*, *Carum carvi*, *Dactylorhiza hatagirea*, *Picrorhiza kurroa*, and *Sinopodophyllum hexandrum* unveils varying densities and threats, with cultivated species facing pressure due to local consumption and trade, and those in specific habitats highly threatened by overexploitation.

The book proceeds with an understanding forest resource scope amidst climate change in the Himalayan region. This entails assessing biodiversity, carbon sequestration potential, and resilience of ecosystems. With rising temperatures and shifting precipitation patterns, proactive management strategies are essential to safeguard these vital resources and mitigate environmental impacts.

In the Himalayan region, Keystone insects' influences on ecosystem structure and their function were elucidated. The chapter underscores the need for research and conservation efforts to safeguard keystone insects and preserve the intricate relationships within these ecosystems.

Using interactive geo-informatics technique, spatio-temporal changes on Rajma crop and citrus fruit cultivation areas in Gori Ganga Watershed, Kumaun Himalaya, were analyzed. That reveals the highest affected villages by temporal changes on Rajma crop and Citrus fruit cultivation areas were observed after 2010, where the crop quality, size, and real local taste disappeared.

The narrative then comprehensively reviews the Conservation Challenges and Adaptation Strategies for Indian Himalayan Biodiversity in a Changing Climate. It emphasizes the imperative of collaborative action, robust policy frameworks, and interdisciplinary approaches to safeguard the invaluable natural and cultural heritage of the IHR and build resilience against ongoing environmental changes.

The book proceeds with status of threatened plants of Uttarakhand that concluded a total of 358 plant species under different threat categories within the state. Moreover, nearly half of the total threatened flora emerged out as endemic (40% bryophytes, 100% pteridophytes, and 40.3% angiosperms). Collectively, the angiosperms dominated the threatened list of total flora of Uttarakhand followed by gymnosperms, bryophytes, and fungi. This may be due to the less attention paid on lower flora to collect the information at different levels.

The exploration then shifts to Bangladesh, where diversity, ecology, and conservation issues of Avifauna in the Ganges Floodplain was assessed. Currently, illegal bird hunting, trade of wild birds, water pollution, agricultural nets, tourism, and existing fishing practices become major threats to birds. This study will help take future conservation initiatives for the conservation of birds in the study area.

In Sikkim, located in the Eastern Himalayas, there has been a shift in traditional cropping patterns, resulting in increased participation in alternative economic activities. This transition in livelihood practices among farmers is accompanied by a decline in agricultural output. The inherent fragility of the region, combined with the growing challenges induced by climate change, has placed significant stress on the livelihoods of the local population.

The investigation delves into the examination of vascular plant endemism in the Western Himalaya region of India, revealing a total of 227 exclusive taxa, encompassing 207 species, 5 subspecies, and 15 varieties, distributed across 115 genera and 43 families. Within the spectrum of life forms, herbaceous taxa were predominant (184), followed by shrubs (34), trees (4), climbers (1), and bamboo (1). This study underscores the imperative for further investigations focusing on population dynamics, conservation assessment, and the formulation of propagation protocols, including reintroduction strategies for endangered endemic species within their natural habitats.

The narrative delves into the repercussions of climate change on the rural areas of the Central Himalayas, examining its effects on local culture, traditions, and ways of life. It also explores the gender dynamics, livelihood challenges, and the phenomenon of out-migration driven by the declining state of agriculture in the research area. This paper sheds light on how developing communities grapple with the challenges posed by climate change.

The discussion focuses on the repercussions of climate change on biodiversity, ecosystem functionality, and services within the Himalayan landscape. It emphasizes the vital role of Himalayan ecosystems in delivering crucial provisioning services like freshwater supply, agricultural support, and forest resources. The narrative illustrates how climate warming disrupts hydrological patterns, impacts agricultural productivity, and alters the distribution of forest flora and fauna. Ultimately, the study underscores the interdependence of ecological systems in the Himalayas and stresses the pressing necessity for collaborative actions to mitigate the effects of climate warming.

To identification of high yield productivity areas, examine of impacts of climate change and analyses of production and productivity of millet crops in different districts of Uttarakhand. Finger, barnyard, and amaranth are major crops of Uttarakhand Himalaya in which finger has the highest area, production, and productivity from 2000 to 2022.

Proceeding further, the book delves into an assessment of the role played by bryophytes in providing various ecosystem services and their function akin to a "Green Brain". This section thoroughly explores the multifaceted contributions of bryophytes across different ecosystems, elucidating their significance in ecological processes and functions. Additionally, it delves into the concept of bryophytes acting as a "Green Brain", implying their ability to serve as vital components in regulating and maintaining ecosystem health and resilience.

In the state of Uttarakhand, the study delves into the examination of lichen species associations and their indicative significance across diverse forest types. It aims to elucidate the importance of lichens as bioindicators, providing valuable insights into the ecological health and conditions of different forest ecosystems. Additionally, the study explores how variations in lichen species composition and abundance can offer valuable clues about environmental conditions, such as air quality, humidity levels, and habitat suitability. Through this research, a deeper understanding of lichen ecology and its relevance to forest management and conservation efforts in Uttarakhand is sought.

The research delves into the effects of climate change on the Himalayan ecosystem, specifically focusing on its impact on traditional crops. This section thoroughly examines how changing climatic conditions are influencing the growth, yield, and distribution of traditional crops cultivated in the Himalayan region. Additionally, the study explores the adaptation strategies employed by local communities to cope with these changes, such as the introduction of resilient crop varieties, adjusting planting schedules, and implementing water management techniques. By understanding the intricate dynamics between climate change and traditional crop cultivation, the research aims to inform sustainable agricultural practices and resilience-building efforts in the Himalayan region.

An extensive investigation into the critical interplay between forest degradation and carbon sequestration on a global scale. The primary aim is to comprehensively assess the current scenario regarding these vital environmental processes.

In the final chapter, our study advocates for the widespread adoption of bamboo-based agroforestry as a transformative strategy for climate change mitigation and livelihood enhancement in the Indian Himalayan Region. By harnessing the ecological resilience of bamboo and the ingenuity of traditional agroforestry practices, we can foster a harmonious relationship between human activities and the natural environment, paving the way for a more resilient and sustainable future for generations to come.

As we traverse the chapters of this book, a poignant narrative unfolds—one underscored by urgency, resilience, and the intricate dance between nature and humanity in the fragile expanses of the Himalayan Region. Through these pages, a compelling tale emerges, transcending mere scientific discourse; it serves as a fervent rallying cry. This anthology passionately implores for action, casting a spotlight on the profound impact of climate change on the biodiversity of the Himalayas. Within these vulnerable landscapes, climate change casts a long shadow, threatening the very fabric of biodiversity that sustains both ecosystems and livelihoods. Each chapter illuminates the pressing need to safeguard our botanical heritage, recognizing it as a cornerstone of resilience in the face of environmental upheaval. This collection serves as a poignant reminder that the survival of Himalayan biodiversity is intricately linked to our collective future. It urges us to heed the call to conservation with a sense of urgency, for these plant legacies are not merely relics of the past but indispensable guides to navigating the challenges of tomorrow. In preserving our botanical heritage, we forge pathways to sustainable futures, where nature and people coexist in harmony, safeguarding the rich tapestry of life that defines the Himalayan landscape.

Dhani Arya
Naveen Chandra
Ravindra Kumar
Manju Lata Upadhayay
Arun Pratap Mishra

Introduction

The Himalayan mountain range is the highest system of mountains on Earth, extending for over 2,400 km. The region displays enormous variation in climatic conditions ranging from subtropical to boreal and features the longest bioclimatic gradient in the world. This means that vascular plants, ranging from 60 to 6,400 m above sea level, can be found across a diverse range of habitats throughout the Himalayas. The confluence of four zoogeographic dominions; the Palearctic, Saharo-Arabian, Sino-Japanese, and Oriental has resulted in the Himalayan region serving as a global biodiversity hotspot. The unique climatic variations, accompanied by an extraordinary range of habitats, have fostered the colonization of several fauna and flora from different territories. As a result, the region contains an exceptionally high level of biodiversity, which is of critical global importance.

The region is subject to both heavy monsoon rains and periodic earthquakes, making it prone to natural disasters. In the next few decades, the mountains of the world are predicted to continue warming at an alarming rate. During the last few decades, the Himalayas have experienced approximately two to three times greater atmospheric temperature warming than the global average. Biodiversity losses can lead to critical environmental crisis. About 9 million types of plants, animals, protists, and fungi, which inhabit the Earth have been influenced directly or indirectly by humans. However, a key question today is, how rapid climate warming intensifying the global biodiversity losses, and why the efforts to protect biodiversity have not been able to address properly. While the anthropogenic disturbances are thought to be the cause of habitat fragmentation and biodiversity losses worldwide, the decline of biodiversity in one of the world's least-disturbed habitats such as the Hindu Kush Himalayan (HKH) region has raised a question of the potential threat of climate warming in regional biodiversity during the twenty-first century. The HKH region hosts 4 of 36 global biodiversity hotspots, 60 of the 200 global ecoregions, and 330 important bird and biodiversity areas (Kotru et al. 2020). The region is also a major source of some of Asia's largest river systems and is regarded as the "global biodiversity hotspot." The large glacier-fed river systems including the Indus, Ganges, Brahmaputra, Salween, Yangtze, Yellow, and Mekong originated from the Himalayan mountains support a vast amount of ecosystem goods and services to more than one billion inhabitants residing in the downstream river basins. However, the Himalayas have been experiencing unprecedented warming at a rate greater than any time in the past.

Major Challenges

As global temperatures continue to rise, the Himalayan region is experiencing drastic changes in its climate patterns. This phenomenon has a direct impact on the biodiversity of the region, posing a significant threat to numerous plant and animal species that are specially adapted to the delicate mountain environment. The accelerated melting of glaciers and snowpacks disrupts the natural water flow, affecting the availability of freshwater for both wildlife and human populations. As a result, many species are facing altered habitats, reduced food sources, and increased competition for limited resources, leading to potential declines in population numbers and even extinction risks for some endemic species.

Moreover, the Himalayas serve as an important biodiversity hotspot, containing numerous endemic and endangered species that are not found anywhere else on Earth. Climate change exacerbates existing challenges related to habitat loss, deforestation, and human encroachment, further pushing these vulnerable species towards the brink of extinction. Additionally, the migration patterns of various wildlife species are shifting due to changing temperature and weather patterns, creating new ecological interactions and potential conflicts with human communities residing in the region. Conservation efforts in the Himalayas are facing mounting complexities as policymakers and stakeholders grapple with finding sustainable solutions that balance the needs of both wildlife and human populations while mitigating the adverse impacts of climate change on this ecologically sensitive area. The book delves into these challenges, offering insights into potential strategies for preserving the unique biodiversity of the Himalayan region in the face of an uncertain climate future.

Magnitude of Threats

(i) ***Rapid Glacier Retreat:*** The Himalayan region contains numerous glaciers that serve as a vital water source for millions of people and support diverse ecosystems. As temperatures rise, glaciers are melting at an accelerated pace, leading to reduced water availability during certain seasons. This affects the biodiversity of plants and animals that rely on specific water-dependent habitats, such as wetlands and freshwater ecosystems.

(ii) ***Altered Plant and Animal Distribution:*** Rising temperatures and changing precipitation patterns are forcing many plant and animal species to migrate to higher altitudes seeking suitable habitats. However, the availability of suitable habitats may be limited at higher elevations, leading to habitat fragmentation and potential loss of species unable to adapt or disperse.

(iii) ***Loss of Alpine Habitats:*** The Himalayan alpine regions are home to unique and specialized species adapted to harsh conditions. As temperatures increase, these sensitive ecosystems are at risk of being disrupted, leading to the loss of specialized species like snow leopards, Himalayan Tahr, and various alpine flora. Additionally, invasive species may move up to higher altitudes, further impacting native biodiversity.

(iv) ***Changing Phenology:*** Climate change is causing shifts in the timing of seasonal events like flowering, breeding, and migration in plants and animals. For example, early flowering of certain plants may lead to mismatches with pollinators, affecting both the plants and the pollinators that depend on them. Such disruptions can have cascading effects on food webs and ecosystem dynamics.

(v) ***Increased Frequency of Extreme Events:*** The Himalayan region is experiencing more frequent and intense extreme weather events, such as droughts, floods, and landslides, as a result of climate change. These events can lead to habitat destruction, soil erosion, and loss of vegetation, which can have devastating impacts on local biodiversity and the communities that depend on it for their livelihoods.

It is essential to address these threats through coordinated efforts, including conservation initiatives, sustainable land-use planning, and global climate action to mitigate the impacts of climate change on Himalayan biodiversity. Local communities, governments, and international organizations must work together to protect this unique and ecologically important region for future generations.

Existing Research Gap in Himalayan Biodiversity

(a) ***Fine-Scale Impact Assessment:*** Despite considerable research on the general impacts of climate change on global biodiversity, there remains a paucity of studies focusing on fine-scale assessments in the Himalayan region. The diverse topography, ranging from low-lying valleys to towering peaks, creates microclimatic variations, influencing the distribution and abundance of species. Identifying these microclimatic refugia and assessing their role in preserving biodiversity is essential to inform targeted conservation efforts. Investigating the local-scale effects of climate change will enable researchers to comprehend the vulnerabilities of specific species and their adaptive potential, ultimately aiding in crafting localized management plans.

(b) ***Community-Level Responses:*** While individual species' responses to climate change are often studied, there is a lack of research on understanding the complex interactions within Himalayan ecosystems and the cascading effects of climate change on biodiversity at the community level. Species interactions, such as predator-prey dynamics, competition, and mutualisms, play a crucial role in shaping ecosystem structure and function. By analyzing these interactions in light of changing climatic conditions, researchers can gain insights into potential disruptions in ecological networks, species assemblages, and ecosystem stability. Filling this research gap will provide a more comprehensive understanding of how climate change-induced alterations could affect the overall resilience of Himalayan biodiversity.

(c) ***Long-Term Monitoring and Predictive Modeling:*** Long-term ecological monitoring initiatives are limited in the Himalayas, hindering researchers' ability to track biodiversity changes accurately. Continuous data collection across various elevational gradients and over extended periods is crucial to discerning the trajectory of species' responses to climate change. Establishing comprehensive monitoring programs will enable researchers to identify trends, potential tipping points, and non-linear responses that may arise as climate change accelerates. Additionally, integrating data into predictive models will allow for more accurate projections of future biodiversity scenarios, aiding policymakers in devising effective conservation and adaptation strategies.

(d) ***Indigenous Knowledge Integration:*** The Himalayan region is home to diverse indigenous communities that have coexisted with its ecosystems for generations. Their traditional knowledge systems hold valuable insights into local biodiversity, climate patterns, and ecosystem dynamics. However, the integration of indigenous knowledge with scientific research on climate change impacts is often lacking. By bridging this gap and collaborating with local communities, researchers can enhance their understanding of climate change-induced biodiversity changes, identify culturally relevant adaptation strategies, and foster more inclusive conservation practices that are respectful of traditional ecological knowledge.

(e) ***Species-Specific Vulnerability:*** Identifying species-specific vulnerabilities to climate change in the Himalayan region will help prioritize conservation efforts and management strategies.

(f) ***Interactions with Invasive Species:*** Research is required to understand how climate change might facilitate the spread of invasive species in the Himalayas and its implications for native biodiversity.

(g) ***Phenological Changes:*** Studying phenological shifts in plant and animal species due to climate change can provide critical information about ecological disruptions in the Himalayan ecosystem.

(h) ***Ecosystem Services:*** Assessing the potential impacts of climate change on ecosystem services provided by the Himalayan biodiversity, such as water provisioning and carbon sequestration, is crucial for sustainable resource management.

(i) ***Community-Based Conservation Strategies:*** Investigating the role of local communities in mitigating the impact of climate change on Himalayan biodiversity can provide valuable insights into the effectiveness of community-based conservation initiatives.

(j) ***Policy and Governance:*** Analyzing existing policies and governance structures related to biodiversity conservation in the Himalayas and identifying gaps and opportunities for integrating climate change considerations is essential for effective conservation planning.

Addressing the research gaps in the context of climate change impacts on Himalayan biodiversity is essential to developing effective conservation and adaptation measures. Fine-scale impact assessments, community-level responses, long-term monitoring, and the integration of indigenous knowledge can collectively enhance our understanding of the intricate ecological dynamics in the Himalayan region. By prioritizing research efforts in these areas, we can better safeguard the unique biodiversity and ecological services that the Himalayas provide to millions of people and species that call this extraordinary region home.

Contents

About the Editors

Dr. Dhani Arya holds an M.Sc. in Botany from Kumaun University, Nainital, and a Ph.D. in Botany from HNB Garhwal University, Srinagar, with a focus on Ecology, Biodiversity Conservation, and Climate Change. Currently serving as an Associate Professor and Head and convener at Botany Department in Soban Singh Jeena University, Almora, Uttarakhand. He brings over 14 years of combined teaching and research expertise to his role. His scholarly contributions include the publication of 44 research papers, 3 book chapters in esteemed national and international journals, and 1 book to his credit. Dr. Arya has been instrumental in mentoring four Ph.D. scholars to successful completion under his guidance. Additionally, he has actively participated as a convener and co-convener in various national and international seminars, fostering academic discourse and collaboration within the scientific community.

Dr. Naveen Chandra is serving as a faculty in Department of Botany, Soban Singh Jeena University Campus, Almora. He did his Post graduation in 2016 and obtained Ph.D. degree in botany from Kumaun University, Nainital in 2022. Prior to joining the university, he served as a Project Fellow at the Uttarakhand Space Application Centre (USAC), Dehradun, for over six years and worked on different aspect of Himalayan biodiversity. His research pursuits encompass diverse domains within the botanical sciences, including plant taxonomy and ecology, vegetation ecology, invasion ecology, high-altitude physiology, medicinal plants, ecological modelling, and the application of Remote Sensing (RS) and Geographic Information System (GIS) in biodiversity and ecological research. His scholarly contributions include 35 research papers, 8 book chapters, 2 edited book, 2 book and 2 scientific reports. Notably, he currently serves as an editor for two books published by Springer Nature on different aspect of Himalayan ecosystem. His research activities involve comprehensive surveys of high-altitude regions in the western Himalaya, focusing on monitoring vegetation patterns and conducting long-term ecological monitoring initiatives.

Dr. Ravindra Kumar is working as a Faculty in the Department of Botany, Soban Singh Jeena University at Almora. He has published 15 research papers and 4 book chapters in different national and international journals. Notably, he currently serves as an editor for one book. He has a teaching experience of more than 6 years. His research interests include plant taxonomy, gymnosperm, high-altitude ecology, and ethnobotany.

Dr. Manju Lata Upadhayay is working as a Faculty at SSJ University, Almora, Uttarakhand. She has more than 15 years of teaching experience for the Department of Botany. She did her Ph.D. on the topic "Studies on leaf surface microfungi of Setaria italic Beauv. with reference to antagonism and disease development in Almora hills" and since then she has been involved in extensive research in the field of Plant Pathology and Ethnobotany. She has published more than 40 research papers in journals of national and international repute and has a patent and one book published in India. She has attended various workshops and seminars and also organized and conducted various webinars in her department. Her research projects have been funded by DST and CSIR, India. Her admiration for nature and efforts to restore the importance of indigenous practices are well mirrored in her research work.

Mr. Arun Pratap Mishra is a passionate environmental researcher with a masters degree in Environmental Science from Hemwati Nandan Bahuguna Garhwal (Central) University in Uttarakhand, India. His dedication to advancing ecological research and sustainability is evident in his extensive body of work, which includes 40 research articles, 15 book chapters, 3 conference papers, and 1 technical report. He also serves as the editor of four notable books that focus on Himalayan plant ecology, natural resource management, climate change in the Himalayas, and the Earth's carbon footprint, utilizing cutting edge techniques and cloud based solutions. With professional experience spanning esteemed institutions like the Uttarakhand Space Application Centre, Botanical Survey of India, Wildlife Institute of India, and RBased Services Private Limited, Mr. Mishra has developed a deep expertise in geospatial technologies. He has further solidified his credentials with a postgraduate diploma in Remote Sensing and GIS from the Indian Institute of Remote Sensing (IIRS, ISRO). Currently, Mr. Mishra holds the position of GIS Analyst at Earthtree Enviro Pvt. Ltd., Shillong, Meghalaya, where he focuses on utilizing innovative computational methods and geospatial analysis to advance ecological conservation and sustainable development. His goal is to address pressing global challenges such as climate change and carbon management through data-driven, practical solutions. Mr. Mishra is always open to collaboration, believing that interdisciplinary efforts are key to driving meaningful change. His work is driven by a passion for science and its potential to create sustainable, real-world impact.

Chapter 1
Research on Medicinal Plants in the Himalaya: Current Knowledge and Future Directions

Bhawana Dangwal, Dhani Arya, Indra D. Bhatt, and Suresh K. Rana

Abstract Medicinal plants have always remained a topic of interest to researchers and policymakers due to their importance in supplying raw materials for modern therapeutics. We assessed the current state of knowledge on Himalayan medicinal plants to understand the gaps and future research opportunities. We reviewed the scientific literature published during 1959–2022 to assess spatial, temporal, and thematic trends in the research on the Himalayan medicinal plants. The number of publications increased continuously nearly fivefold during the last decade in the Himalaya. Ethnobotany, phytochemistry, and ecology are the dominant fields of research. India ranks at the top in research contribution and funding for research on medicinal plants in the Himalaya. More research has been conducted in western Himalayan states and Nepal, whereas eastern Himalayan biodiverse regions are least studied for their medicinal plant wealth. The highest percentage of research has been conducted on themes like conservation, ethnobotany, antioxidants, traditional knowledge, essential oil, ethnomedicine, phytochemicals, etc. The scientific information on the Himalayan medicinal plants provides a powerful resource for pharmaceutical advancement, drug discovery, and economic upliftment of the indigenous communities in the region. Summarized information in this study can further be utilized for data-driven and well-informed policies for the conservation and sustainable utilization of the medicinal plant wealth of the Himalaya.

Keywords Ethnobotany · Himalaya · Medicinal plants · Pharmacology · Phytochemistry

B. Dangwal · I. D. Bhatt · S. K. Rana (✉)
G.B. Pant, National Institute of Himalayan Environment, Almora, Uttarakhand, India
e-mail: envsuresh09@gmail.com

B. Dangwal · D. Arya
Department of Botany, Soban Singh Jeena University, Almora, Uttarakhand, India

D. Arya et al. (eds.), *Climate Change Impact on Himalayan Biodiversity*, Environmental Science and Engineering, https://doi.org/10.1007/978-3-031-77149-1_1

1.1 Introduction

Humans have long relied on medicinal plants to cure ailments and support their livelihood. Medicinal plants play a critical role in providing healthcare solutions worldwide, especially in rural and remote communities where access to modern medicine is limited (Sofowora et al. 2013; Howes et al. 2020). According to the World Health Organization (WHO), 80% of the world's population relies on traditional medicine, and the majority of these remedies are derived from medicinal plants (WHO 2002). The global trade in medicinal plant products is estimated to be worth billions of dollars annually, with many countries relying on the export of medicinal plants as a significant source of income (Vasisht et al. 2016). Moreover, medicinal plants provide a rich source of compounds for the development of new drugs, with over 25% of modern drugs originating from natural products, including medicinal plants (Newman and Cragg 2012). It is estimated that over 50,000 plant species with medicinal properties are used in various medicine systems worldwide, with the Chinese Pharmacopoeia alone listing over 5700 plant-based traditional medicines (Schippmann et al. 2002; Chen et al. 2016). In recent years, the medicinal plant industry has gained significant momentum due to its potential to supply raw materials for modern therapeutics and to serve as precursor molecules for new drug synthesis (Atanasov et al. 2021). The role of medicinal plants in drug discovery is widely recognized, with a significant portion of modern medicines derived directly or indirectly from them (Pešić 2015). The global market for plant-derived drugs was valued at $1.2 billion in the 1990s (Lange 1998) and has increased to $25.6 billion in 2015 (Roosta et al. 2017), with an estimated annual growth of $5 trillion by 2050 (Booker et al. 2012; Roosta et al. 2017). Thus, medicinal plant industry is a promising sector with enormous economic growth potential.

Indigenous knowledge, particularly on the utilization of medicinal plants, when combined with the latest scientific advancements, has the potential to offer new holistic models of sustainable development that are economically viable, environmentally friendly, and socially acceptable (Magni 2017). Traditional herbal medicines are rooted in indigenous knowledge systems and have been used for generations to treat various ailments. However, the increased demand for plant-based products in the medical industry over the past few decades has led to overexploitation of wild habitats and put many species at risk of extinction (Chen et al. 2016; Gowthami et al. 2021). In recent years, there has been a substantial increase in interest in traditional systems of herbal medicine both in developed and developing countries (Newman and Cragg 2020; Theodoridis et al. 2023). This has brought attention to the importance of documenting traditional knowledge related to the use of medicinal plants, particularly in light of the growing market demand for these products and ongoing discussions about access-benefit sharing and bio-piracy (Reyes-García 2010; Ravi and Bharadvaja 2019; Imran et al. 2021).

India is known to have nearly 7,500 species of medicinal plants, with over 2,000 of them listed as threatened according to the IUCN Red List (Gowthami et al. 2021). The Indian Himalayan region alone houses approximately 1,800 of these medicinal plants (Samant et al. 1998). The use of Himalayan plants as medicine dates back to the ancient Rigveda texts, which were written more than 5,000 years ago (Jaiswal and Williams 2016; Ghooi 2022). Recent pharmaceutical research has characterized the important phytochemicals in many Himalayan plant species (Tariq et al. 2015; Saini et al. 2022). The Himalayan region boasts a diverse array of ethnomedicinal plants used to treat various diseases, though only a few have been scientifically validated (Amber et al. 2017). Unfortunately, traditional plant-based healthcare practices and their preservation in the wild are facing a decline due to factors such as lack of interest among the younger generation, over-extraction for economic purposes, and increasing preference for modern medicine (Raut et al. 2012; Nishteswar 2014).

The Himalaya is rich in biological diversity, boasting over 10,000 species of seed plants (Rana and Rawat 2017) and recognized as a global biodiversity hotspot (Myers et al. 2000). This unique mountain system offers unique topographical, ecological, climatic, and physiographical conditions that are home to many valuable medicinal plant species (Singh and Singh 1987; Kala 2006). However, there is limited information regarding the distribution, habitat suitability, demand and supply ratio, and the impacts of climate change on medicinal plants in the Himalaya, leading to a lack of effective monitoring and conservation strategies (Shrestha et al. 2022; Manish 2022). These medicinal plants are under threat from their limited distribution, over-exploitation, and high demand for trade (Ghimire et al. 2005). Human-induced habitat degradation and changing environmental conditions are exacerbating the situation, leading to a rapid decline in wild populations of these plants (Samant et al. 2011). In the Indian Himalayan Region, 70% of the medicinal plant species are over-harvested from their wild habitats, particularly in the sub-alpine and alpine regions (Dhar et al. 2000; Hopping et al. 2018).

Medicinal plant species in the Himalayan region play a crucial role in the rural economy by providing a variety of valuable products for food and medicine (Dhar et al. 2000; Kala 2005; Bhat et al. 2021). They are also significant sources of raw materials for both traditional and modern pharmaceutical industries (Negi et al. 2021). This importance of medicinal plants extends beyond economic benefits and includes social and cultural significance for the indigenous communities of the Himalayan region (Joshi et al. 2016). Research on various aspects of medicinal plants, including ecology, ethnobotany, phytochemistry, agro-techniques, and conservation, has been conducted in the Himalaya for the past six decades (Dhar et al. 2000; Kala et al. 2006). Despite this long history of research, there has yet to be a comprehensive review of the spatial, temporal, and thematic trends in research on Himalayan medicinal plants. The present study aims to fill this gap by conducting a systematic review of the published literature on Himalayan medicinal plants to (a) compile available literature, (b) understand the contributions to research over time and location, (c) identify key areas of research, gaps in knowledge, and their impact, and (d) evaluate the contributions of research institutions and funding agencies.

1.2 Materials and Methods

We used Web of Science (www.webofscience.com) and Scopus (www.scopus.com) citation databases on 24th September 2022 to search published literature in English language on medicinal plants of the Himalaya. We used the "AND" and "OR" Boolean operators to systematically identify peer-reviewed journal articles, book chapters, conference papers, reviews, notes and books on medicinal plants of the Himalaya. We searched the title, abstract, and keywords fields and used a combination of the following keywords in our search string on Scopus: (TITLE-ABS-KEY (Medicinal Plants OR Aromatic Plants) AND (Himalaya OR Himalayas OR Himalayan OR Ladakh OR Kashmir OR Jammu OR Himachal Pradesh OR Uttarakhand OR Uttaranchal OR Nepal OR Sikkim OR Darjeeling OR Bhutan OR Arunachal Pradesh)). The search string connects names of all prominent regions in the Himalaya with the keywords of our interest i.e., medicinal and aromatic plants. A similar strategy was also followed using the same search string on Web of Science. The search strategy yielded a total of 2,422 records on the Scopus and 1,249 records on the Web of Science database. All records from both databases were merged and screened for duplicates resulting in the removal of 1,107 records. The remaining 2,564 records were further screened for their relevance in terms of the subject area of the investigation i.e., medicinal plants, and spatial coverage i.e., the Himalayan region. A careful evaluation of the titles and abstracts of the records by first and lead author resulted in the removal of 627 irrelevant publications; thus a total of 1,937records were selected for the final investigation (Fig. 1.1). We followed the Preferred Reporting Items for Systematic Reviews (PRISMA) guidelines (Moher et al. 2009) for identification and inclusion of studies for the review.

From the final list of the records, we retrieved the following information from each publication: document type, journal/publisher name, affiliation, funding agencies, authors, and year of publication. We further classified all the publications into the following research disciplines, i.e., Ecology, Ethnobotany, Phytochemistry, Molecular biology, Tissue culture, Microbiology, and Cultivation. The first three disciplines were prominent fields covering 84% of the studies; thus, the remaining disciplines with few studies were merged under *Others* category in the analyses. In addition, all the records were checked for the geographical coverage of the study and were classified into 6 Indian Himalayan States/Union Territories, Nepal and Bhutan (Fig. 1.2). Regional studies on multiple states/countries were considered in all the covered regions for spatial coverage except studies on the whole Himalaya which were considered under a separate category. Based on the retrieved information, we conducted comprehensive bibliometric analyses of the research on medicinal plants of the Himalaya. These analyses include spatial research trends, temporal growth, the contribution from different research areas or disciplines, contribution from different nations, collaboration networks, funding countries and agencies, prominent keywords, top contributing organizations, journals, etc. We further analysed the word cloud of author's keywords and the co-occurrence of these keywords in the research on the Himalayan medicinal plant. All analyses were conducted using

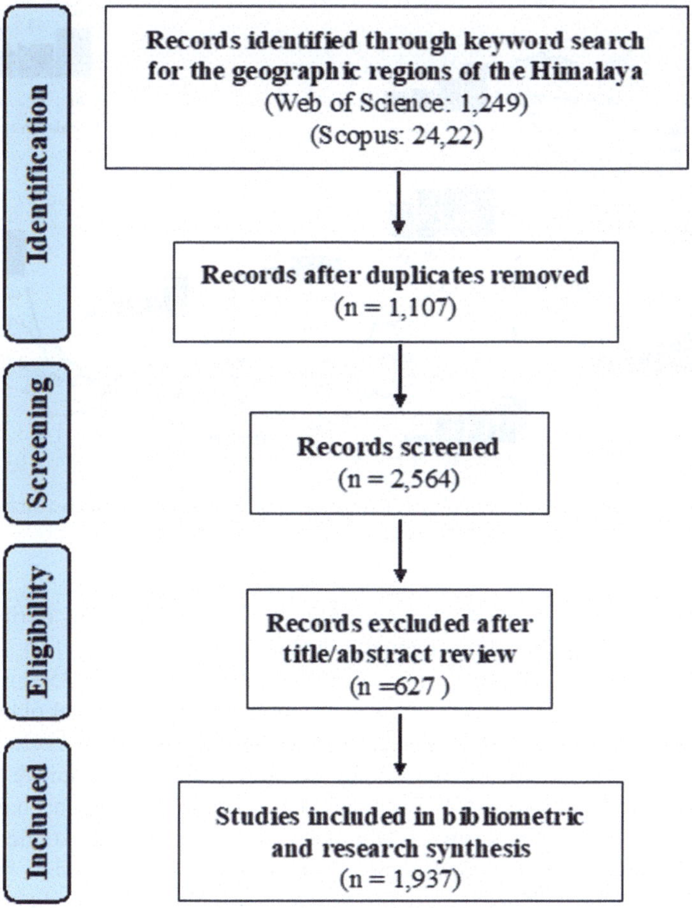

Fig. 1.1 PRISMA flow diagram for the systematic review on medicinal plants of the Himalaya included in the analyses

bibliometrix package (Aria and Cuccurullo 2017) and "ggplot2" (Wickham 2016) in R version 4.1.1 (R Core Team 2021).

1.3 Results

1.3.1 Spatial and Temporal Distribution of Studies

A total of 1,937 publications were analysed on medicinal plants of the Himalaya, including 1,894 journal articles, 34 books, and 9 conference papers published during 1959–2022. The medicinal plant research in the Himalaya shows continuous growth

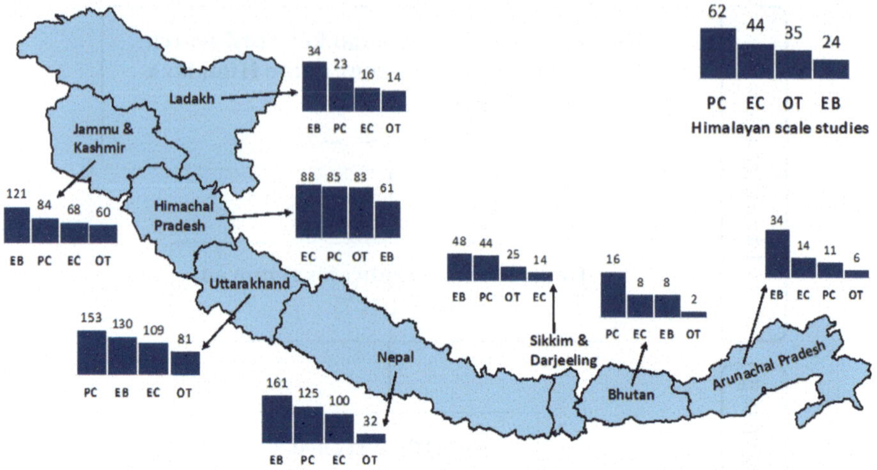

Fig. 1.2 Spatial distribution of the publications on medicinal plants of the Himalaya (abbreviations: EC: Ecology, EB: Ethnobotany, PC: Phytochemistry, OT: Other studies)

with 1–10 publications annually before 2000, up to 50 publications during 2001–2010, and up to 250 publications annually during recent years (Fig. 1.3). Only 143 studies were published till 2000, whereas during 2001–2010 nearly 350 studies and during 2011–2022 nearly 1,500 studies were published on medicinal plants of the Himalaya. In terms of spatial coverage, the maximum number of 473 studies have been conducted in Uttarakhand, followed by 418 in Nepal, 333 in Jammu & Kashmir, 317 in Himachal Pradesh (Fig. 1.2). Eastern Himalaya regions like Arunachal Pradesh and Bhutan and Ladakh in northwest Himalaya are among the least studied regions in terms of research on medicinal plants. Further, a total of 165 studies have been conducted on the whole Himalaya, which mostly cover reviews on medicinal plants, phytochemicals, or the ecology of different groups of plants across the Himalaya.

The majority of the medicinal plants research in the Himalaya has been conducted in the field of Ethnobotany (30.5%) followed by Phytochemistry (30.5), and Ecology (23%). Other less studied disciplines include Molecular biology (6.7%), Tissue culture (5.2%), Microbiology (2.7%), Cultivation (1.4%), etc. Nepal contributed the highest percentage of studies before 2000 in all research disciplines whereas in Uttarakhand, studies are distributed evenly throughout history (Fig. 1.4). Arunachal Pradesh, Bhutan, and Ladakh contributed less than 10% of the research studies in all disciplines across all the time periods (Fig. 1.4). Interestingly, number of studies on medicinal plants during the last decade have increased 3–5 times as compared to the previous decade in all research disciplines in the Himalaya (Fig. 1.4).

Fig. 1.3 Temporal growth in the published literature on medicinal plants of the Himalaya

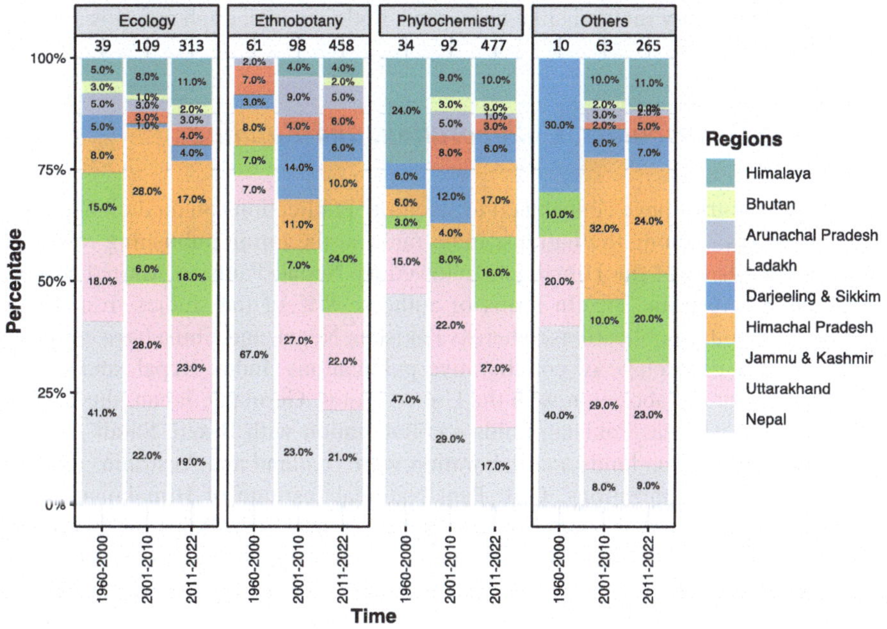

Fig. 1.4 Spatial and temporal contribution of medicinal plant research under different disciplines in the Himalaya. (Numbers on the top of the panel show the total number of studies published under the respective discipline during the period)

1.3.2 Keywords and Thematic Research

Prominent author's keywords used in the publications on medicinal plants of the Himalaya were conservation (10%), antioxidants (8%), ethnobotany (7%), traditional knowledge (5%), ethnomedicine (3%), essential oil (3%), phytochemicals (3%), indigenous knowledge (3%), traditional medicine (3%), antimicrobial (3%), genetic diversity (2%), micropropagation (2%) etc. (Fig. 1.5). Earlier studies on medicinal plants were mainly focused on traditional research topics like non-timber forest products, indigenous uses, herbal remedies, etc., whereas terms like phytocompounds, maxent modelling, phytochemistry, pharmacology, secondary metabolites, propagation, antioxidants, genetic diversity, etc. become more prominent in recent years. The co-occurrence network of keywords shows the association of ethnobotany, ecology, and conservation-related terms with medicinal plants of different regions across the Himalaya (Fig. 1.6). However, terms related to phytochemistry and pharmacology are prominently associated with the names of high value medicinal plant species and laboratory methods like hplc, rapd, hptlc, gc–mc, dpph etc. (Fig. 1.6).

1.3.3 Top Contributors, Collaboration, and Funding

In total, 78 countries have contributed to scientific publications on medicinal plants in the Himalaya. Among the nations, India ranks at the top in publishing research on medicinal plants of the Himalaya, followed by Nepal, Pakistan, United States, China, Japan, Australia, etc. In terms of authors, 95% of the studies from India were single-country publications, whereas Pakistan, Nepal, and China have comparatively a larger percentage of collaborative publications. India, Nepal, and China hold a prominent collaboration with the United States, Germany, Japan, the United Kingdom, etc., whereas Pakistan holds a collaboration with Brazil, Saudi Arabia, Egypt, etc., and Bhutan holds a collaboration with Thailand and Australia. Among various research organizations, G.B. Pant National Institute of Himalayan Environment, India ranks at the top with the highest number of studies on medicinal plants in the Himalaya followed by Tribhuvan University, Nepal, H.N.B. Garhwal University, India, Institute of Himalayan Bioresource Technology India, University of Kashmir, India, and Indian Institute of Integrative Medicine, Jammu, etc. (Fig. 1.7). A total of 160 agencies funded the research on medicinal plants in the Himalaya, among which the Department of Science and Technology, India, funded the highest number of studies followed by the Council of Scientific and Industrial Research, India (CSIR), University Grants Commission, India, Department of Biotechnology, India, etc. (Fig. 1.8).

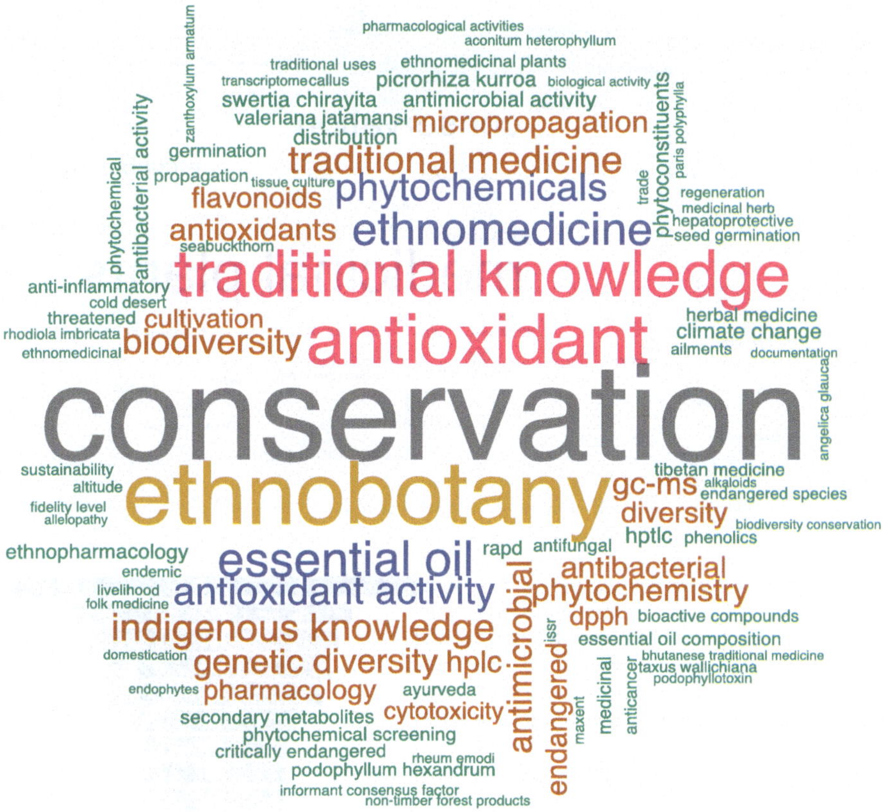

Fig. 1.5 A visualized word cloud of frequently used keywords in the research on medicinal plants in the Himalaya

1.4 Discussion

The current review on the medicinal plants of the Himalaya showed that (1) the majority of the research has been conducted on ethnobotany, phytochemistry and ecology of the medicinal plants (2) the last decade has witnessed 3–5 times increase in the number of published studies as compared to the previous decade (3) Maximum research has been conducted in Uttarakhand, Nepal and Jammu & Kashmir whereas Arunachal Pradesh and Bhutan in the east and Ladakh in the northwest Himalaya are least studied and (4) India ranks at the top both in the number of studies as well as funding for research on the medicinal plants in the Himalaya. Medicinal plants are a globally valuable source of herbal products, and their demand is continuously increasing due to growing interest in alternative healthcare means and systematic progression in the scientific validation of traditional medicines (Lubbe and Verpoorte 2011; Chen et al. 2016). Himalayan medicinal plants play an important role in traditional medicine systems like Ayurveda, Unani, Siddha, Chinese medicine, Tibetan

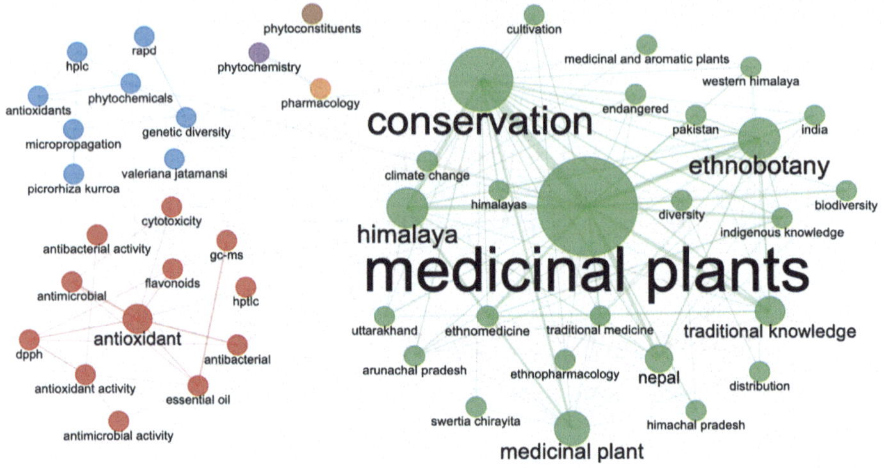

Fig. 1.6 Co-occurrence network of keywords in the research on medicinal plants in the Himalaya

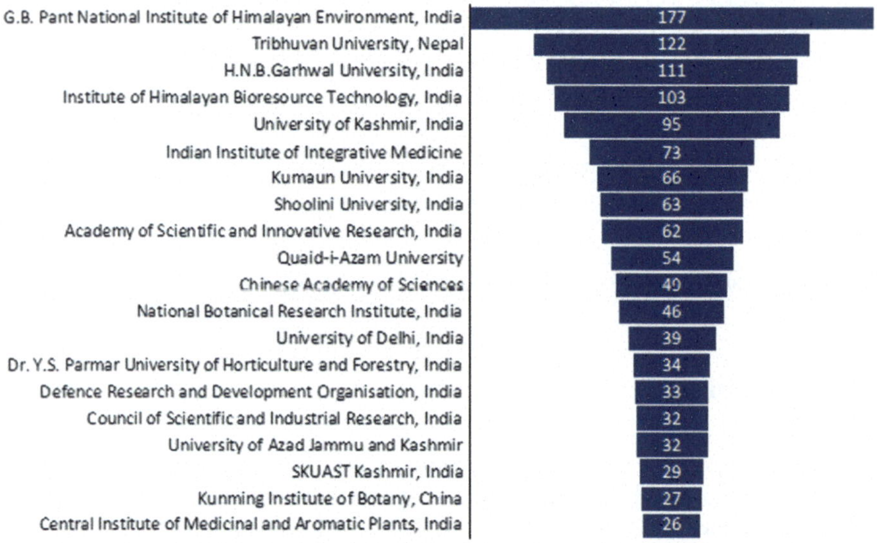

Fig. 1.7 Top organization publishing research on medicinal plants of the Himalaya

medicine, and other indigenous medicine systems (Arji et al. 2019; Pandey et al. 2022). Nearly 10% of rural households are involved in the commercial harvesting of medicinal plants in the Himalaya, with an annual trade of 7000 to 27,000 tons of herbal material corresponding to an export value of 7–30 million USD (Olsen 2005). Conservative estimates predict that about 36% of the volume and 51% of the

Fig. 1.8 Top funding agencies for research on medicinal plants of the Himalaya

value of the traded herbal medicines in the Himalaya is harvested through destructive measures resulting in permanent damage to the regeneration capability of the medicinal plants (Olsen 2005). The increasing demand for herbal products in modern medicine industries and traditional herbal medicine systems is a serious concern for the future sustainability of these valuable herbal bioresources in the Himalaya (Dhar et al. 2000; Manish 2022). Ensuring the sustainability of the Himalayan medicinal plants thus requires an integrative approach by implementing in-situ and ex-situ conservation strategies as well as resource management through sustainable agricultural practices for improvement in yield and potency of herbal bioresources (Hamilton 2004).

In terms of research, recent decades have witnessed an enormous increase in the number of publications. The decade of 2001–2010 witnessed a 60% increase in the number of publications from all research conducted on Himalayan medicinal plants before 2000, which witnessed a further increase of 76% during 2011–2022 as compared to the previous decade (Fig. 1.3). In the research on Himalayan medicinal plants, India stand out from the rest of the countries with more than 1500 publications followed by Nepal, Pakistan, United States and China. Globally, also among the 159 nations publishing research on medicinal plants, India ranks at the top, followed by China, the USA, and Brazil (Salmerón-Manzano et al. 2020). Initially, the research on Himalayan medicinal plants was more focused on themes like distribution (~50%), conservation status (~16%), Cultivation and propagation (~24%), with less than 5% research in the field of phytochemistry (Dhar et al. 2000). However, research in the field of phytochemistry has gained recent momentum with a nearly 74% increase in the number of publications during the last decade as compared to the number of publications during the previous 50 years (Fig. 1.3).

Phytochemistry research in India and the Himalayan region had increased substantially during recent decades (Joshi et al. 2016; Velmurugan and Radhakrishnan 2017). Indian Medicinal Plants, Phytochemistry, and Therapeutics (IMPPAT), a database on phytochemicals of Indian medicinal plants, reports 17,967 phytochemicals from 4,010 species (Mohanraj et al. 2018), many of which are distributed in the Himalayan region. A large number of studies have been conducted on the phytochemical composition of the high-value medicinal plants in the Himalayan region. For example, 606 species with antimicrobial and anti-inflammatory properties are used to cure skin diseases in the Himalaya (Heinrich et al. 2021). Reviews on Himalayan phytochemicals report 64 species with anticancer properties (Tariq et al. 2015), 116 species with essential oils (Joshi et al. 2016), 137 species with ethnomedicines, 59 of which have antimicrobial potential (Adnan et al. 2019) and 45 species with anticancer, anti-inflammatory, and antibacterial properties (Saini et al. 2022). Similarly, many wild edible species with nutraceutical and therapeutic values have also been reported from the Himalayan region (Bhatt et al. 2017; Semwal et al. 2022). Despite the increasing trend of phytochemistry research, very few species have been screened for their phytochemicals as compared to the total number of medicinal plant species reported from the Himalaya. Thus, the region still holds enormous potential for the discovery of plant-based drugs and nutraceuticals.

Although research on medicinal plants in the Himalayan has increased many folds during the last decade but its spatial distribution within the region is still uneven. Eastern Himalayan regions of Arunachal Pradesh and Bhutan, which constitutes biologically the most diverse regions (Price et al. 2014; Rana et al. 2019), contribute the least number of publications on medicinal plants. From the eastern Himalaya region of Arunachal Pradesh and Bhutan with a diversity of 5000–6000 seed plant species (Rana and Rawat 2017), only 500–600 species of medicinal plants have been reported (Wangchuk and Tobgay 2015; Wangpan and Tangjang 2020). In contrast, from the western Himalayan region with seed plant diversity of 2000–3000 species (Rana and Rawat 2017), 643 species from Himachal Pradesh, 1123 species from Jammu & Kashmir, and 1127 species from Uttarakhand have been documented for their medicinal uses (Samant et al. 2007; Kumar et al. 2018; Tali et al. 2019). This indicates that very few efforts have been made toward research and development for bioprospecting medicinal plants in the eastern Himalaya. Thus, with its rich cultural and plant diversity, the region need for intensive research and development efforts to realise its potential for discovering novel pharmaceuticals and nutraceuticals to address regional and global health and nutritional security.

Over the past decades, pharmaceutical companies have shown an increasing interest in exploring new compounds from plants worldwide and in the Himalaya (Katiyar et al. 2012; Atanasov et al. 2021). Many medicinal plants are harvested from their wild habitats to obtain useful substances in medicine and pharmacy. Extensive use and destructive harvesting of medicinal plants are posing a serious threat to population regulation in their wild habitats in different parts of India and the Himalaya (Dhar et al. 2000; Badola and Aitken 2003; Ved et al. 2003). Further, the lack of a rational approach in research on medicinal plants has often limited the success rate compared to the actual potential in drug discovery from plant-based compounds (Kinghorn and Seo 1996; Pešić 2015). The knowledge of medicinal plants is also vanishing rapidly as traditional healthcare systems disappear (Bhat et al. 2013). The traditional knowledge provides important clues for the discovery of important pharmaceuticals; thus, it is very important to document such knowledge for its usage towards human well-being in the future (Arji et al. 2019; Süntar 2020). Due to limited information on the medicinal use of the Himalayan plants, more pharmacological studies are needed to acquire their ethnomedicinal uses (Gaire and Subedi 2011). Similarly, for effective conservation planning, information on the population structure and regeneration rates of medicinal plants are very important in the Himalaya (Bhat et al. 2013).

Currently, the majority of the research on medicinal plants in the Himalaya is focused on ethnobotany and phytochemistry. However, aspects related to the conservation of medicinal plants, like population assessment in the wild habitats, habitat ecology, propagation and cultivation protocols, conservation education and advocacy, etc., are still lacking in the research on Himalayan plants (Bhattarai and Karki 2004; Kala 2006). Although *conservation* is the most frequent keyword in the research on the Himalayan plants (Fig. 1.5), it has often been used only to raise concerns. However, studies related to population and conservation assessment, agroforestry techniques for cultivation and ex-situ conservation, etc., are limited in number and are responsible for implementing medicinal plants conservation on the ground (Phondani et al. 2016; Negi et al. 2018). Most of the funding is focused on phytochemical screening and drug discovery from the medicinal plants, but very few studies have been commissioned for conservation-related aspects of the Himalayan medicinal plants (Kala et al. 2006). The medicinal plant propagation and cultivation sector, which is the key to ex-situ conservation, also provides an employment avenue for indigenous communities (Siwach et al. 2013). Thus, in addition to the documentation of the traditional knowledge (Negi et al. 2021) and phytochemical characterization of the medicinal plants (Natesh et al. 2021), rigorous research is required for populations assessment, regeneration potential and ex-situ cultivation of medicinal plants for their conservation, sustainable utilization and economic upliftment of the traditional communities in the Himalaya.

Traditional herbal medicine is well recognized in the Himalaya and plays an important role in healthcare and partial livelihood in rural areas of the region (Hamilton 2004; Kuniyal et al. 2015; Negi et al. 2018). Despite the great contribution of traditional health care, knowledge of herbal medicines is gradually being lost since this knowledge is mainly restricted to local and indigenous healers; there is no transfer of knowledge across generations, particularly younger generations who are no longer interested. Simultaneously, human-caused biodiversity losses are affecting medicinal plants and associated traditional knowledge too (Tali et al. 2019). Depletion and loss of medicinal plants diversity and associated traditional knowledge base may significantly impact human health and livelihood. Hence, protecting the genetic pool of this valuable wealth in nature is urgently required for long-term sustainability and making it available for future generations (Ezuruike and Prieto 2014; Negi et al. 2021). Further, documentation of indigenous knowledge for future use and benefit sharing mechanism is extremely important as socio-cultural practices contribute significantly towards the conservation of these threatened plants.

The sustainability of the Himalayan herbal bioresources faces challenges like lack of awareness among the primary producers, over-exploitation of the resources, and Illicit trade practices (Chandra 2020). In addition, lack of marketing infrastructure, information system, overlooked value chain, long supply chain, involvement of middlemen, policy complexity and ineffectiveness, etc., are other prominent challenges to sustainable utilization of herbal medicines in the Himalaya (Chandra 2020). Conservation management and planning strategies in the Himalaya shall include medicinal plants as an important component for the designation of new and expansion of existing protected areas and eco-sensitive zones (Shrestha et al. 2022; Manish 2022). Further, conservation reserves designated for the in-situ protection and natural regeneration of medicinal plants will be a strong tool for ensuring sustainability and future supply. Scientific efforts have been made to develop propagation protocols for the cultivation and ex-situ conservation of high-demand medicinal plants in the Himalaya (Phondani et al. 2016; Negi et al. 2018). Cultivation of high-value medicinal plants like *Aconitum heterophyllum, Picrorhiza kurroa, Saussurea costus, Valeriana jatamansi,* and some other important species is being promoted at a commercial scale to meet their growing market demand and ex-situ conservation (Chandra et al. 2006; Rawat et al. 2018; Kuniyal et al. 2019; Jugran et al. 2021). However, many Himalayan medicinal plant species with high market demand and pharmaceutical potential are less studied and under-utilized. Thus, the development of cultivation protocols for such species using biotechnological advancements, along with the promotion of Cultivation and awareness of indigenous communities on access-benefit sharing for medicinal plants, will help in sustainable utilization and conservation of the herbal bioresources of the Himalaya.

Although a large volume of published literature is available on the Himalayan medicinal plants, its comprehensive industrial usage, particularly in the field of drug discovery, is lacking. Himalayan plants are a rich source of pharmaceutical and nutraceutical compounds which, if discovered and used sustainably through ex-situ cultivation, will act as a major source of economy for the region's rural communities. Thus, there is a need for a comprehensive review and summarisation of existing knowledge on the Himalayan medicinal plants for its usage in policy and practice and for informed decision-making by the government, academic and industrial sector. A collaborative framework involving policymakers, academia, and industry, along with the involvement of the local entrepreneurs particularly focusing on the access-benefit sharing of the local communities, will be an important step towards sustainable economic utilization of these important bioresources of the Himalaya. Data-driven and well-informed policies for sustainable utilization of the medicinal plants will be of utmost importance for the conservation of the Himalayan medicinal plant wealth. Such activities and actions will address various components of the SDG-15: Life on Land including conservation of Mountain ecosystems, protection of biodiversity and natural habitats, protection of genetic resources and fair sharing of the benefits to indigenous communities, integrating biodiversity in governmental planning, increase in financial resources to conserve biodiversity and incentivize sustainable forest management.

Recommendations

1. A spatially and culturally explicit comprehensive review on traditional uses of the medicinal plants is important to understand the diversity and distribution of these bioresources in the Himalaya.
2. Documentation of reported phytochemical constituents particularly of medicinal values from all studied species will help in identification of bioresources containing important biomolecules for drug discovery.
3. Comparison of species usage in traditional medicines and constituent phytochemicals will help in identifying alternatives of high value and high demand medicinal plants.
4. Review of less studied but high value species for phytochemical characterisation will help in identification of new bioresources for industrial usage.
5. Preparation of Peoples Biodiversity Registers (PBR's) at village level which is an ongoing scheme of Government of India, will help in comprehensive documentation of medicinal plant wealth of the Himalaya.
6. A policy with well-defined mechanism for use of medicinal plant and Access Benefit Sharing (ABS) will help in sustainable utilization of these important bioresources.

Acknowledgements This research was part of BD's doctoral thesis which is supported by the Council of Scientific & Industrial Research, Government of India under the Senior Research Fellowship (SRF-Direct) programme (Award No: 132-1445-7447-2K23/1). The authors thank Director, G.B. Pant National Institute of Himalayan Environment (NIHE), Almora, Uttarakhand for providing facilities and support to conduct this research.

Conflict of Interest All authors declares that they have no conflict of interest to disclose.

References

Adnan M, Ali S, Sheikh K, Amber R (2019) Review on antibacterial activity of Himalayan medicinal plants traditionally used to treat pneumonia and tuberculosis. J Pharm Pharmacol 71(11):1599–1625. https://doi.org/10.1111/jphp.13156

Amber R, Adnan M, Tariq A, Mussarat S (2017) A review on antiviral activity of the Himalayan medicinal plants traditionally used to treat bronchitis and related symptoms. J Pharm Pharmacol 69(2):109–122. https://doi.org/10.1111/jphp.12669

Aria M, Cuccurullo C (2017) Bibliometrix: an R-tool for comprehensive science mapping analysis. J Informet 11(4):959–975. https://doi.org/10.1016/j.joi.2017.08.007

Arji G, Safdari R, Rezaeizadeh H, Abbassian A, Mokhtaran M, Hossein Ayati M (2019) A systematic literature review and classification of knowledge discovery in traditional medicine. Comput Methods Programs Biomed 168:39–57. https://doi.org/10.1016/j.cmpb.2018.10.017

Atanasov AG, Zotchev SB, Dirsch VM, Supuran CT (2021) Natural products in drug discovery: advances and opportunities. Nat Rev Drug Discov 20(3):200–216. https://doi.org/10.1038/s41573-020-00114-z

Badola HK, Aitken S (2003) The Himalayas of India: a treasury of medicinal plants under siege. Biodiversity 4(3):3–13. https://doi.org/10.1080/14888386.2003.9712694

Bhat MN, Singh B, Surmal O, Singh B, Shivgotra V, Musarella CM (2021) Ethnobotany of the Himalayas: safeguarding medical practices and traditional uses of Kashmir Regions. Biology 10(9):851. https://doi.org/10.3390/biology10090851

Bhat JA, Kumar M, Bussmann RW (2013) Ecological status and traditional knowledge of medicinal plants in Kedarnath Wildlife Sanctuary of Garhwal Himalaya, India. J Ethnobiol Ethnomed 9(1). https://doi.org/10.1186/1746-4269-9-1

Bhatt ID, Rawat S, Badhani A, Rawal RS (2017) Nutraceutical potential of selected wild edible fruits of the Indian Himalayan region. Food Chem 215:84–91. https://doi.org/10.1016/j.foodchem.2016.07.143

Bhattarai NK, Karki MB (2004) Conservation and management of Himalayan medicinal plants in Nepal. General Technical Reports of the US Department of Agriculture, Forest Service, vol 604, pp 45–50

Booker A, Johnston D, Heinrich M (2012) Value chains of herbal medicines—research needs and key challenges in the context of ethnopharmacology. J Ethnopharmacol 140(3):624–633. https://doi.org/10.1016/j.jep.2012.01.039

Chandra P (2020) The medicinal and aromatic plants business of Uttarakhand: a mini review of challenges and directions for future research. Nat Res Forum 44(3):274–285. https://doi.org/10.1111/1477-8947.12208

Chandra B, Palni LMS, Nandi SK (2006) Propagation and conservation of *Picrorhiza kurrooa* Royle ex Benth.: an endangered Himalayan medicinal herb of high commercial value. Biodivers Conserv 15(7):2325–2338. https://doi.org/10.1007/s10531-005-0770-z

Chen S-L, Yu H, Luo H-M, Wu Q, Li C-F, Steinmetz A (2016) Conservation and sustainable use of medicinal plants: problems, progress, and prospects. Chin Med 11:37. https://doi.org/10.1186/s13020-016-0108-7

Dhar U, Rawal RS, Upreti J (2000) Setting priorities for conservation of medicinal plants—a case study in the Indian Himalaya. Biol Cons 95(1):57–65. https://doi.org/10.1016/S0006-3207(00)00010-0

Ezuruike UF, Prieto JM (2014) The use of plants in the traditional management of diabetes in Nigeria: pharmacological and toxicological considerations. J Ethnopharmacol 155(2):857–924. https://doi.org/10.1016/j.jep.2014.05.055

Gaire BP, Subedi L (2011) Medicinal plant diversity and their pharmacological aspects of Nepal Himalayas. Pharmacognosy J 3(25):6–17. https://doi.org/10.5530/pj.2011.25.2

Ghimire SK, McKey D, Aumeeruddy-Thomas Y (2005) Conservation of Himalayan medicinal plants: harvesting patterns and ecology of two threatened species, *Nardostachys grandiflora* DC and *Neopicrorhiza scrophulariiflora* (Pennell) Hong. Biol Conserv 124(4):463–475. https://doi.org/10.1016/j.biocon.2005.02.005

Ghooi RB (2022) Ayurveda research-are we on the right track? J Ayurveda Integr Med 13(3):100564. https://doi.org/10.1016/j.jaim.2022.100564

Gowthami R, Sharma N, Pandey R, Agrawal A (2021) Status and consolidated list of threatened medicinal plants of India. Genet Resour Crop Evol 68(6):2235–2263. https://doi.org/10.1007/s10722-021-01199-0

Hamilton AC (2004) Medicinal plants, conservation and livelihoods. Biodivers Conserv 13(8):1477–1517. https://doi.org/10.1023/B:BIOC.0000021333.23413.42

Heinrich M, Jiang H, Scotti F, Booker A, Walt H, Weckerle C, Maake C (2021) Medicinal plants from the Himalayan region for potential novel antimicrobial and anti-inflammatory skin treatments. J Pharm Pharmacol 73(7):956–967. https://doi.org/10.1093/jpp/rgab039

Hopping KA, Chignell SM, Lambin EF (2018) The demise of caterpillar fungus in the Himalayan region due to climate change and overharvesting. Proc Natl Acad Sci 115(45):11489–11494. https://doi.org/10.1073/pnas.1811591115

Howes M-JR, Quave CL, Collemare J, Tatsis EC, Twilley D, Lulekal E, Farlow A, Li L, Cazar M-E, Leaman DJ, Prescott TAK, Milliken W, Martin C, Canha MND, Lall N, Qin H, Walker BE, Vásquez-Londoño C, Allkin B, Rivers M, Simmonds MSJ, Bell E, Battison A, Felix J, Forest F, Leon C, Williams C, Lughadha EN (2020) Molecules from nature: Reconciling biodiversity

conservation and global healthcare imperatives for sustainable use of medicinal plants and fungi. Plants, People, Planet 2(5):463–481. https://doi.org/10.1002/ppp3.10138
Imran Y, Wijekoon N, Gonawala L, Chiang Y-C, De Silva KRD (2021) Biopiracy: abolish corporate hijacking of indigenous medicinal entities. Sci World J 2021:e8898842. https://doi.org/10.1155/2021/8898842
Jaiswal YS, Williams LL (2016) A glimpse of Ayurveda—the forgotten history and principles of Indian traditional medicine. J Tradit Complement Med 7(1):50–53. https://doi.org/10.1016/j.jtcme.2016.02.002
Joshi RK, Satyal P, Setzer WN (2016) Himalayan aromatic medicinal plants: a review of their ethnopharmacology, volatile phytochemistry, and biological activities. Medicines (Basel) 3(1):6. https://doi.org/10.3390/medicines3010006
Jugran AK, Bhatt ID, Rawal RS (2021) Cultivation and utilization of *Valeriana jatamansi* jones for conservation planning and management. In: Ekiert HM, Ramawat KG, Arora J (eds) Medicinal plants: domestication, biotechnology and regional importance. Springer International Publishing, Cham, pp 113–178
Kala CP (2005) Indigenous uses, population density, and conservation of threatened medicinal plants in protected areas of the Indian Himalayas. Conserv Biol 19(2):368–378
Kala CP (2006) Problems and prospects in the conservation and development of the Himalayan medicinal plants sector. Int J Sustain Dev 9(4):370–389. https://doi.org/10.1504/IJSD.2006.014221
Kala CP, Dhyani PP, Sajwan BS (2006) Developing the medicinal plants sector in northern India: challenges and opportunities. J Ethnobiol Ethnomed 2:32. https://doi.org/10.1186/1746-4269-2-32
Katiyar C, Gupta A, Kanjilal S, Katiyar S (2012) Drug discovery from plant sources: an integrated approach. Ayurveda 33(1):10–19. https://doi.org/10.4103/0974-8520.100295
Kinghorn AD, Seo E-K (1996) Plants as sources of drugs. in: agricultural materials as renewable resources. Am Chem Soc 179–193
Kumar A, Kumar R, Sharma M, Kumar U, Gajula MNVP, Singh KP (2018) Uttarakhand medicinal plants database (UMPDB): a platform for exploring genomic, chemical, and traditional knowledge. Data 3(1):7. https://doi.org/10.3390/data3010007
Kuniyal CP, Bisht VK, Negi JS, Bhatt VP, Bisht DS, Butola JS, Sundriyal RC, Singh SK (2015) Progress and prospect in the integrated development of medicinal and aromatic plants (MAPs) sector in Uttarakhand, Western Himalaya. Environ Dev Sustain 17(5):1141–1162. https://doi.org/10.1007/s10668-014-9595-9
Kuniyal CP, Heinen JT, Negi BS, Kaim JC (2019) Is cultivation of *Saussurea costus* (Asterales: Asteraceae) sustaining its conservation? J Threatened Taxa 11(13):14745–14752. https://doi.org/10.11609/jott.3581.11.13.14745-14752
Lange D (1998) Europe's medicinal and aromatic plants: their use, trade and conservation. Europe's medicinal and aromatic plants: their use, trade and conservation
Lubbe A, Verpoorte R (2011) Cultivation of medicinal and aromatic plants for specialty industrial materials. Ind Crops Prod 34(1):785–801. https://doi.org/10.1016/j.indcrop.2011.01.019
Magni G (2017) Indigenous knowledge and implications for the sustainable development agenda. Eur J Educ 52(4):437–447. https://doi.org/10.1111/ejed.12238
Manish K (2022) Medicinal plants in peril due to climate change in the Himalaya. Eco Inform 68:101546. https://doi.org/10.1016/j.ecoinf.2021.101546
Mohanraj K, Karthikeyan BS, Vivek-Ananth RP, Chand RPB, Aparna SR, Mangalapandi P, Samal A (2018) IMPPAT: a curated database of Indian medicinal plants, phytochemistry and therapeutics. Sci Rep 8(1):4329. https://doi.org/10.1038/s41598-018-22631-z
Moher D, Liberati A, Tetzlaff J, Altman DG, Group TP (2009) Preferred reporting items for systematic reviews and meta-analyses: the PRISMA statement. PLOS Med 6(7):e1000097. https://doi.org/10.1371/journal.pmed.1000097
Myers N, Mittermeier RA, Mittermeier CG, Da Fonseca GA, Kent J (2000) Biodiversity hotspots for conservation priorities. Nature 403(6772):853–858. https://doi.org/10.1038/35002501

Natesh J, Mondal P, Kaur B, Abdul Salam AA, Kasilingam S, Meeran SM (2021) Promising phytochemicals of traditional Himalayan medicinal plants against putative replication and transmission targets of SARS-CoV-2 by computational investigation. Comput Biol Med 133:104383. https://doi.org/10.1016/j.compbiomed.2021.104383

Negi VS, Kewlani P, Pathak R, Bhatt D, Bhatt ID, Rawal RS, Sundriyal RC, Nandi SK (2018) Criteria and indicators for promoting cultivation and conservation of medicinal and aromatic plants in Western Himalaya, India. Ecol Ind 93:434–446. https://doi.org/10.1016/j.ecolind.2018.03.032

Negi VS, Pathak R, Thakur S, Joshi RK, Bhatt ID, Rawal RS (2021) Scoping the need of mainstreaming indigenous knowledge for sustainable use of bioresources in the Indian Himalayan Region. Environ Manage. https://doi.org/10.1007/s00267-021-01510-w

Newman DJ, Cragg GM (2012) Natural products as sources of new drugs over the 30 years from 1981 to 2010. J Nat Prod 75(3):311–335. https://doi.org/10.1021/np200906s

Newman DJ, Cragg GM (2020) Natural products as sources of new drugs over the nearly four decades from 01/1981 to 09/2019. J Nat Prod 83(3):770–803. https://doi.org/10.1021/acs.jnatprod.9b01285

Nishteswar K (2014) Depleting medicinal plant resources: a threat for survival of Ayurveda. Ayurveds 35(4):349–350. https://doi.org/10.4103/0974-8520.158972

Olsen CS (2005) Valuation of commercial central Himalayan medicinal plants. Ambio 34(8):607–610

Pandey J, Singh L, Saxena G, Devkota HP (2022) Environmental challenges for himalayan medicinal plants. In: Aftab T (ed) Environmental challenges and medicinal plants: sustainable production solutions under adverse conditions. Springer International Publishing, Cham, pp 29–47

Pešić M (2015) Development of natural product drugs in a sustainable manner. Brief for united nations global sustainable development report 2013, pp 1–4

Phondani PC, Bhatt ID, Negi VS, Kothyari BP, Bhatt A, Maikhuri RK (2016) Promoting medicinal plants cultivation as a tool for biodiversity conservation and livelihood enhancement in Indian Himalaya. J Asia-Pac Biodivers 9(1):39–46. https://doi.org/10.1016/j.japb.2015.12.001

Price TD, Hooper DM, Buchanan CD, Johansson US, Tietze DT, Alström P, Olsson U, Ghosh-Harihar M, Ishtiaq F, Gupta SK, Martens J, Harr B, Singh P, Mohan D (2014) Niche filling slows the diversification of Himalayan songbirds. Nature 509(7499):222–225. https://doi.org/10.1038/nature13272

R Core Team (2021) R: a language and environment for statistical computing. R Foundation for Statistical Computing, Vienna, Austria

Rana S, Rawat G (2017) Database of Himalayan plants based on published floras during a century. Data 2(4):36. https://doi.org/10.3390/data2040036

Rana SK, Price TD, Qian H (2019) Plant species richness across the Himalaya driven by evolutionary history and current climate. Ecosphere 10(11):e02945. https://doi.org/10.1002/ecs2.2945

Raut N, Adhikari B, Rawat G (2012) Depletion of traditional knowledge of medicinal plants. Curr Sci 103:873–874

Ravi S, Bharadvaja N (2019) Market analysis of medicinal plants in India. Curr Pharm Biotechnol 20(14):1172–1180

Rawat S, Jugran AK, Bhatt ID, Rawal RS (2018) Hedychium spicatum: a systematic review on traditional uses, phytochemistry, pharmacology and future prospectus. J Pharm Pharmacol 70(6):687–712. https://doi.org/10.1111/jphp.12890

Reyes-García V (2010) The relevance of traditional knowledge systems for ethnopharmacological research: theoretical and methodological contributions. J Ethnobiol Ethnomed 6(1):32. https://doi.org/10.1186/1746-4269-6-32

Roosta RA, Moghaddasi R, Hosseini SS (2017) Export target markets of medicinal and aromatic plants. J Appl Res Med Aromatic Plants 7:84–88. https://doi.org/10.1016/j.jarmap.2017.06.003

Saini N, Lather V, Gahlawat SK (2022) Exploring phytochemicals from Himalayan medicinal plants as novel therapeutic agents. Anticancer Agents Med Chem 22(9):1674–1698. https://doi.org/10.2174/1871520621666211015141020

Salmerón-Manzano E, Garrido-Cardenas JA, Manzano-Agugliaro F (2020) Worldwide research trends on medicinal plants. Int J Environ Res Public Health 17(10):3376. https://doi.org/10.3390/ijerph17103376

Samant PS, Singh M, Lal M, Singh A, Sharma A, Bhandari S (2007) Medicinal plants in Himachal Pradesh, north western Himalaya, India. Int J Biodivers Sci Manag 3(4):234–251. https://doi.org/10.1080/17451590709618177

Samant SS, Vidyarthi S, Pant S, Sharma P, Marpa S, Sharma P (2011) Diversity, distribution, indigenous uses and conservation of the medicinal plants of Indian Himalayan region used in cancer. J Biodivers 2(2):117–125. https://doi.org/10.1080/09766901.2011.11884732

Samant SS, Dhar U, Palni LMS, G.B. Pant Institute of Himalayan Environment & Development (1998) Medicinal plants of Indian Himalaya: diversity, distribution, potential values. Gyanodaya Prakashan, Nainital

Schippmann U, Leaman D, Cunningham A (2002) Impact of cultivation and gathering of medicinal plants on biodiversity: global trends and issues. In: Biodiversity and the ecosystem approach in agriculture, forestry and fisheries, pp 142–167

Semwal P, Painuli S, Jamloki A, Rauf A, Rahman MdM, Olatunde A, Hemeg HA, Abu-Izneid T, Naz S, Punia Bangar S, Lorenzo JM, Simal-Gandara J (2022) Himalayan wild fruits as a strong source of nutraceuticals, therapeutics, food and nutrition security. Food Rev Int 1–37. https://doi.org/10.1080/87559129.2022.2121407

Shrestha UB, Lamsal P, Ghimire SK, Shrestha BB, Dhakal S, Shrestha S, Atreya K (2022) Climate change-induced distributional change of medicinal and aromatic plants in the Nepal Himalaya. Ecol Evol 12(8):e9204. https://doi.org/10.1002/ece3.9204

Singh JS, Singh SP (1987) Forest vegetation of the Himalaya. Bot Rev 53(3):80–192

Siwach M, Siwach P, Solanki P, Gill A (2013) Biodiversity conservation of Himalayan medicinal plants in India: a retrospective analysis for a better vision. https://doi.org/10.5897/IJBC2013.0580

Sofowora A, Ogunbodede E, Onayade A (2013) The role and place of medicinal plants in the strategies for disease prevention. Afr J Tradit Complement Altern Med 10(5):210–229

Süntar I (2020) Importance of ethnopharmacological studies in drug discovery: role of medicinal plants. Phytochem Rev 19(5):1199–1209. https://doi.org/10.1007/s11101-019-09629-9

Tali BA, Khuroo AA, Ganie AH, Nawchoo IA (2019) Diversity, distribution and traditional uses of medicinal plants in Jammu and Kashmir (J&K) state of Indian Himalayas. J Herbal Med 17–18:100280. https://doi.org/10.1016/j.hermed.2019.100280

Tariq A, Mussarat S, Adnan M (2015) Review on ethnomedicinal, phytochemical and pharmacological evidence of Himalayan anticancer plants. J Ethnopharmacol 164:96–119. https://doi.org/10.1016/j.jep.2015.02.003

Theodoridis S, Drakou EG, Hickler T, Thines M, Nogues-Bravo D (2023) Evaluating natural medicinal resources and their exposure to global change. Lancet Planet Health 7(2):e155–e163. https://doi.org/10.1016/S2542-5196(22)00317-5

Vasisht K, Sharma N, Karan M (2016) Current perspective in the international trade of medicinal plants material: an update. Curr Pharm des 22(27):4288–4336

Ved D, Kinhal G, Kaliamoorthy R, V P, Ghate, Raman V, JH I (2003) Conservation assessment & management prioritisation for the medicinal plants of Jammu & Kashmir, Himachal Pradesh & Uttaranchal

Velmurugan C, Radhakrishnan N (2017) Phytochemistry research in India: a scientometric profile. Int J Inform Sci Manage (IJISM) 15(2)

Wangchuk P, Tobgay T (2015) Contributions of medicinal plants to the gross national happiness and biodiscovery in Bhutan. J Ethnobiol Ethnomed 11(1):48. https://doi.org/10.1186/s13002-015-0035-1

Wangpan T, Tangjang S (2020) Status of medicinal plants in context of Arunachal Pradesh. In: Shukla V, Kumar N (eds) Environmental concerns and sustainable development: Volume 2: biodiversity, soil and waste management. Springer, Singapore, pp 63–80
WHO (2002) WHO traditional medicine strategy 2002–2005. World Health Organization
Wickham H (2016) Ggplot2: elegant graphics for data analysis. Springer-Verlag, New York

Chapter 2
Deciphering Prehistoric Forts of Uttarakhand, Western Himalaya: An Opportunity for Historical Tourism

Govind Singh Negi, Sonam Bahuguna, Naveen Chandra, Yahya Kazmi, Shashank Lingwal, and Gajendra Singh

Abstract Uttarakhand is among the most sacred places in the Himalaya and has always been the core of enticement from the outlook of tourism. Due to little information available on the historical and ancient places of the state, the growth of tourism is restricted to a few known places. The present study investigated (i) the prehistoric forts (*Garh/Kot*) along with their current status and (ii) exploring possibilities for historical tourism in the Uttarakhand state, western Himalaya. Under SALSA framework, systematic literature review was conducted to collect primary information from online sources, followed by extensive field surveys over 15 years, across the state to collect information related to present status and exact geographical position of *Garh/Kot* using Global Positioning System (GPS). A total of 235 *Garh* (majorly from Garhwal region) and 168 *Kot* (majorly representing Kumaun region) were listed. Maximum number of *Garh/Kot* were reported from Pauri (81/27) district followed by Tehri (55/9) and Uttarkashi (28/6) districts. Similarly, maximum number of *Kot*were inventoried from Pithoragarh district (42) followed by Almora (29), Pauri (27), Nainital (18) and Champawat (17). The information of such historic places would give the riveting surge to the tourism circle to plan, protect and promote the historical heritages for sightseeing and economic purpose, which would help in decentralizing tourists from overcrowded places.

Keywords Heritage tourism · Western Himalaya · Garhwal · Kumaun · Garh

---

G. S. Negi · S. Bahuguna · Y. Kazmi · S. Lingwal
Uttarakhand Space Application Centre, Upper Aamwala, Dehradun, Uttarakhand, India

N. Chandra · G. Singh (✉)
Department of Botany, Soban Singh Jeena University Campus, Almora, India
e-mail: gajendrawat@yahoo.com

D. Arya et al. (eds.), *Climate Change Impact on Himalayan Biodiversity*,
Environmental Science and Engineering, https://doi.org/10.1007/978-3-031-77149-1_2

2.1 Introduction

Historical forts in Uttarakhand are currently known as *Garh* and *Kot.* The *Garh/ Kot* of Uttarakhand embraces historic significance behind them along with scenic ancient structures. These structures were usually built at high mountain ridges for residence and were also act as watch tower (Rawat 2018). The state has divided into two divisions, namely Garhwal and Kumaun (Arora 1996). Both the divisions enfold marked historical and ancient heritages, having historical and religious connotation which attracts pilgrims, tourist and explorers to their doors Gupta 1998, Bagri and Babu 2011). Garhwal and Kumaon regions are mentioned as '*Kedarkhand*' and '*Manaskhand*' respectively in the religious Indian epic 'Skand purana' (Raturi 1928; Uniyal 2014).

The word Garhwal is believed to be derived from the title '*Garh-wala*' (owner of forts/region) given to the 37th ruler of the Garhwal kingdom Ajay Pal (Dabral 1971; Nautiyal 2011). The history of Garhwal started in 823 AD, when King Kanak Pal founded the Parmar dynasty in the Garhwal region and gradually extended his kingdom (Rawat 1983). In the medieval period, the state was fragmented into numerous principalities known as '*Garh*', each one of which was ruled by an emperor (Atkinson 2011). *Garh* usually symbolizes fort, built on hill top is an enclosed building safe from attacks and invasions (Rawat 2018). Over the period of time, some of the small *Garh* got integrated together and aggregated into 52 *Garh* (Ram 1916; Rawat 1989). In the fifteenth century, King Ajay Pal merged all the 52 separate *Garh* into one region and earned the title of '*Garh-wala*' and owing which the current Garhwal division got to be known as 'Garhwal' (Raturi 1928), Whereas, according to Hindu believe, the word Kumaun is derived from the incarnation of the Lord Vishnu as tortoise at present day Kumaun region, earlier known as '*Kurmanchal*' (land of Kurma avatar') (Pandey 1937; Uniyal 2014). The '*Garh*' are termed as '*Kot*' in Kumaun (Kafaltia and Kafaltia 2019). The early medieval history of Kumaun is the history of the Katyur dynasty, later followed by the Chand dynasty. Construction of temples was flourished during the Katyuris, whereas Chand dynasty known for building of the *Kot*and magnificent temples in the Kumaun region (Walton 1910, 1911). Ancient structures have always been center of enticement for travel and expedition that eventually put their demand in tourism sphere.

In Uttarakhand, key entities of tourism affairs are primarily managed by the Department of Tourism, Uttarakhand Tourism Development Board (UTDB), and two regional entities namely: Garhwal Mandal Vikas Nigam (GMVN) and Kumaon Mandal Vikas Nigam (KMVN) (Nautiyal 1969). In spite of having multiple variants of tourism in Uttarakhand, it is usually inclined towards religious, sightseeing and trekking destinations and lacking in historical tourism (Ogra 2014). Historical tourism is a spiritual receiving of richness of culture in their authenticity through travel and excursions (Aukland 2017). The modern-day tourism is represented by internal (people travel to their own country/state) and external tourism (people travel to different country/state) (Ismagilova et al. 2015). The historical and cultural heritage plays an important role in boosting domestic and external tourism, also increasingly appreciated as a positive force for natural and cultural conservation (Ivanova 2016). The sights of the historical enticements of the state can be observe by one in the form of forts and temples, made by the regional emperor. Unfortunately, such places are not into go list of the backpackers because of the lack of historical and ancient link of the state among the people, as it is limited to the books and to the folktales by elderly people who lived that history. There is limited published literature and researches of historical places of the Uttarakhand state to promote its growth in tourism prospects (Ivanova 2016; Kala and Bagri 2018; Rawat 2012; Rawat and Nautiyal 2020). Not only for the health of the tourism circle but also to fulfill the expectations and desires of tourists, it is imperative to include historical tourism to the tourism sphere of Uttarakhand state.

Historical tourism represents a considerable research gap in Uttarakhand. Therefore, the present attempts were made to explore the historic *Garh/Kot* (forts) of the Uttarakhand state along with their present status. There are several *Garh/Kot,* those still have structures either in the form of a temple devoted to local deity or in the form of a fort i.e., *Garh/Kot*, however some of them are ruined. The assessment could be utilized to divaricate the masses from being used to roam around only religious places and give them ancient, cultural and historic link to their travel and excursion.

2.2 Study Area

The *Garh/Kot* of the entire Uttarakhand state were investigated. The state is located between 28° 53' and 31° 27' N latitude and 77° 34' to 81° 02' E longitude, covering an area of 53,483 km^2. It is surrounded by Tibet Autonomous Region (TAR) from the North, Nepal from East, Uttar Pradesh to the South and Himachal Pradesh to the West and Northwest.

The two administrative division of the state, i.e., Kumaun and Garhwal are grouped in 13 districts. The Kumaun division includes six (Pithoragarh, Bageshwar, Champawat, Almora, Nainital and Udham Singh Nagar) and Garhwal division includes seven (Chamoli, Dehradun, Haridwar, Rudraprayag, Pauri Garhwal, Tehri Garhwal, Uttarkashi) districts. As the study seeks current status of past entity, some survey sites fall under present neighboring states of Uttarakhand *i.e.,* Himachal Pradesh and Uttar Pradesh. About 86% of the total geographical area of the Uttarakhand state is mountainous, that engulf scenic mountain peaks, lush green forests, holy rivers, mythological temples and historical places which perpetually lure people to come over here (Fig. 2.1).

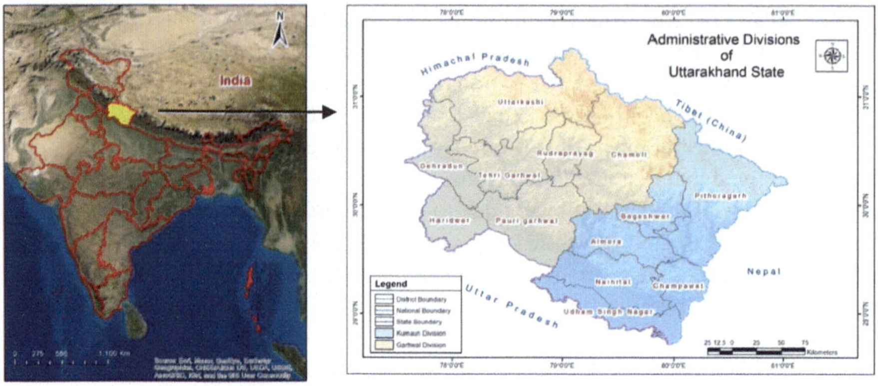

Fig. 2.1 The intensive study area

2.3 Methodology

To collect the primary and ancillary information for qualitative analysis, systematic literature review (SLR) has been performed considering its detailed and precise approach for identifying and synthesizing existing body of recorded information (Rawat 2018). Following SLR approach we conducted 'Search, Appraisal, Synthesis, and Analysis' (SALSA) framework of Grant and Booth (Rawat 2012). The detailed course of actions of the SALSA framework has been explained in Fig. 2.2.

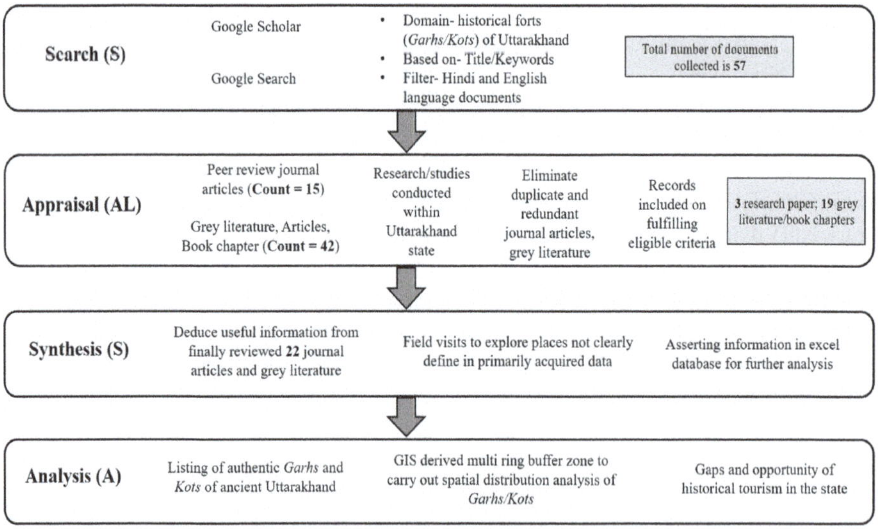

Fig. 2.2 Flow diagram for systematic literature review using SALSA framework

The search databases were Google scholar and Google search and Research gate, where the relevant sources of information were identified using appropriate search strings. Based on the search string, each database was searched to find out relevant literature work reference to Uttarakhand ancient forts within our region of interest, followed by abstract/full body reading of articles has been carried out. Duplicate papers, articles and other literature that lack clear information regarding *Garh/Kot* were removed from the list. Finally, 3 research papers and 19 book literature has been included for Synthesis process. To begin with, well-known and published 52 *Garh* of erstwhile Garhwal kingdom have been identified and listed, followed by information of other *Garh/Kot* of the state have been collected through various published and unpublished sources. The intensive literature survey from finally reviewed 22 journal article and grey literature provides the name of the *Garh/Kot* but not the geo-locations of the same. The information received from online sources after performing SLR was limited, hence, the locations were identified by using the Cartosat and LISS-IV merged high resolution (2.5 m) satellite data (provided by NRSC), available toposheets and Google Earth imagery were also used during SLR synthesis process. Several robust field visits were made to ascertain the information of *Garh* incorporating their geo-location (using Global Positioning System (GPS)) and present status, i.e., availability of ancient structure or ruined. As it is said that 'People are trapped in history and history is trapped in them', locals were also inquired about the authenticity of particular *Garh/Kot,* while performing field visits. Initially, excel database was prepared to store *Garh/Kot* geo-location and their associated attribute information, later, vector layer (in point data format) of recorded locations of *Garh/Kot* has been prepared using ArcMap 10.5 software.

2.4 Results and Discussion

2.4.1 Garh of Ancient Garhwal Kingdom

In the fifteenth century, when all the *Garh*of the erstwhile Garhwal kingdom were brought together, Devalgarh was its capital, which was later shifted to Srinagar (Garhwal). Ajay Pal of Parmar kingdom, built temple of his deity Goddess Raj Rajeshwari at Devalgarh (19 km from Srinagar). Presently, the temple exists and is one of the most visited ancient structures which reflects the archeological riches of that period (Grant and Booth 2009; Sankrityayan 1953; Joshi 2020; Nautiyal 1969). Similarly, several Kot viz., Vanasur, Khagmara and Rajbunga displays the diverse array of archeology of the Kumaun division (Nautiyal 1969). The inventory of acclaimed 52 *Garh* of Garhwal kingdom has been verified through various online and offline sources. Since, all the *Garh* are not enlisted at one place in literature, ancient historical books of regional historians have been glanced to check the authenticity of the recorded 52 *Garh* (Table 2.2 marked as *). Presently, of these about 56% (29) *Garh* are ruined and some of them have no remnant available, however, about 38% (20) *Garh* are still present in the form of temple dedicated to various local deity/goddess.

Local communities on regular basis and tourists occasionally visit these temples to perform various rituals.About six temples viz.,Badhangarhi temple, Chandpur Garhi (Chamoli), Chandrabadani, Kunja Devi temple (Tehri Garhwal), Raj Rajeshwari-Gaura Devi temple, Bhairav temple (Pauri Garhwal), are popular among communities and visitors, where about more than 10,000 people visit every year and seven*Garhviz.,* Asthavakra (Uttarkashi), Kalimath, Lal Bhairav, Nagnath (Chamoli), Gujrumata (Pauri), Someshwar (Uttarkashi) and Kaal Bhairav (Tehri Garhwal) where about more than 5,000 people visit every year.

2.5 *Garh/Kot* of Uttarakhand State

Based on intensive field and literature surveysa total of 235 *Garh* (Table 2.2) and 168 *Kot* (Table 2.3) were recorded including 52 historical *Garh* of erstwhile Garhwal kingdom (Fig. 2.3). Where, maximum number of *Garh/Kot* were recorded in Pauri Garhwal district (81/27) followed by Tehri (55/9) and Uttarkashi (28/6), whereas, Pithoragarh account of maximum *Kot* (42), followed by Almora (29), Pauri (27),), Nainital (18) and Champawat (17) (Table 2.1). Few Garh are located in the fringes of the Uttarakhand state and lies in Himachal Pradesh namely, Dodrakwar, Rami and Ravan *Garh*, and two *Garh* are in Uttar Pradesh, namely, Mordwaj and Parasnath.

Of the total *Garh/Kot* recorded in the state, about 85 are about less than 2 km of distance from road head (national/state/district/others), 57 are between 2–5 km, about 20 are within 5–10 km and 18 have no proper path connecting to the *Garh/Kot* and only can be reached by trekking. Of the total recorded *Garh/Kot* about 43 are frequently visited (> 10,000 people every year) by the local communities and tourists to perform *pooja* and local rituals. Similarly, there are 34 *Garh/Kot* where about 5000 people and 43 *Garh/Kot* where about 1,000 people visit every year and rest have no proper record of visitors or visited only by the surrounded residents. Presently, only 30% of historic *Garh/Kot* are visited by the people, whereas, most of the *Garh/Kot* have limited information, facilities and maintained connecting path.

2.5.1 Elevation Wise Distribution of Garh/Kot in Uttarakhand

The distribution of *Garh* and *Kot* were assessed across various eco-climatic zones of the state i.e., Terai, Bhabhar-Shivalik, Temperate (warm and cool temperate), Sub-alpine and Alpine region covering below 300 m msl to above 3500 m msl. Maximum number of *Garh/Kot* were found distributed along warm temperate (198) and bhabhar-shivalik (98) region below 2000 m msl (Fig. 2.4). Anciently, these

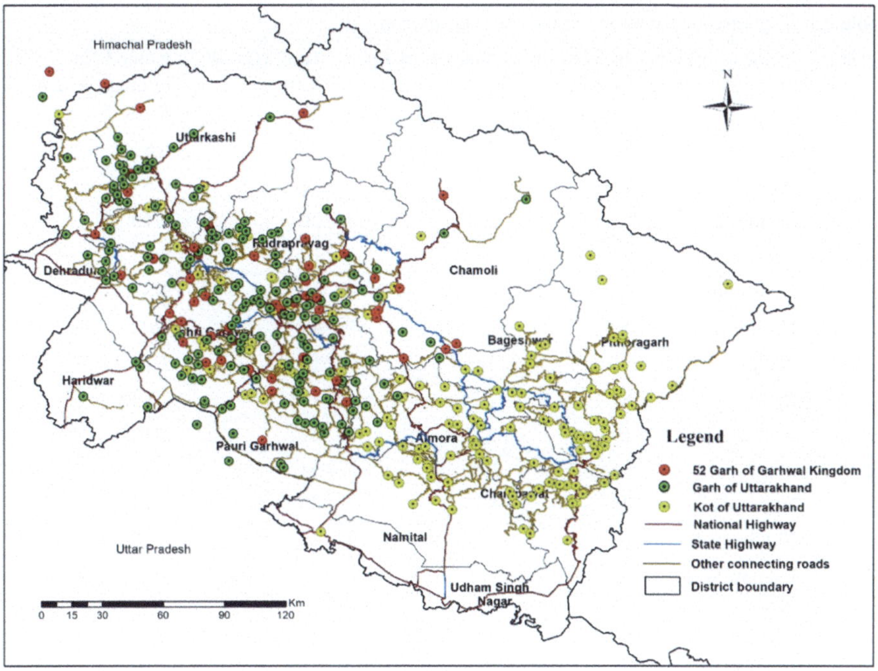

Fig. 2.3 Distribution of *Garh* and *Kot* of Uttarakhand

regions were highly suitable for farming and animal husbandry as the main livelihood and day to daypractices, with plenty of rainfall, long summer and availability of good fuel wood and fodder. The mountainous temperate region also provided the safer places from the attack of invaders.

2.5.2 Historical Importance of Garh/Kot *and Rationale for Their Protection*

Historical places are the cultural heritage of any nation, region or community. In Uttarakhand, Parmar, Katyurand Chand dynasty were prominent in shaping its history. Being a prime necessity of all times since ages, fortification has played an important role for human settlement. In the state, forts are termed as *Garh* and *Kot,* which were the cardinal place of human settlement during medieval period. There were numerous *Garh/Kot* of Uttarakhand known for its historical and cultural riches. The Deval Garhand Chaundkot Garh are ideal exemplar that have prehistoric temples and ancient structures, preserved under Archeological Survey of India protected monument. The Gujru Garh located in Nainidanda block of Pauri Garhwal have complex of temple, historical cavesand man-made ditches that played crucial

Table 2.1 Regional distribution of Garh of Uttarakhand

S. No	Name of region	Number of *Garh*	Number of *Kot*	Total (*Garh/Kot*)	Ruined/No remnant (*Garh/Kot*)
1	Almora	8	29	37	6/8
2	Bageshwar	1	11	12	0/6
3	Chamoli	15	5	20	2/10
4	Champawat	–	17	17	5/5
5	Dehradun	19	–	19	3/8
6	Haridwar	4	–	4	1/0
7	Nainital	–	18	18	3/11
8	Pauri	81	27	108	20/29
9	Rudraprayag	19	3	22	5/10
10	Tehri	55	9	64	8/21
11	Uttarkashi	28	6	34	5/17
12	Pithoragarh	–	42	42	8/17
13	Udham Singh Nagar	–	1	1	1/0
14	Himachal Pradesh	3	–	3	0 /3
15	Uttar Pradesh	2	–	2	0/1
		235	**168**	**403**	**66 (16%)/146 (36%)**

role during Gorkha invasions. Katyur dynasty had built temple of their deity 'Goddess Manila', situated at Manila Garh, Almora district. Till date, the goddess is worshipped by the locals and have great importance from a historical and religious point of view. Vanasur Kot situated at 7 km from Lohaghat in Champawat district was built in medieval-era has unique architecture of Katyur kingdom gambled by vantage view of Himalayan peaks. Likewise, there are countless *Garh/Kot* displaying architectural opulence of that time, while embracing untold glory of Uttarakhand's civic and culture.

India ranks 6th in UNESCO world heritage list. For any community, heritage is an important aspect, to empower its people to participate more fully in social and cultural life. The prehistoric *Garh/Kot* are reflection of Uttarakhand history (Fig. 2.5). Currently, several*Garh/Kot* are ruined and many are at the verge, which need to be recognized in order to save them. Historic environment is a proven source of benefit to local economies, particularly through tourism, by virtue of which it is significant to preserve and protect them (Chandra & Kumar 2021). The explored information of such historic *Garh/Kot* would give riveting surge to the tourism circle, to plan, protect and promote the historical heritages for sightseeing and economic purpose, which would help in decentralizing tourists from overcrowded places.

Table 2.2 List of all historical *Garh* of Uttarakhand

S. No	Garh name	Elevation	Block name	District name	Present status (Temple/Structure Name)
1	Aerasu Garh*	1075	Khirsu	Pauri Garhwal	Astavakra Temple
2	Agar Garh	1721	Bironkhal	Pauri Garhwal	Nanda Garh
3	Ajmeer Garh*	1620	Yamkeshwar	Pauri Garhwal	Shyam Sundari Devi Temple
4	Akaal Garh	1233	Agustmuni	Rudraprayag	No remnant
5	Argarh	1581	Bhilingana	Tehri Garhwal	Raj Rajeshwari Temple
6	Asoour garh	1882	Syaldeh	Almora	Temple
7	Baag Garh	858	Dwarikhal	Pauri Garhwal	No remnant
8	Badalpur Garh/ Raja Gadi*	2036	Jaihrikhal	Pauri Garhwal	Old Ruin
9	Badhaan Garh*	2439	Tharali	Chamoli	Raj Rajeshwari Temple
10	Bagi Garh	1154	Chiniyalisaur	Uttarkashi	Devi Temple
11	Baira Garh	883	Yamkeshwar	Pauri Garhwal	Bhuvneshwari Temple
12	Bairat Garh	2258	Kalsi	Dehradun	Old Ruin
13	Bandar Garh	796	Duggada	Pauri Garhwal	No remnant
14	Bangar Garh*	1997	Jakholi	Rudraprayag	Old Ruin
15	Bariet Garh	1366	Dwarikhal	Pauri Garhwal	Godeshwar Mahadev Temple
16	Barsar Garh	2655	Thalisain	Pauri Garhwal	Gandiyal Dewata Temple
17	Bhadiyar Garh	1662	Kirtinagar	Tehri Garhwal	No remnant
18	Bhagmal Garh	1108	Yamkeshwar	Pauri Garhwal	Old Ruin
19	Bhairav Garh	1761	Rikhanikhal	Pauri Garhwal	Bhairav Temple
20	Bhairon garh	772	Deoprayag	Tehri Garhwal	Bhairav Temple
21	Bhardar Garh*	1487	Jakholi	Rudraprayag	Old Ruin
22	Bharpur Garh*	781	Deoprayag	Tehri Garhwal	Kotha
23	Bhetula Garh	1553	Pratapnagar	Tehri Garhwal	No remnant

(continued)

Table 2.2 (continued)

S. No	Garh name	Elevation	Block name	District name	Present status (Temple/Structure Name)
24	Bhilingana Garh	1572	Bhilingana	Tehri Garhwal	Raj Rajeshwari Temple
25	Bhuj garh	3494	Joshimath	Chamoli	No remnant
26	Bhumiya Garh	1302	Yamkeshwar	Pauri Garhwal	No remnant
27	Bhuwana Garh*	812	Agustmuni	Rudraprayag	Old Ruin
28	Billa Garh	1149	Naugaon	Uttarkashi	Old Ruin
29	Biralta Garh*	2076	Naugaon	Uttarkashi	No remnant
30	Boungi Garh	1440	Rikhanikhal	Pauri Garhwal	Boungi Devi Temple
31	Budkot Garh	1346	Lamgaon	Tehri Garhwal	Garh Devi Temple
32	Bunga Garh	2127	Ukhimath	Rudraprayag	Old Ruin
33	Bungakhal Garh	1677	Thalisain	Pauri Garhwal	Kalinka Devi Temple
34	Bungal garh-Dobar Garh	1731	Rikhanikhal	Pauri Garhwal	Old Ruin
35	Bura Garh	1424	Nanidanda	Pauri Garhwal	No remnant
36	Butola Garh	1179	Agustmuni	Rudraprayag	No remnant
37	Chamba Garh*	1854	Chamba	Tehri Garhwal	No remnant
38	Chandarbadni Garh/Ban Garh	2260	Deoprayag	Tehri Garhwal	Chandarbadni Temple
39	Chandi Garh	513	Bahadrabad	Haridwar	Chandi Devi Temple
40	Chandpur Garh*	1298	Karanprayag	Chamoli	Raj Rajeshwari Temple
41	Char Garhi	1595	Pabho	Pauri Garhwal	Devi Temple
42	Chaunda Garh*	1603	Karanprayag	Chamoli	No remnant
43	Chaundan Garh	1474	Jakhinidhar	Tehri Garhwal	No remnant
44	Chaundkot Garh*	2085	Pokhra	Pauri Garhwal	Old Ruin
45	Chhamrai Garh	1802	Chakrata	Dehradun	No remnant
46	Chhetra Garh	2035	Pokhra	Pauri Garhwal	Old Ruin

(continued)

Table 2.2 (continued)

S. No	Garh name	Elevation	Block name	District name	Present status (Temple/Structure Name)
47	Chil Garh	708	Kot	Pauri Garhwal	No remnant
48	Chorgarh	2113	Jaunpur	Tehri Garhwal	No remnant
49	Chula Garh/ Chilla Garh	1725	Bhilingana	Tehri Garhwal	Raj Rajeshwari Temple
50	Dadur Garh	1556	Chamba	Tehri Garhwal	Old Ruin
51	Dasholi Garh*	1392	Pokhri	Chamoli	Old Ruin/Kalimath Temple
52	Deewa Garh	2506	Nanidanda	Pauri Garhwal	DewaMata Temple
53	Deo Garh	2022	Pokhra	Pauri Garhwal	No remnant
54	Dev Garhi	2082	Kirtinagar	Tehri Garhwal	Devi Temple
55	Deval Garh*	1075	Khirsu	Pauri Garhwal	Raj Rajeshwari-Goradevi
56	Dhamma Garh	657	Raipur	Dehradun	Jhanda SahaibGurudwara
57	Dhamod Garh/ Ratan Garh*	1112	Narendranagar	Tehri Garhwal	Bageshwari Devi Temple
58	Dhangarh	934	Dwarikhal	Pauri Garhwal	Temple
59	Dhangu Garh*	683	Yamkeshwar	Pauri Garhwal	Old Ruin
60	Dhantari	1595	Purola	Uttarkashi	No remnant
61	Dhargarh	1490	Chiniyalisaur	Uttarkashi	No remnant
62	Dharmal Garh	1214	Agustmuni	Rudraprayag	No remnant
63	Dhol Sungda Garh	2249		Uttarkashi	No remnant
64	Dhona Garh*	2128	Jaunpur	Tehri Garhwal	No remnant
65	Dhung Garh	1365	Jakhinidhar	Tehri Garhwal	Old Ruin
66	Dordraa Kwar Garh*	3565		Himachal	No remnant
67	Dungar Garh	2708	Narayan Bagad	Chamoli	No remnant
68	Dwar Garh	2184	Jaunpur	Tehri Garhwal	Bhadrraj Temple

(continued)

Table 2.2 (continued)

S. No	Garh name	Elevation	Block name	District name	Present status (Temple/Structure Name)
69	Erasu Garh/ Badagadi	1270	Bhatwari	Uttarkashi	Kuteti Devi Temple
70	Farasu Garh	1225	Khirsu	Pauri Garhwal	No remnant
71	Gahad Garh	957	Khirsu	Pauri Garhwal	No remnant
72	Garh	2227	Raipur	Dehradun	No remnant
73	Garh	769	Kritinagar	Tehri Garhwal	No remnant
74	Garh Khatal	2089	Chakrata	Dehradun	No remnant
75	Garhkot Garh*	1052	Dwarikhal	Pauri Garhwal	Old Ruin/ Mandlushwer Mahadev Temple
76	Garhthaang Garh*	4252	Bhatwari	Uttarkashi	Old Ruin
77	Garlani Garh	2059	Chakrata	Dehradun	No remnant
78	Gatu Garh	1202	Naugaon	Uttarkashi	Temple
79	Gawana Garh	2981	Thalisain	Pauri Garhwal	No remnant
80	Gawni Garh	2309	Bironkhal	Pauri Garhwal	Old Ruin
81	Ghindayal garh	2118	Agustmuni	Rudraprayag	Ghindayal Temple
82	Ghughti Garh	1328	Yamkeshwar	Pauri Garhwal	Old Ruin
83	Gon Garh	1303	Naugaon	Uttarkashi	Tateshwar Mahadev Temple
84	Gona Garh	2319	Bhilingana	Tehri Garhwal	No remnant
85	Gujru Garh*	2405	Nanidanda	Pauri Garhwal	Mata Temple
86	Guraad Garh	1391	Ekeshwar	Pauri Garhwal	Kotha Telu Roteli
87	Gurkeda Garh	1460	Yamkeshwar	Pauri Garhwal	Old Ruin
88	Guti Garh	1527	Purola	Uttarkashi	No remnant
89	Hansora Garh	1662	Naugaon	Uttarkashi	No remnant
90	Hariyali Devi Garh	2856	Agustmuni	Rudraprayag	Hariyali Devi
91	Hidyani Garh	923	Chiniyalisaur	Tehri Garhwal	Old Ruin

(continued)

Table 2.2 (continued)

S. No	Garh name	Elevation	Block name	District name	Present status (Temple/Structure Name)
92	Hindau Garh	1646	Bhilingana	Tehri Garhwal	Kotha
93	Idwal Garh	1263	Pauri	Pauri Garhwal	Temple
94	Iriya Garh	1248	Naugaon	Uttarkashi	No remnant
95	Iriya Garh	1108	Karanprayag	Chamoli	Temple
96	Iriya Garh/ Kandara garh	1573	Pauri	Pauri Garhwal	Devi Temple
97	Jagdev Garh	466	Duggada	Pauri Garhwal	Jagdev Asram
98	Jaman Garh	851	Duggada	Pauri Garhwal	Near By Sidhawali Temple
99	Jamar Garh	1210	Duggada	Pauri Garhwal	No remnant
100	Jaspal Garh	1111	Yamkeshwar	Pauri Garhwal	No remnant
101	Jaunt Garh*	803	Junpur	Tehri Garhwal	No remnant
102	Jaunya Garh	1063	Yamkeshwar	Pauri Garhwal	No remnant
103	Jaurasi Garh*	2233	Narendranagar	Tehri Garhwal	Ghantakarna Temple
104	Jayal Garh	616	kritinagar	Tehri Garhwal	No remnant
105	Julam Garh	1309	Rikhanikhal	Pauri Garhwal	No remnant
106	Jundalyun Garh	1405	Salt	Almora	Bhaun Devi Temple
107	Juni Sera Garh	1395	Pokhara	Pauri Garhwal	Old Ruin
108	Juniya Garh	2047	Syaldeh	Almora	Devi Temple
109	Kaala Garh	665	Duggada	Pauri Garhwal	Near By Kalagarh Dam
110	Kaani Garh	1650	Chamba	Tehri Garhwal	No remnant
111	Kachora garh / Mukhba Garh	2722	Bhatwari	Uttarkashi	Old Ruin
112	Kalinka Garh	2100	Bironkhal-Syaldeh	Pauri -Almora	Kalinka Devi Temple
113	Kalsi Garh/ Kailarkot	635	Kalsi	Dehradun	No remnant

(continued)

Table 2.2 (continued)

S. No	Garh name	Elevation	Block name	District name	Present status (Temple/Structure Name)
114	Kanda Garh	1475	Bironkhal	Pauri Garhwal	Telu Roteli Temple
115	Kanda Garh*	1107	Kot	Pauri Garhwal	Kalinka Devi Temple
116	Kandara Garh*	1277	Agustmuni	Rudraprayag	Raj Rajeshwari Temple
117	Kandrala Garh	1940	Purola	Uttarkashi	No remnant
118	Kathoor Garh	1348	Bhilingana	Tehri Garhwal	Raj Rajeshwari Temple
119	Kathuli Garh	1540	Khirsu	Pauri Garhwal	Old Ruin
120	Kaula Garh	688	Sahaspur	Dehradun	Tapkeshwar Temple
121	Kaunpur Garh	2969	Narayan Bagad	Chamoli	Old Ruin
122	Khaira Garh*	893	Duggada	Pauri Garhwal	Old Ruin
123	Khait Garh	2277	Jakhinidhar	Tehri Garhwal	Khait Temple
124	Khamlekh garh	2154	Nanidanda	Pauri Garhwal	No remnant
125	Khand Garh	942	Chamba	Tehri Garhwal	Kali Temple
126	Khatli Garh	2408	Bironkhal	Pauri Garhwal	Old Ruin
127	Khola Garh	2198	Pratapnagar	Tehri Garhwal	Kotha
128	Kimgari-Garh	1484	Pokhra	Pauri Garhwal	Old Ruin
129	Kishan Garh/ Raja Garhi	2266	Garur	Bageshwer	No remnant
130	Koli Garh*	1968	Agustmuni	Rudraprayag	No remnant
131	Kotli Garh	1349	Bhilingana	Tehri Garhwal	Old Ruin
132	Kujnee Garh*	1648	Narendranagar	Tehri Garhwal	Kunja Devi Temple
133	Kungarh	2064	Ukhimath	Rudraprayag	No remnant
134	Kweeli Garh*	1026	Narendranagar	Tehri Garhwal	Narshinh Temple
135	Langur Garh/ Bhairav Garh*	1897	Dwarikhal	Pauri Garhwal	Bhairav Temple
136	Lohbaag Garh*	1877	Gairsain	Chamoli	Lal Bhairav Temple

(continued)

Table 2.2 (continued)

S. No	Garh name	Elevation	Block name	District name	Present status (Temple/Structure Name)
137	Lodan Garh	1434	Naugaon	Uttarkashi	No remnant
138	Lodi Garh*	1249	Jaiharikhal	Pauri Garhwal	Old Ruin
139	Lwai Garh	1090	Agustmuni	Rudraprayag	No remnant
140	Lukundar Garh/Rani Garh	2143	Kot	Pauri Garhwal	Adeshwer Mahadev
141	Madwal Garh	608	Kot	Pauri Garhwal	Dhaneshwar Temple
142	Mahaab Garh	1724	Duggada	Pauri Garhwal	Bholenath Temple
143	Mandaar Garh	1694	Jakhinidhar	Tehri Garhwal	Kotha
144	Manila Garh	1952	Syaldeh	Almora	Manila Mata Temple
145	Mansa Garh	448	Bahadrabad	Haridwar	Mansa Devi Temple
146	Manur	1432	Salt	Almora	Old Ruin
147	Massur Garh	1646	Salt	Almora	No remnant
148	Mathiyana Garh	2079	Jakholi	Rudraprayag	Mathiyana Mata Temple
149	Mavakot Garh	801	Duggada	Pauri Garhwal	Near By Kanvashrama
150	Mol Garh	868	Kot	Pauri Garhwal	No remnant
151	Moliya Garh*	1536	Pratapnagar	Tehri Garhwal	Raj Rajeshwari Temple
152	Mordhwaj Garh	303		Uttar Pradesh	Mordhwajeshwar Mahadev Temple
153	Mor Garh	1270	Nanidanda	Pauri Garhwal	NO
154	Mounda Garh	1240	Chakrata	Dehradun	Lakhamandal
155	Mugra Garh*	1109	Naugaon	Uttarkashi	Old Ruin
156	Mukhem Garh	1838	Pratapnagar	Tehri Garhwal	Mukhem Temple
157	Mulya Garh	660	Deoprayag	Tehri Garhwal	Old Ruin
158	Mundeswer Garh	1921	Kalgikhal	Pauri Garhwal	Mundeswer Mahadev Temple
159	Munor Garhi	1676	Naugaon	Uttarkashi	No remnant
160	Naala Garh	951	Raipur	Dehradun	Old Ruin

(continued)

Table 2.2 (continued)

S. No	Garh name	Elevation	Block name	District name	Present status (Temple/Structure Name)
161	Nagdev Garhi	1945	Dwarikhal	Pauri Garhwal	Nagdevta Temple
162	Nagnee Garh	2702	Dunda	Uttarkashi	Nagni Temple
163	Nagpur Garh*	2128	Pokhri	Chamoli	Nagnath Temple
164	Nainthna Garh	1726	Chaukhutia	Almora	Naithana Devi
165	Nakot Garh	1687	Agustmuni	Rudraprayag	Old Ruin
166	Nakuri Garh/ Renuka Garh	1407	Dunda	Uttarkashi	Renuka Temple
167	Nanur Garh	712	Raipur	Dehradun	Kalinga Memorial
168	Nau Garhi	1525	Pabho	Pauri Garhwal	Near By Jabroli Mahadev Temple
169	Naugarh	1787	Jaunpur	Tehri Garhwal	Old Ruin
170	Nawasu Garh	1713	Agustmuni	Rudraprayag	Devi Temple
171	Nayal Garh*	1004	Khirsu	Pauri Garhwal	NO
172	Nayan Garh	2101	Khirsu	Pauri Garhwal	Bhairav Temple
173	Nimga Garh	1509	Chakrata	Dehradun	Old Ruin
174	Nun Garh	330	Duggada	Pauri Garhwal	Old Temple
175	Onchirgarh	1559	Pokhri	Chamoli	NO
176	Painkhanda Garh	1871	Joshimath	Chamoli	Arpana Devi Temple
177	Paino Garh	1245	Rikhanikhal	Pauri Garhwal	NO
178	Palkot Garh	989		Tehri Garhwal	NO
179	Parasnath Fort	278	Kalagarh road	Uttar Pradesh	-
180	Patou Garh	1949	Thalisain	Pauri Garhwal	Old Ruin
181	Phaliyaan Garh*	1255	Yamkeshwar	Pauri Garhwal	Bhuvneshwari Temple
182	Puran Garhi/ Hanuman garhi	1893	Dwarikhal	Pauri Garhwal	Hanuman Temple
183	Quali Garh	1566	Rikhanikhal	Pauri Garhwal	Devi Temple

(continued)

Table 2.2 (continued)

S. No	Garh name	Elevation	Block name	District name	Present status (Temple/Structure Name)
184	Raika Garh*	795	Pratapnagar	Tehri Garhwal	NO
185	Raj Garh	1540	Khirsu	Pauri Garhwal	Old Ruin
186	Raj Garhi	1804	Naugaon	Uttarkashi	Kotha
187	Rami Garh*	2935	Rohru	Himachal	-
188	Rana Garh	623	Dwarikhal	Pauri Garhwal	-
189	Rang Mahal	267	Narsan	Haridwar	Old Ruin
190	Rani Garh*	1592	Agustmuni	Rudraprayag	NO
191	Rani Garh-2	1369	Bhilingana	Tehri Garhwal	Raj Rajeshwari Temple
192	Rani Garh-3	894	Vikasnagar	Dehradun	NO
193	Rawad Garh*	2588	Joshimath	Chamoli	NO
194	Rawan Garh	1520		Himachal	-
195	Rawat Garh	1528	Chakrata	Dehradun	NO
196	Ringal Garh	1900	Junpur	Tehri Garhwal	NO
197	Roulakot-Garh	932	Pratapnagar	Tehri Garhwal	Nag Temple
198	Sabal Garh	256	Bhadrabad	Haridwar	Siddh Peeth
199	Saitan Garh	2208	Bironkhal	Pauri Garhwal	NO
200	Sangela/ Nailchami Garh*	1448	Bhilingana	Tehri Garhwal	Old Ruin
201	Sankri Garh*	1898	Mori	Uttarkashi	Someshwer Devta Temple
202	Santala Garh	935	Sahaspur	Dehradun	Santala Devi Temple
203	Santour Garh	1828	Sahaspur	Dehradun	Santala Devi Temple
204	Santura Garh	935	Sahaspur	Dehradun	Santura Devi Temple
205	Sari/Garhkot Garh	1100	Dwarikhal	Pauri Garhwal	Mata Temple
206	Sashastra Bahu Garh	1299	Naugaon	Uttarkashi	Old Ruin
207	Sateragarh	1457	Agustmuni	Rudraprayag	Temple
208	Sawali Garh*	1974	Bironkhal	Pauri Garhwal	Old Ruin

(continued)

Table 2.2 (continued)

S. No	Garh name	Elevation	Block name	District name	Present status (Temple/Structure Name)
209	Sentula Pani Garh	1613	Deoprayag	Tehri Garhwal	Temple
210	Shakeri-Raghri Garh	1483	Bhilingana	Tehri Garhwal	Devi Temple
211	Shirgur Garh*	1367	Tharali	Chamoli	No remnant
212	Shrikot Garh	2156	Purola	Uttarkashi	Mahadev Temple
213	Sil Garh*	1084	Jakholi	Rudraprayag	No remnant
214	Sindwal Garh	1555	Pratapnagar	Tehri Garhwal	Temple
215	Sirwal Garh	780	Raipur	Dehradun	No remnant
216	Surkanda Garh	2760	Junpur	Tehri Garhwal	Surkanda Devi Temple
217	Syalda Garh/ Saula Garhi	1801	Purola	Uttarkashi	No remnant
218	Taidi Garh	1390	Kot	Pauri Garhwal	Old Ruin
219	Tamkhani Garh	3013	Naugaon	Uttarkashi	Nagdevta Temple
220	Tara Garh	2425	Thalisain	Pauri Garhwal	Tarakund
221	Tegad	1321	Kirtinagar	Tehri Garhwal	Temple
222	Thadung Garh	1406	Purola	Uttarkashi	No remnant
223	Thander Garh	1263	Pablio	Pauri Garhwal	Devi Temple
224	Thati Garh	2078	Bhilingana	Tehri Garhwal	Thati Mata Temple
225	Thati Ramoli Garh	1252	Pratapnagar	Tehri Garhwal	Raj Rajeshwari Temple
226	Top Garh*	1343	Karanprayag	Chamoli	Raj Rajeshwari Temple
227	Udipur Garh	1869	Thaludhar	Tehri Garhwal	Devi Temple
228	Ulkha Garh	2070	Khirsu	Pauri Garhwal	Ulkha Devi Temple
229	Umrasu Garhi	609	Kot	Pauri Garhwal	No remnant
230	Umta Garh	1400	Nanidanda	Pauri Garhwal	Bungi Devi Temple
231	Unchra Garh	1910	Deoprayag	Tehri Garhwal	No remnant

(continued)

Table 2.2 (continued)

S. No	Garh name	Elevation	Block name	District name	Present status (Temple/Structure Name)
232	Uppu Garh*	999	Thaludhar	Tehri Garhwal	Kaal BhairavTemple
233	Vajri Garh	1746	Naugaon	Uttarkashi	Raj Rajeshwari Temple
234	Van Garh*	1134	Deoprayag	Tehri Garhwal	Old Ruin
235	Virbhadra	338	Doiwala	Dehradun	Virbhadra

* 52 known *Garh* of erstwhile Garhwal Kingdom

2.6 Conclusion

Uttarakhand is well known for its religious tourist destinations and a few places of sightseeing, adventure and trekking. Other than the religious places, other destinations have limited tourist visit due to arduous geography, limited access and lack of desired promotion and basic facilities. Similarly, historical tourism of the state is not much promoted. The little information about the 52 forts (*Garh*) is found commonly in the literature. The present study inventories more than 403 *Garh/Kot* in the state having 235 *Garh* and 168 *Kot* including acclaimed 52 *Garh* of Garhwal kingdom. Unfortunately, of the total, more than 50% of historical places (214 *Garh/Kot*) are either old ruin or have no structure in the ground. There is only 120 *Garh/Kot* where > 1000 people visit annually. However, there is an opportunity to promote at least about 142 *Garh/Kot* situated < 5 km distance from road. The excursion to *Garh/Kot* is a cracking way to proportion the seasonal fluctuations and evenly distributing tourist streams in the state domain. Moreover, this would encourage people to know the heroic history of the state and would probably make historical trips of the state.

Table 2.3 List of *Kot* of Uttarakhand

S. No	Kot name	Elevation	Block name	District name	Mandir name
1	Amsar Kot	960	Bageshwar	Bageshwar	Bhagwati Temple
2	Aara Kot	1341	Mori	Uttarkashi	No remnant
3	Asar Kot	1442	Pokhra	Pauri Garhwal	No remnant
4	Askot	1365	Kanalichina	Pithoragarh	Mallikarjun Temple
5	Badiyakot	2253	Kapkot	Bageshwar	Bhagwati Temple
6	Bai Kot	1053	Dharchula	Pithoragarh	No remnant
7	Balara Kot	1430	Vin	Pithoragarh	No remnant
8	Baluwa Kot	770	Dharchula	Pithoragarh	Kalsin Temple
9	Ban Kot	1547	Gangolihat	Pithoragarh	No remnant
10	Banskot	1340	Vin	Pithoragarh	No remnant
11	Bar Kot	1373	Kot	Pauri Garhwal	Temple
12	Bara Kot	1453	Barakot	Champawat	No remnant
13	Bara Kot	1563	Vin	Pithoragarh	No remnant
14	Bari Kot	876	Bhikyasain	Almora	No remnant
15	Bel Kot /Karki Kot	1408	Barinag	Pithoragarh	Old Ruin
16	Bhains Kot	1432	Khirsu	Pauri Garhwal	Temple
17	Bhan Kot	824	Jaiharikhal	Pauri Garhwal	Temple
18	Bhat Kot	941	Kot	Pauri Garhwal	Temple
19	Bhatt Kot	1675	Vin	Pithoragarh	No remnant
20	Bhins Kot	1272	Munsayari	Pithoragarh	Bhakli Temple
21	Bich Kot	1748	Vin	Pithoragarh	No remnant
22	Bigra Kot	1283	Pati	Champawat	Temple
23	Bil Kot	936	Nanidanda	Pauri Garhwal	No remnant
24	Bin Kot	1402	Betalghat	Nainital	No remnant
25	Bisoud Kot	1590	Lamgara	Almora	Old Ruin
26	Boyal Kot	1025	Gangolihat	Pithoragarh	No remnant
27	Budhla Kot	1886	Betalghat	Nainital	No remnant
28	Bunga Kot	1665	Munakot	Pithoragarh	Old Ruin
29	Bura Kot	953	Pauri	Pauri Garhwal	Temple

(continued)

Table 2.3 (continued)

S. No	Kot name	Elevation	Block name	District name	Mandir name
30	Bura Kot	1431	Bironkhal	Pauri Garhwal	No remnant
31	Bursum Kot	776	Gangolihat	Pithoragarh	No remnant
32	Chal Kot	1387	Didihat	Pithoragarh	No remnant
33	Cham Kot	1551	Agustmuni	Rudraprayag	No remnant
34	Chandal Kot	2005	Champawat	Champawat	Old Ruin
35	Chara Kot	1398	Champawat	Champawat	No remnant
36	Chhot Kot	952	Pauri	Pauri Garhwal	Temple
37	Chipla Kot	4042	Munsyari	Pithoragarh	No remnant
38	Chirpat Kot	2010	Kapkot	Bageshwar	Bhagwati Temple
39	Chitkot	1506	Lohaghat	Champawat	Old Ruin
40	Choura Kot	1446	Pati	Champawat	No remnant
41	Chulkot	2228	Munsyari	Pithoragarh	Temple
42	Dan Kot	1248	Agustmuni	Rudraprayag	No remnant
43	Dewal Kot	1223	Karanprayag	Chamoli	No remnant
44	Dhaniakote	1084	Betalghat	Nainital	Old Ruin
45	Dhaniya Kot	792	Jaiharikhal	Pauri Garhwal	Temple
46	DhanpurKot	2245	Kapkot	Bageshwar	No remnant
47	Dhar Kot	1092	Khirsu	Pauri Garhwal	Temple
48	Dhar Kot	1341	Yamkeshwar	Pauri Garhwal	Temple
49	Dhar Kot	1755	Bhilingana	Tehri Garhwal	Temple
50	Dhar Kot	1750	Chiniyalisaur	Uttarkashi	No remnant
51	Dhar Kot	1237	Jakholi	Rudraprayag	Rajeshwari Temple
52	DharPayan Kot	1743	Kirtinagar	Tehri Garhwal	No remnant
53	Dharchula Kot	2110	Dharchula	Pithoragarh	Temple
54	Drona Kot	1734	Champawat	Champawat	Old Ruin
55	Duna Kot	1250	Didihat	Pithoragarh	No remnant
56	Falda Kot	1914	Tarikhet	Almora	Temple
57	GangarKot	1264	Ramgarh	Nainital	No remnant
58	Gar Kot	974	Yamkeshwar	Pauri Garhwal	Temple

(continued)

Table 2.3 (continued)

S. No	Kot name	Elevation	Block name	District name	Mandir name
59	Garh Kot	1593	Kalzikhal	Pauri Garhwal	Temple
60	Ghach kot	1387	Salt	Almora	Bhuvaneshwari Devi
61	Ghat Garh	1046	Bhimtal	Nainital	Mata Temple
62	Golachura Fort	1624	Champawat	Champawat	GoluDevta Temple
63	Gopal Kot	2709	Takula	Almora	BudhaPinath
64	Gorkha Gadi	2115	Takula	Almora	MallikaTemple
65	Gujru Kot	1538	Salt	Almora	Raja HaruheetTemple
66	Gwasikot	1445	Gangolihat	Pithoragarh	Temple
67	Har Kot	1378	Ekeshwar	Pauri Garhwal	Temple
68	Harkot	1594	Munsyari	Pithoragarh	Bharadi Devi Temple
69	Jai Kot	998	Rikhanikhal	Pauri Garhwal	No remnant
70	Jaltha Kot	1674	Bageshwar	Bageshwar	No remnant
71	Jaman Kot	1125	Gangolihat	Pithoragarh	No remnant
72	JatiyaKot	1142	Gangolihat	Pithoragarh	Temple
73	Jeyoli Kot	1348	Bhimtal	Nainital	No remnant
74	Kaira Kot	1382	Okhalkanda	Nainital	No remnant
75	Kamali Kot	1118	Bhimtal	Nainital	Old Ruin
76	Kana Kot	1526	Pati	Champawat	Temple
77	Kapkot	1214	Kapkot	Bageshwar	KashilBubu Temple
78	Kara Kot	1325	Bhikyasain	Almora	Temple
79	Kara Kot	1490	Dhauladevi	Almora	GoljyuTemple
80	Kashipur Fort/Govishan	248	Kashipur	U S Nagar	Old Ruin
81	Kathhal Kot	1342	Salt	Almora	No remnant
82	Kathyur Kot	2076	Chaukhutiya	Almora	Mansa Devi Temple
83	KhadiKot	1677	Vin	Pithoragarh	Temple
84	Khagmara Fort	1507	Hawalbag	Almora	Khagmara Temple
85	Khai Kot	1533	Lohaghat	Champawat	Temple
86	Kot Bhramari—Ranchulahat	1035	Garur	Bageshwar	Nanda Devi Temple

(continued)

Table 2.3 (continued)

S. No	Kot name	Elevation	Block name	District name	Mandir name
87	Kota	1020	Dharikhal	Pauri Garhwal	No remnant
88	Kotdhar	2535	Dasholi	Chamoli	Old Ruin
89	Kotyura Kot	1298	Hawalbag	Almora	No remnant
90	Kumarkot	1789	Dunda	Uttarkashi	Temple
91	Lakhanpur Kot	1172	Chaukhutiya	Almora	Dham Deval Temple
92	Lalmandi Fort/Moira Fort	1690	Hawalbag	Almora	Temple
93	Lam Kot	1390	Chamba	Tehri Garhwal	No remnant
94	Lat Kot	1578	Munsayari	Pithoragarh	Temple
95	Loha Kot	1240	Kapkot	Bageshwar	No remnant
96	Loha Kot	1272	Kanalichina	Pithoragarh	Old Ruin
97	Luwa Kot	1453	Barakot	Champawat	Temple
98	Madkot	1417	Munsyari	Pithoragarh	Kali Temple
99	Mall Kot	1305	Betalghat	Nainital	No remnant
100	Malla Kot Fort	1612	Hawalbag	Almora	Devi Temple
101	Malya Kot	1177	Jakhinidhar	Tehri Garhwal	No remnant
102	Man Kot	1625	Bageshwar	Bageshwar	No remnant
103	Mandal Kot	1405	Tarikhet	Almora	Temple
104	Mani Kot	1785	Gangolihat	Pithoragarh	Mankeshwar Mahadev Temple
105	Mani Kot	1398	Yamkeshwar	Pauri Garhwal	Temple
106	Manj Kot	1102	Thaludhar	Tehri Garhwal	No remnant
107	Manur Kot	1432	Salt	Almora	Old Ruin
108	Markot	1346	Ekeshwar	Pauri Garhwal	Temple
109	Milam Kot	3471	Munsyari	Pithoragarh	Old Ruin
110	MusKot	1603	Kanalichina	Pithoragarh	No remnant
111	Nakot	1466	Gangolihat	Pithoragarh	Nagimal Devta Temple
112	Nakot	1082	Karanprayag	Chamoli	No remnant
113	Nir Kot	1300	Bhikyasain	Almora	No remnant
114	Nira Kot	1518	Bhatwari	Uttarkashi	No remnant

(continued)

Table 2.3 (continued)

S. No	Kot name	Elevation	Block name	District name	Mandir name
115	Pan Kot	1466	Hawalbag	Almora	GoluDevta Temple
116	Pandav Fort	3810	Dharchula	Pithoragarh	Old Ruin
117	Par Kot	1362	Dwarahat	Almora	No remnant
118	Par Kot	1552	Kirtinagar	Tehri Garhwal	Temple
119	Pat Kot	934	Ramnagar	Nainital	No remnant
120	Pattharkot	1412	Kanalichina	Pithoragarh	Temple
121	Pina Kot	1604	Thalisain	Pauri Garhwal	No remnant
122	Pipal Kot	1178	Jaiharikhal	Pauri Garhwal	Temple
123	Pithoragarh fort/Loudon Fort	1626	Vin	Pithoragarh	Old Ruin
124	Puna Kot	1665	Pati	Champawat	Temple
125	Raila Kot/Gorkha Garh	1548	Hawalbag	Almora	No remnant
126	Raj Kot	1481	Didihat	Pithoragarh	No remnant
127	Rajbunga fort	1665	Champawat	Champawat	Nagnath Temple
128	Rala Kot	1894	Lamgara	Almora	Old Ruin
129	Ran Kot	1662	Gangolihat	Pithoragarh	No remnant
130	Ranakot	1584	Yamkeshwar	Pauri Garhwal	Temple
131	Rani Kota	746	kotabagh	Nainital	No remnant
132	Ranikot	1445	Okhalkanda	Nainital	Old Ruin
133	Rat Kot	1433	Bironkhal	Pauri Garhwal	No remnant
134	Rautela Kot	1290	Okhalkanda	Nainital	Temple
135	Rilkot	3252	Munsyari	Pithoragarh	Old Ruin
136	Routakot	720	Munakot	Pithoragarh	Old Ruin
137	SakniyaKot	1463	Takula	Almora	No remnant
138	Satniyan Kot	1735	Salt	Almora	SatyalokTemple
139	Sera Kot	2062	Didihat	Pithoragarh	Malaynath Temple
140	Sirkot	1215	Pokhri	Chamoli	No remnant
141	Sirkot	872	Pauri	Pauri Garhwal	Temple
142	Sirkot	981	Yamkeshwar	Pauri Garhwal	No remnant
143	Sirmohi Fort	1615	Pati	Champawat	Old Ruin

(continued)

Table 2.3 (continued)

S. No	Kot name	Elevation	Block name	District name	Mandir name
144	Sita Kot	1078	Bhilingana	Tehri Garhwal	No remnant
145	Siyan Kot	1386	Kapkot	Bageshwar	Temple
146	Sui Kot	1871	Lohaghat	Champawat	No remnant
147	Sumgarh	1784	Kapkot	Bageshwar	No remnant
148	Sun Kot	1398	Okhalkanda	Nainital	Sun Kot Temple
149	Suwa Kot	1304	Okhalkanda	Nainital	No remnant
150	Suwalek Kot	1102	Kanalichina	Pithoragarh	Temple
151	Syunar Kot	1346	Chamba	Tehri Garhwal	Temple
152	Syunra Kot	1374	Hawalbag	Almora	Old Ruin
153	Tall Kot	1456	Betalghat	Nainital	No remnant
154	Tara Kot	1487	Chiniyalisaur	Uttarkashi	No remnant
155	TikhonaKot	2120	Takula	Almora	Airadiyo Temple
156	Tikhun Kot	1662	Hawalbag	Almora	Hanuman Temple /Old Ruin
157	Ucha Kot	1492	Betalghat	Nainital	No remnant
158	Ucha Kot	1827	Vin	Pithoragarh	Temple
159	Ucha Kot	1472	Ekeshwar	Pauri Garhwal	Temple
160	Ucha Kot	1186	Yamkeshwar	Pauri Garhwal	Temple
161	Udai Kot	1556	Chamba	Tehri Garhwal	No remnant
162	Udaipur Kot	2090	Vin	Pithoragarh	No remnant
163	Uma Garh	2345	Ramgarh	Nainital	Uma Devi Temple
164	Umrakot	982	Pokhri	Chamoli	No remnant
165	Upri Kot	1547	Chiniyalisaur	Uttarkashi	No remnant
166	Usha Kot	1912	Lohaghat	Champawat	No remnant
167	Vanasur's Fort/ Markot	1905	Lohaghat	Champawat	Old Ruin
168	Vimal Kot	2145	Bhisiyachana	Almora	Vimalkot Devi Temple

Acknowledgements The authors would like to thank Director, Uttarakhand Space Application Centre (USAC), Dehradun, for providing all the necessities to conduct the field investigations. The financial support provided by the GB Pant National Institute of Himalayan Environment (GBPNIHE), Ministry of Environment, Forest and Climate Change (MoEF& CC) for financial assistance under Himalayan Knowledge Network (HKN) Program and Department of Information and Science Technology, Government of Uttarakhand is dully acknowledged. We thank National

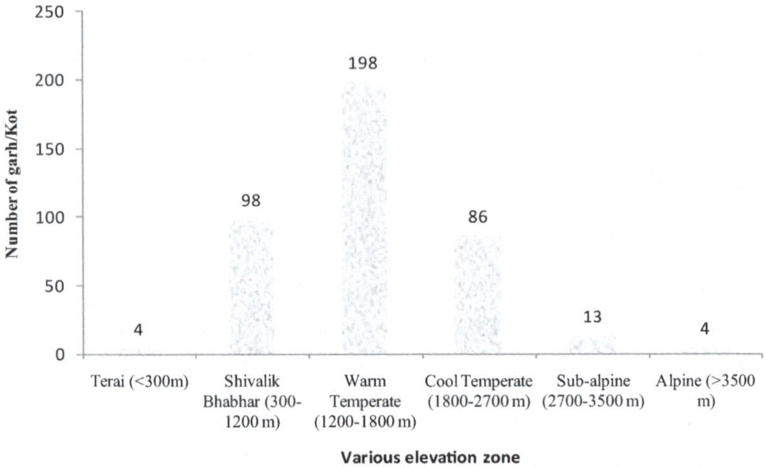

Fig. 2.4 Elevation wise distribution of *Garh/Kot* of Uttarakhand

(a) Rajrajeshwari temple of Dewal Garh, Pauri Garhwal

(b) Kaalbhairav Temple of Uppu Garh, Tehri Garhwal

(c) Nagnath Temple of Nagpur Garh, Chamoli

(d) Partially ruined Chandpur Garh, Chamoli

(e) Ruined Bunga Garh, Rudraprayag

Fig. 2.5 Glimpse of some historical *Garh* of Uttarakhand

Remote Sensing Centre (NRSC), Hyderabad for providing Cartosat-1 & LISS-IV merged satellite data under SISDP-update project. The assistance received from researchers at USAC are gratefully acknowledged.

The research meets al apllicable standards with regards to ethics of experiments and research integrity and being cetified true. On the behalf of co-authors, I declare that, there is no duplicate publication, fraud, plagiarism, or other malicious concerns involved while submitting the manuscript.

Compliance with Ethical Standards

Conflict of Interest Authors and funding agencies have no conflict of interest.

References

https://archive.org/details/in.ernet.dli.2015.94182/page/n5/mode/2up; https://archive.org/details/in.gov.ignca.35289/page/n3/mode/2up; https://panotbook.com/kumau-ka-itihas/
Arora A (1996) Administrative History of Uttarakhand
Atkinson ET (1981) The Himalayan Gazetter
Aukland K (2017) Pilgrimage expansion through tourism in contemporary India: the development and promotion of a Hindu pilgrimage circuit. J Contemp Relig 32(2):283–298
Bagri S, Babu A (2011) Historical development of tourism education in India: The case of the Himalayan state of Uttarakhand. J Tour 12(1):39–59
Chandra P, Kumar J (2021) Strategies for developing sustainable tourism business in the Indian Himalayan Region: Insights from Uttarakhand, the Northern Himalayan State of India. J Destin Mark Manag 19:100546
Dabral SP (1971) Uttarakhand kaitihaas. Veer Gatha press, Duggada, Pauri Garhwal, India
Grant MJ, Booth A (2009) A typology of reviews: An analysis of 14 review types and associated methodologies. Health Inf Libr J 26(2):91–108. https://doi.org/10.1111/j.1471-1842.2009.00848
Gupta SK (1998) Tourism in Garhwal Himalaya: Strategy for sustainable development. Domestic tourism in India, 199–218. https://archive.org/details/in.ernet.dli.2015.181556
Ismagilova G, Safiullin L, Gafurov I (2015) Using historical heritage as a factor in tourism development. Procedia Soc Behav Sci 188:157–162
Ivanova P (2016) Cultural tourism-development, scope and prospects. In: The Priority Directions of National Economy Development: Proceedings of the International Scientific Conference-Nis (pp 341–347)
Joshi G (2020) Uttarakhand karajnitik, samajikevamsanskritikitihaas, IV edn
Kafaltia H, Kafaltia GS (2019) A comprehensive study of Uttarakhand. Notion Press, Vanagaram, Chennai, India
Kala D, Bagri SC (2018) Barriers to local community participation in tourism development: Evidence from mountainous state Uttarakhand India. Tour: Int Interdiscip J 66(3):318–333
Nautiyal KP (1969) The Archeology of Kumaon
Nautiyal RM (2011) Pauranik Uttarakhand
Ogra A (2014) Tourism Web Analytics-A Case Study of Garhwal and Kumaon Regions of Uttarakhand
Pandey BD (1937) Kumaon Ka Itihaas
Ram P (1916) Garhwal: Ancient and modern
Raturi HK (1928) Garhwal kaltihas, Bhagirathi Prakashan Tehri, Uttarakhand, India
Rawat NS, Nautiyal V (2020) Exploring chaundkot fort in Garhwal, central Himalaya, India. Antiquity 94:(376)
Rawat AS (1983) Garhwal Himalaya: A historical survey: Political and administrative history of garhwal, 1815–1947
Rawat AS (1989) History of Garhwal, 1358–1947: An erstwhile kingdom in the Himalayas
Rawat NS (2012) Investigation of Garh/Hill forts (fortress) using Geo-informatics tools in district Pauri of Garhwal Himalaya. J HumIties Soc Sci 4(4):10–18. ISSN: 2279–0837, ISBN: 2279–0845
Rawat NS (2018) An archaeological and geospatial investigation of lost forts and fortalices in Yamuna Valley with special reference to vairat and mungra forts in garhwal (Central) Himalaya, Uttarakhand, India. Herit: J Multidiscip Stud Archaeol 6:872–886
Sankrityayan R (1953). Himalaya Parichay (1) Garhwal. Allahabad Law J Press

Uniyal H (2014) Manaskhand.Uttara books B-4/310C, Keshavpuram, Delhi. ISBN 978–81–924138–3–9
Walton HG(1910) Gazetteer of British Garhwal
Walton HG (1911) Gazetteer of Dehradun

Chapter 3
Investigating the Richness and Therapeutic Applications of Indigenous Medicinal Flora in Arunachal Pradesh, Northeastern India

Suman Halder, Subhajit Lahiri, Arijit Ghosh, and Mriganka Shekhar Sarkar

Abstract Ethnomedicine, a holistic approach integrating plants, spirituality, and the natural world, has long served as a fundamental source of healing. In the Eastern Himalaya region, particularly Arunachal Pradesh, medicinal plants play a pivotal role in cultural, spiritual, and healthcare domains. Despite extensive documentation of traditional medicinal plant use by various tribal communities in North-East India, comprehensive exploration in Arunachal Pradesh remains scarce. This study aims to fill this gap by compiling indigenous knowledge of medicinal plants in the region, focusing on traditional uses, plant parts utilized, medicinal properties, and their richness distribution across elevational gradients. Data compilation from 199 sources facilitated the identification of 508 medicinal plant species spanning 158 families. Leaves, fruits, and roots emerged as primary plant parts utilized in traditional medicine. The study employed a Generalized Additive Model (GAM) to analyze species richness across elevation gradients. The model featuring an automatic smoothing term, emerged as the best-fit model, effectively elucidating the relationship of medicinal plant richness over elevation gradients of the state. The richness map revealed concentrations of medicinal flora in the tropical Himalayan region, emphasizing the ecological significance of conservation efforts. The findings underscore the region's ecological uniqueness and the importance of preserving its rich medicinal plant biodiversity for future generations.

S. Halder · A. Ghosh
Botanical Survey of India, Arunachal Pradesh Regional Centre, Itanagar 791111, India

S. Lahiri
Botanical Survey of India, Head Quarter, Kolkata 700064, India

M. S. Sarkar (✉)
GB Pant National Institute of Himalayan Environment, North Eastern Regional Centre, Itanagar 791113, India
e-mail: mriganka.sarkar@gbpihed.nic.in

D. Arya et al. (eds.), *Climate Change Impact on Himalayan Biodiversity*, Environmental Science and Engineering, https://doi.org/10.1007/978-3-031-77149-1_3

Keywords Medicinal flora · Richness map · Generalized additive model · Eastern Himalaya · North East India

3.1 Introduction

Ethnomedicine comprises a multifaceted interdisciplinary approach involving the utilization of plants, spirituality, and the natural environment, serving as an ancient source of healing through the ages. Traditionally, medicinal plants have held significant sway in the socio-cultural, spiritual, and medicinal spheres among the local communities of the Eastern Himalaya region (Arya et al. 2017). Multiple studies in North-East India document the utilization of medicinal plant species in traditional healthcare practices by diverse tribal communities for addressing minor and major ailments (Chakraborty et al. 2012). In the past decade, there has been a remarkable surge in research interest and activity within the field of the use of Medicinal Plants and as a cure. Presently, approximately 80% of the global population primarily depends on traditional medicinal plants and their extracts for healthcare. Data from the World Health Organization (WHO 2002) indicates that ethnomedicine continues to maintain its popularity across all regions of the developing world, while its significance is swiftly growing in industrialized nations. The systematic examination of medicinal and aromatic plants has been increasingly recognized and prioritized, with a notable emphasis on systematic botanical and ethno-medicobotanical research. This includes comprehensive documentation, conservation efforts, and sustainable utilization on a global scale. Medicinal plants have been pivotal in shaping contemporary healthcare systems. For numerous decades, humanity has depended on traditional medicinal plants to treat and prevent illnesses, fostering improvements in both physical and spiritual well-being. India, encompassing merely 2.4% of the land area, stands among the twelve mega biodiversity nations worldwide. Within its borders thrive approximately 500,000 species, a fraction of the estimated 10–30 million species on Earth. Notably, India has bestowed at least 166 plant species to global agriculture. Furthermore, it hosts four of the world's 25 biodiversity hotspots: the Western Ghats, the Eastern Himalayas, the Indo-Burma region, and Sundaland. The floral diversity found in India's Northeastern region accounts for approximately 43% of the nation's total flora. The state of Arunachal Pradesh is in the junction of two biodiversity hotspots—The Eastern Himalaya and Indo-Burma region making the state cradle of life (Chowdhury 2008). Across the state of Arunachal Pradesh, 28 distinct ethnic groups inhabit various regions, each possessing unique indigenous knowledge of medicinal plants and having crafted diverse methodologies for harnessing nature's abundant resources. The utilization of these resources varies significantly from one tribe to another, with distinct approaches and understandings. Through the utilization of medicinal plants, their confidence in traditional folk medicine has flourished. This knowledge and belief system have been orally transmitted from one generation to the next. While some researchers have documented the diverse uses of medicinal plants by indigenous tribes of the region (Rawat & Chowdhury 1998;

Tag & Das 2004; Kala 2005; Das & Tag 2006; Sarmah et al. 2006; Nimachow et al. 2008; Sarmah et al. 2008; Tiwari et al. 2009; Sarmah 2010; Khongsai et al. 2011; Namsa et al. 2011; Das et al. 2013; Doley et al. 2014; Ghosh et al. 2014; Perme et al. 2015; Tiling et al. 2015; Omem et al. 2016; Ayam 2017; Bharali et al. 2017; Chowlu et al. 2017; Kalita et al. 2017; Tripathi & Limasenla 2017; Singh 2017; Kamum et al. 2018; Rinyo et al. 2018; Wangpan et al. 2019; Hage et al. 2020; Arya et al. 2020; Ngomle et al. 2020; Talukdar 2020; Tag et al. 2021; Gogoi & Sharma 2022; Lungphi et al. 2022), but there remains a significant need for comprehensive exploration in the field of ethno-medico-botany in Arunachal Pradesh. The primary objective of this study is to compile comprehensive information regarding the medicinal plants of Arunachal Pradesh, emphasizing their traditional uses as documented by indigenous tribes. Specifically, the study aims to (1) Document the traditional uses of medicinal plants by indigenous communities in Arunachal Pradesh. (2) Identify the parts of plants utilized for medicinal purposes. (3) Assess the medicinal significance and properties of each documented plant species. (4) Develop richness maps of medicinal plants across different elevational gradients in Arunachal Pradesh.

3.2 Study Area

Arunachal Pradesh has been designated as a "Biodiversity Hotspot" by Myers (1988), owing to its distinctive physiographical position, topography, and abundant precipitation. These factors play a crucial role in facilitating the presence of numerous medicinal plants within the state's forests. The indigenous tribes residing in the region recognize over 500 plant species for their medicinal properties (Anon 2003). These plants thrive across various altitudes and vegetation types. Arunachal Pradesh, often referred to as the "Land of Dawn-Lit Mountains," is also recognized as the botanically under-explored State of India or the Botanist's Paradise. Arunachal Pradesh spans a total geographical area of 83,743 km^2, situated between latitude 26°30'N and 29°30'N, and longitude 91°30'E and 97°30'E (Fig. 3.1). The state shares its southern borders with Assam and Nagaland, while its western boundary abuts Bhutan. To the east lies Myanmar, and the north stretches the international border with the People's Republic of China. According to other available literature, Kaul and Haridasan (1987) have extensively classified the forest types in Arunachal Pradesh. They have identified six major types of forests, namely tropical semi-evergreen, subtropical broad-leaved, subtropical pine, alpine, and secondary forest/bamboo breaks. These are categorized within four climatic categories and are compared with the classical types proposed by Champion and Seth (1968). When discussing land cover mapping of Arunachal Pradesh through RS-GIS, it has been conducted by various entities including the Forest Survey of India (FSI), the State Remote Sensing Agency (Government of Arunachal Pradesh), and additional experts such as Roy and Behera (2005). Similar to the forest classifications, the medicinal plants found in the state can be categorized as tropical, subtropical, temperate, or alpine species. The relief in Arunachal Pradesh ranges from low-lying plains, standing only a few hundred meters above sea

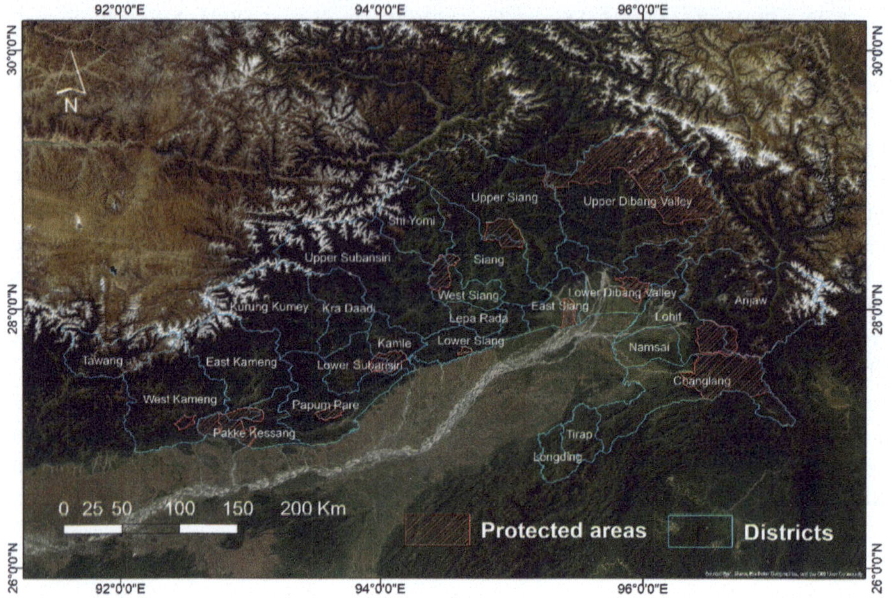

Fig. 3.1 The study area depicts the geographical extent of Arunachal Pradesh, India with its protected areas

level, to towering mountains exceeding 7,000 m (23,000 feet) in elevation (Sharma & Shukla 1992). The towns of Naharlagun, Pasighat, and Tezu in the southern region boast elevations of 290 m, 155 m, and 210 m respectively. Conversely, Kangto, Nyegi Kangsang, and the Gorichen mountain group represent some of the loftiest peaks in this Himalayan region (Sharma & Shukla 1992). Major hills found in this region include *Aka*, *Daphla*, *Miri*, *Abor*, *Mishmi*, and *Patkai* hills. The water and wetland cover span 154,609 hectares (1,546.09 km^2), constituting 1.91% of the total area. Of this, rivers comprise 86% of the wetlands. Lohit district and Dibang Valley district host the largest number of wetlands in the state (NWA 2013). The primary river systems, listed clockwise from west to east, include *Kameng*, *Subansiri*, *Siang*, *Dibang*, and *Lohit* (Singh 2005). The lower regions of the state undergo hot and humid climates, with temperatures in the foothills soaring up to 40 °C during the summer. Winter temperatures in this area typically range from 15 °C to 21 °C, while during the monsoon season, they fluctuate between 22 °C and 30 °C while the high-altitude regions remain snow-covered most of the parts of the year (Mizuno & Tenpa 2015). Arunachal Pradesh receives significant rainfall from May to September. The average annual rainfall in the region is recorded at 300 cm, with fluctuations ranging from 80 cm to 450 cm. This kind of diverse climatic condition across the different altitude gradients makes the state one of the most medicinally plant-rich states.

3.3 Materials and Methods

3.3.1 Literature Review and Data Compilation

Data compiled from published literature on medicinal plants in Arunachal Pradesh, sourced from Google Scholar, Scopus, Web of Science and the Botanical Survey of India (BSI) publications, totalled 199 literatures. These resources were consulted to create a list of medicinally important plants, detailing their usage and the parts used in treating various ailments. Authors also referenced additional multiple Botanical Survey of India publications (Singh et al. 2019; Sinha et al. 2019; Hajra et al. 1996; Giri et al. 2008; Chowdhury et al. 2009) for altitudinal information and distribution in Arunachal Pradesh. Furthermore, recent discoveries of new species among medicinal plants were cross-checked with the Botanical Survey of India's publications spanning from 2008 to 2020. The current nomenclature of each species was verified by consulting Plant Names (POWO 2024) and the World Flora Online (WFO 2024) databases.

3.3.2 Developing Richness Map of Medicinal Flora

We established the relationship between elevation and species richness within a given geographic region of Arunachal Pradesh. To achieve this, we employed a methodological approach involving the computation of species counts across various elevation levels. Initially, we determined the elevation range of each species by generating a sequence spanning from the minimum to the maximum elevation values recorded in our dataset (Table 3.2). Subsequently, we constructed a data frame, to capture species richness information corresponding to each elevation value within the defined range. Species richness was initialized to zero for all elevation levels. Next, we iteratively processed the medicinal plant data of 508 species, incrementing the species richness count for each elevation value covered by the range of a particular species. This iterative process enabled us to quantify species richness at discrete elevation intervals, providing insights into the distribution patterns of plant species along elevation gradients. Subsequently, we fitted the Generalized Additive Model (GAM) with four different model parameterizations (Table 3.1) using the *gam* Package in R (). The *Model 1* is the default smooth term with automatic smoothing. This model employs a default smooth term for the predictor variable Elevation in a GAM. The smooth term is automatically adjusted by the GAM algorithm to capture the nonlinear relationship between Elevation and Species Richness without specifying a specific functional form. It represents a basic approach to modeling the relationship between the predictor and response variables. *Model 2* specifies a larger basis dimension for more flexibility. In this model, a larger basis dimension ($k = 10$) is specified for the smooth term associated with Elevation. By increasing the basis dimension, the

Table 3.1 Model specifications and their corresponding AIC and BIC values

Models	Model specification	AIC	BIC
Model 1	s(Elevation)	42,431	42,504
Model 2	s(Elevation, k = 10)	42,431	42,504
Model 3	s(Elevation, sp = 0.5)	46,647	46,605
Model 4	te(Elevation)	56,480	56,520

model gains more flexibility to capture complex nonlinear patterns in the relationship between Elevation and Species Richness. Further, this model allows for a more detailed representation of the underlying relationship compared to *Model 1*, which may be beneficial for capturing fine-scale variability in the data. *Model 3* specifies a smaller smoothing parameter for more smoothing. Here, a smaller smoothing parameter ($sp = 0.5$) is specified for the smooth term of Elevation. Decreasing the smoothing parameter results in a smoother curve, which can help to reduce noise and capture broad trends in the relationship between Elevation and Species Richness. This particular model prioritizes overall trend estimation over capturing fine-scale variability in the data. Model 4 usually uses a different type of smooth term (e.g., thin plate regression spline). Unlike the previous models, this model employs a different type of smooth term, specifically a thin plate regression spline *te()*, for modeling the relationship between Elevation and Species Richness. Thin plate regression splines are a type of smoother that balances flexibility and smoothness, making them suitable for capturing complex nonlinear relationships. The best-fitted model was then used to predict over-elevation raster to get an estimate of Species Richness across the spatial scale of 1 km.

3.4 Results

3.4.1 Literature Review and Data Compilation

The analysis of the medicinal plant dataset reveals a rich diversity of species with various medicinal uses, shedding light on traditional medicine practices and presenting opportunities for further research and conservation efforts. With 339 unique genera and 508 species spanning 158 families, this comprehensive dataset underscores the importance of biodiversity conservation in the region. The top 10 families showcased in Fig. 3.2 highlight the significant contribution of these plant families to medicinal practices. This localization emphasizes the unique ecological niche and the importance of conservation efforts in this region. Among the different plant parts utilized for various ailments, leaves emerge as the most frequently used part, constituting 28% of the total, followed by fruits (16%) and roots (15%). Understanding the prevalence of these plant parts in medicinal practices can guide conservation strategies to ensure sustainable utilization while preserving biodiversity.

Table 3.2 Descriptions of medicinal Flora of Arunachal Pradesh, India

No	Plant species	Family	Altitude (mtr.)	Local name	Parts used	Medicinal uses	References
	Artocarpus heterophyllus Lam	Moraceae	700	Belang	Roots, latex, leaves	Skin disease, ringworm infestation, cracking of feet (Hazarika and Tayeng 2020), fever, boils, wounds (Prakash et al. 2009), asthma (Jagtap and Bapat 2010)	Hazarika and Tayeng 2020; Prakash et al. 2009; Jagtap and Bapat 2010
	Ardisia crenata Sims	Myrsinaceae	1400–2000	So-Bumlo	Leaf	Skin disease (Hazarika and Tayeng 2020), chicken pox, ringworm, infestation (Hazarika et al. 2022)	Hazarika and Tayeng 2020; Hazarika et al. 2022
	Averrhoa carambola L.	Oxalidaceae	500–1000	Tador, Kordoi, Kadung, Namsu (A), Maakphung	Leaves, seed, fruits	Chicken pox, ringworms, fever, throat inflammation (Hazarika and Tayeng 2020), jaundice (Namsa and Tangjang 2011), hepatoprotective agent (Asha and Singh 2020), fruits edible (Ronald et al. 2019), used in Jaundice (Hage Yanka and Tag 2020), stomach problem, check bleeding and environment purifier (Ayam 2017), decoction of leaf, root, bark and twig is given in jaundice (Benniamin 2011), ripen fruits are eaten and preserved as chips (Shankar et al. 2015)	Ayam 2017; Hazarika and Tayeng 2020; Namsa and Tangjang 2011; Asha and Singh 2020; Ronald et al. 2019; Hage Yanka and Tag 2020; Shankar et al. 2015; Benniamin 2011
	Antidesma bunius (L.) Spreng	Phyllanthaceae	1000–1600	Kompong-Perong	Bark, leaves, fruits	Indigestion, cough, stomachache, insomnia (Hazarika and Tayeng 2020), wounds (Hazarika et al. 2022)	Hazarika and Tayeng 2020; Hazarika et al. 2022
	Bischofia javanica Blume	Phyllanthaceae	270–2000	Sitiil, Mebu-changne (N)	Leaves, bark	Cholera, skin diseases (Hazarika and Tayeng 2020), gastric and jaundice (Tiwari et al. 2009), cancer, inflammation, tuberculosis, diarrhoea, sore throat, burns and different allergic conditions (Rajbongshi et al. 2014)	Tiwari et al. 2009; Hazarika and Tayeng 2020; Rajbongshi et al. 2014
	Bauhinia variegata L.	Fabaceae	Upto 1800	Ogok	Leaves, flower	Stomachache, diarrhoea, dysentery (Hazarika and Tayeng 2020; Bushi et al. 2021), antidiarrhoeal, antidiabetic, antioxidant, anti-hyperlepidemic activity (Sharma et al. 2021)	Hazarika and Tayeng 2020; Sharma et al. 2021; Bushi et al. 2021
	Baccaurea ramiflora L.	Phyllanthaceae	100–1300	Bureng	Bark, fruit	Constipation (Hazarika and Tayeng 2020), analgesic, anthelmintic, antioxidant, anti-diarrheal, anti-inflammatory, cytotoxic, haemolytic, hpoglycemic, hypolipidemic, insecticidal, neuropharmacological, thrmobocytic, anti-fungal and antimicrobial activities (Goyal et al. 2022)	Hazarika and Tayeng 2020; Goyal et al. 2022
	Citrus grandis (L.)	Rutaceae	500–1200	Rebab	Fruit, leaf	Swelling, constipation (Hazarika and Tayeng 2020)	Hazarika and Tayeng 2020

(continued)

Table 3.2 (continued)

No	Plant species	Family	Altitude (mtr.)	Local name	Parts used	Medicinal uses	References
	Citrus medica L.	Rutaceae	500–1200	Sipa-Yegra, Jipin, Nimu Tiinga, Chachhori, Jipin, Kaginemu, Makhmuyum	Root, leaves, fruits	Constipation, backpain (Hazarika and Tayeng 2020), ripe fruits (sweet) are edible (Ronald et al. 2019), anticancer (Hage Yanka and Tag 2020), treatment of scurvy, intestinal ailments, antidote, anticancer, weak eyesight, vomiting, skin diseases, haemorrhoids (Tshering et al. 2018), treatment of scurvy, intestinal ailments, antidote, anti cancer, weak eyesight, vomiting, skin diseases, haemorrhoids (Gogoi and Sharma 2022b), gastric and tumors (Shankar and Rawat 2008), fruit juice used for common cold is also applied on head for dandruff (Benniamin 2011)	Hazarika and Tayeng 2020; Ronald et al. 2019; Hage Yanka and Tag 2020; Tshering et al. 2018; Gogoi and Sharma 2022b; Shankar and Rawat 2008; Benniamin 2011
	Calamus tenuis Roxb	Arecaceae		Esong	Shoot	Stomach disorder (Hazarika and Tayeng 2020)	Hazarika and Tayeng 2020
	Castanopsis indica Roxb	Fagaceae	Upto 2000	Koyu–Korang Raaye, Hirang, Khe-Shing (M), Sirang, Kaeshing, Hirang (A)	Leaves, bark, fruit	Stomach disorder, skin disease, indigestion, headache (Hazarika and Tayeng 2020), applied to wounds to stop bleeding (Hage Yanka and Tag 2020), cough and cold (Kalita et al. 2017), ripe fruits are edible (Ronald et al. 2019), fish poison and raw seeds are eaten (Kamum et al. 2018), nuts are taken as edible fruits (Shankar et al. 2015)	Kalita et al. 2017; Kamum et al. 2018; Hazarika and Tayeng 2020; Ronald et al. 2019; Hage Yanka and Tag 2020; Shankar et al. 2015
	Citrus limon Burm.f	Rutaceae	200–500	Singkom, Hatta	Fruit, leaves and stem	Dysentery (Tiwari et al. 2009; Hazarika and Tayeng 2020), cuts, indigestion (Hazarika and Tayeng 2020), dehydration, stomachic trouble and liver problem (Tshering et al. 2018), leg paralysis (Das and Tag 2006b)	Tiwari et al. 2009; Hazarika and Tayeng 2020; Tshering et al. 2018; Das and Tag 2006b
	Canarium strictum Roxb	Burseraceae	1000–1200	Hilum/Tikok, Schellum, Saan-ri	Leaves, bark, root	Insect bite, fever, cough, chicken pox (Hazarika and Tayeng 2020), juice from bark used against insect bite (Murtem and Chaudhry 2016), eczema in pigs, cough and cold (Tiwari et al. 2009)	Tiwari et al. 2009; Hazarika and Tayeng 2020; Murtem and Chaudhry 2016

(continued)

Table 3.2 (continued)

No	Plant species	Family	Altitude (mtr.	Local name	Parts used	Medicinal uses	References
	Dillenia indica L.	Dilleniaceae	100–500	SompaJaopa, Masangsi, Trogochocina (Mem/Khamb), Akka (I), Sampak, Ahutenga, Jampa, Bhavya, Chaltatanga, Champa	Fruits, leaf, flower	Dysentery, diarrhoea, fever (Hazarika and Tayeng 2020), stomachache (Saikia et al. 2021), antimalaria (Kalita and Khan 2013), diabetes (Tiwari et al. 2009), cough and cold, cholera, diarrhea and abnormal obstruction (Sarmah et al. 2008b), to improve appetite, heart fever, cough and mouth disease (Tshering et al. 2018), curing dandruff, wound healing, bone fracture (Gogoi and Sharma 2022b), hair tonic (Shankar and Rawat 2008), edible and medicinal (Rama 2013), fruit decoction is applied to scalp for curing dandruff, wound healing, bone fracture (Gogoi and Sharma 2022b), gummy substance from fruit used in hair fall problem. fruits also edible (Murtem and Chaudhry 2016), malnutrition (Bharali et al. 2017), fruit is used to improve appetite, heart fever, cough and mouth disease (Tripathi and Limasenla 2017), calyx is boiled and takes as vegetable to relieve cough and fever; jelly substance of fruit used for washing hair; flowers taken orally for treating dysentery (Rethy et al. 2010), fruits along with fishes are used during stomach disorders (Wangpan et al. 2019)	Tiwari et al. 2009; Bharali et al. 2017; Tripathi and Limasenla 2017; Wangpan et al. 2019; Hazarika and Tayeng 2020; Tshering et al. 2018; Murtem and Chaudhry 2016; Saikia et al. 2021; Kalita and Khan 2013; Sarmah et al. 2008b; Gogoi and Sharma 2022b; Shankar and Rawat 2008; Rethy et al. 2010; Rama 2013
	Emblica officinalis G.	Phyllanthaceae	350–1700	Amloki, Phatmakri	Bark, fruits	Scurvy (Hazarika and Tayeng 2020), antidiarrhoeal (Namsa and Tangjang 2011), juice of bark is used for treatment of gonorrhea; fruits eaten for liver trouble, diarrhea; and dysentery, young fruit extract with water is used as hair dye (Rethy et al. 2010)	Hazarika and Tayeng 2020; Namsa and Tangjang 2011; Rethy et al. 2010
	Elaeocarpus floribundus Blume	Leguminoceae	100–1500	Jolpai	Leaves, bark	Inflamed gums (Hazarika and Tayeng 2020), bark and leaf infusion is used as mouth wash for inflamed gums. fruit is rich source of vitamin C and digestive (Tshering et al. 2018)	Hazarika and Tayeng 2020; Tshering et al. 2018
	Entada pursaetha DC.	Elaeocarpaceae		Pyopa	Leaves, bark	Fever, cold (Hazarika and Tayeng 2020)	Hazarika and Tayeng 2020

(continued)

Table 3.2 (continued)

No	Plant species	Family	Altitude (mtr.)	Local name	Parts used	Medicinal uses	References
	Ficus auriculata Lour	Moraceae	400–2000	Tapang, Hote (A), Bangoh (B), Choma (M), Takuk (N)	Stem, fruits, inflorescence, leaf, bark	Diarrhoea, dysentery, cuts, wounds, vomiting, cholera (Hazarika and Tayeng 2020), fruit is edible as a food supplement, leaf as fodder, used as a laxative, to relieve bee sting and during diarrhea (Asha and Singh 2020), ripe fruits are edible (Ronald et al. 2019), mature inflorescence are sweet in taste which is consumed to removed fish bone hooked up in throat (Kalita et al. 2017), fruits used as pig feed (Murtem and Chaudhry 2016)	Kalita et al. 2017; Hazarika and Tayeng 2020; Asha and Singh 2020; Ronald et al. 2019; Murtem and Chaudhry 2016
	Ficus semicordata Buch.-Ham	Moraceae	500–1700	Takuk, Bangoh (B), Tekuk, Tukusen (Adi/N/M)	Latex, fruits, leaf, root, bark	Boils, constipation, headache, indigestion (Hazarika and Tayeng 2020), ripe fruits are edible (Tiwari et al. 2009), root medicinal, fruits edible (Rama 2013), toothache and diarrhea (Tiwari et al. 2009)	Tiwari et al. 2009; Tiwari et al. 2009; Hazarika and Tayeng 2020; Rama 2013
	Ficus hirta Reinw.ex Blume	Moraceae	Upto 1200	Siipop, Bangoh (B)	Root	General weakness (Hazarika and Tayeng 2020)	Hazarika and Tayeng 2020
	Garcinia cowa Roxb	Clusiaceae	Upto 1200	Tabi-Tarak, Kowthekera	Fruits	Dysentery, fever, constipation (Hazarika and Tayeng 2020), dysentery and diarrhea (Tiwari et al. 2009), digestive disorder (Bharali et al. 2017), sliced dried fruit juice drank for treatment of dysentery	Tiwari et al. 2009; Bharali et al. 2017; Hazarika and Tayeng 2020
	Gynocardia odorata Roxb	Achariaceae	100–800	Sibeng-Tulpii	Seeds, young shoot	Sprains, skin diseases, jaundice (Hazarika and Tayeng 2020)	Hazarika and Tayeng 2020
	Melastoma malabathricum L.	Melastomataceae	100–800	Kesi-Yayi, Pudiraju, Acha-kaya, Dai dasa (N), Pepu yoyu	Flowers, fruits and leaves	Dysentery, diarrhoea, cuts, wound (Hazarika and Tayeng 2020), ripe fruits are edible (Ronald et al. 2019), fruits edible, leaves medicinal (Rama 2013), edible fruit; fruit used in dying. Fruit used in dry lips, wound healing and toothache (Asha and Singh 2020), used for the treatment of Jaundice (Kalita et al. 2017)	Kalita et al. 2017; Hazarika and Tayeng 2020; Asha and Singh 2020; Ronald et al. 2019; Rama 2013
	Myrica esculenta Buch.-Ham	Myricaceae	1200–2100	Ketir	Bark, flower	Diarrhoea (Hazarika and Tayeng 2020)	Hazarika and Tayeng 2020
	Morus alba L.	Moraceae	2700	Eeyum-Tayin, Talu/ Tuda, Nuni (As)	Leaves, fruit, bark	Fever (Hazarika and Tayeng 2020), fruit is used as remedy for throat sour and fever (Tripathi and Limasenla 2017), bark used as anthelmentic and wound healing (Shankar et al. 2015)	Tripathi and Limasenla 2017; Hazarika and Tayeng 2020; Shankar et al. 2015

(continued)

Table 3.2 (continued)

No	Plant species	Family	Altitude (mtr.)	Local name	Parts used	Medicinal uses	References
	Maesa japonica (Thunb.) Moritzi and Zoll	Primulaceae	Upto 1000	Ko-Kebung	Leaves	Itching (Hazarika and Tayeng 2020)	Hazarika and Tayeng 2020
	Mangifera sylvatica Roxb	Anacardiaceae	1200–1400	Monaam-Tagung, Haisotri	Seed, Bk, jungli neem (A)	Diarrhoea (Hazarika and Tayeng 2020), nail infection, diseases of pigs (Tiwari et al. 2009), diseases spread (Sarmah et al. 2008b)	Tiwari et al. 2009; Hazarika and Tayeng 2020; Sarmah et al. 2008b
	Litsea cubeba (Lour.) Pers	Lauraceae	500–2500	Raayil, Nyangshing (M), Tayer Rajil/ Taier/rayil (A), Santero (Ap), Tayer/Ischi tak—ke ame (G), Earking, Jayar/tanam (N), Santero (Wild), Tasmu (A), Tair, Minzay, Nengshing	Seed, fruit, leaf, wood, tender leaves and fruits, stem	Cough, indigestion (Hazarika and Tayeng 2020), fruits are used as spice. Leaves and wood are bum to purify the air (Kalita et al. 2017), curries, chutiny and pickle (Ayam 2017), cough, cold, hair tonic, indigestion, good sleep (Saikia et al. 2021), pounded tender leaves and fruits mixed with water are taken for dysentery, indigestion, stomach ache, cold and cough. It also provides sound sleep (Ratan et al. 2016), worm parasite (Sarmah et al. 2008b), blood clotting agent (Bharali et al. 2017), fruits edible (Rama 2013), food (Bajpay et al. 2019), essential oils and antibacterial (Namsa and Tangjang 2011), condiments, eczema, heart disease and stomach disorder (Tshering et al. 2018), condiments, eczema, heart disease and stomach disorder (Kamum et al. 2018)	Ayam 2017; Bharali et al. 2017; Kalita et al. 2017; Kamum et al. 2018; Hazarika and Tayeng 2020; Namsa and Tangjang 2011; Tshering et al. 2018; Saikia et al. 2021; Sarmah et al. 2008b; Ratan et al. 2016; Rama 2013; Bajpay et al. 2019
	Livistona jenkinsiana Griff	Arecaceae	100–2500	Taek	Seed, fruits, leaves, petiole	Stomach ailment (Hazarika and Tayeng 2020), ripe fruits are edible (Ronald et al. 2019), fruits edible, leaf as roofing material	Hazarika and Tayeng 2020; Ronald et al. 2019; Rama 2013
	Leea indica (Burm.f.) Merr	Leeaceae	780–2100	Talu	Leaves, roots	Dysentery, fever, headache, sprain (Hazarika and Tayeng 2020)	Hazarika and Tayeng 2020
	Phoebe cooperiana P.C. Kanj. and Das	Lauraceae	400–1300	Tapil, Tapir	Bark	Itching (Hazarika and Tayeng 2020), wounds and boils, itching (Hage Yanka and Tag 2020)	Hazarika and Tayeng 2020; Hage Yanka and Tag 2020
	Pandanus odoratissimus L.	Pandanaceae	294	Taku	Leaf	Skin diseases (Hazarika and Tayeng 2020)	Hazarika and Tayeng 2020

(continued)

Table 3.2 (continued)

No	Plant species	Family	Altitude (mtr.)	Local name	Parts used	Medicinal uses	References
	Rhus chinensis Mill	Anacardiaceae	100–2800	Tagmo, Mithrek (B) Phsilun (S), Basang	Fruit, leaf, branch, seeds	Cough, fever, dysentery (Hazarika and Tayeng 2020), the branches are used in rituals. Seeds are crushed into paste and use as chutney. Also used as anti-emetic and helps in cleansing of stomach. The stems are also used as a boundary fencing demarcating the area of two communities (Kalita et al. 2017), high blood pressure, sysentery, indigestion (Tiwari et al. 2009)	Tiwari et al. 2009; Kalita et al. 2017; Hazarika and Tayeng 2020
	Rubus calycinoides Hayata ex KoidZ	Rosaceae	780–2100	Pasi-Payi	Leaf	Wound (Hazarika and Tayeng 2020)	Hazarika and Tayeng 2020
	Rubus ellipticus Smith	Rosaceae	2000–3000	Tayin, Pakom tinne, Mipya jilyung (Ap), Raspberry, Nyinkhe jilying, Tap tara (A)	Fruit, roots, young shoot	Gastric, fever, wound, dysentery (Hazarika and Tayeng 2020), fruit edible (Ronald et al. 2019), indigestion (Saikia et al. 2021), fruit is used as a food supplement, used as laxative during constipation (Asha and Singh 2020), jaundice, chronic abdominal pair, arthritis, rheumatic, inflammatory disorder (Singh et al. 2012), used for the treatment of cough (Kalita et al. 2017), leaves as substitute for pepper betel; sweet ripen fruits are eaten (Shankar et al. 2015)	Kalita et al. 2017; Hazarika and Tayeng 2020; Asha and Singh 2020; Ronald et al. 2019; Saikia et al. 2021; Shankar et al. 2015; Singh et al. 2012
	Rubus niveus Thunb	Rosaceae	2000–3000	Inmin-Tayin, Miliang-sai (B) Mulang (S), Yikhe jilyung (Ap)	Fruits, roots	Dysentery (Hazarika and Tayeng 2020), fruits are sweet which is eaten raw (Kalita et al. 2017), fruit is used as a food supplement, root in wound healing and anti-inflammatory agent (Asha and Singh 2020)	Kalita et al. 2017; Hazarika and Tayeng 2020; Asha and Singh 2020
	Saurauia roxburghii Wall	Actinidiaceae	200–1500	Taan	Leaves, bark	Fever, boils (Hazarika and Tayeng 2020)	Hazarika and Tayeng 2020
	Syzygium cumini Lam	Myrtaceae	2000–3000	Jongkeng, Jamun	Bark and fruits	Dysentery, diarrhoea (Hazarika and Tayeng 2020), medicinal and edible fruits (Rama 2013)	Hazarika and Tayeng 2020; Rama 2013
	Spondias pinnata Kurz	Anacardiaceae	500–800	Belum/Dorgeng, Ansiari (G)	Leaves, fruits, bark	Cough, joints pain, diarrhoea, sour throat (Hazarika and Tayeng 2020), gastric, jaundice and dysentery (Tiwari et al. 2009), skin diseases (Bharali et al. 2017)	Tiwari et al. 2009; Bharali et al. 2017; Hazarika and Tayeng 2020
	Sterculia lanceifolia Roxb	Malvaceae	500–1200	Aanggom-Bayom	Pod	Fever (Hazarika and Tayeng 2020)]	Hazarika and Tayeng 2020

(continued)

Table 3.2 (continued)

No	Plant species	Family	Altitude (mtr.)	Local name	Parts used	Medicinal uses	References
	Trichosanthes dunniana Levl	Cucurbitaceae	900–1900	Pak-Sung gal	Fruits	Wound, diarrhoea (Hazarika and Tayeng 2020)	Hazarika and Tayeng 2020
	Terminalia citrina Roxb.ex	Combretaceae	500–800	Hilika	Seed, bark, fruits	Fever, stomach aches, kidney problem (Hazarika and Tayeng 2020), fruits used in constipation and gastric (Murtem and Chaudhry 2016)	Hazarika and Tayeng 2020; Murtem and Chaudhry 2016
	Tamarindus indica L.	Fabaceae	100–300	Imli, Tetuli (A)	Fruits, leaf, bark	Cough, cold, asthma, fever (Hazarika and Tayeng 2020), chutney and pickles (Ayam 2017), tonic, jaundice and diarrhea (Tiwari et al. 2009)	Tiwari et al. 2009; Ayam 2017; Hazarika and Tayeng 2020
	Zanthoxylum rhetsa (Roxb) DC.	Rutaceae	1200–2000	Onger, honyor (N)	Bark, leaves, seeds, fruits	Wart, jaundice, stomach pain, snake bite, asthma, toothache (Hazarika and Tayeng 2020), leaves used as spice with meat, also eaten as chutney (Ronald et al. 2019), used in Jaundice, fever, headache (Hage Yanka and Tag 2020), boiled foods (Ayam 2017), jaundice, wart (Tshering et al. 2018), leaves are used constipation and other stomach disorders. leaves mixed with pork help to avoid tapeworm infection (Wangpan et al. 2019)	Ayam 2017; Wangpan et al. 2019; Hazarika and Tayeng 2020; Ronald et al. 2019; Hage Yanka and Tag 2020; Tshering et al. 2018
	Zanthoxylum armatum DC.	Rutaceae	1000–2000	Ombey, Brasing, Zabrang (M), Suiji (S), Trimal onger (A), Yakhung/ Yorkhung (Ap), Rikon (G), Singlu/ sente (N), Timbru, Honyur, Yarkhung, Kundu (L), Nangri (B), Timur or Timmur (N), Khagi (M)	Leaves, bark, fruit, seed	Cuts, wounds, stomach pain (Hazarika and Tayeng 2020), leaves used as spice with meat, also eaten as chutney (Ronald et al. 2019), Powdered dry fruits are use as spice along with stable food (Kalita et al. 2017), boiled foods and chutney (Ayam 2017), anti diabetic (Bam et al. 2015), seeds and bark aromatic tonic, dental disorders and also in rheumatism (Balkrishna et al. 2019), seed and bark are used as tonic during fever and cholera, stomach disorder (Tshering et al. 2018), cold, cough, fever, appetizer (Saikia et al. 2021), seed and bark are used as tonic during fever and cholera, stomach disorder (Gogoi and Sharma 2022b), parts are finely crushed and thrown into stream (Tiwari et al. 2009), used for the treatment of cough (Kalita et al. 2017), leech repellent (Namsa et al. 2011b)	Tiwari et al. 2009; Ayam 2017; Kalita et al. 2017; Kalita et al. 2017; Hazarika and Tayeng 2020; Ronald et al. 2019; Tshering et al. 2018; Saikia et al. 2021; Gogoi and Sharma 2022b; Bam et al. 2015; Balkrishna et al. 2019; Namsa et al. 2011b
	Ziziphus mauritiana Lam	Rhamnaceae	300–500	Bogori	Bark, root, fruits	Fever, old wounds, indigesion (Hazarika and Tayeng 2020)	Hazarika and Tayeng 2020
	Alsophila gigantea Wall. ex Hook	Cyatheaceae	200–1200		Fronds, rhizome	Inflammation, leucorrhoea and used against snake bite and in leucorrhoea (Tsering and Tag 2015)	Tsering and Tag 2015

(continued)

Table 3.2 (continued)

No	Plant species	Family	Altitude (mtr.)	Local name	Parts used	Medicinal uses	References
	Alsophila henryi Baker	Cyatheaceae	120–500		Whole plant	Ornamental (Tsering and Tag 2015)	Tsering and Tag 2015
	Angiopteris crassipes Wall. ex C. Presl	Marattiaceae	1000–1600		Stem, fronds	Leucoderma (Tsering and Tag 2015)	Tsering and Tag 2015
	Angiopteris helferiana C. Presl	Marattiaceae	600–1000		Whole plant	Indigestion, dysentery and hairloss (Tsering and Tag 2015)	Tsering and Tag 2015
	Asplenium finlaysonianu m Wall. ex Hook	Aspleniaceae	50–600		Roots	Dysentery (Tsering and Tag 2015)	Tsering and Tag 2015
	Asplenium nidus L.	Aspleniaceae	600–1600		Whole plant	Skin disease, jaundice, malaria, fever, bone fracture and urinary disorders (Tsering and Tag 2015)	Tsering and Tag 2015
	Asplenium nitidum Sw	Aspleniaceae	50–1000		Whole plant	Ornamental (Tsering and Tag 2015)	Tsering and Tag 2015
	Asplenium phyllitidis D.Don	Aspleniaceae	600–1600		Fronds	Jaundice and fever (Tsering and Tag 2015)	Tsering and Tag 2015
	Blechnum orientale L.	Blechnaceae	50–600		Fronds, rhizome	Urinary bladder complaints, worm infection and diaphoretic, also used as aperitif (Tsering and Tag 2015)	Tsering and Tag 2015
	Bolbitis heteroclita (C.Presl) Ching	Dryopteridacea	600–1000		Whole plant	Ornamental for aquarium (Tsering and Tag 2015)	Tsering and Tag 2015
	Ceratopteris thalictroides (L.) Brongn	Pteridaceae	50–1600		Fronds	Skin diseases, wounds, stomachache and piles (Tsering and Tag 2015)	Tsering and Tag 2015
	Cibotium barometz (L.) J.Sm	Cyatheaceae	50–1600		Whole plant	Health tonic and lumbago (Tsering and Tag 2015)	Tsering and Tag 2015
	Diplazium dilatatum Blume	Athyriaceae	600–1000		Fronds	Diuretic (Tsering and Tag 2015)	Tsering and Tag 2015
	Diplazium esculentum (R etz.) Sw	Athyriaceae	50–2200	Taka, Hiikahamang, Kajingna, Dangsom (S), Pamsam (B)	Rhizome, fronds, young shoot, leaves, whole plant	Blood disorders, gout, constipation, earache, malaria fever, cough and cold (Tsering and Tag 2015), vegetable (Rama 2013), Young shoots are eaten (Ronald et al. 2019), anti diabetic (Bam et al. 2015), constipation (Saikia et al. 2021), it relieves indigestion (Gangwar and Ramakrishnan 1990), digestive disorder (Bharali et al. 2017), decoction prepared from rhizome and young leaves are used for haemophtysis and cough (Yomgam et al. 2017), tender frond are cooked and consumed as vegetable along with fish (Kalita et al. 2017)	Bharali et al. 2017; Ronald et al. 2019; Saikia et al. 2021; Rama 2013; Bam et al. 2015; Tsering and Tag 2015; Gangwar and Ramakrishnan 1990; Yomgam et al. 2017

(continued)

Table 3.2 (continued)

No	Plant species	Family	Altitude (mtr.)	Local name	Parts used	Medicinal uses	References
	Dipteris wallichii (R. Br.) T. Moore	Dipteridaceae	1000–1600		Whole plant	Stomachache, jaundice (Tsering and Tag 2015)	Tsering and Tag 2015
	Equisetum diffusum D. Don	Equisetaceae	600–1000		Whole plant	Urinary problems, haemostatic, acidity, dyspepsia, bone fracture and hydrophobia (Tsering and Tag 2015)	Tsering and Tag 2015
	Equisetum ramosissimu m subsp. debile (Roxb. ex Vaucher) Hauke	Equisetaceae	50–3000		Stem, rhizome, cones	Diuretic, bleeding, gonorrhoea, joint pain, fungal infection and kidney disorders (Tsering and Tag 2015)	Tsering and Tag 2015
	Leptochilus pteropus (Blume) Fraser-Jenk	Polypodiaceae	600–1000		Whole plant	Cut and wounds (Tsering and Tag 2015)	Tsering and Tag 2015
	Lycopodiella cernua (L.) Pic. Serm	Lycopodiaceae	600–1600		Rhizome	Nervous disorders, rheumatism, itching, cough and cold, leucorrhoea, fever and dropsy (Tsering and Tag 2015)	Tsering and Tag 2015
	Macrothelypteris torresiana (Gaudich.) Ching	Thelypteridaceae	50–1600		Frond	Inflammation, pain, fever (Tsering and Tag 2015)	Tsering and Tag 2015
	Microlepia rhomboidea (Wall. ex Kunze) Prantl	Dennstaedtiaceae	600–1600		Whole plant	Ornamental (Tsering and Tag 2015)	Tsering and Tag 2015
	Microsorum punctatum (L.) Copel	Polypodiaceae	50–600		Fronds	Constipation, urinary disorders, snakebite, dysentery and for healing wounds (Tsering and Tag 2015)	Tsering and Tag 2015
	Nephrolepis cordifolia (L.) C. Presl	Nephrolepidaceae	1000–2200		Rhizome, fronds	Bacterial infection, cough and cold, fungal infection, jaundice, rheumatism, liver and skin disorder (Tsering and Tag 2015)	Tsering and Tag 2015
	Odontosoria chinensis (L.) J. Sm	Lindsaeaceae	50–2200		Fronds	Intestine infection, urinary problems, cut and wounds (Tsering and Tag 2015)	Tsering and Tag 2015
	Onychium siliculosum (D esv.) C. Chr	Pteridaceae	50–1600		Whole plant	Dysentery, hair fall, dysentery, skin diseases (Tsering and Tag 2015)	Tsering and Tag 2015
	Pityrogramma calomelanos (L.) Link	Pteridaceae	50–3000		Whole plant	Kidney trouble, body insects, fever, hypertension, cough, measles, boils in mouth and nose, wounds, gastric disorders, fever, cold, asthma, anthelmintic, cancer (Tsering and Tag 2015)	Tsering and Tag 2015
	Pseudodrynaria coronan s (Wall. ex Mett.) Ching	Polypodiaceae	50–1000		Rhizomes	Herpes (Tsering and Tag 2015)	Tsering and Tag 2015

(continued)

Table 3.2 (continued)

No	Plant species	Family	Altitude (mtr.)	Local name	Parts used	Medicinal uses	References
	Pteris biaurita L.	Pteridaceae	50–1600		Fronds	Sores and wounds, body pain, dysentery, body pain (Tsering and Tag 2015)	Tsering and Tag 2015
	Pteris ensiformis Burm. f	Pteridaceae	600–1000		Rhizome, young fronds	Dysentery, bone fracture, glandular swelling of neck and menstruation (Tsering and Tag 2015)	Tsering and Tag 2015
	Pteris vittata L.	Pteridaceae	50–600		Whole plant, rhizome	High blood pressure, diarrhoea and dysentery, fever (Tsering and Tag 2015)	Tsering and Tag 2015
	Pyrrosia lanceolata (L.) Farw	Polypodiaceae	600–1000		Fronds	Cough and cold and sore throats (Tsering and Tag 2015)	Tsering and Tag 2015
	Pyrrosia piloselloides (L.) M.G. Price	Polypodiaceae	Upto 1000		Leaves	Bone fracture; fronds pounded with Gypsum is applied to irritating rashes and poultice for headaches (Tsering and Tag 2015)	Tsering and Tag 2015
	Selaginella uncinata (Desv. ex Poir.) Spring	Selaginellaceae	Upto 400		Whole plant	Bacterial infection, liver infection, brain tumor (Tsering and Tag 2015)	Tsering and Tag 2015
	Selaginella wallichii (Hoo k. and Grev.) Spring	Selaginellaceae	100–1500		Whole plant	Protective medicine after child birth and cough (Tsering and Tag 2015)	Tsering and Tag 2015
	Sphaeropteris brunoniana (Wall. ex Hook.) R.M.Tryon	Cyatheaceae	600–1000		Whole plant	Pus (Tsering and Tag 2015)	Tsering and Tag 2015
	Stenochlaena palustris (Burm. f.) Bedd	Blechnaceae	50–600		Fronds	Burns, ulcer and body cooling, fever, skin diseases, throat infection and gastric ulcer, fever, stomach ache (Tsering and Tag 2015)	Tsering and Tag 2015
	Tectaria fuscipes (Wall.) C. Chr	Tectariaceae	600–1000		Fronds	Food (Tsering and Tag 2015)	Tsering and Tag 2015
	Tectaria polymorpha (W all. ex Hook.) Copel	Tectariaceae	50–2200		Whole plant	Eczema and scabies, fever, worm infection (Tsering and Tag 2015)	Tsering and Tag 2015
	Thelypteris arida (D.Don) Morton	Thelypteridacea	600–1600		Whole plant	Wounds and cuts (Tsering and Tag 2015)	Tsering and Tag 2015
	Thelypteris dentata (Forssk.) E.P. St.John	Thelypteridacea e	50–2200		Fronds	Swellings, rheumatism, blood vomiting, urinary disorders, insect repellent, wounds and cuts (Tsering and Tag 2015)	Tsering and Tag 2015
	Thelypteris nudata (Roxb.) C.V.Morton	Thelypteridacea e	50–2200		Whole plant	Tooth trouble (Tsering and Tag 2015)	Tsering and Tag 2015

(continued)

Table 3.2 (continued)

No	Plant species	Family	Altitude (mtr.	Local name	Parts used	Medicinal uses	References
	Thelypteris procera (D.Don) Fraser-Jenk	Thelypteridacea e	50–2200		Whole plant	Skin diseases (Tsering and Tag 2015)	Tsering and Tag 2015
	Vittaria elongata Sw	Pteridaceae	50–600		Whole plant	Knee pain, therapeutic pain, cuts, wounds, rheumatism and joint pains (Tsering and Tag 2015)	Tsering and Tag 2015
	Aconitum ferox Seringe	Ranunculaceae	3000–5000	Niyong, Emo, Tsando, Shaga-manshing, Wild, Omli, Bish	Roots, rhizome/leaf, tuber	Small amount of dry root powder used with water in fever and digestive problem; also used as a poison (Ngomle et al. 2020), antidote for snake bite, boil, body pain and swelling. Used as arrow poison for hunting large animals like bear, tiger (Hage Yanka and Tag 2020), dried tuber boiled in alcohol to detoxify; a very small peace used as antipyretic, analgesic and anti-inflammatory; also against common ailments like high fever, stomachic, tonsillitis and throat pain, and cheilitis (Sen et al. 2008), underground roots and tubers are used in arrow poisoning by local hunters (Tshering et al. 2018), poison to kill rats and wild animals (Kamum et al. 2018), underground roots and tubers are used in arrow poisioning by local hunters (Gogoi and Sharma 2022b), antipyretic, rheumatism and skin diseases (Shankar and Rawat 2008)	Kamum et al. 2018; Ngomle et al. 2020; Hage Yanka and Tag 2020; Tshering et al. 2018; Gogoi and Sharma 2022b; Shankar and Rawat 2008; Sen et al. 2008
	[Aconitum heterophyllum Wallich ex Royle	Ranunculaceae	2800–4500	Chando, Emo, Gonga-karpu, Atish, Bonga-Karpho, Mra (A)	Root, rhizome/leaf, tuber	Small amount of powder of dried root taken with water in digestive problem dysentery and constipation (Ngomle et al. 2020), antidote for snake bite, boil, body pain and swelling. Used as arrow poison for hunting large animals like bear, tiger (Hage Yanka and Tag 2020), dried tuber consumed during toxicities (snake bite, aconite contamination), stomachic, high fever, and for all unknown diseases (Sen et al. 2008), root aphrodisiac, tonic (Balkrishna et al. 2019), antimalaria (Kalita and Khan 2013), root used as arrow poisoning and in snakebite; rhizome is applied on	Ngomle et al. 2020; Hage Yanka and Tag 2020; Kalita and Khan 2013; Shankar et al. 2015; Balkrishna et al. 2019; Sen et al. 2008
	Anemone vitifolia DC.	Ranunculaceae	1400–1600	Vine-leaved anemone	Roots and leaves	Skin infection mainly for ring worms and eczema (Ngomle et al. 2020)	Ngomle et al. 2020
	Artemisia annua L.	Asteraceae	2000–3700	Newglum	Leaves	Dermatological problem and eye pain (Ngomle et al. 2020)	Ngomle et al. 2020

(continued)

Table 3.2 (continued)

No	Plant species	Family	Altitude (mtr.)	Local name	Parts used	Medicinal uses	References
	Centella asiatica (L.) Urban	Apiaceae	200–1900	Sher-Gor-Go, Ngilyang Khiko haman, Loram/ Kipum Yayum, Manimuni, Cult, Barang, Kedoh keroh, Kaso Kruna or sewal ing, Nguri, Mendooh pinri, Mawanro, Yatunemih, Ningkhoi, Ngiilyan akho hamang (Wild), Enge (A).Son a (I), Mandookapami/ Brahmi, Mandukapami, Manimuni, Panaang, Bengpata Tita sak	Leaves and young shoot, whole plant, shoots	Dry leaf is used against fever; fresh leaf juice is used against dysentery (Ngomle et al. 2020), antimicrobial, antiinflammatory, anticancer, antioxidant (Hage Yanka and Tag 2020), used in gastric problem, constipation, blood purification (Hage Yanka and Tag 2020), antibacterial and essential oil (Namsa and Tangjang 2011), plant is used in arthritis, diabetes, blood disorders and brain tonic (Tshering et al. 2018), constipation, gastritis, blood purification (Saikia et al. 2021), stomach disorder, cuts, wounds, inflammations& common vegetable (Kamum et al. 2018), fresh plant juice with honey is given in stomach ulcer, leprosy, the leaves are warm on fire and applied on ringworm affected areas (Tiwari et al. 2009), cures skin disorder, syphilis, rheumatism, leprosy, epilepsy, nervous and immune system disorders (Gangwar and Ramakrishnan 1990), leaves used in stomach trouble and as brain tonic (Murtem and Chaudhry 2016), fever, amenorrhea, rheumatism (Singh et al. 2012), antimalaria (Kalita and Khan 2013), chronic digestion (Tag et al. 2005), gastric ulcer, weak eyesight, liver problems, liver problems, purification of blood (Tiwari et al. 2009), leaf juice is taken against gastric, jaundice and stomach ache. Raw leaves or juice is also taken for diarrhea and dysentery (Ratan et al. 2016), used for the treatment of toothache/dental worms (Kalita et al. 2017), check bleeding (Sarmah et al. 2008b), digestive disorder and dysentery (Bharali et al. 2017), plant is used in arthritis, diabetes, blood disorders and brain tonic (Tripathi and Limasenla 2017), plant juice mixed with stephania hentandifolia plant juice is used in vomiting (Tripathi and Limasenla 2017), plant extract mixed with honey is is taken in empty stomach every morning in the treatment of chronic dysentery, high blood pressure and for enriching memory (Benniamin 2011), taken as boiled vegetable in stomach disorders, good in treating skin diseases, given to mother after childbirth (Rethy et al. 2010)	Tiwari et al. 2009; Tiwari et al. 2009; Bharali et al. 2017; Kalita et al. 2017; Tripathi and Limasenla 2017; Kamum et al. 2018; Ngomle et al. 2020; Namsa and Tangjang 2011; Hage Yanka and Tag 2020; Hage Yanka and Tag 2020; Tshering et al. 2018; Murtem and Chaudhry 2016; Saikia et al. 2021; Kalita and Khan 2013; Sarmah et al. 2008b; Gogoi and Sharma 2022b; Rethy et al. 2010; Ratan et al. 2016; Singh et al. 2012; Gangwar and Ramakrishnan 1990; Tag et al. 2005; Benniamin 2011

(continued)

Table 3.2 (continued)

No	Plant species	Family	Altitude (mtr.)	Local name	Parts used	Medicinal uses	References
	Gaultheria trichophylla Royle	Ericaceae	3000–3900	Shsksima	Fruit	Curing cough and cold (Ngomle et al. 2020)	Ngomle et al. 2020
	Houttuynia cordata Thunberg	Saururaceae	200–1000	Honya, Mumarang, Matring (B) Chumong (S), Roram, Reram, Zizi baying (A), Siiya hamang (Ap), Moyum-kneme (G), Hongyea (N), Momarengpa, Wild, Nekir name, Amkhalai, Thalai-naam, Siyan hamang (Wild), Sia-hamang (Ap), Nyengma (Kham), Mosondori, Masandari (As)	Leaves, roots, whole plant, rhizome, shoot, stem, tender leaves	Leaves are used in measles, dysentery, gonorrhea, eye and skin troubles; anti-inflammatory, anti-viral, anti-fungal, and anti-microbial in action (Ngomle et al. 2020), calf diarrhea (Namsa and Tangjang 2011), whole plant including root, stem and leaves are crushed to paste and consumed as salad along with rice. Juice is useful against dysentery (Kalita et al. 2017), chutney making (Ayam 2017), stomach ache, diarrhoea and deworming (Tshering et al. 2018), freshness, good sleep, heart disorders (Saikia et al. 2021), stomachache and diarrhoea (Kamum et al. 2018), measles, gonorrhoea, skin troubles, anti-tumour, anti-cancer, pneumonia, bronchitis, stomach ulcer (Gogoi and Sharma 2022b), fresh wounds for stopping bleeding and fast healing (Gangwar and Ramakrishnan 1990), blood impurity, high blood pressure (Tiwari et al. 2009), juice is extracted and taken to get relief from dysentery, indigestion, loose motion, and stomach ache. It is also known to provide sound sleep (Ratan et al. 2016), used for the treatment of dysentery, toothache/dental worms (Kalita et al. 2017), plant is consumed raw for heart diseases (Goswami et al. 2009), tape and ring worm parasite (Sarmah et al. 2008b), diarrhoea and digestive disorder (Bharali et al. 2017), plant extract is drank for treatment of liver, digestive problems and urinary disorder. Juice of the plant drank to remedy (Rethy et al. 2010), shoot/leaf used as salad, chatni, vegetables, decoction in stomach problem (Shankar et al. 2015), leaves are considered good for lactating mothers (Wangpan et al. 2019)	Tiwari et al. 2009; Ayam 2017; Bharali et al. 2017; Kalita et al. 2017; Kalita et al. 2017; Kamum et al. 2018; Wangpan et al. 2019; Ngomle et al. 2020; Namsa and Tangjang 2011; Tshering et al. 2018; Saikia et al. 2021; Sarmah et al. 2008b; Gogoi and Sharma 2022b; Rethy et al. 2010; Shankar et al. 2015; Ratan et al. 2016; Gangwar and Ramakrishnan 1990; Goswami et al. 2009

(continued)

Table 3.2 (continued)

No	Plant species	Family	Altitude (mtr.)	Local name	Parts used	Medicinal uses	References
	Nardostachys jatamansi (D. Don) DC.	Valerianaceae		Pang-posh, Pangpos. Jatamansi	Root-stock, rhizome	General tonic; also used in skin diseases, fever, cough, cold and dysentery (Ngomle et al. 2020), rhizome used in respiratory related diseases; water decoction given during insomnia (Sen et al. 2008), root, aromatic bitter tonic (Balkrishna et al. 2019), antimalaria (Kalita and Khan 2013), female disorders, Urinary tract infection and indigestion (Shankar and Rawat 2008)	Ngomle et al. 2020; Kalita and Khan 2013; Shankar and Rawat 2008; Balkrishna et al. 2019; Sen et al. 2008
	Neopicrorhiza scrophulariiflora (Pennell) Hong	Scrophulariaceae	3500–4880	Khongleng	Root-stock, rhizome	Root-stock used to cure fever, dysentery, cold, influenza, chronic fever and jaundice; also a cardiac tonic (Ngomle et al. 2020), decoction of rhizome consumed to treat stomach worms, fevers and constipation; prescribed during liver complaints, anemia, jaundice and respiratory tract disorders (Sen et al. 2008)	Ngomle et al. 2020; Sen et al. 2008
	Rubia manjith Roxburgh ex Fleming	Rubiaceae	700–3600	Langye, Manjishtha, Tamin	Root, leaf, fruits and stem	Mixed with oil and used for body pain and headache mainly for joints and the stem of the plant used for local dye (Ngomle et al. 2020), root tonic, astringent (Balkrishna et al. 2019), used to cure headache, cough, cold, locally used as a textile dye (Tshering et al. 2018), Used to cure headache, cough, cold, locally used as a textile dye (Gogoi and Sharma 2022b), diarrhea, fever (Singh et al. 2012)	Ngomle et al. 2020; Tshering et al. 2018; Gogoi and Sharma 2022b; Singh et al. 2012; Balkrishna et al. 2019
	Satyrium nepalense D. Don	Orchidaceae	1500–3000	Peepee	Tuberous root	Root-tubers are used as tonic and to cure cough and cold (Ngomle et al. 2020)	Ngomle et al. 2020
	Saussurea gossipiphora D. Don	Asteraceae	2000–3000	Ruiposh	Root-stock	Root-stock used to treat pre-menstruation cycle (Ngomle et al. 2020)	Ngomle et al. 2020
	Swertia chirayita (Roxburgh) Karsten	Gentianaceae	7350–2440	Gonga Marpo, Chiraita, Chirata	Whole plant	Decoction of plant used against malaria, cough and cold (Ngomle et al. 2020), whole plant bitter tonic, laxative in fever, malaria, jaundice (Balkrishna et al. 2019), plant decoction is taken in fever, antihepatitis B (Tshering et al. 2018), plant decoction is taken in fever, anti-hepatitis B (Gogoi and Sharma 2022b), antimalaria (Kalita and Khan 2013), decoction of the dried plant is taken in malarial fever and cold. It is also used as anthelmintic	Ngomle et al. 2020; Tshering et al. 2018; Kalita and Khan 2013; Gogoi and Sharma 2022b; Ratan et al. 2016; Balkrishna et al. 2019

(continued)

Table 3.2 (continued)

No	Plant species	Family	Altitude (mtr.)	Local name	Parts used	Medicinal uses	References
	Swertia nervosa (G. Don) Clarke	Gentianaceae	1500–3000	Yanshi-pong-yong (Monpas)	Whole plant	Decoction of plants used to cure fever, malaria and also used as blood purifier (Ngomle et al. 2020), antimalaria (Kalita and Khan 2013)	Ngomle et al. 2020; Kalita and Khan 2013
	Swertia speciosa D. Don	Gentianaceae	3353–4266	Sherkuto	Root	Decoction of plants used against fever, cough, dysentery and cold (Ngomle et al. 2020)	Ngomle et al. 2020
	Thalictrum chelidonii DC.	Ranunculaceae		Krekpa	Leaf	Leaf-paste is applied on skin irritation or any infectious diseases (Ngomle et al. 2020)	Ngomle et al. 2020
	Viola biflora L.	Violaceae	1400–2200	Sher-Minto	Whole plant	Used against constipation, cough and skin diseases (Ngomle et al. 2020)	Ngomle et al. 2020
	Acacia pennata (L.) Wild	Leguminosae	Upto 1700 m	Ramli, Tappa pad	Whole plant	Pisicidaland detergent (Rama 2013), fish poison (Baruah and Bhattacharyya 2019)	Rama 2013; Baruah and Bhattacharyya 2019
	Acacia caesia (L.) Wild	Leguminosae	Throughout	Talik, Ragra	Bark	Medicinal and Pisicidal (Rama 2013), crushed leaves applied on head for killing lice (Murtem and Chaudhry 2016)	Murtem and Chaudhry 2016; Rama 2013
	Acmella paniculata (Wall. ex DC..) R.K. Jansen	Asteraceae	Upto 1500	Marsa, Pajong-naam, Marsang, Marsha (Galo), Marsang (Adi), Yakho hamang (Ap), Maresa	Leaves and flower, Inf, Ls, Lf, leaves and young twigs, inflorescence, young shoot, inflorescence and dried flower	Tender leaves and flowers (cooked) are edible as vegetable. The fresh flowers are used as fishes poison. Flowers are used for toothache (Ronald et al. 2019), stop bleeding, skin infections and gastric, fish poison (Chowdhery et al. 2009), used in toothache (Hage Yanka and Tag 2020), bioactive metabolites (Namsa and Tangjang 2011), boiled and making chutney (Ayam 2017), antimalarial, antipyretic, analgesic, flowers are chewed during toothache (Tshering et al. 2018), the most common and widespread use is to treat toothache, throat and gum infections. The entire plant (root, stem, leaf and flower) is medicinally active and non-toxic to humans. The leaves and flower heads contain analgesic, antifungal, anthelminthic, and antibacterial agents (Gangwar and Ramakrishnan 1990)	Tiwari et al. 2009; Rama 2013

(continued)

Table 3.2 (continued)

No	Plant species	Family	Altitude (mtr.)	Local name	Parts used	Medicinal uses	References
	Acorus calamus L.	Acoraceae	500–1000	Buch, Vacha, Wok-kak-hing, Talyo, Vocha, Kiile tolyo (Wild), Daruga/Boch, Boj, Shueta (M), Zingche (S), Vacha	Rhizome, root, leaves, stems	Medicinal (Rama 2013), rhizome in respiratory diseases, nerve tonic (Balkrishna et al. 2019), rhizome is used in respiratory disorders (Tshering et al. 2018), cut, wounds, skin diseases, bone fracture (Saikia et al. 2021), sedative, laxative, carminative, stroke, insecticidal activities, also in making perfume (Gogoi and Sharma 2022b), rhizome used in treating diarrhea, dysentery. Juice of rhizome also used in asthma and bronchitis (Murtem and Chaudhry 2016), crushed rhizomes mixed with water, given to hen and goats for treating loose motion (Murtem and Chaudhry 2016), leprosy, asthma (Singh et al. 2012), gastric problems (Singh et al. 2012), antimalaria (Kalita and Khan 2013), a paste of rhizome is applied to cure headache, joint pain, wounds and skin rashes. Juice is also taken for stomach ache (Ratan et al. 2016), carminative, general Tonic, diarrhea and dysentery (Shankar and Rawat 2008), digestive disorder (Bharali et al. 2017), rhizome juice mixed with Phologacanthus thyrsifbt leaves is used in bodyache and fever. Pills prepared fr < paste of rhizome, onion and kalajira are used for dysentery (Tripathi and Limasenla 2017), rhizome extract is used for treatment of sinus problems, fever and spondolytis (Rethy et al. 2010), rhizome is used in child birth as nerve tonic (Shankar et al. 2015), paste of rhizome are applied on knee to relieve inflammation. Rhizomes are used against dysentery and also used in rituals (Kalita et al. 2017), rhizome is used in respiratory disorders (Tripathi and Limasenla 2017)	Bharali et al. 2017; Tripathi and Limasenla 2017; Tshering et al. 2018; Murtem and Chaudhry 2016; Saikia et al. 2021; Kalita and Khan 2013; Gogoi and Sharma 2022b; Shankar and Rawat 2008; Rethy et al. 2010; Shankar et al. 2015; Ratan et al. 2016; Singh et al. 2012; Rama 2013; Balkrishna et al. 2019

(continued)

Table 3.2 (continued)

No	Plant species	Family	Altitude (mtr.	Local name	Parts used	Medicinal uses	References
	Ageratum conyzoides L.	Asteraceae	Upto 1000	Eae-Namya, Bonger, Ngonamshu, Wild, Pashpaya, Nyeam ei, Sahadevi bhed, Enepu, Einamya/ Eirego, Thiingriing, Borbe tami (Wild), Pashpai (Ni), Pakku (A); Jackok (A). Dilap (A). Elompra (I), Deem ehh, Gundhuabon, Ingkir/Nangkir/ Bongir (A), Botpa ngon (M), Eww Namya (G)	Leaves or whole plant, stem	Medicinal and Pisicidal (Rama 2013), paste is used in the Cuts and wounds (Hage Yanka and Tag 2020), blood flow, dysentery and diarrhoea, wound healer, veterinary, fish poison (Tshering et al. 2018), cut, wounds (Saikia et al. 2021), wound healer, Veterinary, fish poison (Kamum et al. 2018), wound healing, antihelmintic (Gogoi and Sharma 2022b), paste of leaves applied on cuts and wounds (Murtem and Chaudhry 2016), diarrhea, fever, pneumonia (Singh et al. 2012), antimalaria (Kalita and Khan 2013), blood clotting (Tag et al. 2005), cuts and bruises (Tiwari et al. 2009), paste of leaves is applied to stop bleeding in minor cuts (Ratan et al. 2016), crushed leaves applied on cuts (Goswami et al. 2009), check bleeding (Sarmah et al. 2008b), fish poison (Baruah and Bhattacharyya 2019), crude leaf extract is applied on fresh cut to stop bleeding (Rethy et al. 2010), whole plant is used for blood clotting and wound healing (Shankar et al. 2015), to stop bleeding from leech bite site (Namsa et al. 2011b)	Tiwari et al. 2009; Kamum et al. 2018; Hage Yanka and Tag 2020; Tshering et al. 2018; Murtem and Chaudhry 2016; Saikia et al. 2021; Kalita and Khan 2013; Sarmah et al. 2008b; Gogoi and Sharma 2022b; Rethy et al. 2010; Shankar et al. 2015; Ratan et al. 2016; Singh et al. 2012; Rama 2013; Namsa et al. 2011b; Tag et al. 2005; Goswami et al. 2009; Baruah and Bhattacharyya 2019
	Angiopteris evecta (G. Forster) G.F. Hoffmann	Marattiaceae	50–600	Bakum, Fadey (M), Taba (A), Ghora Top/Bom, Tach, Giant fern, Chahmah	Tuber, leaf, stem, rhizome, spore	Edible (Rama 2013), leaves are edible as vegetable (Ronald et al. 2019), rhizome in dysentery (Balkrishna et al. 2019), health tonic (Saikia et al. 2021), powdered rhizome used in dysentery and diarrhea, stem cut into pieces and used as pig feed (Murtem and Chaudhry 2016), cough, skin disease (Singh et al. 2012), stings of spider or snake bites (Tiwari et al. 2009), substitute for staple food; underground parts and base of rachis stem is eaten after boiling and through washing (Shankar et al. 2015), leaf extract is used in the treatment of dysentery and diseases of blood ulcers. Spores are said to be effective in the treatment of leprosy and other skin diseases (Yomgam et al. 2017)	Tiwari et al. 2009; Ronald et al. 2019; Murtem and Chaudhry 2016; Saikia et al. 2021; Shankar et al. 2015; Singh et al. 2012; Rama 2013; Balkrishna et al. 2019; Yomgam et al. 2017
	Ardisia solanacea (Poir.) Roxb	Primulaceae	400–1000	Tamak/choochak	Berries, fruit	Edible (Rama 2013), ripe fruits (sweet) are edible (Ronald et al. 2019)	Ronald et al. 2019; Rama 2013
	Artemisia nitagirica (C.B. Clarke) Pampanini	Asteraceae	1000–2500	Tape-Nyobeng	Leaves	Medicinal	Rama 2013

(continued)

Table 3.2 (continued)

No	Plant species	Family	Altitude (mtr.)	Local name	Parts used	Medicinal uses	References
	Begonia roxburghii A.DC	Begoniaceae	500–800	Bukusurbu, Babeyaying, Bekhoo, Lukhu, Choiwang	Whole plant, petiole, shoot and rhizome, leaf	Edible, skin care (Rama 2013), edible as pickle. Petiole is eaten raw (Ronald et al. 2019), used in stomach problem, dysentery, diarrhoea (Hage Yanka and Tag 2020), anti diabetic (Bam et al. 2015), indigestion (Saikia et al. 2021), rashes, stings of spider (Tiwari et al. 2009)	Tiwari et al. 2009; Ronald et al. 2019; Hage Yanka and Tag 2020; Saikia et al. 2021; Rama 2013; Bam et al. 2015
	Bryophyllum pinnatum (Lamarck) Okcn	Crassulaccace	2000–7600	Pate-goja	Leaves	Medicinal (Rama 2013)	Rama 2013
	Buddleja asiatica Loureiro	Scrophulariaceae	500–1700	Komki-Hiri, Phamshing, Wild, Faab Shang, Feb-shang, Nigranda	Whole plant, leaves and young twigs, root, bark	Pisicidal (Rama 2013), diarrhoea, beverages fermentation (Tshering et al. 2018), inflammation (Saikia et al. 2021), diarrhoea, beverages fermentation (Kamum et al. 2018), flowers used in sinus (Murtem and Chaudhry 2016), antimalaria (Kalita and Khan 2013), bark extract is drank for treatment of fever and sinus remove hepatic problem (Rethy et al. 2010)	Kamum et al. 2018; Tshering et al. 2018; Murtem and Chaudhry 2016; Saikia et al. 2021; Kalita and Khan 2013; Rethy et al. 2010; Rama 2013
	Calamus leptospadix Griff	Arecaceae	Upto 1200	Tahh	Stem	House, bridge construction and handicraft (Rama 2013)	Rama 2013
	Calllcarpa arbórea Roxb	Verhenaceae	300–800	Penlo	Leaves	Condiments (Rama 2013)	Rama 2013
	Chromolaena odorata (L.) R.M.King and H.Rob	Asteraceae	100–1000	Daglin, Heling, Eeneh pudeh, Gocham Nemi, Malinga-jumpak, Thangba, Daglin (G)	Leaves	Medicinal (Rama 2013), used in major cuts and wounds (Hage Yanka and Tag 2020), cut, wounds, headache, fever (Saikia et al. 2021), paste of leaves are directly applied on cut and wounds (Tiwari et al. 2009), paste of leaves used on wounds infested by worms (Murtem and Chaudhry 2016), antimalaria (Kalita and Khan 2013), cuts (Tiwari et al. 2009), digestive disorder (Bharali et al. 2017), to stop bleeding from leech bite site (Namsa et al. 2011b)	Tiwari et al. 2009; Tiwari et al. 2009; Bharali et al. 2017; Hage Yanka and Tag 2020; Murtem and Chaudhry 2016; Saikia et al. 2021; Kalita and Khan 2013; Rama 2013; Namsa et al. 2011b
	Cinnanioniuni tañíala iBuch.-Ham.) T. Ness and Eberm	Lauraceae	1000–1500	Nalok	Leaves	Spice and condiments (Rama 2013)	Rama 2013
	Clerodendrunt glandulosunt Lindl	Verbenaceae	500–800	Oin	Leaves	Vegetable and medicinal (Rama 2013)	Rama 2013

(continued)

Table 3.2 (continued)

No	Plant species	Family	Altitude (mtr.)	Local name	Parts used	Medicinal uses	References
	Colocasia esculenta (L.) Schott	Araceae	100–400	Riksin/Rikyu, Ngaglin (M), Chowk-pema (B), Chak (S), Clung/ Otenga/Soh-kriban (As)	Tuber, fruit	Medicinal (Rama 2013), corm are cooked and eaten as staple food (Kalita et al. 2017), fever (Sarmah et al. 2008b), diabetes and dandruff (Bharali et al. 2017)	Bharali et al. 2017; Kalita et al. 2017; Sarmah et al. 2008b; Rama 2013
	Crassocephalum crepidioides (Berth)S Moore	Asteraceae	300–1800	Jugen, Gende, Manchala (B), Gendahamang	Leaves, tender shoots	Vegetable and medicinal (Rama 2013), leaves (cooked) are edible as vegetable (Ronald et al. 2019), consumed as vegetable during constipation. The cooked leaf is also used in insomnia (Hage Yanka and Tag 2020), leaves are cooked and consumed as vegetable (Kalita et al. 2017), anti diabetic (Bam et al. 2015), antimalarial, analgesic, epileptic, wound bleeding, headache (Tshering et al. 2018), indigestion, headache, stomachache, cut, wounds (Saikia et al. 2021), grinded leaves paste is applied on minor cuts to stop bleeding (Ratan et al. 2016)	Kalita et al. 2017; Ronald et al. 2019; Hage Yanka and Tag 2020; Tshering et al. 2018; Saikia et al. 2021; Ratan et al. 2016; Rama 2013; Bam et al. 2015
	Croton tiglium L.	Euphorbiaceae	800–1200	Pakmo/Mosir, Saklang, Aug aag sin	Leaves, flower and fruits, seeds and inflorescence	Bio weapons (Rama 2013), antimalaria (Kalita and Khan 2013), fish poison (Baruah and Bhattacharyya 2019), Malaria (Das and Tag 2006b)	Das and Tag 2006b; Kalita and Khan 2013; Rama 2013; Baruah and Bhattacharyya 2019
	Cyclosorus parasiticus (L.) Farw	Thelypteridaceae	Upto 1500	Rukdik	Leaves, rhizome	Medicinal (Rama 2013), gout and rheumatism, anthelmintic, antifungal and antibacterial (Tshering et al. 2018), gout and rheumatism, anthihelmintic, antifungal and antibacterial	Tshering et al. 2018; Gogoi and Sharma 2022b; Rama 2013
	Dendrocalamus giganteus Munro	Poaceae	800–1200	Diba, Ehh herom	Culm and tender shoot, stem and leaves	Medicinal and cultural (Rama 2013), fish poison (Baruah and Bhattacharyya 2019)	Rama 2013; Baruah and Bhattacharyya 2019
	Dendrocnide sinnata (Blume) Chew	Urticaceae	800–1500	Reteh	Root	Medicinal (Rama 2013)	Rama 2013

(continued)

Table 3.2 (continued)

No	Plant species	Family	Altitude (mtr.)	Local name	Parts used	Medicinal uses	References
	Drymaria cordata (L.) Willd ex Schultes	Caryophyllaceae	200–1900	Keji-kero, Kaira, Kadokairon, Pipep-jaak, Akka/ Aabra (A), Yatikhoi	Whole plant, leaf	Anti inflammatory (Rama 2013), applied to ringworm and other skin diseases; bone fracture; leech bite (Hage Yanka and Tag 2020), Vomiting (Tshering et al. 2018), antimalaria (Kalita and Khan 2013), abscess (Tiwari et al. 2009), joint pain. cut. wound and check bleeding (Sarmah et al. 2008b), whole herb wrapped in banana leaf is burned in a chulha and the smoke is inhaled for relief from headache (Benniamin 2011)	Tiwari et al. 2009; Hage Yanka and Tag 2020; Tshering et al. 2018; Kalita and Khan 2013; Sarmah et al. 2008b; Rama 2013; Benniamin 2011
	Ficus hispida Linnaeus f	Moraceae	300–1200	Kukcoo-looloo, Mukonpong, Cheyum, Takpiya (A)	Fruits, bark, roots, sap of stem	Fruits edible (Rama 2013), tuberculosis (Das and Tag 2006b), Burn injury (Bora et al. 2007), fruits are edible (Shankar et al. 2015)	Das and Tag 2006b; Shankar et al. 2015; Rama 2013; Bora et al. 2007
	Garcinia pedunculata Roxburgh ex Buchanan-Hamilton	Clusiaceae	100–1200	Porok-Tumgalo, Liba (A), Mibia, Tabing-asing, Haipum, Kaw, Mhahau, Thekera	Fruits, leaves	Fruits edible (Rama 2013), antibacterial (Kamum et al. 2018), chutney (Ayam 2017), leaves used in dysentery and cough (Tshering et al. 2018), leaves used in dysentery and cough (Murtem and Chaudhry 2016), antimalaria (Kalita and Khan 2013), dysentery, stomach-ache (Tiwari et al. 2009) Diarrhoea and dysentery (Shankar and Rawat 2008), digestive disorder (Bharali et al. 2017), fruit pulp extract is used as a treatment against blood dysentery (Benniamin 2011), slices dried fruit juice drank for treatment of dysentery (Rethy et al. 2010), fruits are used as salad and pickles; in stomach problems (Shankar et al. 2015)	Tiwari et al. 2009; Ayam 2017; Bharali et al. 2017; Kamum et al. 2018; Tshering et al. 2018; Murtem and Chaudhry 2016; Kalita and Khan 2013; Shankar and Rawat 2008; Rethy et al. 2010; Shankar et al. 2015; Rama 2013; Benniamin 2011
	Gonostegia hirta (Blume ex Hassle) Miquel	Urticaceae	500–2500	Oyik, Masaipuk-irap (B) Dankhuchu (S)	Leaves and stem	Vegetable and medicinal (Rama 2013), leaves (cooked) are edible as vegetable; ripe fruits are edible (Ronald et al. 2019), leaves and young stem are boiled and consumed as a vegetable (Kalita et al. 2017), gastric problem, constipation (Tshering et al. 2018), food (Bajpay et al. 2019), acidity, gastric, appetizer, constipation (Saikia et al. 2021), increasing breast milk/lactation (Tag et al. 2005)	Kalita et al. 2017; Ronald et al. 2019; Tshering et al. 2018; Rama 2013; Bajpay et al. 2019
	Gymnocladus burmanicus C.E Parkinson	Fabaceae	160–200	Dika	Fruit, bark, and leaves	Soap, shampoo and detergent And pisicidal (Rama 2013)	Rama 2013

(continued)

Table 3.2 (continued)

No	Plant species	Family	Altitude (mtr.)	Local name	Parts used	Medicinal uses	References
	Hedyotis scandens Roxburgh	Rubiaceae	500–2400	Airofaik/Asirpaye, Phamshing, Taja hoor, Yakauka	Leaves and roots, young twigs	Vegetable and medicinal (Rama 2013), leaves (cooked) are edible as vegetable; ripe fruits are edible (Ronald et al. 2019), gastritis, beverages fermentation (Tshering et al. 2018), gastritis, beverages fermentation (Kamum et al. 2018), dried root powder used orally in stomach pain (Murtem and Chaudhry 2016), leaf extract is used for the treatment of diabetes (Benniamin 2011)	Kamum et al. 2018; Ronald et al. 2019; Tshering et al. 2018; Murtem and Chaudhry 2016; Rama 2013; Benniamin 2011
	Hydrocotyle sibthorpioides Lamarck	Araliaceae	2300–2400	Nyinyo-Tamo, Manni Ao (G), Hah-naam	Whole plant, leaf	Medicinal, pisicidal and vegetable (Rama 2013), leaves edible as vegetable (Ronald et al. 2019), curries (Ayam 2017), dysentery (Tshering et al. 2018), rashes due to caterpillar sting (Tiwari et al. 2009)	Tiwari et al. 2009; Ayam 2017; Ronald et al. 2019; Tshering et al. 2018; Rama 2013
	Lasia spinosa (L.) Thawaites	Araceae	100–1500	Pimibro/ fiikru-Epreh, Hamwey	Petiole, leave/stem	Medicine for human and animal (Rama 2013), joint pain, healing wounds and kills intestinal worm (Tag et al. 2005), jaundice, intestinal worms in pigs (Tiwari et al. 2009)	Tiwari et al. 2009; Rama 2013; Tag et al. 2005
	Molineria capitulata (Lour,) Herb	Hypoxidaceace	800–2200	Royek	Leaves, root and shoot	Wound-healing and spiritual healing (Rama 2013), digestive disorder (Bharali et al. 2017)	Bharali et al. 2017; Rama 2013
	Musa x paradisiaca Linnaeus	Musaccace	305–914	Kodok, Musung (S) (Banana), Kopak, Nyoro-kopa	Fruits and watery exudates from the shoot part	Vegetable and medicinal (Rama 2013), fruit edible also as vegetable (Ronald et al. 2019), ripe fruit is consumed as source of carbohydrate. Unripe fruits are used against dysentery (Kalita et al. 2017), diarrhoea, dysentery (Hage Yanka and Tag 2020), dysentery, urinary problems, boiled unripe fruits are given during dysentery, diabetes, anaemia (Tshering et al. 2018), boiled unripe fruits are given during dysentery, diabetes, in anaemia (Gogoi and Sharma 2022b)	Kalita et al. 2017; Ronald et al. 2019; Rama 2013
	Mussaenda roxburghii Hooker f	Rubiaceae	450–1500	Taksap, Tangmeng	Leaves, whole plant	Vegetable and medicinal (Rama 2013), leaves (cooked) are edible as vegetable (Ronald et al. 2019), detoxify mushroom poison, antipyretic, diuretic, treat blemishes on tongue, acute gastroenteritis (Tshering et al. 2018), detoxify mushroom poison, anti-pyretic, diuretic, treat blemishes on tongue, acute gastroenteritis (Gogoi and Sharma 2022b), the leaf paste is applied to freshly cut wound in order to enable blood clotting; and it is also edible as vegetable (Tiwari et al. 2009)	Tiwari et al. 2009; Ronald et al. 2019; Tshering et al. 2018; Gogoi and Sharma 2022b; Rama 2013

(continued)

Table 3.2 (continued)

No	Plant species	Family	Altitude (mtr.)	Local name	Parts used	Medicinal uses	References
	Oxyspora paniculata (D. Don) DC.	Melastomaceae	500–2000	Pekn-jalo, Porkijale	Berries and young shoots, Leave, whole plant	Fruits and young shoot edible (Rama 2013), treatment of various liver disorder, stomachic, antidote against snake poisoning (Tshering et al. 2018), treatment of various liver disorder, stomachic, antidote against snake poisioning (Gogoi and Sharma 2022b)	Tshering et al. 2018; Gogoi and Sharma 2022b; Rama 2013
	Pandanus tectorius Parkinson ex Du	Pandanaccae	200–400	Tako	Leaves	Traditional handicrafts (Rama 2013)	Rama 2013
	Piper longum L.	Piperaceae	500–1200	Dolo-pan, Saturikki Ahoma, Riidii, Ere ro:ri (A)	Leaves and fruits, stem, roots	Anti inflammatory (Rama 2013), Malaria (Tshering et al. 2018), treat joints pain, gout, paralysis, improve immune and digestive system, arthritis (Gogoi and Sharma 2022b), antimalaria (Kalita and Khan 2013), used for the treatment of cough, dysentery (Kalita et al. 2017), joints pain, arthritis (Bora et al. 2007), tender leaves are used as vegetable (Shankar et al. 2015)	Kalita et al. 2017; Tshering et al. 2018; Kalita and Khan 2013; Gogoi and Sharma 2022b; Shankar et al. 2015; Rama 2013; Bora et al. 2007
	Piper pedicellatum C.DC	Piperaceae	300–1800	Rar	Leaves	Vegetable (Rama 2013), leaved are edible as vegetable (Ronald et al. 2019), food (Bajpay et al. 2019), stomach ache, dysentery (Tshering et al. 2018)	Ronald et al. 2019; Tshering et al. 2018; Rama 2013; Bajpay et al. 2019
	Psidium guajava Linnaeus	Myrtaceae	2000–3000	Mudori	Leaves and fruits	Edible fruits and medicinal (Rama 2013)	Rama 2013
	Rhynchotechum ellipticum (Wallich ex D. Dietr.) A. DC.	Gesneriaceae	500–1000	Joko	Leaves	Vegetable (Rama 2013), leaves (cooked) are edible as vegetable (Ronald et al. 2019)	Ronald et al. 2019; Rama 2013
	Ricinus communis L.	Euphorbiaceae	300–1800	Porok-Ak-Ane	Leaves	Porok-Ak-Ane (Rama 2013)	Rama 2013
	Saurauia napaulensis DC.	Actinidiaceae	400–2500	Enehi, Eti-aou (B)	Berries and leaves	Edible fruits and socio-cultural (Rama 2013), fruits are eaten by human, and leaves are eaten by animals (cows) (Kalita et al. 2017)	Kalita et al. 2017; Rama 2013
	Solanum americanum P. Miller	Solanaceae	200–2800	Or-re, Dhangbu (S)	Seed and leaves, fruit	Vegetable and medicinal (Rama 2013), fruits (cooked) are edible as vegetable (Ronald et al. 2019), food (Bajpay et al. 2019), leaves are eaten raw with maize, and also cooked and consumed with staple food. It is also used as medicine for treatment of dysentery (Kalita et al. 2017)	Kalita et al. 2017; Ronald et al. 2019; Rama 2013; Bajpay et al. 2019

(continued)

Table 3.2 (continued)

No	Plant species	Family	Altitude (mtr.	Local name	Parts used	Medicinal uses	References
	Solanum torvum Swartz	Solanaceae	200–1250	Kope, Singkotey, Baka (Galo), Byako (Adi), Byakta (Nyishi), Sathi Byako (Apatani), Aying tanyum, Borang Kharangjeh, Tutusik, ShoatBeyak, Vodhata/Bri hati, Byakta, Mehengchang, Sukkang, Titabhekuri (As), Kopi	Fruits, seed and roots, immature fruits	Edible fruits (Rama 2013), fruits (cooked) are edible as vegetable (Ronald et al. 2019), infection in gums and tootache; stomachache, high blood pressure fever, cough, toothache malaria and stomach pain (Chowdhery et al. 2009), used in Toothache (Hage Yanka and Tag 2020), stomach ache, spleen problem and anthelmintic (Tshering et al. 2018), cough, skin diseases (Saikia et al. 2021), anthelmintic (Kamum et al. 2018), the fruits are useful against liver and spleen enlargement (Gangwar and Ramakrishnan 1990), fruits used in stomach ache and high blood pressure, fresh roots known as oral contraceptive (Murtem and Chaudhry 2016), cancer, addison's disease, rheumatic arthritis (Singh et al. 2012), antimalaria (Kalita and Khan 2013; Srivastava 2010), high blood pressure (Tiwari et al. 2009), Malaria, stomach pain (Das and Tag 2006b), capsule are used as vegetable; used in toothache (Shankar et al. 2015), roasted fruits are eaten in diarrhoea (Wangpan et al. 2019)	Tiwari et al. 2009; Kamum et al. 2018; Wangpan et al. 2019; Chowdhery et al. 2009; Ronald et al. 2019; Hage Yanka and Tag 2020; Tshering et al. 2018; Das and Tag 2006b; Murtem and Chaudhry 2016; Saikia et al. 2021; Kalita and Khan 2013; Shankar et al. 2015; Singh et al. 2012; Rama 2013; Gangwar and Ramakrishnan 1990; Srivastava 2010
	Stemona tuberosa Loureiro	Stemonaceae	300–2300	Inginn, Waakkao-thah	Rhizome, root tuber	Anti inflammatory (Rama 2013), appendix; cuts and wounds (Tiwari et al. 2009)	Tiwari et al. 2009; Rama 2013
	Urtica ardens Link	Urticaccae	Upto 2500	Peji-pomo	Leaves	Vegetable and medicinal (Rama 2013)	Rama 2013
	Zanthoxylum acanthopodium DC.	Rutaceae	1000–2500	Onyer, Ombe or ombeng, Ongar (A), Yakhung (Ap), Onyor (G), Mekat	Leaves, bark and fruits	Vegetable and pisicidal (Rama 2013), boiled foods (Ayam 2017), dysentery (Saikia et al. 2021), stomach pain (Das and Tag 2006b)	Ayam 2017; Das and Tag 2006b; Saikia et al. 2021; Rama 2013
	Zanthoxylum oxyphyllum Edgeworth	Rutaceae	1500–2700	Rikom, Honyor (N), Nemba yorkhung (Ap), Mathana or manjang, Gagra	Root, leaves and stem, fruit, tender shoots	Medicinal, vegetable and pisicidai (Rama 2013), boiled foods (Ayam 2017), stomach ache (Tshering et al. 2018), stomach disorder (Saikia et al. 2021), Used against curing are carminative, depurative, rheumatism, dyspepsia and asthma (Gangwar and Ramakrishnan 1990), liver tonic (Shankar and Rawat 2008)	Ayam 2017; Tshering et al. 2018; Saikia et al. 2021; Shankar and Rawat 2008; Rama 2013; Gangwar and Ramakrishnan 1990
	Amomum subulatum Roxb	Zingiberaceae	500–1000	Jepo (A)	Inflorescence, fruit, flower	Inflorescence are eaten as vegetables. Fruits are eaten raw (Ronald et al. 2019), Meat curry (Ayam 2017), fruit is used in cough and stomachic disorders (Tshering et al. 2018)	Ayam 2017; Ronald et al. 2019; Tshering et al. 2018

(continued)

Table 3.2 (continued)

No	Plant species	Family	Altitude (mtr.)	Local name	Parts used	Medicinal uses	References
	Ardisia crispa (Thunb.) A.DC	Primulaceae	100–2500		Leaf	Leaves are edible as chutney (Ronald et al. 2019)	Ronald et al. 2019
	Balakata baccata (Roxb.) Esser	Euphorbiaceae	400–6400	Shigum, Yaoulu	Fruit, bark	Fruits (sweet) are edible (Ronald et al. 2019), antimalaria (Kalita and Khan 2013)	Ronald et al. 2019; Kalita and Khan 2013
	Baliospermum calycinum Müll.-Arg	Euphorbiaceae	600–1600	Gilgal	Leaf, whole plant	Leaves (cooked) are edible as vegetable (Ronald et al. 2019), Purgative, stimulant, antidote in snakebite, asthma, jaundice, gastric problem, gout and rheumatism, toothache (Tshering et al. 2018), Purgative, stimulant, antidote in snake bite, asthmajaundice, gastric problem, gout and rheumatism, toothache (Gogoi and Sharma 2022b)	Ronald et al. 2019; Tshering et al. 2018; Gogoi and Sharma 2022b
	Begonia griffithiana (A.DC.) Warb	Begoniaceae	750–1400	Sudum Meku	Leaf	Leaves (cooked) are edible as vegetable and sour in taste	Ronald et al. 2019
	Clerodendrum glandulosum Lindl	Lamiaceae	500–800	Tapen, Oin, Khangjela-Shing (M), Taapin, Khahmao	Leaf, Ts	Leaved (cooked) are edible as vegetable (Ronald et al. 2019), food (Bajpay et al. 2019), Young leaves are cooked and nsumed as vegetable. Soup are taken to cure hypertension (Kalita et al. 2017), Malaria (Tshering et al. 2018), blood pressure, fever, cough (Saikia et al. 2021), leaves, young twigs used as vegetables to treat low blood pressure (Tiwari et al. 2009), high blood pressure (Tiwari et al. 2009), to control high blood pressure (Bharali et al. 2017)	Tiwari et al. 2009; Tiwari et al. 2009; Bharali et al. 2017; Kalita et al. 2017; Ronald et al. 2019; Tshering et al. 2018; Saikia et al. 2021; Bajpay et al. 2019
	Debregeasia longifolia (Burm.f.) Wedd	Urticaceae	400–2100	Jirepole, khajung Nyajumna, Jirpollee	Leaf, fruit, tender leaves	Tender leaves (cooked) are edible as vegetable, fruits edible (Ronald et al. 2019), antitumours, rheumatism; juice is applied to the areas of the skin affected by scabies (Tshering et al. 2018), anti-tumours, rheumatism, juice is applied to the areas of the skin affected by scabies (Gogoi and Sharma 2022b), helps indigestion (Gangwar and Ramakrishnan 1990), antimalaria (Kalita and Khan 2013)	Ronald et al. 2019; Tshering et al. 2018; Kalita and Khan 2013; Gogoi and Sharma 2022b; Gangwar and Ramakrishnan 1990
	Elatostema sessile J.R.Forst. and G.Forst	Urticaceae	400–2600	Sakobadha, Chulukpa (M)	Leaf	Whole plant is consumed as vegetable (Ronald et al. 2019), Young shoot and leaves are cooked and Stem is used for fencing of kitchen garden (Kalita et al. 2017)	Kalita et al. 2017; Ronald et al. 2019

(continued)

Table 3.2 (continued)

No	Plant species	Family	Altitude (mtr.)	Local name	Parts used	Medicinal uses	References
	Fagopyrum esculentum Moench	Polygonaceae	2000–3700	Amintatek, Mamey (B), Chhika, Tabing-asing (A), Yong jiba (M)	Leaf, fruit, seed, grain	Leaves (cooked) are edible as vegetable; ripe fruits are edible (Ronald et al. 2019), leaves are cooked and consumed as vegetable. Seeds are crushed into flour, cooked and consumed (Kalita et al. 2017), stomach ache (Tshering et al. 2018), antimalaria (Kalita and Khan 2013), diarrhea, dysentery and stomach pain (Sarmah et al. 2008b), rheumatic pain and Fever (Shankar and Rawat 2008), tender shoots are used as vegetable (Shankar et al. 2015)	Kalita et al. 2017; Ronald et al. 2019; Tshering et al. 2018; Kalita and Khan 2013; Sarmah et al. 2008b; Shankar and Rawat 2008; Shankar et al. 2015
	Fragaria nubicola (Lindl. ex Hook.f.) Lacaita	Rosaceae	1800–2500	Pasi, Aki tayin Ziro (Ap)	Fruit	Ripe fruits are edible (Ronald et al. 2019), edible fruit, treatment of cough, 56 fever, and profuse menstruation (Asha and Singh 2020)	Asha and Singh 2020; Ronald et al. 2019
	Glochidion lanceolarium (Roxb.) Voigt	Phyllanthaceae	200–1500		Fruit	Ripe fruits are edible (Ronald et al. 2019)	Ronald et al. 2019
	Hydrangea robusta Hook.f. and Thoms	Hydrangeaceae	Upto 2500	Takmi	Leaf	Leaves (cooked) are edible as vegetable; ripe fruits are edible (Ronald et al. 2019)	Ronald et al. 2019
	Juglans regia L.	Juglandaceae	1800–2550	Tirkhuh, Man-gou (B), Walnut, Akrot	Fruit	Fruit edible (Ronald et al. 2019), Nut of the fruit is eaten. Bark of stem is chewed to treat toothache and gum infection, constipation, asthma, diarrhea, dyspepsia (Singh et al. 2012), anthelminthic and Rheumatisms (Shankar and Rawat 2008)	Kalita et al. 2017; Ronald et al. 2019; Shankar and Rawat 2008; Singh et al. 2012
	Murraya paniculata (L.) Jack	Rutaceae	400–2100	Nyibumtarum	Fruit	Ripe fruits are edible (Ronald et al. 2019), analgesic, antidiarrhoea, antiinflammatory (Tshering et al. 2018)	Ronald et al. 2019; Tshering et al. 2018
	Musa aurantiaca G.Mann ex Baker	Musaceae	400–600	Kodok	Fruit	Inflorescence is edible; cooked unripe fruits eaten as vegetable (Ronald et al. 2019)	Ronald et al. 2019
	Mycetia longifolia (Wall.) Kuntze	Rubiaceae	700–1800	Tangmge	Leaf	Leaves (cooked) are edible as vegetable (Ronald et al. 2019), pain relief, ulcer, wound healing, inflammation, antinociceptive, pain relief, ulcer, wound healing, inflammation, antinociceptive (Gogoi and Sharma 2022b)	Ronald et al. 2019; Tshering et al. 2018; Gogoi and Sharma 2022b
	Osbeckia nepalensis Hook.f	Melastomataceae	200–1500	Pudirasa, tree paeony, Branglaga sing (M)	Fruit	Ripe fruits are edible (Ronald et al. 2019), skin disorders, eye problems, gastro intestinal disorder, tender leaves are used as vegetable	Ronald et al. 2019; Shankar et al. 2015; Singh et al. 2012
	Osbeckia nutans Wall. ex C.B.Clarke	Melastomataceae	300–2000	Jikenemi	Fruit	Ripe fruits are edible (Ronald et al. 2019)	Ronald et al. 2019

(continued)

Table 3.2 (continued)

No	Plant species	Family	Altitude (mtr.)	Local name	Parts used	Medicinal uses	References
	Paris polyphylla Sm	Melanthiaceae	1800–3000	Mithi Vacha, Kala Katchu or Jungli Katchu, Dipogoiak	Fruit, leaf, rhizome	Leaves (cooked) are edible as vegetable, tuber medicinal (Ronald et al. 2019), used in Piles, constipation and high fat (Hage Yanka and Tag 2020), root stimulant, tonic (Balkrishna et al. 2019), fruit is eaten raw. Initially, never used as in the herbal treatment of ailments. Exploited and sold to neighboring countries, illegally, despite the ban by the govt. for its uses in brain tumor, snake bite and as aphrodisiacs, analgesic, antipyretic, antispasmodic, antitussive, narcotics (Asha and Singh 2020), antimalaria (Kalita and Khan 2013), bone fracture (Sarmah et al. 2008b)	Asha and Singh 2020; Ronald et al. 2019; Hage Yanka and Tag 2020; Kalita and Khan 2013; Sarmah et al. 2008b; Balkrishna et al. 2019
	Perilla frutescens (L.) Britton	Lamiaceae	500–1000	Namdung (A), Tining (Ap), Tanam-namdu (N),	Leaf, seed, stem	Seed paste is used to enhance the taste of curry-soup; also used as a substitute of mustard oil or spices. Leaves (cooked) are edible as vegetables (Ronald et al. 2019), curries, soup, meat preservation, sweet dish, rice cakes and boiled foods (Ayam 2017), fever, headache (Saikia et al. 2021), antimalaria (Kalita and Khan 2013)	Ayam 2017; Ronald et al. 2019; Saikia et al. 2021; Kalita and Khan 2013
	Physalis minima L.	Solanaceae	Upto 1800	Bodopati. Mamang	Fruit, leaf	Fruit edible (Ronald et al. 2019), gastric trouble, laxative, diuretic, anticancer, in hypertension, antiinflammatory (Tshering et al. 2018), gastric trouble (Saikia et al. 2021), leaves are considered good stomach disorders (Wangpan et al. 2019)	Wangpan et al. 2019; Ronald et al. 2019; Tshering et al. 2018; Saikia et al. 2021
	Pilea bracteosa Wedd	Urticaceae	600–2700	Guge	Leaf	Leaves (cooked) are edible as vegetable (Ronald et al. 2019)	Ronald et al. 2019
	Piper mullesua Buch.-Ham. ex D.Don	Piperaceae	1000–3000	Lo-ree, Odor (A), Pinee (G). Rer Edik, Nemar/Pipli/ Pan, Ishapgul (A)	Leaf, fruit, inflorescence	Young tender leaves are used as flavour, fruits edible (Ronald et al. 2019), antidiarrhoeal (Namsa and Tangjang 2011), chutney (Ayam 2017), fruits used in cough and cold problem (Murtem and Chaudhry 2016), antimalaria (Kalita and Khan 2013), whole plant is used in wound healing. (Shankar et al. 2015)	Ayam 2017; Namsa and Tangjang 2011; Ronald et al. 2019; Murtem and Chaudhry 2016; Kalita and Khan 2013; Shankar et al. 2015
	Piper sylvaticum Roxb	Piperaceae	300–1800	Rari, Ahanna	Leaves, tender shoot	Leaved are edible as vegetable (Ronald et al. 2019), the green pods are mildly diuretic and contain a substance that reduces the blood sugar level. It is used in the treatment of diabetes	Ronald et al. 2019; Gangwar and Ramakrishnan 1990
	Persicaria chinensis (L.) H.Gross	Polygonaceae	200–2000	Amintaktabe, Choihai-jaak	Fruit, Ls	Ripe fruits (sweet) are edible (Ronald et al. 2019), weakness (Tiwari et al. 2009)	Tiwari et al. 2009; Ronald et al. 2019

(continued)

Table 3.2 (continued)

No	Plant species	Family	Altitude (mtr)	Local name	Parts used	Medicinal uses	References
	Persicaria hydropiper (L.) Delarbre	Polygonaceae	270–2900	Kotsamang	Leaf	Leaves are used as spice with meat, also eaten as chutney (Ronald et al. 2019)	Ronald et al. 2019
	Persicaria nepalensis (Meisn.) Miyabe	Polygonaceae	1500–3900	Ruri (A), Yarung (N), Roli hamang (Ap)	Whole plant, young shoot	The entire plant is crushed and eaten as chutney (Ronald et al. 2019), chutney (Ayam 2017)	Ayam 2017; Ronald et al. 2019
	Rhynchotechum calycinum C.B. Clarke	Gesneriaceae	300–600		Leaf	Leaves (cooked) are edible as vegetable (Ronald et al. 2019)	Ronald et al. 2019
	Rubus insignis Hook. f	Rosaceae		Taptere	Fruit	Leaves are edible as substitute of betel (Piper betel) leaves; Ripe fruits (sweet) are edible (Ronald et al. 2019)	Ronald et al. 2019
	Rubus paniculatus	Rosaceae	300–800	Taptara, Kebo-lepum (A)	Fruit	Leaves are edible by children as substitute for betel leaves. Ripe fruits (sweet) are edible (Ronald et al. 2019), stomach disorder (Saikia et al. 2021), ripen fruits are taken raw, sour in taste (Shankar et al. 2015)	Ronald et al. 2019; Saikia et al. 2021; Shankar et al. 2015
	Rubus rosifolius Sm	Rosaceae	200–600	Kibo-lepum, Jigang (M), Mipir (A)	Fruit	Fruits (sour) are edible (Ronald et al. 2019),) Ripen fruits are taken raw (Shankar et al. 2015)	Ronald et al. 2019; Shankar et al. 2015
	Rubus treutleri Hook.f	Rosaceae	600–1100		Fruit	Ripe Fruits (sweet) are edible (Ronald et al. 2019)	Ronald et al. 2019
	Sauropus androgynus (L.) Merr	Phyllanthaceae	400–1000		Leaf	Tender leaves and shoots (cooked) are edible as vegetable (Ronald et al. 2019)	Ronald et al. 2019
	Silene herminii (Welw. ex Rouy) Welw. ex Rouy	Caryophyllaceae	900–2300	Kubumeku	Fruit	Fruits (sweet) are edible (Ronald et al. 2019)	Ronald et al. 2019
	Solanum anguivi Lam	Solanaceae	500–1000	Bangko, Brahati	Leaf, fruit	Half ripe fruits are edible as vegetable, paste or pickles. Leaves are edible as vegetable. Fruits are digestive but more quantity is considered toxic (Ronald et al. 2019), root carminative, expectorant (Balkrishna et al. 2019)	Ronald et al. 2019; Balkrishna et al. 2019
	Spiradiclis bifida Wall. ex Kurz	Rubiaceae	300–1000	Sokho	Leaf	Leaves (cooked) are edible as vegetable (Ronald et al. 2019)	Ronald et al. 2019
	Stixis suaveolens (Roxb.) Pierre	Capparaceae	300–800	Titegille	Fruit	Fruits (cooked) are edible as vegetable (Ronald et al. 2019)	Ronald et al. 2019
	Pteridium aquilinum	Pteridaceae		Khang, duwai (M), Okang (A)	Leaf	Leaves (cooked) are edible as vegetable (Ronald et al. 2019)	Ronald et al. 2019

(continued)

Table 3.2 (continued)

No	Plant species	Family	Altitude (mtr.)	Local name	Parts used	Medicinal uses	References
	Xanthophyllum burkillii J.R.Drumm. and Dunn	Polygalaceae	800–1400		Fruit	Fruits edible (Ronald et al. 2019)	Ronald et al. 2019
	Youngia japonica (L.) DC.	Asteraceae	Upto 2000	Rum Dum	Leaf	Entire plant (cooked) edible as vegetable (Ronald et al. 2019)	Ronald et al. 2019)
	Zanthoxylum nitidum (Roxb.) DC.	Rutaceae	400–500	Honior	Leaf, bark and fruits	Leaves used as spice with meat (Ronald et al. 2019), gastric problem (Tshering et al. 2018), fish poison (Baruah and Bhattacharyya 2019)	Ronald et al. 2019; Tshering et al. 2018; Baruah and Bhattacharyya 2019
	Clerodendrum colebrookianum Walp	Verbenaceae	500–2000	Oen-tapen Galo), ongin (Nyishi), Khangjela-shing, Wild, tuplickna or li pha fu, Pato hamang, Poto-ow (Ni), Bhati, Helea sak Titapata, Papa tsitsu (M)	Leaves, bark and tender shoots	Blood pressure; stomach trouble, malarial and bronchitis treatment (Chowdhery et al. 2009), high blood pressure, stomach disorder, headache (Tshering et al. 2018), high blood pressure, stomach disorder, headache (Kamum et al. 2018), consumption of leaves and young shoots controls high blood pressure. Also applied in wounds for blood clotting (Gangwar and Ramakrishnan 1990), boiled or steamed leaves are taken to control high blood pressure (Ratan et al. 2016), used for the treatment of blood pressure (Kalita et al. 2017), water extract of fresh leaves is taken for cough, diarrhoea and stomach pain (Goswami et al. 2009), blood pressure, cough and cold and liver disorders (Shankar and Rawat 2008), leaves used in fever; bark extract is drank for malaria treatment; boiled leaves taken for treatment of high blood pressure (Rethy et al. 2010), Tender leaves are used as vegetable; leaf decoction is given in high blood pressure (Shankar et al. 2015), leaves are used in curing diabetes, and high blood pressure (Wangpan et al. 2019)	Kalita et al. 2017; Kamum et al. 2018; Wangpan et al. 2019; Chowdhery et al. 2009; Tshering et al. 2018; Shankar and Rawat 2008; Rethy et al. 2010; Shankar et al. 2015; Ratan et al. 2016; Gangwar and Ramakrishnan 1990; Goswami et al. 2009

(continued)

Table 3.2 (continued)

No	Plant species	Family	Altitude (mtr.)	Local name	Parts used	Medicinal uses	References
	Coptis teeta Wall	Ranunculaceae	1500–2500	Mishmi teeta/ Rinko, Mamira, Rinko, idu-aro, Rinke, Mameera, Aku-taiim (Kham), Ringgo (A), Dung tsa	Fresh and dried rhizomes, leaves and roots, seeds	Antimicrobial and Antibacterial, anti-diarrheal (Verma 2018), used in blood clotting, stomach trouble, malaria (Hage Yanka and Tag 2020), root tonic in jaundice and diabetes (Balkrishna et al. 2019), Malaria, fever, jaundice, stomach ache, liver diseases, hypertension and diabetes (Tshering et al. 2018), fever, headache, gastric trouble, dysentry, ulcer, insomnia, vomitting, stimulant to heart, antibacterial (Gogoi and Sharma 2022b), a multiple usages whole plant for gastric, diarrhea, dysentery, fever (Murtem and Chaudhry 2016), antimalaria (Kalita and Khan 2013; Srivastava 2010), rhizome used in eye diseases and good appetizer, curing the digestive system. Anti-inflammatory and is useful in skin disorders (Hazarika and Tayeng 2020), leach bite (Sarmah et al. 2008b), cough and Cold, diarrhea and Dysentery and any liver disorders (Shankar and Rawat 2008), root/Rhizome is used in fever, liver diseases, hypertension and diabetes, root extract is taken twice daily in worm infestation, diabetes and hypertension (Tripathi and Limasenla 2017), dried rhizome is taken in fever, gastric, cold, cough, malaria, dysentery (Shankar et al. 2015)	Tripathi and Limasenla 2017; Hazarika and Tayeng 2020; Hage Yanka and Tag 2020; Tshering et al. 2018; Murtem and Chaudhry 2016; Kalita and Khan 2013; Sarmah et al. 2008b; Gogoi and Sharma 2022b; Shankar and Rawat 2008; Shankar et al. 2015; Balkrishna et al. 2019; Srivastava 2010; Verma 2018
	Allium cepa L.	Amaryllidaceae	Upto 750	Piyaji, Chong, Mong (B) Bzap (S), Byaku (Cultivated)	Leaf and bulb, rhizome	Anticancer, antimicrobial (Hage Yanka and Tag 2020), bulb used against cold, cough, flu and bronchitis (Sen et al. 2008), leaves are eaten raw as salad. Bulb are consumed as spice (Kalita et al. 2017), anti diabetic (Bam et al. 2015), eye pain (Saikia et al. 2021), Used to cures high blood cholesterol, blood sugar, asthma, liver problems, intestinal problems and locally cures fever (Gangwar and Ramakrishnan 1990), leaves juice is taken orally for cold and cough (Ratan et al. 2016)	Kalita et al. 2017; Hage Yanka and Tag 2020; Saikia et al. 2021; Ratan et al. 2016; Bam et al. 2015; Sen et al. 2008; Gangwar and Ramakrishnan 1990

(continued)

Table 3.2 (continued)

No	Plant species	Family	Altitude (mtr.)	Local name	Parts used	Medicinal uses	References
	Allium sativum L.	Amaryllidaceae	Upto 1400	Lossung. Jilap (A), Lasun (Ap), Losun, Chong, Cult, Pulou, Dibechengki (A), Jilap (A)	Whole plant, leaf, rhizome, cloves	Anticancer (Hage Yanka and Tag 2020), chutney and boiled food (Ayam 2017), anti diabetic (Bam et al. 2015), bone fracture and malaria (Tshering et al. 2018), bone fracture (Kamum et al. 2018), antimalaria (Kalita and Khan 2013), diarrhea and stomach problems (Sarmah et al. 2008b), tender leaves are used as vegetable; tubers as spice (Shankar et al. 2015)	Ayam 2017; Kamum et al. 2018; Hage Yanka and Tag 2020; Tshering et al. 2018; Kalita and Khan 2013; Sarmah et al. 2008b; Shankar et al. 2015; Bam et al. 2015
	Eryngium foetidum L.	Apiaceae	770–1500	Dunia, Hariyo, Ori, Ori-ritak, migom ori (A), dhaniyapatta (Ap), Rithak (G), Jongli Dhaniya (N), Damgra/ Tanum-tananang, Ori/Bandhaniya (A)	Leaf, seed, flower	Anticancer, antioxidant, antimicrobial (Hage Yanka and Tag 2020), chutney, salad and boiled food (Ayam 2017), dysentery (Tshering et al. 2018), Madness, headache (Saikia et al. 2021), anti-epileptic, headache, scorpion sting, anti-diabetic, anti-bacterial, analgesic, fever, arthritis (Gogoi and Sharma 2022b), helps in curing burns, earache, fevers (Gangwar and Ramakrishnan 1990), antimalaria (Kalita and Khan 2013), antidote for Scorpion sting (Bharali et al. 2017), leaf is used as spice (Shankar et al. 2015)	Ayam 2017; Bharali et al. 2017; Hage Yanka and Tag 2020; Tshering et al. 2018; Saikia et al. 2021; Kalita and Khan 2013; Gogoi and Sharma 2022b; Shankar et al. 2015; Gangwar and Ramakrishnan 1990
	Daucus carota L.	Apiaceae	2000–3500	Gajor	Root	Anticancer (Hage Yanka and Tag 2020)	Hage Yanka and Tag 2020
	Catharanthus roseus (L.) G.Don	Apocynaceae	90	Periwinkle, Dhuvephool	Whole plant, leaves	Anticancer (Hage Yanka and Tag 2020), anti diabetic (Bam et al. 2015), antimalaria	Hage Yanka and Tag 2020; Kalita and Khan 2013; Bam et al. 2015
	Panax ginseng C.A.Mey	Araliaceae	1500–2000	Zinseng	Root	Antioxidant, anti-stress, antineoplastic, anticancer (Hage Yanka and Tag 2020)	Hage Yanka and Tag 2020
	Aloe vera (L.) Burm.f	Asphodelaceae	1200–1600	Alo bera, Aloe vera, Ghikunvar	Whole plant, leaf pulp	Antioxidant, anticancer (Hage Yanka and Tag 2020), anti diabetic (Bam et al. 2015), acid reflux (Tshering et al. 2018), leaves used in stomach trouble, purgative, pulp in menstrual disorders, dried juice in constipation, root in colic disorders (Hazarika and Tayeng 2020)	Hazarika and Tayeng 2020; Hage Yanka and Tag 2020; Tshering et al. 2018; Bam et al. 2015
	Helianthus annuus L.	Asteraceae	Cultivated throughout	Danyi apu	Leaf, seed, flower	Anticancer (Hage Yanka and Tag 2020)	Hage Yanka and Tag 2020
	Sonchus oleraceus L.	Asteraceae	2000–3500	Kochi Hamang	Leaf	Anticancer (Hage Yanka and Tag 2020)	Hage Yanka and Tag 2020

(continued)

Table 3.2 (continued)

No	Plant species	Family	Altitude (mtr.	Local name	Parts used	Medicinal uses	References
	Cannabis sativa L.	Cannabaceae	300–3650	Bangw, Ganja, Namku	Leaves and seeds, flower, resin	Anticancer (Hage Yanka and Tag 2020), anti diabetic (Bam et al. 2015), Plant leaf is used in digestion and dysentery, stomach disorder, hypnotic, sedative, anti-inflammatory, analgesic, nausea, vomiting, hallucinogenic (Tshering et al. 2018), Veterinary (Kamum et al. 2018), dry leaves are taken as smoking cigarette for relieving body pain; a pinch of seed powder is used in dysentery, diarrhoea (Tiwari et al. 2009), Paste of leaves used against ringworms, leaves given to cattle for flatulence and indigestion (Murtem and Chaudhry 2016), pain relief (Bharali et al. 2017)	Bharali et al. 2017; Kamum et al. 2018; Hage Yanka and Tag 2020; Tshering et al. 2018; Murtem and Chaudhry 2016; Bam et al. 2015
	Glycine max (L.) Merr	Fabaceae	100–3000	Pottung Perung, Mawsung (B), Suk (S), Rontung (A), Yanni (Ap), Yagya (G), ammiiperung	Fruit, seed, stem	Antioxidant, anticancer (Hage Yanka and Tag 2020), seed are cooked, fermented and and consumed as vegetable along with rice (Kalita et al. 2017), fermented seeds make chutney and boiled food preparation (Ayam 2017), anti diabetic (Bam et al. 2015)	Ayam 2017; Kalita et al. 2017; Hage Yanka and Tag 2020; Bam et al. 2015
	Oxalis corniculata L.	Oxalidaceae	500–2800	O khui Hamang, Phakep, Nanchirikmu (S) Pulungsukhu (M), amrul, PakHukku, Kanjal (small), Pawa hiyub (Ap), Dhundi, Yasompi, Chageri	Leaf, root, fruit, shoot, whole plant	Anticancer (Hage Yanka and Tag 2020), appetizer, headache, relieves intoxication from wine; Diarrhoea (Hage Yanka and Tag 2020), leaves and roots as consumed as vegetable after boiled or raw. Leaves are against jaundice (Kalita et al. 2017), dyspepsia, bowel disorder, anaemia, scurvy, cure opacity of cornea, dysentery (Tshering et al. 2018), appetizer, headache (Saikia et al. 2021), dyspepsin, bowel disorder, aneamia, scurvy, daturapoisioning, cure opacity of cornea (Gogoi and Sharma 2022b), Paste of plant applied on fire burn (Murtem and Chaudhry 2016), antimalaria (Kalita and Khan 2013), leaves are grinded into paste and are applied in cuts and wounds. Raw leaves are used as an appetizer. Leaves are also chewed for foul smell of mouth (Ratan et al. 2016), used for the treatment of Dysentery (Kalita et al. 2017), Plant eaten raw for headache (Goswami et al. 2009), wormicide and liver tonic (Shankar and Rawat 2008), headache, indigestion, loss of appetite (Das and Tag 2006b), Plant for burning sensation, digestion and hyperacidity (Tripathi and Limasenla 2017)	Kalita et al. 2017; Kalita et al. 2017; Tripathi and Limasenla 2017; Hage Yanka and Tag 2020; Hage Yanka and Tag 2020; Tshering et al. 2018; Murtem and Chaudhry 2016; Saikia et al. 2021; Kalita and Khan 2013; Gogoi and Sharma 2022b; Ratan et al. 2016; Goswami et al. 2009

(continued)

Table 3.2 (continued)

No	Plant species	Family	Altitude (mtr.)	Local name	Parts used	Medicinal uses	References
	Plantago asiatica subsp. erosa (Wall.) Z.Yu Li	Plantaginaceae	600–1500	Mepi Hɛmang, Donihana-khang	Whole plant, leaves	Anticancer, antioxidant (Hage Yanka and Tag 2020), antimalaria (Kalita and Khan 2013)	Hage Yanka and Tag 2020; Kalita and Khan 2013
	Nicotiana tabacum L.	Solanaceae	305–2500	Dui Muku	Leaf	Anticancer (Hage Yanka and Tag 2020), leech repellent (Namsa et al. 2011b)	Hage Yanka and Tag 2020; Namsa et al. 2011b
	Solanum nigrum L.	Solanaceae	200–2800	Hiroh Hamang, Byako, Oryange, Nakana, Hor, Bangko kopi (A)	Berries, leaves, roots, stem, tender shoots	Anticancer (Hage Yanka and Tag 2020), diarrhoea (Tshering et al. 2018), liver tonic, indigestion (Saikia et al. 2021), Vomiting, diarrhoea, also used to cure tuberculosis, reduce mild abdominal pain (Gogoi and Sharma 2022b), the powdered dry leaves are used in the treatment of malaria. The fresh leaves are boiled and juice is taken during gastritis problem and in constipation (Tiwari et al. 2009), the leaves improve appetite and used against asthma, skin diseases, ringworms and urinary discharges (Gangwar and Ramakrishnan 1990), leaves used in gastric problem (Murtem and Chaudhry 2016), antimalaria (Kalita and Khan 2013), fresh leaves are chewed to cure boils in mouth and tongue. Boiled leaves and tender shoots are taken to cure high blood pressure and stomach ache (Ratan et al. 2016), used for the treatment of Dysentery, blood pressure, toothache/dental worms (Kalita et al. 2017), fruits are used as vegetable, chatni, challad, used in stomachache (Shankar et al. 2015)	Tiwari et al. 2009; Kalita et al. 2017; Hage Yanka and Tag 2020; Tshering et al. 2018; Murtem and Chaudhry 2016; Saikia et al. 2021; Kalita and Khan 2013; Gogoi and Sharma 2022b; Shankar et al. 2015; Ratan et al. 2016; Gangwar and Ramakrishnan 1990
	Taxus wallichiana Zucc	Taxaceae	100–3500	Tale Noori	Flower	Anticancer (Hage Yanka and Tag 2020)	Hage Yanka and Tag 2020
	Curcuma longa L.	Zingiberaceae	200–2000	Haldi, Hɛlodi, Zuing (S), Kaya Haldi, Khohmin, Khumein Nak	Root, rhizome	Anti-inflammatory, antioxidant, anticancer (Hage Yanka and Tag 2020), antidiarrhoeal (Namsa and Tangjang 2011), rhizome is consumed to relieve stomach pain or sore throat (Kalita et al. 2017), stomach ache (Tshering et al. 2018), rhizomes used in fracture and pain relieving (Murtem and Chaudhry 2016), cuts and wounds (Tiwari et al. 2009), rhizomes are chewed for relief from asthma (Benniamin 2011), rhizome is used as spice and in body pain (Shankar et al. 2015)	Tiwari et al. 2009; Kalita et al. 2017; Namsa and Tangjang 2011; Hage Yanka and Tag 2020; Tshering et al. 2018; Murtem and Chaudhry 2016; Shankar et al. 2015; Benniamin 2011

(continued)

Table 3.2 (continued)

No	Plant species	Family	Altitude (mtr)	Local name	Parts used	Medicinal uses	References
	Zingiber officinale Roscoe	Zingiberaceae	300–1500	Taki, Ching, Mongchu (S), Tatsua (B), Jakeng (A), Sing taki (Ap), Saagha, Cult, Kekir, Takeeh, Take, Deing, Chiingsa, Chhiing, Sinka, ardraka, Kiting	Leaf, rhizome	Antioxidant, anti-inflammation, anticancer (Hage Yanka and Tag 2020), antidiarrhoeal (Namsa and Tangjang 2011), Paste of rhizome is used for treatment of blockage in nostril, clear sore throat (Kalita et al. 2017), boiled food and curries (Ayam 2017), cough and stomach ache (Kamum et al. 2018; Tshering et al. 2018), cough (Saikia et al. 2021), stomach pain, carminative, stimulant, rhizome juice mixed with honey is used for cough (Gogoi and Sharma 2022b), rhizomes are consumed raw or cooked for immediate cure from cough and cold (Tiwari et al. 2009), rhizomes used in cough, cold and tonsillitis (Murtem and Chaudhry 2016), antimalaria (Kalita and Khan 2013), cough and cold (Tiwari et al. 2009), juice mixed with honey is taken for cold and cough. It is also taken with hot water against asthma and indigestion. Raw rhizome is chewed to cure tooth ache (Ratan et al. 2016), used for the treatment of cough (Kalita et al. 2017), extract of rhizome and black pepper (Imphitlom) is taken as remedy for ulcer (Tripathi and Limasenla 2017)	Tiwari et al. 2009; Tiwari et al. 2009; Ayam 2017; Kalita et al. 2017; Kalita et al. 2017; Tripathi and Limasenla 2017; Kamum et al. 2018; Namsa and Tangjang 2011; Hage Yanka and Tag 2020; Tshering et al. 2018; Murtem and Chaudhry 2016; Saikia et al. 2021; Kalita and Khan 2013; Gogoi and Sharma 2022b; Ratan et al. 2016
	Artemisia nilagirica C.B.Clarke	Asteraceae	1000–2500	Pamp, Masiang (B), Nei (S), Hyeman (S), Merangma, Namiperi, Ili mu (IM)	Leaf	Cough, headache, sores, wound and nose bleeding (Hage Yanka and Tag 2020), water of boiled leaves are used for bathing to cure body itching. Use as Antiseptic for freshly cut wound. Also used as air purifier during rituals and is used as an insecticides (Kalita et al. 2017), in headache and stomach pain, used as vegetable, to get relief from asthma (Tshering et al. 2018), cough, headache, sores (Saikia et al. 2021), wounds, cuts, scabies, and inflammations (Kamum et al. 2018), antimalaria (Kalita and Khan 2013), leaf paste is used in wound healing, fresh leaf decoction is used during nose bleeding (Shankar et al. 2015)	Kalita et al. 2017; Kamum et al. 2018; Hage Yanka and Tag 2020; Tshering et al. 2018; Saikia et al. 2021; Kalita and Khan 2013; Shankar et al. 2015

(continued)

Table 3.2 (continued)

No	Plant species	Family	Altitude (mtr.)	Local name	Parts used	Medicinal uses	References
	Bidens pilosa L.	Asteraceae	Upto 2000	Gomnik, Rhobashing (M), Hou-bak, Phutium, Tagam, Tangum	Leaf, seeds and roots, whole plant	Cuts and wounds, ulcers, ear and eye problems; Skin inflammation; hepatitis, urinary tract infection, malaria (Hage Yanka and Tag 2020), tender shoots are boiled and consumed as vegetable along with rice (Kalita et al. 2017), wounds and skin inflammations (Kamum et al. 2018; Tshering et al. 2018), wound healing, ulcer, ear and eye problem, influenza, hepatitis, urinary tract infection, anti-malaria, anti-pathogenic (Gogoi and Sharma 2022b), leprosy, eye disorder (Singh et al. 2012), antimalaria (Kalita and Khan 2013), chronic indigestion (Tag et al. 2005), leaves are considered good for eyes and stomach (Wangpan et al. 2019)	Kalita et al. 2017; Kamum et al. 2018; Hage Yanka and Tag 2020; Tshering et al. 2018; Kalita and Khan 2013; Gogoi and Sharma 2022b; Singh et al. 2012; Tag et al. 2005
	Blumea fistulosa (Roxb.) Kurz	Asteraceae	Upto 1500	Rumdum	Leaf	Used in Diarrhoea (Hage Yanka and Tag 2020)	Hage Yanka and Tag 2020
	Campylandra aurantiaca Baker	Asparagaceae	1800–2900	Kekong kelong	Root	Used in vomiting and headache (Hage Yanka and Tag 2020), indigestion (Tshering et al. 2018), antimalaria (Kalita and Khan 2013)	Hage Yanka and Tag 2020; Tshering et al. 2018; Kalita and Khan 2013
	Carica papaya L.	Caricaceae	Upto 1700	Kadongri, Omiir, Omita schein, Omri, Maak sangfo	Mature seeds, fruits, whole plant	Used for fertility control in Women (Hage Yanka and Tag 2020), antisecretory activity (Namsa and Tangjang 2011), Malaria and gastric (Tshering et al. 2018), anti-malarial, treatment of cuts, rashes, burns, stings, digestive problem, improve hearing capacity and improve lactation (Gogoi and Sharma 2022b), Unriped fruits used in stomach problems (Murtem and Chaudhry 2016), antimalaria (Kalita and Khan 2013), fruits cut open at one end, seeds are taken out and filled with common salt and tightly plugged with bamboo pins. This is burned in a chulha and the charred material is then cooled and mixed in water and filtered. Filtrate used is administered for the treatment of stubborn and dry cough (Benniamin 2011)	Namsa and Tangjang 2011; Hage Yanka and Tag 2020; Tshering et al. 2018; Murtem and Chaudhry 2016; Kalita and Khan 2013; Gogoi and Sharma 2022b; Benniamin 2011)

(continued)

Table 3.2 (continued)

No	Plant species	Family	Altitude (mtr.)	Local name	Parts used	Medicinal uses	References
	Chromolaena odorata (L.) R.M.King and H.Rob	Asteraceae	100–1000	Heling, Eeneh pudeh, Gocham Nemi, Malinga-jumpak, Thangba, Daglin (G)	Leaf	Used in major cuts and wounds (Hage Yanka and Tag 2020), cut, wounds, headache, fever (Saikia et al. 2021), Paste of leaves are directly applied on cut and wounds (Tiwari et al. 2009), Paste of leaves used on wounds infested by worms (Murtem and Chaudhry 2016), antimalaria (Kalita and Khan 2013), cuts (Tiwari et al. 2009), digestive disorder (Bharali et al. 2017), to stop bleeding from leech bite site (Namsa et al. 2011b)	Tiwari et al. 2009; Tiwari et al. 2009; Bharali et al. 2017; Hage Yanka and Tag 2020; Murtem and Chaudhry 2016; Saikia et al. 2021; Kalita and Khan 2013; Namsa et al. 2011b
	Embelia ribes Burm.f	Myrsinaceae	400–1500	Hingkong, Vai bidang, Onior, Vidanga, Biakol-lata	Tender leaves, fruits and shoots	Vegetable and is said to act as stomachic (Hage Yanka and Tag 2020), fruits antihelmintic, tonic (Balkrishna et al. 2019), fruit and root used in worm infestation, liver disorders and as tonic (Tshering et al. 2018), anti-diarrhoea, also used against intestinal worm infection (Gogoi and Sharma 2022b), skin disease, bronchitis, urinary discharge, dyspepsia, jaundice (Singh et al. 2012), antimalaria (Kalita and Khan 2013), fruits used in Liver disorder, worm infestation (Hazarika and Tayeng 2020), fruit and root used in worm infestation, liver disorders and as tonic (Tripathi and Limasenla 2017), diarrhoea (Bora et al. 2007)	Tripathi and Limasenla 2017; Hazarika and Tayeng 2020; Hage Yanka and Tag 2020; Tshering et al. 2018; Kalita and Khan 2013; Gogoi and Sharma 2022b; Singh et al. 2012; Balkrishna et al. 2019; Bora et al. 2007
	Entada rheedei Spreng	Leguminosae	Upto 2500	Pio-pia/pyopa, Prthvika, Gilgachh, Kolokpa sing tse (M), Po: piya (A)	Seeds	Crushed seeds are used in hair wash and as anti-dandruff (Hage Yanka and Tag 2020), cancer, gastrointestinal disorder, diarrhea, wounds, fever, skin disease (Singh et al. 2012), seeds are used as tonic and in worm infestation (Tripathi and Limasenla 2017), seed used for edible oil, spiritual use, used in Mandup (Shankar et al. 2015)	Tripathi and Limasenla 2017; Hage Yanka and Tag 2020; Shankar et al. 2015; Singh et al. 2012
	Erigeron bonariensis L.	Asteraceae	Upto 1600	Daglentado	Leaf	Vapour of the leaf is used in Sinusitis problem (Hage Yanka and Tag 2020), Vapour of leaves is inhaled in sinus problem (Tshering et al. 2018), Nose block (Saikia et al. 2021), Vapour of leaves is inhaled in sinus problem (Gogoi and Sharma 2022b)	Hage Yanka and Tag 2020; Tshering et al. 2018; Saikia et al. 2021; Gogoi and Sharma 2022b
	Gymnocladus assamicus Kanjilal ex P.C. Kanjilal	Mimosaceae	Upto 1500	Dekang, Minangmashing	Pods	Used as hair wash to remove dandruff (Hage Yanka and Tag 2020), detergent (soap), religious and veterinary (Tshering et al. 2018), detergent (soap), religious and veterinary (Kamum et al. 2018)	Kamum et al. 2018; Hage Yanka and Tag 2020; Tshering et al. 2018

(continued)

Table 3.2 (continued)

No	Plant species	Family	Altitude (mtr.)	Local name	Parts used	Medicinal uses	References
	Mikania micrantha Kunth	Asteraceae	1500–2500	Chaahpan-jaak, Mantami, Japanilota (A), Hipini, Chakpan, Tare Nemi, Bebli lata	Leaf, tender leaves and stem, flower	Used in itching, skin diseases, headache; cut and wound healing (Hage Yanka and Tag 2020), itching, skin diseases, headache (Saikia et al. 2021), burns, cuts (Tiwari et al. 2009), tender leaves and stem are grinded into paste and applied on cuts and wounds to stop bleeding. Juice is also taken orally to cure nose bleeding (Ratan et al. 2016), used for the treatment of skin disease (Bharali et al. 2017), leaf used in blood clotting, cuts, and wound healing (Shankar et al. 2015), blood clotting, insect bites and sting, antifungal, gastric ulcer, locally used as ornamental plant (Tshering et al. 2018), blood clotting, insect bites and sting, anti-fiingal, gastric ulcer, locally used as ornamental plant (Gogoi and Sharma 2022b), Paste of leaves used on cuts and wounds (Murtem and Chaudhry 2016), stomach ulcers, scabies, cholic, gastric ulcer, cardiac disorder (Singh et al. 2012)	Tiwari et al. 2009; Bharali et al. 2017; Kalita et al. 2017; Hage Yanka and Tag 2020; Saikia et al. 2021; Shankar et al. 2015; Ratan et al. 2016
	Paederia foetida L.	Rubiaceae	300–1300	Yepetere, Epitari, Phadobas lodi/ Paritaru/gandhali (Apatani), Aape-tare (G), Prasarini, Yepe-tere, Eppehtareh, Raunam, Gandhali, Taksirma momnang	Stem, leaves, whole plant	Used in gastritis, diarrhoea, dysentery, stomach disorder, paste of leaves applied to skin diseases (Hage Yanka and Tag 2020), antidiarrhoeal activity (Namsa and Tangjang 2011), salad and boiled foods (Ayam 2017), stomach ache, gastric, indigestion (Tshering et al. 2018), boils, stomach pain (Singh et al. 2012), antimalaria (Kalita and Khan 2013), jaundice and gastric (Tag et al. 2005), joint pains, dysentery, gastric, Pneumonia (Tiwari et al. 2009), juice of leaves and stem is used to cure dysentery, diarrhea, gastric, indigestion and stomach ache (Ratan et al. 2016), blood pressure control (Bharali et al. 2017), whole plant extract taken for treatment of urinary disorder, kidney stone and digestive problems (Rethy et al. 2010), extract of boiled mixture of inflorescence is used in the crack of chick and (Shankar et al. 2015)	Tiwari et al. 2009; Ayam 2017; Namsa and Tangjang 2011; Hage Yanka and Tag 2020; Tshering et al. 2018; Kalita and Khan 2013; Rethy et al. 2010; Shankar et al. 2015; Ratan et al. 2016; Singh et al. 2012; Tag et al. 2005
	Persicaria barbata (L.) H.Hara	Polygonaceae	Upto 1300	Rukji	Leaf	Ringworm and other skin diseases (Hage Yanka and Tag 2020)	Hage Yanka and Tag 2020

(continued)

Table 3.2 (continued)

No	Plant species	Family	Altitude (mtr.)	Local name	Parts used	Medicinal uses	References
	Pothos scandens L.	Araceae	100–1000	Lomangloset, Wiise-pyari, Madure (A)	Stem and leaf	Used in treatment of Bone fracture (Hage Yanka and Tag 2020), joint and waist pain (Tiwari et al. 2009), tender shoots are taken raw in dysentery, fruits are edible (Shankar et al. 2015)	Tiwari et al. 2009; Hage Yanka and Tag 2020; Shankar et al. 2015
	Pouzolzia bennettiana Wight	Urticaceae	300–1500	Oyik, Chandermuli	Leaves, stem	Vegetable is used in Dysentery (Hage Yanka and Tag 2020), stomach disorder (Kamum et al. 2018; Tshering et al. 2018), eaten as vegetable by expecting mothers to increase secretion of milk (Rethy et al. 2010), cooked tender leaves are consumed as vegetable in case of indigestion, burning sensation in chest and constipation. Paste of whole plant is applied in case of skin burn with inflammation (Wangpan et al. 2019)	Kamum et al. 2018; Wangpan et al. 2019; Hage Yanka and Tag 2020; Tshering et al. 2018; Rethy et al. 2010
	Rauvolfia verticillata (Lour.) Baill	Apocynaceae	Upto 1700	Ruki	Root, bark and seeds	Administered orally to cure malaria (Hage Yanka and Tag 2020)	Hage Yanka and Tag 2020
	Trichosanthes cordata Roxb	Cucurbitaceae	100–500	Dongkoriyong, Kangkulung (M), Dongkyong riyong (A)	Root	Used in Dysentery, diarrhoea (Hage Yanka and Tag 2020), edible problem; fruits are eaten (Shankar et al. 2015)	Hage Yanka and Tag 2020; Shankar et al. 2015
	Zingiber zerumbet (L.) Roscoe ex Sm	Zingiberaceae	200–1500	Kekir	Rhizome	Used in Stomach-ache, diarrhoea, cough (Hage Yanka and Tag 2020), stomach ache, vomiting, diarrhoea, cough	Hage Yanka and Tag 2020; Tshering et al. 2018
	Aconitum hookeri Stapf	Ranunculaceae	4200–4800	Sza-tsando	Tuber	Root used as substitute of A. ferox (Bajpay et al. 2019)	Bajpay et al. 2019
	Ajania tibetica (Hooker f. and Thomson) Tzvelev	Asteraceae	3900–4700	Otung-karpo	Whole plant	Used against respiratory related ailments (Sen et al. 2008)	Sen et al. 2008

(continued)

Table 3.2 (continued)

No	Plant species	Family	Altitude (mtr.)	Local name	Parts used	Medicinal uses	References
	Allium hookeri Thwaites	Amaryllidaceae	1900–3050	Maan, Tongpsi (B) Tschong (M), Disang-talap (A), Lepi/Taley (Ap), Dilap (G), Talap (N), Lam, Cult, Alo/Pula, Nyishi Talap, Lepi	Whole plant, seeds, leaves, rhizome, bulb	Whole plant consumed as vegetable to relieve fatigue; also eaten during insomnia (Sen et al. 2008), leaves and rhizome are crushed and used as salad. Rhizome is boiled and water is used for bathing among the Monk to clean the body before proceeding for ritual ceremony (Kalita et al. 2017), fritters salad and chutney, curries, fried egg (Ayam 2017), skin diseases, veterinary, bone fracture (Tshering et al. 2018), eruption of skin, cough, cold, wounds (Saikia et al. 2021), skin diseases, Veterinary, bone fracture (Kamum et al. 2018), to cure ulcer and stomach ailments (Gangwar and Ramakrishnan 1990), bulb used in cough and cold problem (Murtem and Chaudhry 2016), Pounded leaves mixed with oil is warmed and massaged on body to get relief from body ache. It is also used against cold and cough (Ratan et al. 2016)	Ayam 2017; Kalita et al. 2017; Kamum et al. 2018; Tshering et al. 2018; Murtem and Chaudhry 2016; Saikia et al. 2021; Ratan et al. 2016; Sen et al. 2008; Gangwar and Ramakrishnan 1990
	Fritillaria cirrhosa D.Don	Liliaceae	3000–4500	Loshu	Bulb	Bulb used in bronchitis and respiratory related ailments; applied to reduce swelling; eaten cooked as health tonic (Sen et al. 2008)	Sen et al. 2008
	Gentiana kurroo Royle	Gentianaceae	1500–3000	Panggin-mento	Whole plant	Water decoction used during back pain; consumed as vegetables to reduce cough, cold and fever (Sen et al. 2008)	Sen et al. 2008
	Gentiana stylophora C.B. Clarke	Gentianaceae	3048–4266	Shermen	Root	Root paste applied as poultice to wounds and swellings (Sen et al. 2008)	Sen et al. 2008
	Oxyria digyna (Linnaeus) Hill	Polygonaceae	2400–5800	Lugsho	Leaf	Leaf decoction used as purgative, diuretic, digestive and anti-inflammatory; paste applied to treat wounds and cuts (Sen et al. 2008)	Sen et al. 2008
	Potentilla peduncularis D. Don	Rosaceae	3000–4000	Gugae	Rhizome	Rhizome used in the treatment of diarrhea and pre-menstruation disorders (Sen et al. 2008)	Sen et al. 2008
	Rheum acuminatum Hooker f. and Thomson	Polygonaceae	3000–4100	Chutzsa	Fruit, root	Pounded fruit used to treat herpes; root used as diuretic and digestive (Sen et al. 2008)	Sen et al. 2008
	Rheum nobile Hooker f. and Thomson	Polygonaceae	3000–4200	Dugu-mento	Root	Water decoction of root consumed orally to treat of common cold, cough and throat pain (Sen et al. 2008)	Sen et al. 2008

(continued)

Table 3.2 (continued)

No	Plant species	Family	Altit▪de (mtr.	Local name	Parts used	Medicinal uses	References
	Sinopodophyllum hexandrum (Royle) T.S. Ying	Berberidaceae	2500–4500	Vomosi, Papra, Himalayan Mayapple	Whole plant	Extract of whole plant used in various female reproductive related diseases including irregular menstruation, birth difficulties, fluid discharge, and urinary track problems (Sen et al. 2008), root hepatic stimulant, purgative (Balkrishna et al. 2019)	Balkrishna et al. 2019; Sen et al. 2008
	Swertia hookeri C.B. Clarke	Gentianaceae	300C–3500	Tiktah	Whole plant	Root of matured plant used as anti-poison (Sen et al. 2008)	Sen et al. 2008
	Silene himalayensis (Rohrbach) Majumdar Lungara	Caryophyllaceae	200C–5000		Whole plant	Whole plant used against skin infections (Sen et al. 2008)	Sen et al. 2008
	Aegle marmelos (Corr.) L.	Rutaceae	Uptc 1000	Bel (A), Bilva	Unripe fruit pulp and bark	Antidiarrhoeal (Namsa and Tangjang 2011), fruit is used in diarrhoea and dysentery (Tshering et al. 2018), fruits used as digestive (Murtem and Chaudhry 2016), tonic and diarrhea (Tiwari et al. 2009), fruit is used in diarrhoea and dysentery (Tripathi and Limasenla 2017)	Tiwari et al. 2009; Tripathi and Limasenla 2017; Namsa and Tangjang 2011; Tshering et al. 2018; Murtem and Chaudhry 2016
	Bauhinia vahlii Wight. Arn	Apiaceae	Uptc 2300	Lingchirijong	Seeds	Antimicrobial activity (Namsa and Tangjang 2011)	Namsa and Tangjang 2011
	Rotheca serrata (L.) Steane and Mabb	Lamiaceae	500–1500	Kharam	Leaves		Prakash et al. 2009
	Cocos nucifera Linn	Arecaceae	Upt▪ 400	Nekuan	Nut water	Antimicrobial and phenolics comp (Namsa and Tangjang 2011)	Namsa and Tangjang 2011
	Dalbergia sissoo Roxb	Leguminosae	Upt▪ 2000	Sissoo	Mature leaves	Infectious diarrhoea (Namsa and Tangjang 2011)	Namsa and Tangjang 2011
	Manihot esculenta Crantz	Euphorbiaceae	200–1200	Banghakhoan	Tubers		Namsa and Tangjang 2011

(continued)

Table 3.2 (continued)

No	Plant species	Family	Altitude (mtr.)	Local name	Parts used	Medicinal uses	References
	Ocimum tenuiflorum L.	Lamiaceae	Upto 500	Mahnjor, Tilosi, Cult, Eulochi, Tulsi	Leaves, whole plant	Activity against enteric pathogens (Namsa and Tangjang 2011), stomach disorder, inflammations, wounds, cuts (Tshering et al. 2018), stomach disorder, inflammations, wounds, cuts (Kamum et al. 2018), bronchitis, cough, in several heart disease, believe to promote longevity (Gogoi and Sharma 2022b), leaf extract is swallowed to cure cough (Rethy et al. 2010)	Kamum et al. 2018; Namsa and Tangjang 2011; Tshering et al. 2018; Gogoi and Sharma 2022b; Rethy et al. 2010
	Oroxylum indicum Vent	Bignoniaceae	500–1000	Mano, Sayonaka, Namkalingshing, Panokni, Orok Tak Shein, Arlu, Tapatale (M), Sona Patha, Bhatghila	Leaves, seed, bark and root	Flavanoids (Namsa and Tangjang 2011), bark bitter tonic, in rheumatism (Balkrishna et al. 2019), cancer, antimalarial, jaundice, anti-arthritic, diarrhoea, fever, ulcer, antiinflammatory, Malaria and jaundice (Tshering et al. 2018), Purgative, headache (Tilling et al. 2015), rituals (Kamum et al. 2018), cancer, anti-malarial, jaundice, anti-arthritic, diarrhoea, fever, ulcer, anti-inflammatory (Gogoi and Sharma 2022b), bark powder used in skin itching, swelling. Also used in liver and stomach problem (Murtem and Chaudhry 2016), dysentery, leucoderma, asthma, bronchitis (Singh et al. 2012), diabetes, dysentery and cough (Tiwari et al. 2009), antimalaria (Kalita and Khan 2013), stem bark powder used in breast cancer (Hazarika and Tayeng 2020), hair tonic, anti-helminthic and carminative (Shankar and Rawat 2008), digestive disorder (Bharali et al. 2017), root bark is astringent, tonic; useful in dysentery Stem bark is bitter, tonic, useful in chronic rheumatism (Tripathi and Limasenla 2017)	Tiwari et al. 2009; Tilling et al. 2015; Bharali et al. 2017; Tripathi and Limasenla 2017; Kamum et al. 2018; Hazarika and Tayeng 2020; Namsa and Tangjang 2011; Tshering et al. 2018; Murtem and Chaudhry 2016; Kalita and Khan 2013; Gogoi and Sharma 2022b; Shankar and Rawat 2008; Singh et al. 2012; Balkrishna et al. 2019
	Papaver somniferum L.	Papaveraceae	2285–3050	Kanyi	Fruits	Diarrhoea (Namsa and Tangjang 2011)	Namsa and Tangjang 2011
	Piper betle Linn	Piperaceae	100–1500	Palu, Ghi (M)	Leaves	Beverages fermentation (Kamum et al. 2018), flowers are used in religious medicine–Mandup (Shankar et al. 2015)	Kamum et al. 2018; Namsa and Tangjang 2011; Shankar et al. 2015
	Sonchus wightianus DC.	Asteraceae	1200–2500	Balakhar	Roots		Namsa and Tangjang 2011

(continued)

Table 3.2 (continued)

No	Plant species	Family	Altitude (mtr)	Local name	Parts used	Medicinal uses	References
	Terminalia chebula Retz	Combretaceae	900–1500	Hilikha, Ontyal, Hellica, Bumura, Logyo, Harir, Laakchaa-ri, Harra/ Hilika	Fruits	Antidysentric (Namsa and Tangjang 2011), Malaria (Tshering et al. 2018), cough (Tilling et al. 2015), dehydration, cure blindness, cardiotonic, constipation, tumours, antihelmintic (Gogoi and Sharma 2022b), fruit used as a food supplement. Fruits are chewed for the treatment of cough, stomach and renal disorder (Asha and Singh 2020), fruits used in constipation and gastric (Murtem and Chaudhry 2016), antimalaria (Kalita and Khan 2013), bark diuretic, cardiotonic, fruit astringent, laxative, alterative used externally as local application to chronic ulcer (Hazarika and Tayeng 2020), gastric, cough, indigestion (Tiwari et al. 2009), Purgative, skin ulcer, cardio tonic, tooth ache and asthma in children's (Shankar and Rawat 2008), diarrhoea and digestive disorder (Bharali et al. 2017), cough, chest pain, stomach pain (Das and Tag 2006b)	Tiwari et al. 2009; Tilling et al. 2015; Bharali et al. 2017; Hazarika and Tayeng 2020; Namsa and Tangjang 2011; Asha and Singh 2020; Tshering et al. 2018; Das and Tag 2006b; Murtem and Chaudhry 2016; Kalita and Khan 2013; Gogoi and Sharma 2022b; Shankar and Rawat 2008
	Terminalia myriocarpa Heurck and Muell	Combretaceae	750–1150	Hilika	Fruit, leaves, bark	Antidiarrhoeal (Namsa and Tangjang 2011), bark extract is given in chest pain and as cardiac stimulant (Tshering et al. 2018), bark extract is given in chest pain and as cardiac stumulant (Gogoi and Sharma 2022b)	Namsa and Tangjang 2011; Tshering et al. 2018; Gogoi and Sharma 2022b
	Acmella oleracea (L.) R.K. Jansen	Asteraceae	Upto 1200	Mersang, Marsa	Young stem and leaf, fruit	Young shoots are boiled and eaten. Raw leaf is chewed to cure toothache (Kalita et al. 2017), the fresh flower is used in the treatment of toothache and the leaves are edible as vegetable (Tiwari et al. 2009)	Tiwari et al. 2009; Kalita et al. 2017
	Alnus nepalensis D. Don	Betulaceae	1200–3300	Hincha (S), Taram sin, Piak	Wood, branches, bark, leaves	Wood is used as household firewood and local fencing materials (Kalita et al. 2017), disinfectant, diuretic, reduce swelling, prevent excessive sweating, also used for carpentry (Tshering et al. 2018; Gogoi and Sharma 2022b), cancer, hepatitis, uterine cancer (Singh et al. 2012)	Kalita et al. 2017; Tshering et al. 2018; Singh et al. 2012
	Alpinia galanga (L.) Willd	Zingiberaceae	1000–1300	Skeh (B), Pipa, King Pang	Fruit, rhizome	Ripe fruits are sweet and edible (Kalita et al. 2017), the rhizome is an antibacterial agent and is used during constipation (Tiwari et al. 2009), bone Fracture (Das and Tag 2006b), rhizome powder mixed with Hemidesmus indicus leaves is tied over the fractured part with strips of bamboo internodes (Tripathi and Limasenla 2017)	Tiwari et al. 2009; Kalita et al. 2017; Tripathi and Limasenla 2017; Das and Tag 2006b

(continued)

Table 3.2 (continued)

No	Plant species	Family	Altitude (mtr.)	Local name	Parts used	Medicinal uses	References
	Arenga micrantha C.F. Wei	Arecaceae	2000–2150	Nap (B), Nuek (S)	Leaf and stem	The leaves are used for construction of roof and wall and fencing. The midrib of the leaves are used as broom. The stems are beaten and grounded to flour, roasted, cooked and consumed (Kalita et al. 2017)	Kalita et al. 2017
	Bambusa tulda Roxb	Poaceae	Upto 1200	Mayui (B), Me-Meyɔ (S), Hej	Tender shoot and culm, stem and leaves	Stem is used to make traditional hat. Leaves are used in rituals (Kalita et al. 2017), bamboo shoots are consumed as integral part of diet (Tshering et al. 2018), fish poison (Baruah and Bhattacharyya 2019)	Kalita et al. 2017; Tshering et al. 2018; Baruah and Bhattacharyya 2019
	Brugmansia suaveolans (Humb. and Bonpl. ex Willd.) Berch and Presl. Rostl	Solanaceae	1000–2500	Zanlin minthu (S), Dhatura	Whole plant	The plant is grown as barricade in the kitchen garden to prevent trespassing of animal (Kalita et al. 2017), skin infections, respiratory tract disorder, gastro intestinal tract disorders (Singh et al. 2012)	Kalita et al. 2017; Singh et al. 2012
	Capsicum annuum L.	Solanaceae	1200–3000	Chilliau (B), Nyamdak/pichak nyamday (N), Tero (Ap)	Fruit	Fruit is used as spice with rice and vegetable (Kalita et al. 2017), boiled food, chutneys and also used in making pickle (Ayam 2017)	Ayam 2017; Kalita et al. 2017
	Chenopodium album L.	Amaranthaceae	300–2600	Kungzo (S), Blamon (M), Taye, Khanaja	Young shoot and grains, Leaves, seed	Young shoot and grain are used as vegetable and food stuff (Kalita et al. 2017), locally used in preparing local wine and also eat as a vegetable (Tshering et al. 2018), locally used in preparing local wine and also eat as a vegetable (Gogoi and Sharma 2022b), used as carminative, laxative, anthelmintic, diuretic and as tonic (Gangwar and Ramakrishnan 1990)	Kalita et al. 2017; Tshering et al. 2018; Gogoi and Sharma 2022b; Gangwar and Ramakrishnan 1990
	Citrus aurantiifolia (Christm.) Swingle	Rutaceae	Upto 900	Chulukzapa (M)	Fruit	Fruits are processed into pickle and consumed along with rice (Kalita et al. 2017)	Kalita et al. 2017
	Cornus capitata Wall	Cornaceae	1000–1800	Krama (M)	Fruit	Ripe fruits are sweet and edible in raw after removing the peel (Kalita et al. 2017)	Kalita et al. 2017
	Cucumis sativus L.	Curcurbitaceae	7300–3400	Bizie (B). Meku, Gil	Fruit, leaf	Fruit are eaten raw as salad (Kalita et al. 2017), the leaf juice is given orally to children during stomach disorder. It is mainly used as wormicide. The fleshy fruit is edible and is used to cure indigestion (Tiwari et al. 2009), fresh or cooked, it is a main ingredient for salads. It is also consumed for its watery contents (Gangwar and Ramakrishnan 1990)	Tiwari et al. 2009; Kalita et al. 2017; Gangwar and Ramakrishnan 1990

(continued)

Table 3.2 (continued)

No	Plant species	Family	Altitude (mtr)	Local name	Parts used	Medicinal uses	References
	Cucurbita maxima Duchesne	Cucurbitaceae	2000–4000	Mirasami (B), Miniature pumpkin/Kupra	Tender leaves and fruit	Leaves and fruits are cooked and consumed as vegetable (Kalita et al. 2017), used in burns and inflammation (Gangwar and Ramakrishnan 1990)	Kalita et al. 2017; Gangwar and Ramakrishnan 1990
	Daphne papyracea Wall, ex G.Don	Thymelaeaceae	1000–2500	Shugu-sheng (M)	Bark	Fibre extracted from bark is processed into paper and used for writing purpose (Kalita et al. 2017)	Kalita et al. 2017
	Dendrocalamus hamiltonii Nees and Arn. ex Munro	Poaceae	300–1000	Muai-mu (B), Igzome (S), Yayibyapu/Eni (A), Bamboo (N), eepom, Ae Heroom, Hiku (Ap)	Stem, leaf, young shoot	Stem is used for crafting baskets. Leaves are used during rituals and also used as an accessory in traditional hat (Kalita et al. 2017), boiled food (Ayam 2017), the liquid of fermented young bamboo shoots are used against bee bite and have anti-inflammatory property (Tiwari et al. 2009), green peel of culm is effective in cuts (Murtem and Chaudhry 2016), water extract of fermented tender shoots is consumed for dysentery (Goswami et al. 2009)	Tiwari et al. 2009; Ayam 2017; Kalita et al. 2017; Murtem and Chaudhry 2016; Goswami et al. 2009
	Dioscorea alata L.	Dioscoreaceae	Upto 2500	Jiring (B), Rangthangong, Mayong (A)	Tuber	The tuber is consumed either roasted or boiled (Kalita et al. 2017), gastritis (Kamum et al. 2018; Tshering et al. 2018), indigestion (Saikia et al. 2021), tubers boiled with salt are used as vegetable (Shankar et al. 2015)	Kalita et al. 2017; Kamum et al. 2018; Tshering et al. 2018; Saikia et al. 2021; Shankar et al. 2015
	Diploknema butyracea (Roxb.) H.J. Lam	Sapotaceae	820–2500	Finsheng (M)	Fruit	Ripe fruits edible; oil extracted from seeds used for cooking (Kalita et al. 2017)	Kalita et al. 2017
	Erythrina stricta Roxb	Leguminosae	Upto 1400	Charow (S), Sit-pan, Tanyumta sen (N)	Stem, flower, root, bark	Stem is used for fencing of kitchen garden to prevent trespassing of animal (Kalita et al. 2017), scorpion sting, gout, anti-inflammatory, anxiolytic property (Tshering et al. 2018), scorpion sting, gout, ntiinflammatory, anxiolytic property (Gogoi and Sharma 2022b), headache (Tiwari et al. 2009), bone fracture (Bharali et al. 2017)	Tiwari et al. 2009; Bharali et al. 2017; Kalita et al. 2017; Tshering et al. 2018; Gogoi and Sharma 2022b
	Girardinia diversifolia (Link) Friis	Urticaceae	400–3000	Oguma (M), Posh Phon	Root	Roasted leaves used for relieving muscular pain (Murtem and Chaudhry 2016)	Kalita et al. 2017; Murtem and Chaudhry 2016

(continued)

Table 3.2 (continued)

No	Plant species	Family	Altitude (mtr.)	Local name	Parts used	Medicinal uses	References
	Gynura cusimbua (D.Don) S. Moore	Asteraceae	1300–3400	Munnyo-rap (B) Machilow (S), Aglaina, Kochi hamang	Leaf, tender shoot	Leaves are heated under fire until warm and is applied at the site of fractured bone anddislocated joint to fastening healing (Kalita et al. 2017), fine silky fibre is obtained from stems used in making ropes, twine and coarse cloth. Juice of stem and leaves are applied (Gangwar and Ramakrishnan 1990), raw juice is taken against stomach ache and worms. It is also used as blood purifier and is applied against allergy (Ratan et al. 2016)	Kalita et al. 2017; Ratan et al. 2016; Gangwar and Ramakrishnan 1990
	Illicium griffithii Hook.f. and Thomson	Schizandraceae	1200–1800	Lisi (M), Taihelang (Ap), Domlishi	Leaf, fruit	Dried fruit is used as spice. It is also used against cough and cold (Kalita et al. 2017), tea and local drink (Ayam 2017), fruits stimulant, carminative (Balkrishna et al. 2019), stimulant and Carminative (Shankar and Rawat 2008)	Ayam 2017; Kalita et al. 2017; Shankar and Rawat 2008; Balkrishna et al. 2019
	Ipomoea batatas (L.) Lam	Convolvulaceae	500–1200	Alau (B)	Tuber	Tuber are sweet in taste, consumed after Fruit boiled or roasted (Kalita et al. 2017)	Kalita et al. 2017
	Mahonia napaulensis DC.	Berberidaceae	2000–3500	Tsot Sheng (M), Taming	Bark, fruit	Bark is used as dye stuff. Ripe fruit is sweet and edible (Kalita et al. 2017), bark juice is applied against boils in lips, itching, wounds and skin rashes. Juice is also used for preparation of dye (Ratan et al. 2016)	Kalita et al. 2017; Ratan et al. 2016
	Nasturtium officinale R.Br	Brassicaceae	Upto 3700	Si lang-sag (M), Merek Mare (B) Asak doho (S)	Young shoot	Young shoots are cooked and eaten as vegetable and also used as fodder for livestock such as Pig and Cow (Kalita et al. 2017)	Kalita et al. 2017
	Oenanthe javanica (Blume) DC.	Apiaceae	1000–1200	Dhamphuk (S) Dhangbing (S) Szhing-hru (M), Kebunamul/Babon (N), Hugu hamang	Stem, leaf, young shoot	Leaves are consumed in raw as salad or chutney after mixing with salt and chilli (Kalita et al. 2017), boiled food and chutney (Ayam 2017), indigestion (Saikia et al. 2021), raw tender leaves and stem is taken against stomach ache (Ratan et al. 2016)	Ayam 2017; Kalita et al. 2017; Saikia et al. 2021; Ratan et al. 2016
	Opuntia dillenii (Ker Gawl.) Haw	Cactaceae	Upto 1200	Digbazo (S)	Inflorescence	Inflorescence parts are fleshy and sweet which are edible. Planted as barricade in kitchen garden to prevent trespassing of animal (Kalita et al. 2017)	Kalita et al. 2017
	Oryza sativa L.	Poaceae	1300	Chu (S)	Grain	Grain are eaten as staple food. Consumed with varieties of vegetables (Kalita et al. 2017)	Kalita et al. 2017
	Pennisetum glaucum (L.) R.Br	Poaceae	100–1500	Chow (B)	Seed	Seeds are grinded into flour, cooked, ermented and prepared local liquor (Kalita et al. 2017)	Kalita et al. 2017

(continued)

Table 3.2 (continued)

No	Plant species	Family	Altitude (mtr.)	Local name	Parts used	Medicinal uses	References
	Phaseolus vulgaris L.	Leguminosae	Upto 1000	Lape-chiri (B), Mawai-tuk (B), Pachung	Pod, fruits	Pods are cooked and consumed as vegetable (Kalita et al. 2017), it helps in poor digestion, nausea, abdominal cramps, gastro-enteritis, migraine, insomnia, depression and exhaustion (Gangwar and Ramakrishnan 1990)	Kalita et al. 2017; Gangwar and Ramakrishnan 1990
	Plantago major L.	Plantaginaceae	2200–3950	Tsa Shing (M), Germon (S), Mepihamang, Tsashing, Arisapana (\). Achodina (Kham), Kotsomang (M)	Root, leaves stem, whole plant	Leaves are eaten raw as vegetable salad. Leaves paste is used as skin ointment to cure bruises (Kalita et al. 2017), anti diabetic (Bam et al. 2015), wounds, inflammations, veterinary (Tshering et al. 2018), constipation (Saikia et al. 2021), wounds, inflammations, veterinary (Kamum et al. 2018), bone fracture (Sarmah et al. 2008b), used for fish poisoning; whole crushed plants are used to stupefying fishes in small running water. Tender leaves as vegetable, ripening fruits edible (Shankar et al. 2015)	Kalita et al. 2017; Kamum et al. 2018; Tshering et al. 2018; Saikia et al. 2021; Sarmah et al. 2008b; Shankar et al. 2015; Bam et al. 2015
	Plumbago indica L.	Plantaginaceae	600–810	Jaundice man (S), Kensumi, Nanchuni	Leaf, whole plant, root	Leaf is used to treat skin diseases (Kalita et al. 2017), bone pai (Das and Tag 2006b), root extract is applied on mouth ulcers; leaves are used as vegetables (Benniamin 2011)	Kalita et al. 2017; Das and Tag 2006b; Benniamin 2011
	Polygonum molle D. Don	Polygonaceae	1200–3500	Sa-pheung (B), Chan-fun	Young stem and leaf	Paste of leaves is consumed with food (Kalita et al. 2017)	Kalita et al. 2017
	Malus sikkimensis (Wenz.) Koehne	Rosaceae	2500–3500	Kya toh (M), Piita-ahi (Ap)	Fruit	Fruit are used as major ingredient for preparation of traditional pickle (Kalita et al. 2017), fruit is used as a food supplement, against gastrointestinal, constipation and digestive problems (Asha and Singh 2020)	Kalita et al. 2017; Asha and Singh 2020
	Quercus griffithii Hook.f. and Thomson ex Miq	Fagaceae	1300–2500	HingPong (B)	Wood	Wood are dried and use as firewood among the household (Kalita et al. 2017)	Kalita et al. 2017
	Quercus semecarpifolia Sm	Fagaceae	2600–4000	Kethcheng (M)	Fruit	Nuts from acorn are rich in fats which are edible (Kalita et al. 2017)	Kalita et al. 2017
	Rhododendron arboreum Sm	Ericaceae	2700–3700	Samu (M), Woodongmento	Flower, leaf and stem	Tender leaves are cooked and consumed. Flowers are eaten raw, and prepare juice (squash) from raw flower which is used as health tonic (Kalita et al. 2017), dysentery, diarrhoea, throat clearance when fish bones get stuck in the gullet (Kamum et al. 2018; Tshering et al. 2018)	Kalita et al. 2017; Kamum et al. 2018; Tshering et al. 2018
	Ricinus cummunis L.	Euphorbiaceae	300–800	Fopo Lasa (S)	Leaf	Leaves are wrapped over inflamed part of body which relieves pain and inflammation (Kalita et al. 2017)	Kalita et al. 2017

(continued)

Table 3.2 (continued)

No	Plant species	Family	Altitude (mtr.)	Local name	Parts used	Medicinal uses	References
	Rubia cordifolia L.	Rubiaceae	2000–4000	Niyeo-rap (B), Chou (S). Tamen, Manjista	Shoot, root and leaf	Roots are used to make dye (maroon colour). Leaves are used in rituals (Kalita et al. 2017), stomachache (Saikia et al. 2021), powdered root applied on forehead in headache (Murtem and Chaudhry 2016), to check vomiting, female disorders (Shankar and Rawat 2008)	Kalita et al. 2017; Murtem and Chaudhry 2016; Saikia et al. 2021; Shankar and Rawat 2008
	Rubus pedunculosus D. Don	Rosaceae	2600–3700	Bla-mrep (M)	Fruit	Ripe fruits are sweet which is eaten raw (Kalita et al. 2017)	Kalita et al. 2017
	Rumex maritimus L.	Polygonaceae	Upto 1800	Dham-champarak (S)	Roots, leaf	Leaves are used as animal fodder. Root is cut in small pieces and placed in the fishing hook to attract and catch the fishes (Kalita et al. 2017)	Kalita et al. 2017
	Cyphomandra betacea (Cav.) Sendtn	Solanaceae	1250–6900	Japarn-manthow (B), Masang tomator	Fruit	Fruits are used as substitute to tomato and also used as chutney with main course food (Kalita et al. 2017), the fruit is rich in Vitamin C and its soup is said to be good remedy for patients suffering from constipation (Gangwar and Ramakrishnan 1990)	Kalita et al. 2017; Gangwar and Ramakrishnan 1990
	Solanum viarum Duna	Solanaceae	100–1250	Tangchuk (B), Brok-bojo (S), Al tao, Katakari bhed, Khebyong chang (B), Boksikara (N)	Fruit	Mature fruit is yellow and is burnt to produce smoke which is inhaled under the mouth to clear out worms (Kalita et al. 2017), the smoke of burn fruit is sucked through a pipe made of bamboo and retained the smoke in mouth to expel tooth worm from the tooth (Tiwari et al. 2009), fever, diarrhea, pulmonary infection (Singh et al. 2012), leech repellent (Namsa et al. 2011b)	Tiwari et al. 2009; Kalita et al. 2017; Singh et al. 2012; Namsa et al. 2011b
	Thysanolaena latifolia (Roxb. ex Hornem.) Honda	Poaceae	500–1700	Chanchiri (B)	Inflorescence	The inflorescence are harvested and dried which is used as broom (Kalita et al. 2017)	Kalita et al. 2017
	Triticum aestivum L.	Poaceae	Upto 500	Pheuk (B)	Seed	Flour prepared from corn are consumed after cooked (Kalita et al. 2017)	Kalita et al. 2017
	Zea mays L.	Poaceae	Upto 3000	Spaw (B), Pchi (S)	Corn, leaf	Leaves are used as animal fodder and the corn is consumed in varieties of forms (Kalita et al. 2017)	Kalita et al. 2017
	Allium chinense G. Don	Amaryllidaceae	100–1200	Dilab/mirrong (A), Talap (N), Dilaap/ Alo	Bulb, leaves, rhizome	Boiled food and salad (Ayam 2017), help reduce blood cholesterol levels, act as a tonic to the digestive system and circulatory system (Gangwar and Ramakrishnan 1990)	Ayam 2017; Gangwar and Ramakrishnan 1990

(continued)

Table 3.2 (continued)

No	Plant species	Family	Altitude (mtr.)	Local name	Parts used	Medicinal uses	References
	Brassica juncea (L.) Czern	Brassicaceae	300–700	Tulang/tuka (A), Petto, Ka jung na	Leaf, stem, seed	Fermented seeds make chutney and leaves make salad (Ayam 2017), seed are bitter in taste and used in digestive problem. The seed paste is applied to the person for relief in mild fever or headache; the leaves are edible as vegetable (Tiwari et al. 2009), brown mustard is a folk remedy for arthritis, foot ache and rheumatism (Gangwar and Ramakrishnan 1990)	Tiwari et al. 2009; Ayam 2017; Gangwar and Ramakrishnan 1990
	Cardamine hirsuta L.	Brassicaceae	500–1000	Pade-hama (Ap)	Leaf, whole plant	Chutney and salads (Ayam 2017), indigestion (Saikia et al. 2021), whole plant is consumed raw for indigestion, cold and cough (Ratan et al. 2016)	Ayam 2017; Saikia et al. 2021; Ratan et al. 2016
	Cinnamomum bejolghota (Buch.-Ham.) Sweet	Lauraceae	800–1500	Pode (G), Chinchayong, Bontezpata	Leaf, bark, roots	Tea and meat curry (Ayam 2017), liver disorders, fever (Shankar and Rawat 2008), root extract is drank for treatment of liver troubles (Rethy et al. 2010)	Ayam 2017; Shankar and Rawat 2008; Rethy et al. 2010
	Cinnamomum camphora (L.) J. Presl	Lauraceae	1500–2500	Bambe (N)	Leaf, fruit	Tea and meat curry (Ayam 2017), leaf is useful in diarrhoea and skin diseases (Tshering et al. 2018)	Ayam 2017; Tshering et al. 2018
	Cinnamomum tamala (Buch.-Ham.) T. Nees and Eberm	Lauraceae	1000–1500	Rapi ising/ Jongkeng asing/ Sepiriang (A), Tej pat	Leaf, stem, bark, root	Meat curry (Ayam 2017), anti diabetic (Bam et al. 2015), leaf is used in cough, digestion and diabetes (Tshering et al. 2018), carminative, colic pain and diarrhea (Shankar and Rawat 2008), diarrhoea (Bharali et al. 2017)	Ayam 2017; Bharali et al. 2017; Tshering et al. 2018; Shankar and Rawat 2008; Bam et al. 2015
	Coriandrum sativum L.	Apiaceae	2000–3500	ori (A), Dhania	Leaf, stem	Chutney (Ayam 2017), fruits used in many medicinal preparations (Murtem and Chaudhry 2016)	Ayam 2017; Murtem and Chaudhry 2016
	Etlingera linguiformis (Roxb.) R. M. Sm	Zingiberaceae	1200–6600	Zakir (G)	Rhizome	Chutney (Ayam 2017)	Ayam 2017
	Gynura procumbens (Lour.) Merr	Asteraceae	Upto 1500	Ogin tamin (G), Tamin (N)	Young shoot	Chutney (Ayam 2017)	Ayam 2017
	Hibiscus sabdariffa L	Malvaceae	Upto 1500	Amta (A)	Leaf, fruit	Chutney and jam making (Ayam 2017), leaves and fruits are used in diarrhoea and dysentery (Wangpan et al. 2019)	Ayam 2017; Wangpan et al. 2019
	Litsea lancifolia (Roxb. ex Nees) Fern.-Vill	Lauraceae	1500–2000	Take more (G)	Stem, leaf	Chutney and salad (Ayam 2017)	Ayam 2017
	Magnolia oblonga (Wall. ex Hook.f.& Thomson) Figlar	Magnoliaceae	1200–1800	Serio-asing/Salyo (Ap),	Fruit	Chutney and boiled food (Ayam 2017)	Ayam 2017

(continued)

Table 3.2 (continued)

No	Plant species	Family	Altitude (mtr.)	Local name	Parts used	Medicinal uses	References
	Murraya koenigii (L.) Spreng	Rutaceae	500–6600	Narasingha (Ap)	Leaf	Salad, chutney and boiled foods (Ayam 2017), stomach trouble (Saikia et al. 2021), indigestion of food (Tiwari et al. 2009), leaf extract is taken orally to stop vomiting (Rethy et al. 2010)	Tiwari et al. 2009; Ayam 2017; Saikia et al. 2021; Rethy et al. 2010
	Neocinnamomum caudatum (Nees) Merr	Lauraceae	1500–1800	Tej pat (N)	Bark	Meat curry (Ayam 2017)	Ayam 2017
	Ocimum basilicum L.	Lamiaceae	200–1200	Tulsi (G), Sowsang, Bontulsi	Leaves and seeds	Salad (Ayam 2017), helps in urinary infection and discharges (Gangwar and Ramakrishnan 1990), seed powder and leaves used in cough and cold (Murtem and Chaudhry 2016), use for treatment of ringworm (Rethy et al. 2010)	Ayam 2017; Murtem and Chaudhry 2016; Rethy et al. 2010; Gangwar and Ramakrishnan 1990
	Phyllostachys bambusoides Siebold and Zucc	Poaceae	1000–1500	Bije (Ap), Bamboo (N)	Young shoot	Boiled food, pickles (Ayam 2017)	Ayam 2017
	Piper nigrum L.	Piperaceae	500–1200	Jaluk (G), Marcha, Rari	Leaves, fruits, seed	Boiled food (Ayam 2017), fruits used in cough, bronchitis, tonsillitis (Murtem and Chaudhry 2016), antimalaria (Kalita and Khan 2013), fever, tonsillitis (Bora et al. 2007)	Ayam 2017; Murtem and Chaudhry 2016; Kalita and Khan 2013; Bora et al. 2007
	Sesamum indicum L.	Pedaliaceae	500–1500	Tanam-namcha (N)	Stem	Boiled food, curries and rice cake (Ayam 2017)	Ayam 2017
	Zingiber montanum (J. Koenig) Link ex A. Dietr	Zingiberaceae	300–1500	Kekir (A)	Rhizome	Chutney (Ayam 2017)	Ayam 2017
	Azadirachta indica A. Juss	Meliaceae	300–700	Neem, Dokam Belam	Leaf, bark	Anti diabetic (Bam et al. 2015), stomach disorder, diarrhoea (Kamum et al. 2018; Tshering et al. 2018) and malaria (Tshering et al. 2018), leaves used in skin disease and boiled water extract taken for stomach ache (Murtem and Chaudhry 2016), itching, coughing and against hook worm	Kamum et al. 2018; Tshering et al. 2018; Murtem and Chaudhry 2016; Bam et al. 2015
	Bauhinia purpuea L.	Caesalpiniaceae	Upto 500	Kachnar	Bark	Stem bark is used in throat disorder, worm infestation (Tripathi and Limasenla 2017; Tshering et al. 2018)	Tripathi and Limasenla 2017; Tshering et al. 2018
	Begonia josephii A. DC.	Begoniaceae	6600–2900			Dysentery (Tshering et al. 2018), tender leaves and petioles are used as vegetable and chatni. Leaves are crushed and spread in paddy field to repel insects (Shankar et al. 2015)	Tshering et al. 2018; Shankar et al. 2015

(continued)

Table 3.2 (continued)

No	Plant species	Family	Altitude (mtr)	Local name	Parts used	Medicinal uses	References
	Bombax ceiba L.	Bombacaceae	100–650		Root, stem	Root and stem bark are aphrodisiac, stimulant (Tripathi and Limasenla 2017; Tshering et al. 2018)	Tripathi and Limasenla 2017; Tshering et al. 2018
	Brassica rapa L.	Brassicaceae	1800–2500		Seed	Bladder inflammation,	Tshering et al. 2018
	Kalanchoe pinnata (Lam.) Pers	Crassulaceae	2000–7600		Leaf	Leaf juice is used in kidney stone and urinary disorders (Tshering et al. 2018), burns (fire/hot water), stomach inflammation, dysentery, urinary problems (Tiwari et al. 2009)	Tiwari et al. 2009; Tshering et al. 2018
	Callicarpa macrophylla Vahl	Verbenaceae	Upto 1300	Priyangu,	Fruit	Fruit is used in blood dysentery and skin diseases (Tripathi and Limasenla 2017), headace (Saikia et al. 2021), atrophy (Singh et al. 2012),	Tripathi and Limasenla 2017; Saikia et al. 2021; Singh et al. 2012
	Calotropis gigantea (L.) R.Br. ex Ait	Asclepiadaceae	Upto 1000	Tupawknung, Kawenexait	Flower, root, leaf	Flowers are used in cough; root as rasayana (Tripathi and Limasenla 2017; Tshering et al. 2018), dog bite (Murtem and Chaudhry 2016; Saikia et al. 2021), antimalaria (Kalita and Khan 2013), leaves are used in sexual weakness (Swapna dosha) (Tripathi and Limasenla 2017)	Tripathi and Limasenla 2017; Tshering et al. 2018; Murtem and Chaudhry 2016; Saikia et al. 2021; Kalita and Khan 2013
	Senna alata (L.) Roxb	Caesalpiniaceae	Throughpout tropics		Leaf	Leaf is used in ring worm; leaf decoction is used in bronchitis and asthma (Tshering et al. 2018)	Tshering et al. 2018
	Senna tora (L.) Roxb	Caesalpiniaceae	Throughpout tropics		Leaf paste	Low blood pressure (Tripathi and Limasenla 2017), liver problem (Tshering et al. 2018)	Tripathi and Limasenla 2017; Tshering et al. 2018
	Hellenia speciosus (J.Koenig) S.R.Dutta	Zingberaceae	500–1500		Rhizome	Rhizome is used as worm repellent and blood purifier (Tshering et al. 2018)	Tshering et al. 2018
	Crotolaria juncea L.	Fabaceae	Throuhout		Seeds	Seeds, leaves are used in insanity, fever with catarrhal (Tshering et al. 2018)	Tshering et al. 2018
	Cyclosorus parasiticus	Thelypteridaceae				Gout and rheumatism, anthelmintic, antifungal and antibacterial (Tshering et al. 2018)	Tshering et al. 2018
	Datura stramonium L.	Solanaceae	2000–3000		Leaf	Leaves are used as narcotic, sedative and diuretic (Tshering et al. 2018)	Tshering et al. 2018
	Dendrocalamus strictus Nees	Poaceae				Wound or cut (Tshering et al. 2018)	Tshering et al. 2018

(continued)

Table 3.2 (continued)

No	Plant species	Family	Altitude (mtr.)	Local name	Parts used	Medicinal uses	References
	Drymaria cordata subsp. diandra (Blume) J.A.Duke	Caryophyllaceae	600–1300		Juice	Plant juice is laxative and ant febrile (Tshering et al. 2018)	Tshering et al. 2018
	Eleusine coracana (L.) Gaertn	Poaceae	300–1000			Cough, cold, congestion, antimicrobial, anti-inflammatory, food preservative (Tshering et al. 2018)	Tshering et al. 2018
	Ficus elastica Roxb. ex Hornem	Moraceae	350–1500			Dysentery (Tshering et al. 2018)	Tshering et al. 2018
	Gerbera piloselloides	Asteraceae	1500–3000			Treat cold, fever, acute conjunctivitis, rheumatic pain (Tshering et al. 2018)	Tshering et al. 2018
	Laphangium affine (D.Don) Tzvelev	Asteraceae	Upto 2500			Treatment of common cold, antimicrobial (Tshering et al. 2018)	Tshering et al. 2018
	Hedychium gracile Roxb	Zingiberaceae	800–1800			Mosquito repellent, antifungal, also used as spies (Tshering et al. 2018)	Tshering et al. 2018
	Hedychium aureum C.B. Clarke and H. Mann ex Baker	Zingiberaceae	1000–2000			Liver problem (Tshering et al. 2018)	Tshering et al. 2018
	Hedychium coccineum	Zingiberaceae	400–2000			Cure asthma and indigestion, antimicrobial, also used for local ornamental purposes (Tshering et al. 2018)	Tshering et al. 2018

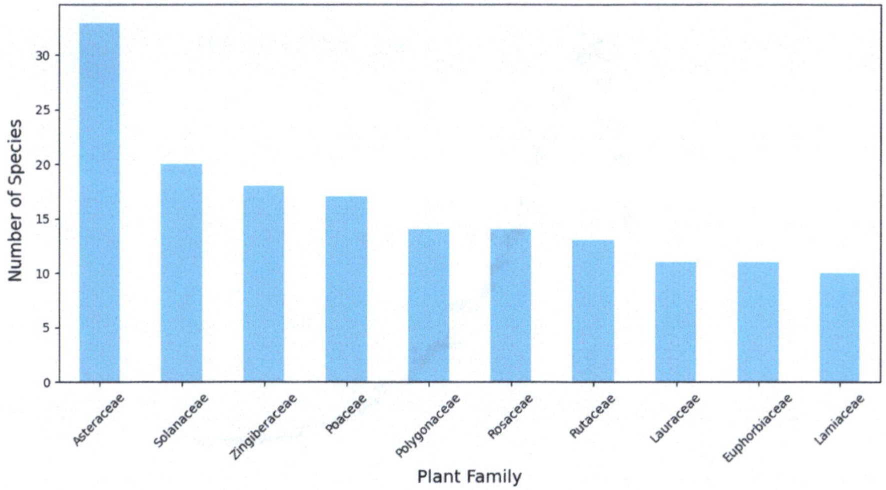

Fig. 3.2 Number of medicinal flora belonging to the top 10 families within Arunachal Pradesh, India

3.4.2 Richness Map of Medicinal Flora

Model 1 featuring the automatic smoothing term has been identified as the most suitable model according to both the Akaike Information Criterion (AIC) and the Bayesian Information Criterion (BIC). It effectively elucidates the variation in medicinal plant richness along the elevational gradients of Arunachal Pradesh (Table 3.1). The best-fitted curve (Model 1) of GAM is represented in the scatter plot representing the relationship of medicinal flora richness across different elevations (Fig. 3.3). The predicted map of the richness of Medicinal flora is presented in Fig. 3.4. These medicinal plants are found across diverse altitude gradients but are primarily concentrated in the tropical Himalayan region, as illustrated in Fig. 3.4.

3.5 Discussion

Traditional Knowledge and Practices of Medicinal Plant Use among Indigenous Communities play a crucial role in the cultural heritage of indigenous peoples, offering a treasure trove of plant-based remedies that have been passed down through generations. In Arunachal Pradesh, Northeastern India, these practices are integral to the healthcare systems of various tribes such as the Nyishi, Adi, and Galo. Indigenous healers, known as 'bono' or 'mori,' possess a profound understanding of the local flora and its medicinal properties, using plants to treat a myriad of ailments ranging from common colds to more serious diseases. These traditional healing methods reflect a deep connection between nature and health, emphasizing the holistic approach to

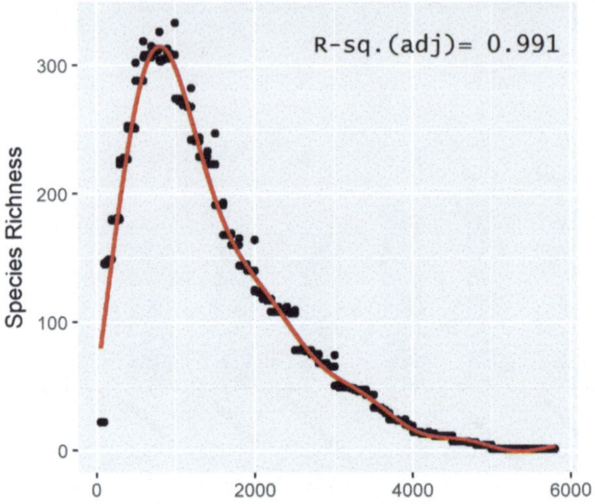

Fig. 3.3 The relationship of medicinal flora richness across different elevations and best-fitted GAM curve

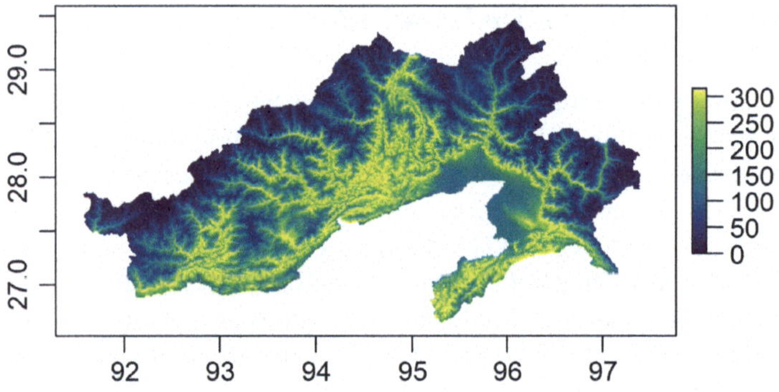

Fig. 3.4 Predicted spatially explicit medicinal flora richness map of Arunachal Pradesh

well-being among indigenous communities in the region. The documentation and preservation of such knowledge are essential not only for the conservation of biodiversity but also for the sustainable management of natural resources and cultural practices (Mary Zacharia Charwi et al., 2023). The results of the study provide valuable insights into the ethno-medico-botanical knowledge of Arunachal Pradesh. By documenting the traditional uses of medicinal plants and identifying the parts utilized for various ailments, this study contributes to the preservation and dissemination of indigenous knowledge systems.

Furthermore, the assessment of the medicinal significance and properties of each documented plant species will serve as a valuable resource for healthcare practitioners, researchers, and policymakers. Understanding the medicinal properties of these plants can aid in the development of new drugs, as well as the conservation and sustainable utilization of medicinal plant resources. Modern scientific research has played a pivotal role in unraveling the potential of indigenous medicinal flora, particularly in Arunachal Pradesh, Northeastern India. By combining traditional knowledge with contemporary scientific methodologies, researchers have identified numerous plant species with pharmacological properties (Devkota, 2014). These findings have not only validated the efficacy of traditional medicinal practices but have also opened avenues for the commercialization of these botanical resources. The commercialization of indigenous medicinal flora can lead to economic opportunities for local communities, conservation incentives for rare plant species, and the development of novel pharmaceutical products derived from natural sources. However, the process of commercialization must be approached with caution to ensure the sustainable utilization of these valuable resources without compromising the traditional knowledge systems or the ecological balance of the region. Thus, modern scientific research must be coupled with ethical considerations to harness the potential of indigenous medicinal flora for the benefit of all stakeholders.

The richness maps developed as part of this study will highlight the distribution patterns of medicinal plant species across different elevations in Arunachal Pradesh. These maps will serve as valuable tools for conservation planning and management, enabling local policymakers and stakeholders to prioritize areas for biodiversity conservation efforts. Biodiversity and conservation efforts regarding medicinal plants in Arunachal Pradesh, Northeastern India, are of paramount importance due to the region's rich flora and the traditional knowledge held by indigenous communities. The diverse range of medicinal plants found in this region has been a subject of interest for researchers and conservationists alike. Arunachal Pradesh is recognized as a biodiversity hotspot, highlighting the need for sustainable practices to protect these plant species. Conservation efforts must consider the indigenous communities' role in preserving and utilizing these plants, acknowledging their valuable traditional knowledge. By integrating scientific research with indigenous practices, a holistic approach to biodiversity conservation and sustainable use of medicinal plants can be achieved, ensuring the preservation of traditional knowledge and the protection of valuable plant species (Sumita Jha et al., 2023). Efforts in this area are crucial for maintaining ecological balance and promoting the health and well-being of both the environment and the local communities.

References

Anon (2003) State forest research institute, Information Bulletin No. 16. SFRI, Itanagar, Arunachal Pradesh

Arya OP, Pandey A, Samal PK (2017) Ethnobotany and nutritional importance of four selected medicinal plants from Eastern Himalaya, Arunachal Pradesh. J Med Plants Stud 5(1):45–49

Arya OP, Mylliemngap W, Pandey A (2020) Ethnomedicinal plants used by Adi community of upper siang district of Arunachal Pradesh in North-East India. Pleione 14(2):265–276

Asha H, Singh AV (2020) Diversity of medicinal wild fruits in the Lower Subansiri district of Arunachal Pradesh in Northeast India. Asian J Conserv Biol 9(1):113–122

Ayam VS (2017) Ethnomedicine of wild plants of Ziro, Arunachal Pradesh. Int J Res Stud Biosci 5(7):1–12

Bajpay A, Nainwal RC, Singh D (2019) Coptis teeta: a potential endemic and endangered medicinal plant of Eastern Himalayas. J Pharmacogn Phytochem 8(4):245–248

Balkrishna A, Joshi B, Srivastava A, Shankar R, Tiwari S, Tyagi SS, Mishra RK (2019) Some economic aspects of ferns and fern-allies of Seijosa forest area of Pakke-Kessang district, Arunachal Pradesh. Int J Adv Res Bot 5(3):14–20

Bam J, Rai S, Bhattacharya D, Maiti S, Islam S, Pathak P, Bera AK, Deb SM (2015) Indigenous curative and prophylactic traditional practices used against haematophagous leeches in Arunachal Pradesh and Sikkim. IJTK 14(3):493–497

Baruah U, Bhattacharyya R (2019) Diverse ethnic food practices of the Galo tribe in Arunachal Pradesh. Food Sci 10(2):197–202

Benniamin A (2011) Medicinal ferns of North Eastern India with special reference to Arunachal Pradesh. IJTK 10(3):516–522

Bharali P, Sharma M, Sharma CL, Singh B (2017) Ethnobotanical survey of spices and condiments used by some tribes of Arunachal Pradesh. J Med Plants Stud 5(1):101–109

Bora U, Sahu A, Saikia AP, Ryakala VK, Goswami P (2007) Medicinal plants used by the people of Northeast India for curing malaria. Phytother Res 21(8):800–804

Bushi D, Bam K, Mahato R, Nimasow G, Dai Nimasow O, Tag H (2021) Ethnomedicinal plants used by the indigenous tribal communities of Arunachal Pradesh, India: a review. Ethnobot Res Appl 22:1–40

Chakraborty R, De B, Devanna N, Sen S (2012) North-East India an ethnic storehouse of unexplored medicinal plants. J Nat Prod Plant Resour 2(1):143–152

Champion HG, Seth SK (1968) Revised survey of forest types ofIndia. Govt. of India publisher, Nasik, p 404

Charwi MZ, Mogha NG, Muluwa JK, Bostoen K (2023) Indigenous knowledge and use of medicinal plants among the kuria communities in the Tarime and Serengeti Districts of Mara Region, Tanzania. J Herbs Spices Med Plants 29(3):288–307

Chowdhery HJ (2008) 12.0 Arunachal Pradesh–The cradle of flowering plants. Special habitats and threatened plants of India 11(1):77

Chowdhery HJ, Giri GS, Pramanik A (2009) Materials for the Flora of Arunachal Pradesh, vol 3. Government of India, Botanical Survey of India, Kolkata

Chowlu K, Mahar KS, Das AK (2017) Ethnobotanical studies on orchids among the Khamti Community of Arunachal Pradesh, India. Indian J Nat Prod Resour (IJNPR) [Formerly Nat Prod Radiance (NPR)] 8(1):89–93

Das M, Anju J, Sarma NH (2013) Traditional medicines of herbal origin practice by the adi tribe of east siang district of Arunachal Pradesh, India. Global J Res Med Plants & Indigen Med 2(5):298–310

Das A, Tag H (2006a) Ethnomedicinal studies of Khamti tribe of Arunachal Pradesh Indian. J Tradit Knowl 5:317–322

Das AK, Tag H (2006b) Ethnomedicinal studies of the Khamti tribe of Arunachal. Indian J Tradit Knowledge IJTK 5(3):317–322

Devkota SR (2014) Ethno-medicinally used anti-inflammatory and pain relieving plants in the Kaski district, western Nepal. World J Pharm Sci 1240–1248

Doley B, Gajurel PR, Rethy P, Buragohain R (2014) Uses of trees as medicine by the ethnic communities of Arunachal Pradesh India. J Med Plant Res 8(24):857–863

Gangwar AK, Ramakrishnan PS (1990) Ethnobiological notes on some tribes of Arunachal Pradesh Northeastern India. Economic Botany 44(1):94–105

Ghosh G, Ghosh DC., Melkania U, Majumdar U (2014) Traditional medicinal plants used by the Adi, Idu and Khamba tribes of Dehang-Debang biosphere Reserve in Arunachal Pradesh. Int J Agric, Environ Biotechnol 7(1):165–171

Giri GS, Pramanik A, Chowdhery HJ (2008) Materials for the flora of Arunachal Pradesh. Vol. 2. Government of India, Botanical Survey of India, Kolkata

Gogoi J, Sharma M (2022a) Medicinal uses of wild edible plants by the Wancho tribe in Longding district of Arunachal Pradesh. Int J Pharm Res Allied Sci 11(4–2022):107–114

Gogoi J, Sharma MB (2022b) Medicinal uses of wild edible plants by the Wancho tribe in Longding district of Arunachal Pradesh. Int J Pharm Res Allied Sci 11(4):107–114

Goswami P, Soki D, Jaishi A, Das M, Sarma HN (2009) Traditional healthcare practices among the Tagin tribe of Arunachal Pradesh. IJTK 8(1):127–130

Goyal AK, Middha SK, Usha T (2022) BaccaurearamifloraLour.: a comprehensive review from traditional usage to pharmacological evidence. Adv Tradit Med 22(2):231–249

Hage Y, Das AP, Tag H (2020) Local understanding of some anticancer plants found in the Ziro Valley of Arunachal Pradesh India. Pleione 14(2):255–264

Hage Yanka AP, Tag H (2020) Local understanding of some anticancer plants found in the Ziro Valley of Arunachal Pradesh India. Pleione 14(2):255–264

Hajra PK, Verma DM, Giri GS (1996) Materials for the flora of Arunachal Pradesh vol 3. Government of India, Botanical Survey of India, Kolkata

Hazarika TK, Tayeng B, Ngurthankhumi R, Lalruatsangi E, Upadhyaya K, Lyngdoh N (2022) Unlocking Wild Edible Fruits of Indo-Burma Biodiversity Hot Spot, Arunachal Pradesh, India, to Support Food Security and Sustainable Rural Livelihood. Sustainability 14(23):16088

Hazarika T, Tayeng B (2020) Wild edible fruits of Arunachal Pradesh, India: Ethno-medicinal uses, socio-economic importance and role in rural economy and livelihood security. Research Square 1–30

Jagtap UB, Bapat VA (2010) Artocarpus: A review of its traditional uses, phytochemistry and pharmacology. J Ethnopharmacol 129(2):142–166

Jha S, Halder M (eds) (2023) Medicinal plants: Biodiversity, biotechnology and conservation. Springer Nature

Kala CP (2005) Ethnomedicinal botany of the Apatani in the Eastern Himalayan region of India. J Ethnobiol Ethnomed 1:1–8

Kalita BC, Yanka H, Gogoi G, Tag H, Das AK (2017) Anti-diabetic plants used by apatani tribe of Arunachal Pradesh India. J Bioresour 4(2):73–79

Kalita J, Khan ML (2013) Medicinal plants from the high altitudes of the western part of Arunachal Pradesh, India and their trade. Int J Conserv Sci 4(3)

Kamum G, Kanwal KS, Lod Y (2018) Ethnomedicinal plants used by Galo community of West Siang district, Arunachal Pradesh. Int J Res Appl Sci & Eng Technol 6(1):438–444

Kaul, Haridasan (1987) Forest types of Arunachal Pradesh, a prelim-inary study. J Econ Taxon Bot 9:383–389

Khongsai M, Saikia SP, Kayang H (2011) Ethnomedicinal plants used by different tribes of Arunachal Pradesh. Indian J Tradit Knowl 10(3):541–546

Lungphi P, Das AP, Ayam VS (2022) Ethnomedicinal Knowledge of Tangsa Community from Patkai Hills Region of Changlang District in Arunachal Pradesh, India. Non-Wood Forest Products of Asia: Knowledge, Conservation and Livelihood. Springer International Publishing, Cham, pp 71–101

Mayers N (1988) Threatened biotas: Hotspots in tropical forests. Environmentalist 8:187–208

Kazuharu M, Lobsang T (2015) Himalayan nature and tibetan buddhist culture in Arunachal Pradesh, India: A study of monpa. Springer. ISBN 9784431554929

Murtem G, Chaudhry P (2016) An ethnobotanical study of medicinal plants used by the tribes in Upper Subansiri district of Arunachal Pradesh India. Am J Ethnomedicine 3(3):35–49

Namsa ND, Tangjang S (2011) Natural pharmacopoeia used in traditional system of medicine for the treatment of diarrhoea in Arunachal Pradesh. Med Plants-Int J Phytomedicines Relat Ind 3(3):191–201

Namsa ND, Mandal M, Tangjang S, Mandal SC (2011a) Ethnobotany of the Monpa ethnic group at Arunachal Pradesh, India. J Ethnobiol Ethnomed 7:1–15

Namsa ND, Mandal M, Tangjang S, Mandal SC (2011b) Ethnobotany of the Monpa ethnic group at Arunachal Pradesh India. J Ethnobiol Ethnomedicine 7(1):1–15

Ngomle S, Eko R, Kanwat M, Kalita H, Moyon NN (2020) Eating from the wild: an insight into the indigenous wild edible plants consumed by the Digaru Mishmi tribe of Arunachal Pradesh. Indian J Tradit Knowl (IJTK) 19(2):360–369

Nimachow G, Taga T, Tag H, Dai O (2008) Linkages between bio-resources and human livelihood: a case study of Adi tribes of Mirem Village, Arunachal Pradesh (India). Initiation 2(1):183–198

Omem R, Rajiv M, Tag H (2016) Ethnobotany of the Galo community of Arunachal Pradesh India. Pleione 10(2):248–261

Perme N, Choudhury SN, Choudhury R, Natung T, De B (2015) Medicinal plants in traditional use at Arunachal Pradesh India. Int J Phytopharm 5(5):86–98

POWO (2024) Plants of the World Online. Facilitated by the royal botanic gardens, kew. Published on the Internet; http://www.plantsoftheworldonline.org/ Retrieved 25 March 2024

Prakash O, Kumar R, Mishra A, Gupta R (2009) Artocarpus heterophyllus (Jackfruit): an overview. Pharmacogn Rev 3(6):353

Rajbongshi P, Zaman K, Boruah S, Das S (2014) A review on traditional use and phytopharmacological potential of Bischofiajavanica Blume. Int J Pharm Sci Rev Res 24(2):24–29

Rama S (2013) Conservation and cultivation of threatened and high valued medicinal plants in North East India. Int J Biodivers Conservn 5(9):584–591

Ratan O, Mili R, Tag H (2016) Ethnobotany of the Galo community of Arunachal Pradesh, India. East Himal Soc Spermatophyte Taxon 10(2):248–261

Rawat MS, Chowdhury S (1998) Ethno medico botany of Arunachal Pradesh: Nishi & Apatani tribes. Dehra Dun, Bishen Singh Mahendra Pal Singh

Rethy P, Singh B, Kagyung R, Gajurel PR (2010) Ethnobotanical studies of Dehang-Debang Biosphere Reserve of Arunachal Prad esh with special reference to Memba tribe. IJTK 9(1):61–37

Rinyo R, Momang T, Pallabi KH, Tag H (2018) Ethnobotanical resources and traditional skills prevalent among the Tagin community of Arunachal Pradesh India. Pleione 12(2):265–274

Ronald K, Lodhi MS, Singha R, Kumari S, Kanwal KS, Arya SC (2019) Wild edible plants used by the ethnic communities of Shi Yomi District of Arunachal Pradesh, India. Pleione 13:247–257

Roy PS, Behera MD (2005) Assessment of biological richness indifferent altitudinal zones in the Eastern Himalayas, Arunachal Pradesh India. Curr Sci 88(2):250–257

Saikia S, Begum RA, Buragohain A (2021) Comprehensive list of anti-malarial plants used by different communities of Assam and Arunachal Pradesh, India. Int J Mosq Res 8:63–69

Sarmah R (2010) Commonly used non-timber forest products (NTFPs) by the lisu tribe in changlang district of Arunachal Pradesh, India. J Front Res 5:68–77

Sarmah R, Arunachalam A, Adhikari D (2006) Indigenous technical knowledge and resource utilization of Lisus in the South eastern part of Namdapha National Park Arunachal Pradesh. Indian J Trad Know. 5:51–56

Sarmah R, Adhikari D, Majumder M, Arunachalam A (2008a) Traditional medicobotany of Chakma community residing in the northwestern periphery of Namdapha national park in Arunachal Pradesh. Indian J. Tradit. Knowl. 7(4):587–593

Sarmah R, Adhikari D, Majumder M, Arunachalam A (2008b) Traditional medicobotany of Chakma community residing in the Northwestern periphery of Namdapha National Park in Arunachal Pradesh. IJTK 7(4):587–593

Sen P, Dollo M, Choudhury MD, Choudhury D (2008) Documentation of traditional herbal knowledge of Khamptis of Arunachal Pradesh. IJTK 7(3):438–442

Shankar R, Rawat MS (2008) Medicinal plants used in traditional medicine in Lohit and Dibang valley districts of Arunachal Pradesh. IJTK 7(2):288–295

Shankar R, Tripathi AK, Neyaz S, Anku G, Rawat MS, Dhiman KS (2015) Distribution of medicinal plants in Kurung Kumey, East Kameng and West Kameng districts of Arunachal Pradesh: their systemic conservation. World J Pharm Res 4(9):977–993

Sharma N, Shukla SP (1992) Geography and Development of Hill Areas: A Case Study of Arunachal Pradesh. Mittal Publications, ISBN, p 9788170993834
Sharma K, Kumar V, Kumar S et al (2021) *Bauhinia variegata*: a comprehensive review on bioactive compounds, health benefits and utilization. Adv Tradit Med (ADTM) 21:645–653
Singh RS (2005) Paths of Development in Arunachal Pradesh. Northern Book Centre. ISBN 9788172111830
Singh AV (2017) Ethnomedicinal studies on the Galo tribe of Nari-Koyu, East Siang, Arunachal Pradesh. Int J Eng, Sci Math 6(3):17–24
Singh A, Singh RK, Bhardwaj R, Singh AK (2012) Adaptations of culturally and nutritionally important traditional foods in Eastern Himalaya: A case study with Adi women of Arunachal Pradesh. IJTK 11(4):623–633
Singh P, Dash SS, Sinha BK (2019) Plants of Indian Himalayan region: an annotated checklist & pictorial guide Part I. Botanical Survey of India
Sinha BK, Dash SS, Singh P (2019) Plants of Indian Himalayan Region: an annotated checklist & pictorial guide Part II. Botanical Survey of India
Srivastava RC (2010) Traditional knowledge of Nyishi (Daffla) tribe of Arunachal Pradesh. IJTK 9(1):26–37
Tag H, Das AK (2004) Ethnobotanical notes on Hill Miri Tribe of Arunachal Pradesh. Indian Jour Trad Knowledge 3:80–85
Tag H, Das AK, Kalita P (2005) Plants used by the Hill Miri tribe of Arunachal Pradesh in ethnofisheries
Tag H, Kalita P, Singh RK, Das AK (2021) Biocultural resources and traditional food systems of nyishi tribe of Arunachal Pradesh (India): An Empirical Learning on the Role of Mythology and Folklore in Conservation. In: Social-Ecological Diversity and Traditional Food Systems (pp 155–198). CRC Press
Talukdar S (2020) Traditional herbal healing practices for the treatment of Malaria and associated symptoms by Miju-Mishmi tribe of Lohit district, Arunachal Pradesh, Northeast India. Research Square
Tilling R, Bharali P, Dutta P, Gogoi G, Paul A, Das AK (2015) Ethnomedicinal plants used by Apatani tribe of Ziro Valley of Arunachal Pradesh. Int J Conserv Sci 6(3)
Tiwari UL, Kotia A, Rawat GS (2009) Medico-ethnobotany of the Monpas in Tawang and West Kameng districts of Arunachal Pradesh India. Pleione 3(1):1–8
Tripathi AK, Limasenla SR (2017) Ethno-medicinal plants used by Nyishi tribe of Arunachal Pradesh India. World J Pharm Pharm Sci 6(5):1246–1253
Tsering J, Tag H (2015) High altitude ethnomedicinal plants of Western Arunachal Himalayan landscape. Pleione 9(1):18–25
Tshering D, Das S, Tsering J, Hui PK, Tag H (2018) Rapid ethnobotanical appraisal on Bugun, Sartang and Monpa communities of West Kameng and Tawang sectors of Arunachal Himalayan Region India. Pleione 12(2):283–297
Verma S (2018) Phytochemical and Pharmacological investigation of Vernonia cinerea: Asteraceae. Pharma Innov J 7(6):519–521
Wangpan T, Tasar J, Taka T, Giba J, Tesia P, Tangjang S (2019) Traditional use of plants as medicine and poison by Tagin and Galo Tribe of Arunachal Pradesh. J Appl Pharm Sci 9(9):098–104
National Wetland Atlas: Arunachal Pradesh (PDF), SAC/RESA/AFEG/NWIA/ATLAS/06/2009, Space Applications Centre (ISRO), Ahmedabad, India, p 98
WFO (2024) World Flora Online. Published on the Internet; http://www.worldfloraonline.org. Accessed on: 25 Mar 2024
World Health Organisation (2002) Traditional medicine strategy 2002–2005. WHO, Geneva
Yomgam Y, Sharma R, Arya OP, Yomgam Y (2017) Phytochemical screening and antifungal activities of three medicinal plants from Arunachal Pradesh India. J Med Plants 5(3):278–283

Chapter 4
Population Status of High-Value Medicinal Plants and Decadal Climate Variability in Niti Valley, Nanda Devi Biosphere Reserve, Western Himalaya

Arun Pratap Mishra, Himanshu Sahu, and Amit Kumar

Abstract This study investigates the impact of climate variability on threatened medicinal plants, using Landsat-8 for NDVI analysis in the Niti valley, a cold-arid region of Nanda Devi Biosphere Reserve, Decadal temperature trends, derived from TERRACLIMATE data, reveal consistent warming, particularly in summer, while precipitation patterns from CHIRPS exhibit both increasing and decreasing trends. Correlation analysis emphasizes complex interactions, highlighting a strong positive correlation between temperature and precipitation. NDVI trends showcase vegetation health dynamics, indicating favorable conditions and stress periods. Population assessment of medicinal plant species namely *Allium stracheyi, Carum carvi, Dactylorhiza hatagirea, Picrorhiza kurroa* and *Sinopodophyllum hexandrum* unveils varying densities and threats, with cultivated species facing pressure due to local consumption and trade, and those in specific habitats highly threatened by overexploitation. Results demonstrate a strong positive correlation between temperature and precipitation, along with a weak negative correlation between NDVI and temperature/precipitation. The study aligns observed trends with global climate change patterns, underscoring the interconnected nature of climate-vegetation relationships. Urging

The original version of the chapter has been revised: Chapter title has been updated. A correction to this chapter can be found at
https://doi.org/10.1007/978-3-031-77149-1_24

A. P. Mishra (✉)
Department of Forestry and Remote Sensing, Earthtree Enviro Private Limited, Shillong, Meghalaya 793012, India
e-mail: arunpratap7371@gmail.com

A. Kumar (✉)
Department of Habitat Ecology, Wildlife Institute of India, Dehradun, Uttarakhand 248001, India
e-mail: amit@wii.gov.in

H. Sahu
Bhomya Foundation, Monal Enclave, Banjarawala, Dehradun, Uttarakhand 248001, India

Department of Animal Ecology and Conservation Biology, Wildlife Institute of India, Dehradun, Uttarakhand 248001, India

D. Arya et al. (eds.), *Climate Change Impact on Himalayan Biodiversity*, Environmental Science and Engineering, https://doi.org/10.1007/978-3-031-77149-1_4

for sustainable harvesting and conservation, the findings stress the need for adaptive strategies. The study contributes to the discourse on environmental challenges, advocating for collaborative efforts to ensure a sustainable future for the Himalayan region and its diverse plant species.

Keywords Adaptive strategies · Climate variability · Medicinal Plants · Cold-arid region · Trans-Himalaya

4.1 Introduction

Globally, the pervasive impacts of climate change extend to diverse ecosystems, and the Himalayan region, often dubbed the 'Third Pole', stands out as a hotspot of vulnerability (Anees et al. 2022). This region is especially vulnerable to the far-reaching effects of climate change because of the complex interaction between its varied terrain and increased sensitivity to temperature changes (Aggarwal et al. 2023). The discernible rise in global temperatures has been more pronounced in the Himalayas than the average observed worldwide (Bolch et al. 2019). This accelerated warming trend highlights the urgent need to comprehend the climate dynamics in ecologically fragile zones such as the trans-Himalayan region (Kattel 2022; Mishra et al. 2023). Nowhere is this need more apparent than in the potential repercussions on the rich biodiversity, including threatened medicinal plant species that are integral to both traditional healing practices and global pharmaceutical interests (Sen and Samantha 2015).

Climate change, with its attendant shifts in temperature, precipitation patterns, and extreme weather events, exerts a profound impact on the distribution, phenology, and abundance of medicinal plants (Prasad et al. 2023). The effects of rising temperatures, which change the altitudinal ranges that these plants may grow in, are felt across the complex network of ecosystems (Lloretr et al. 2012). The melting of glaciers, erratic precipitation, and extreme weather events disrupt the delicate ecological balance, posing a direct threat to the survival of these vital plant species (Upadhyay 2020).

On a global scale, the ramifications extend beyond biodiversity loss, encompassing disruptions in the availability of bioactive compounds crucial for medicinal properties (Chivian 2002). The implications are not confined to ecological consequences alone; they extend to traditional knowledge systems, as indigenous communities grapple with the diminishing abundance of plants integral to their healthcare practices (Zimmerman 2005; Pearson et al. 2021).

Climate change's global impact on vulnerable medicinal plants in the trans-Himalayan region is a sad reminder of the interconnection of ecosystems and human well-being (Gagné 2019). The urgency lies in developing comprehensive strategies that address the complex relationship between climate dynamics, biodiversity conservation, and the preservation of indigenous knowledge in the face of this formidable global challenge (Seddon et al. 2020; Angmor 2024).

At the national level, India grapples with discernible shifts in climate patterns, ushering in profound impacts across multiple sectors, ranging from agriculture to water resources and biodiversity (Srivastava et al. 2024). The reports of the Intergovernmental Panel on Climate Change (IPCC) underscore the escalating frequency and intensity of extreme weather events in the country, sounding a clarion call for urgent and exhaustive climate studies (Change 2014).

Within this context, the Himalayan region emerges as a pivotal indicator and a focal point for monitoring climate-related changes (Chettri et al. 2020). The intricate ecological tapestry of this region, home to a myriad of plant species, including threatened medicinal plants, magnifies the significance of these climatic shifts (Johansson et al. 2018; Pandey et al. 2019). As temperatures rise and precipitation patterns become increasingly unpredictable, the delicate balance that sustains the biodiversity, particularly medicinal plant populations, is disrupted (Ahmadi-Lahijani and Moori 2022).

The Himalayas, often referred to as the 'pharmacy of the world', harbors a wealth of plant species with significant medicinal properties (Samant et al. 2011; Malik et al. 2015). The escalating climate-related changes pose a dual threat to these invaluable resources (Scheffran 2011; Dow et al. 2013). Firstly, the altered climatic conditions directly impact the distribution and abundance of these medicinal plants, potentially pushing them beyond their accustomed habitats (Kumar et al. 2023). Secondly, the perturbations in climate patterns can influence the bioactive compound composition of these plants, altering their medicinal properties and, consequently, impacting traditional healthcare practices deeply rooted in indigenous cultures (Kumar et al. 2015).

For India, the stakes are high, given the reliance on traditional medicine and the burgeoning pharmaceutical industry that draws extensively from the rich biodiversity of medicinal plants (Sheldon et al. 1997; Rao 2015). The evolving climate scenario in the Himalayan region necessitates a nuanced understanding of these shifts and their cascading effects on threatened medicinal plants, intertwining ecological resilience with national healthcare and economic considerations (Jha et al. 2021; Meetei et al. 2023).

In essence, the national urge for comprehensive climate studies coincides with the need to protect the richness and therapeutic value of plant species in the Himalayas (Kunwar et al. 2013; Kunwar 2018). The relationship between climate patterns and threatened medicinal plants is crucial for developing adaptive strategies to ensure ecosystem sustainability and the preservation of valuable traditional knowledge (Das et al. 2016; Aswani et al. 2018).

Nestled in the Western Himalaya, Uttarakhand stands as a region of rich ecological diversity facing environmental challenges from deforestation, tourism, and unsustainable land practices compounded by the threat of climate change (Ogra and Badola 2015). In this scenario, the lesser-explored Niti Valley in the Nanda Devi Biosphere Reserve stands out as a key area for an in-depth study on climate variability over the past decade. Motivated by the valley's critical role in ecological balance, the study integrates climatic parameters such as temperature, precipitation, and the Harmonic Normalized Difference Vegetation Index using Google Earth Engine. The research aims to unravel the complicated relationships between climate, vegetation, and environmental sustainability, offering insights to inform conservation strategies amidst the ecological challenges faced by this unique Himalayan ecosystem. Beyond climate and vegetation dynamics, the study provides a holistic perspective on Uttarakhand's environmental landscape, incorporating the complex relationship between climate change and threatened medicinal plant species. This multifaceted approach highlights the necessity of environmental research for sustaining delicate ecosystems amid the unfolding narrative of a changing climate.

4.2 Study Area

The current study was conducted in the buffer zone of Nanda Devi Biosphere Reserve (NDBR), encompassing seven villages viz., Mahargaon, Kailashpur, Gurgutti, Pharkia, Bampa, Gamsali, and Niti in Chamoli District of Uttarakhand (Fig. 4.1). The Niti Valley, with an average elevation ranging from 3500 to 5000 m, spans approximately 727.7 km^2 and includes three sub-watersheds; Amrit Ganga, Satyagad, and Ganesh Ganga in the Trans-Himalayan region of NDBR. This picturesque valley, situated in the rain-shadow zone, experiences dryness, particularly in the upper reaches, which remain snowbound for over six months annually. The vegetation in the region comprises dry temperate to dry alpine types, featuring a steppe-like flora. Inhabited by the Bhotiya community, specifically Tolchhas and Marchhas, the Niti Valley houses seven migratory villages out of the 47 in the NDBR buffer zone. The Bhotiya people, of Indo-Mongoloid origin, follow unique customs and beliefs, with dual dwellings one in the cold-arid region of Niti Valley and the other in the lesser Himalaya. The region faces biotic pressure from heavy livestock grazing, and the locals rely on natural resources from nearby forests and alpine pastures for sustenance through transhumant pastoralism.

Methodology: The study employed a comprehensive methodology beginning with the delineation of the study area through reconnaissance. A preliminary checklist was compiled, documenting locally and commercially used plants, vegetation types, distribution patterns, trade routes of highly medicinal plants, and the size of the area. Local communities, shepherds, and frontline forest department staff played a crucial role in providing insights. Population assessment of high-use value and threatened medicinal plant species was conducted through a combination of primary and secondary sources. Primary data collection involved semi-structured open-ended and closed-ended questions, individual interactions, and group discussions in seven selected villages and local markets. Secondary information was gathered from various offline and online sources, including scientific research articles, reports, books, dissertations, and PhD theses, utilizing databases such as Google Scholar, ResearchGate, Taylor and Francis Online, SpringerLink, and Elsevier Online. The study further integrated climatic parameters, including temperature, precipitation, and the Harmonic Normalized Difference Vegetation Index, employing advanced

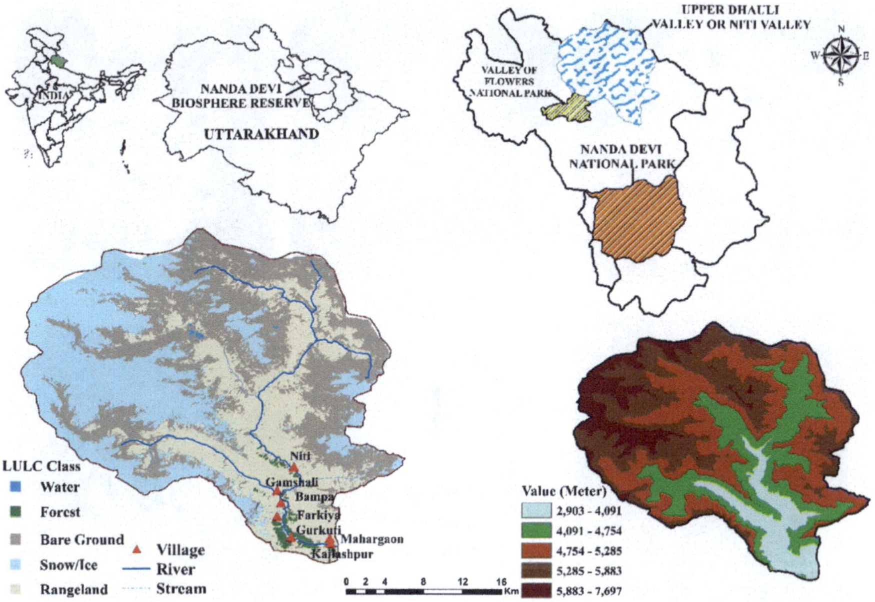

Fig. 4.1 Upper Dhauli Valley or Niti valley, a cold arid region of Nanda Devi Biosphere Reserve, Western Himalaya

geospatial tools and technologies such as Google Earth Engine (GEE). Specific datasets, namely TERRACLIMATE for decadal temperature analysis, CHIRPS for precipitation, and Landsat-8 for Harmonic Normalized Difference Vegetation Index, contributed to a comprehensive and holistic assessment of the study area.

4.3 Results

4.3.1 Selection of Focal Medicinal Plant Species

In Himalayan regions, the increased demand viz., commercial and local consumption has caused the decline in the wild population of several medicinal plant species. Unfortunately, unorganized, over harvesting and premature harvesting are the major threats to the survival of threatened medicinal plant species in the present scenario, which eventually has led to the decline in their natural population, regeneration and survival.

Based on extensive interactions in the form of group discussions, individual communications with younger generations, elder people (men and women) including

Fig. 4.2 High value medicinal and aromatic plant species in Niti valley

local healers, plant collectors, and local traders, a total of five high-value (with significant economic end usage medicinal and aromatic plant species namely *Allium stracheyi, Carum carvi, Dactylorhiza hatagirea, Picrorhiza kurroa* and *Sinopodophyllum hexandrum* were selected in Niti valley, NDBR (Fig. 4.2). The selection of threatened medicinal plant species was based mainly on two focal categories, (i) highest quantum of collection (HQC) and (ii) high threat (HT) due to removal and usages.

4.3.2 Population Status of the Focal Species

According to IUCN Red List guidelines, and Goraya and Ved (2017), population size, threats and geographic range are essential parameters to evaluate the threat or population status of a species. In the present study, based on the vegetation sampling, the population of medicinal plant species was estimated to understand its habitat specificity, operative threats, existing pressure, potential source areas and impact on their survival (Tables 4.1, 4.2 and 4.3). Of the selected focal species, *Allium stracheyi* and *Carum carvi* were cultivated by the villagers of Niti, Gamshali, Bampa, Pharkiya, Gurgutti, Kailashpur and Mahargaon, whereas *Sinopodophyllum hexandrum* was reported to cultivate at micro-scale in Gamshali and Niti. *A. stracheyi* has been brought under cultivation by the ethnic Bhotiyas of the Niti valley as a kitchen garden crop for a long time for their local consumption as well as trade.

Table 4.1 Population of focal species in Niti valley, NDBR

Name of species	Density (ind. m^{-2} ± SE)	Individuals recorded	Associate species
Allium stracheyi Baker	1.8 ± 0.855	186	*Juniperus* spp., *Rosa* spp., *Danthonia cachemyriana, Thymus linearis, Lotus corniculatus, Trigonella emodi* and *Origanum vulgare*
Dactylorhiza hatagirea (D.Don) Soó	**	05	*Sibbaldia parviflora, Euphrasia himalayica, Gaultheria trichophylla* and *Thymus linearis*
Picrorhiza kurroa Royle ex Benth	0.75 ± 1.03	148	*Bistorta affinis, Potentilla argyrophylla, Thymus linearis, Anaphalis* spp., and *Sibbaldia parviflora*
Sinopodophyllum hexandrum (Royle) T.S. Ying	0.91 ± 0.827	65	*Berberis* spp., *Lonicera* spp., *Arisaema tortuosum, Potentilla* spp., *Geranium wallichianum* and *Juniperus* spp.

**Recorded in one locality

Table 4.2 Population size, habitat specificity, threat and level of pressure in Niti valley, NDBR

Species	Local population size	Habitat specificity	Threat	Pressure level
Allium stracheyi	Small, sparse	Wide	Over exploitation for consumption and trade	High
Carum carvi	Large, dominant in and around villages and agricultural fields	Wide	High demand for consumption and trade	High
Dactylorhiza hatagirea	Small, non-dominant	Narrow	Over exploitation for medicine and trade	High
Picrorhiza kurroa	Small, non-dominant	Narrow	Over exploitation for medicine and trade	High
Sinopodophyllum hexandrum	Large, dominant at few locations	Wide	Over exploitation for medicine and trade	High

Table 4.3 Potential source areas, collection pressure and impact on survival of the focal species

Species	Potential source areas	Distribution (m)	Collection pressure	Threat status		Impact on the survival
				IUCN	CAMP	
Allium stracheyi	Adjoining forested areas, bouldary slopes, grassy slopes	2800–3500	Highly destructive and selective	–	–	Threat to its wild population
Carum carvi	Domestic in villages and agricultural areas	2500–3500	High trade and local consumption	Least concern	–	No reports from wild
Dactylorhiza hatagirea	Riverine areas and herbaceous meadows	3000–3800	Highly destructive, unselective and unmanaged	Endangered	Critically endangered	Threat to its wild population, regeneration and survival
Picrorhiza kurroa	Moraine, rocky alpine slopes	3000–4000	Highly trade	Endangered	Critically endangered	Threat to its wild population, regeneration and survival
Sinopodophyllum hexandrum	Adjoining forested areas, bouldary slopes, grassy slopes	2800–4000	Highly destructive, unselective and unmanaged	–	Endangered	Threat to its wild population

4.3.3 Decadal Temperature Trends

The investigation of temperature trends spanning from 2001 to 2022 reveals dynamic patterns in both increasing and decreasing trends, with due consideration to standard errors (Fig. 4.3). In the period from 2001 to 2004, temperatures exhibited a consistent rise, with notable peaks during the summer months. For instance, July 2004 recorded a temperature of 16.5 °C, surpassing the mean, with a standard error of ± 1.2 °C. This upward trajectory persisted from 2005 to 2008, showcasing significant increases, such as the peak of 17.0 °C in August 2007, with a standard error of ± 1.5 °C. The years from 2009 to 2012 continued the trend, marked by fluctuations, and July 2012 displayed a notable high of 17.8 °C, accompanied by a standard error of ± 1.4 °C. Despite occasional decreases in certain winter months, the overall trend remained positive. From 2013 to 2016, temperatures experienced further growth, with a distinct upward trend in the summer and early fall. August 2015 reached a high of 15.5 °C, accompanied by a standard error of ± 1.3 °C. The years 2017 to 2020 witnessed

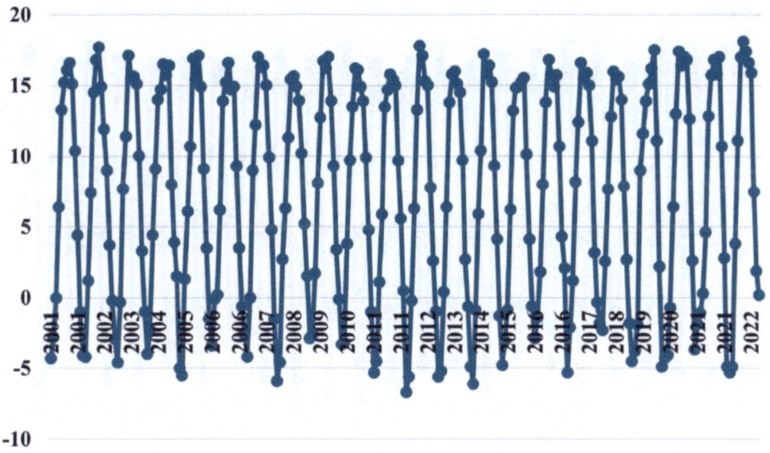

Fig. 4.3 Decadal temperature trends (2001–2022), in the Niti Valley, NDBR

consistent temperature increases, particularly in the late spring and early summer. July 2018 stood out with a temperature of 16.0 °C, and its standard error was ± 1.6 °C. The upward trend continued in 2021 and 2022, with July 2022 registering a high of 18.1 °C, with a standard error of ± 1.9 °C.

4.3.4 Decadal Precipitation Trends

The analysis of precipitation trends from 2001 to 2022 reveals intriguing patterns in both increasing and decreasing trends, with consideration for standard error (Fig. 4.4). In the period from 2001 to 2004, a gradual increase in precipitation is noted, particularly in the summer months. For instance, August 2001 saw a significant positive deviation from the mean, recording a surplus of 118 mm, with a standard error of ± 15 mm. A similar trend continues from 2005 to 2008, with a consistent upward pattern. Notably, January 2007 experienced an unexpected decrease, recording − 5 mm, deviating from the general trend. In the subsequent years from 2009 to 2012, a continued rise in precipitation is observed, and despite occasional fluctuations, the overall trend remains positive. September 2010, however, exhibits a notable decrease of -25 mm, deviating from the increasing pattern, with a standard error of ± 18 mm. From 2013 to 2016, the trend continues with some fluctuations, and September 2014 stands out with a decrease of −33 mm, accompanied by a standard error of ± 20 mm. The years 2017 to 2020 witness a consistent increase in precipitation, with standard errors accounting for the variability. Noteworthy is the deviation in February 2018, where the precipitation is −46 mm below the mean, with a standard error of ± 25 mm. The upward trend persists in 2021 and 2022, with January 2022 showing a deviation of −97 mm below the mean, indicating a substantial decrease, with a standard error of ± 23 mm.

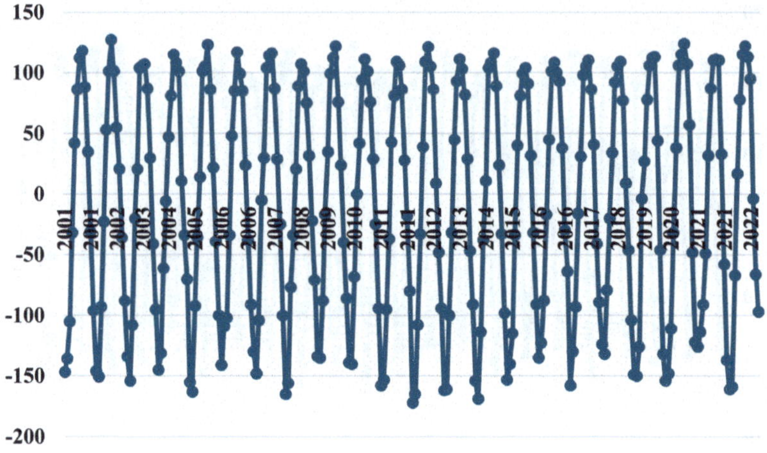

Fig. 4.4 Decadal precipitation trends (2001–2022), in Niti Valley

4.3.5 *Vegetation Health Trends*

The dataset illustrates the Normalized Difference Vegetation Index (NDVI) spanning from April 20, 2013, to December 22, 2021, providing insights into vegetation health over time (Fig. 4.5). Initial missing values on April 20 and May 6, 2013, precede the recorded NDVI values, with a notable initiation at 0.294 on May 22, 2013, indicating moderate vegetation density. Fluctuations follow, showcasing peaks in June 2015 (0.315) and July 2017 (0.327), indicating periods of heightened vegetation growth. Intermittent decreases, observed in March 2014 (−0.029) and March 2019 (−0.058), suggest potential stress or reduced vegetation during those periods. Consistently, higher NDVI values are evident in the summer months of June and July, indicative of favorable conditions for vegetation. Remarkably, June 2020 records a peak of 0.276, highlighting optimal vegetation health during that period.

The Harmonic Normalized Difference Vegetation Index with the fitted model enhances the dataset, offering a smoothed representation of vegetation trends, allowing for a clearer interpretation of long-term patterns and cyclical variations in vegetation health over the specified timeframe.

4.3.6 *Correlation Analysis*

A Pearson correlation analysis was conducted to examine the relationships among Temperature, Precipitation, and NDVI (Fig. 4.6). The results indicate a strong positive correlation between Temperature and Precipitation ($r = 0.99$, $p \leq 0.05$) and a moderate negative correlation between Temperature and NDVI ($r = -0.15$, $p \leq 0.05$). Similarly, Precipitation demonstrates a robust positive correlation with Temperature

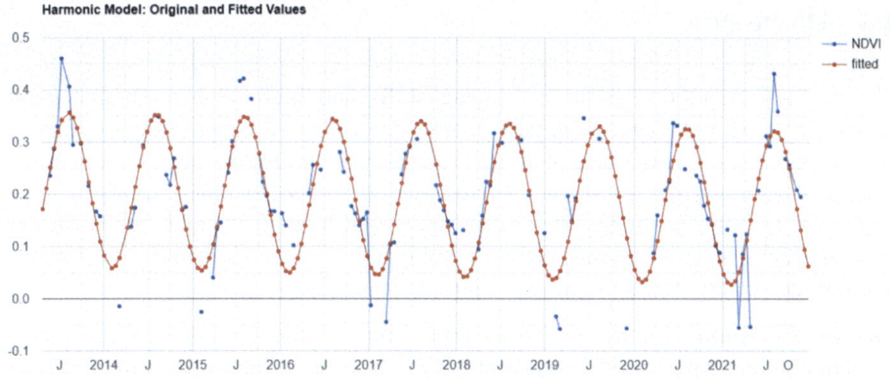

Fig. 4.5 NDVI trends (2013–2021) in Niti Valley, peaks and fluctuations with harmonic NDVI

($r = 0.99$, $p \leq 0.05$) and a modest negative correlation with NDVI ($r = -0.2$, $p \leq 0.05$). NDVI exhibits a weak negative correlation with both temperature ($r = -0.15$, $p \leq 0.05$) and Precipitation ($r = -0.2$, $p \leq 0.05$). All correlations are reported at the 95% confidence level, ensuring statistical significance.

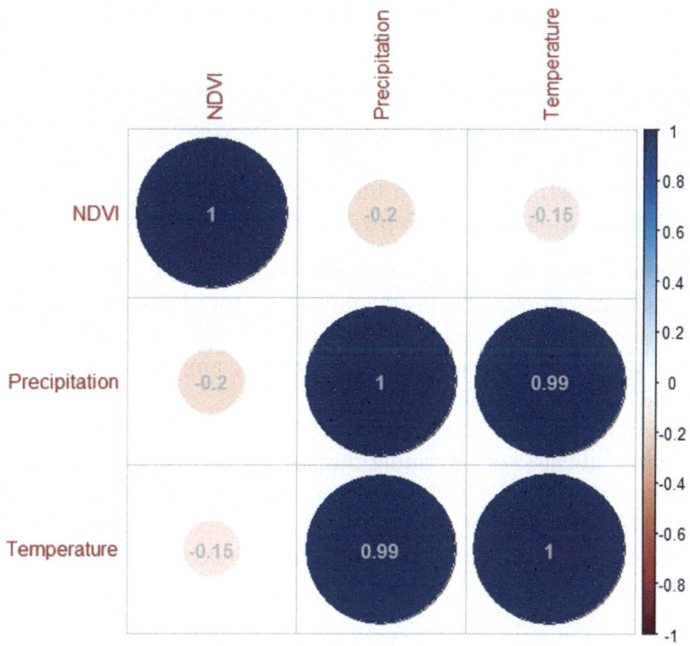

Fig. 4.6 Showing strong positive correlation between temperature and precipitation

4.4 Discussion

The delicate interaction between climate shifts and high-value medicinal plant species in the Himalayan region, particularly Niti Valley, reveals a complex story of ecological vulnerability. These plant species, namely *Allium stracheyi, Carum carvi, Dactylorhiza hatagirea, Picrorhiza kurroa, and Sinopodophyllum hexandrum,* play vital roles in both local ecosystems and traditional medicinal practices. However, their existence is under threat due to a combination of factors, primarily driven by anthropogenic activities and climate-induced variations.

Increased Demand: The surge in demand for these medicinal plant species, driven by both commercial and local consumption needs, poses a substantial threat. *Allium stracheyi* and *Carum carvi,* cultivated in and around villages, face intensified pressure due to their proximity to human settlements. The heightened demand for these plants, often driven by traditional medicine practices and economic considerations, puts local populations at risk.

Unorganized Harvesting: The unorganized and unsustainable harvesting practices further exacerbate the challenges faced by these plant species. The lack of structured management and regulation leads to overharvesting, particularly in the case of *D. hatagirea* and *P. kurroa,* which have smaller populations and specific habitats. Unchecked harvesting can result in the depletion of these species, disrupting the delicate ecological balance.

Habitat Specificity: The specific habitat requirements of *D. hatagirea* and *P. kurroa* make them susceptible to habitat degradation. The pressures exerted by overharvesting, combined with habitat loss due to land-use changes or climate-induced alterations, create a perilous situation for these species. Their smaller populations and restricted distribution make them particularly vulnerable.

Dominance and Threat: *S. hexandrum,* while dominant in certain locations, is not immune to the threats posed by overexploitation. The dominance of this species may attract more harvesting, leading to localized declines in its wild population. Even species with larger populations face threats, emphasizing the need for a comprehensive conservation approach.

The decadal temperature trends observed in Niti Valley from 2001 to 2022 exhibit dynamic patterns indicative of a warming climate. The consistent rise in temperatures, particularly during the summer months, aligns with global climate change trends (IPCC 2021). This warming trend can have profound impacts on the local ecosystem, affecting vegetation, water resources, and overall biodiversity. The positive correlation between temperature and precipitation is a noteworthy finding, suggesting a complex interaction within the hydrological cycle. The observed positive correlation implies that as temperatures increase, the hydrological cycle may intensify, leading to alterations in precipitation patterns. This aligns with projections indicating changes in precipitation distribution and intensity under a warming climate (Trenberth et al. 2003). The implications for vegetation health are substantial, as shifts in precipitation patterns can influence water availability and soil moisture, crucial factors for plant growth and survival (Dai 2013).

The weak negative correlation between NDVI and temperature/precipitation suggests that extreme climatic conditions, such as excessive heat or rainfall, might stress vegetation, leading to a reduction in NDVI values. This aligns with studies highlighting the sensitivity of NDVI to climate extremes and the potential implications for ecosystem functioning (Li et al. 2018). Stress-induced reduction in NDVI values may indicate periods of decreased photosynthetic activity, impacting overall vegetation health.

Decadal precipitation trends reveal intriguing patterns, with both increasing and decreasing trends. Anomalies, such as unexpected decreases in January 2007 and February 2018, underline the complexity of precipitation dynamics. These anomalies could result from regional climatic variations, affecting the hydrological balance and posing challenges to local ecosystems (Fu et al. 2007). The overall positive trend, especially from 2017 to 2020, aligns with global climate change projections indicating shifts in precipitation patterns (IPCC 2021).

The strong positive correlation between temperature and precipitation further emphasizes the interconnected nature of these climatic variables. As temperatures rise, the atmosphere's capacity to hold moisture increases, potentially leading to more intense rainfall events (Held and Soden 2006). However, negative correlations with NDVI suggest that extreme climatic conditions might have adverse effects on vegetation, highlighting the delicate balance between water availability and excess under changing climatic conditions. NDVI trends provide valuable insights into vegetation health over time. Peaks in June 2015 and July 2017 coincide with periods of favorable conditions for robust vegetation growth, indicating the significance of climatic conditions during these months. Intermittent decreases in March 2014 and March 2019 suggest periods of stress, possibly induced by climatic anomalies. The Harmonic NDVI model's smoothed representation enhances the dataset, allowing for a clearer interpretation of long-term vegetation trends.

In the global context, the vulnerability of the Himalayan region to climate change is well-documented (Dimri et al. 2018). The observed trends in Niti Valley contribute to the broader discourse on regional and global environmental challenges. Climate change impacts, including shifts in temperature, precipitation, and vegetation dynamics, have far-reaching consequences for ecosystems and biodiversity. The implications for medicinal plant species highlight the urgent need for sustainable harvesting practices and conservation efforts to mitigate the threats posed by climate variability.

4.5 Conclusion

The climatic analysis of Niti Valley from 2001 to 2022 reveals significant trends with potential implications for the local ecosystem and medicinal plant species. Decadal temperature trends show a consistent warming pattern, particularly during summer, aligning with global climate change projections. The positive correlation between temperature and precipitation implies complex hydrological interactions, while the weak negative correlation with NDVI suggests potential stress on vegetation.

Precipitation trends exhibit both increasing and decreasing patterns, with unexpected anomalies. The overall positive trend, alongside a strong positive correlation between temperature and precipitation, emphasizes the interconnected nature of these climatic variables. However, negative correlations with NDVI highlight potential adverse effects on vegetation health.

NDVI trends provide valuable insights into vegetation dynamics, with peaks indicating robust growth and intermittent decreases signaling periods of stress. The study emphasizes the Himalayan region's vulnerability to climate change, calling for sustainable practices and conservation efforts, especially for high-value medicinal plant species facing overexploitation.

In the global context, the findings contribute to the understanding of climate-vegetation interactions, emphasizing the urgency of adaptive strategies and conservation measures. As climate change impacts ecosystems and biodiversity, informed decision-making and sustainable resource management become paramount for the well-being of both local communities and the environment.

4.6 Way Forward

The study highlights the urgency of implementing adaptive strategies and conservation measures in Niti Valley. A holistic approach involving local communities, policymakers, and researchers is essential. Sustainable harvesting practices for high-value medicinal plant species, coupled with climate-resilient agriculture, can enhance ecosystem resilience. Community-based initiatives, informed by the study's findings, should prioritize education on climate change impacts and promote sustainable resource management. Collaborative efforts, guided by scientific insights, are crucial for mitigating climate-induced threats and fostering a resilient future for both the Himalayan region and its diverse plant species.

Authors Contribution Arun Pratap Mishra: Conceptualized, Visualized, Field Data Collection, Computational Handling, Original draft Writing, Review and Editing; Himanshu Sahu: Computational Handling, Statistics and Writing; Amit Kumar: Review, Edit and Technical Inputs.

References

Aggarwal R, Kaur S, Din Dar MU, Kuriqi A (2023) Uncertainties in climate change scenarios for determining temperature and rainfall patterns in regions with mixed climate conditions. Acta Scientiarum Polonorum. Formatio Circumiectus 22(1)

Ahmadi-Lahijani MJ, Moori S (2022) Current status of medicinal plants in perspective of environmental challenges and global climate changes. Environmental challenges and medicinal plants:

sustainable production solutions under adverse conditions. Springer International Publishing, Cham, pp 1–28

Anees MM, Sharma R, Joshi PK (2022) Urbanization in Himalaya-an interregional perspective to land use and urban growth dynamics. Mountain landscapes in transition: effects of land use and climate change, pp 517–538

Angmor GDM (2024) Climate change flooding and diseases in Sub Sahara Africa: trends and adaptions strategies (a review). Korean J Physiol Pharmacol 28(1):8–21

Aswani S, Lemahieu A, Sauer WH (2018) Global trends of local ecological knowledge and future implications. PLoS ONE 13(4):e0195440

Bolch T et al (2019) Status and change of the cryosphere in the extended Hindu Kush Himalaya region. In: Wester P, Mishra A, Mukherji A, Shrestha A (eds) The Hindu Kush Himalaya assessment. Springer, Cham. https://doi.org/10.1007/978-3-319-92288-1_7

Change IPOC (2014) AR5 climate change 2014: mitigation of climate change. Switzerland, Geneva

Chettri N, Shrestha AB, Sharma E (2020) Climate change trends and ecosystem resilience in the Hindu Kush Himalayas. Himalayan weather and climate and their impact on the environment, pp 525–552

Chivian E (2002) Biodiversity: its importance to human health. Center for Health and the Global Environment, Harvard Medical School, Cambridge, MA, p 23

Dai A (2013) Increasing drought under global warming in observations and models. Nat Clim Chang 3(1):52–58

Das M, Jain V, Malhotra SK (2016) Impact of climate change on medicinal and aromatic plants. Indian J Agric Sci 86(11):1375–1382

Dimri AP, Kumar D, Choudhary A, Maharana P (2018) Future changes over the Himalayas: mean temperature. Global Planet Change 162:235–251

Dow K, Berkhout F, Preston BL (2013) Limits to adaptation to climate change: a risk approach. Curr Opin Environ Sustain 5(3–4):384–391

Fu X, Wang B, Waliser DE, Tao L (2007) Impact of atmosphere–ocean coupling on the predictability of monsoon intraseasonal oscillations. J Atmos Sci 64(1):157–174

Gagné K (2019) Caring for glaciers: land, animals, and humanity in the Himalayas. University of Washington Press

Held IM, Soden BJ (2006) Robust responses of the hydrological cycle to global warming. J Clim 19(21):5686–5699

IPCC (2021) Climate change 2021: the physical science basis. In: Contribution of working group i to the sixth assessment report of the intergovernmental panel on climate change. Cambridge University Press

Jha SK, Negi AK, Alatalo JM, Negi RS (2021) Socio-ecological vulnerability and resilience of mountain communities residing in capital-constrained environments. Mitig Adapt Strat Glob Change 26(8):38

Johansson MU, Frisk CA, Nemomissa S, Hylander K (2018) Disturbance from traditional fire management in subalpine heathlands increases Afro-alpine plant resilience to climate change. Glob Chang Biol 24(7):2952–2964

Kattel GR (2022) Climate warming in the Himalayas threatens biodiversity, ecosystem functioning and ecosystem services in the 21st century: is there a better solution? Biodivers Conserv 31(8–9):2017–2044

Kumar A, Mitra M, Adhikari BS, Rawat GS (2015) Depleting indigenous knowledge of medicinal plants in cold-arid region of Nanda Devi Biosphere Reserve, Western Himalaya. Med Aromatic Plants 4(195):2167–2178

Kumar A, Adhikari BS, Mishra AP (2023) Status of threatened medicinal and aromatic plants and their use by the Bhotiya community in Niti valley, Nanda Devi Biosphere Reserve, Uttarakhand: a report (TR No./2022/35) submitted to International Union for Conservation of Nature, India, p 103

Kunwar RM, Mahat L, Acharya RP, Bussmann RW (2013) Medicinal plants, traditional medicine, markets and management in far-west Nepal. J Ethnobiol Ethnomed 9:1–10

Kunwar RM (2018) Ethnobotany in the Kailash sacred landscape, Nepal: implications for conservation through interactions of plants, people, culture and geography. Doctoral dissertation, Florida Atlantic University

Li J, Wang Z, Lai C, Wu X, Zeng Z, Chen X, Lian Y (2018) Response of net primary production to land use and land cover change in mainland China since the late 1980s. Sci Total Environ 639:237–247

Lloret F, Escudero A, Iriondo JM, Martínez-Vilalta J, Valladares F (2012) Extreme climatic events and vegetation: the role of stabilizing processes. Glob Change Biol 18(3):797–805

Malik ZA, Bhat JA, Ballabha R, Bussmann RW, Bhatt AB (2015) Ethnomedicinal plants traditionally used in health care practices by inhabitants of Western Himalaya. J Ethnopharmacol 172:133–144

Meetei KB, Tsopoe M, Giri K, Mishra G, Verma PK, Rohatgi D (2023) Climate-resilient pathways and nature-based solutions to reduce vulnerabilities to climate change in the Indian Himalayan Region. In: Climate change in the Himalayas. Academic Press, pp 89–119

Mishra AP, Kumar A, Yadav SN (2023) Ecology and conservation of threatened medicinal plants in the Trans-Himalayan region of Nanda Devi biosphere reserve, Western Himalaya. Trees, For People 14:100451

Ogra MV, Badola R (2015) Gender and climate change in the Indian Himalayas: global threats, local vulnerabilities, and livelihood diversification at the Nanda Devi Biosphere reserve. Earth Syst Dyn 6(2):505–523

Pandey V, Bhatt ID, Nandi SK (2019) Environmental stresses in Himalayan medicinal plants: research needs and future priorities. Biodivers Conserv 28:2431–2455

Pearson J, Jackson G, McNamara KE (2021) Climate-driven losses to indigenous and local knowledge and cultural heritage. Anthropocene Rev 10(2)

Prasad R, Timothy R, Malakar S, Dayma V (2023) Measurable impacts of climate change on biodiversity and livelihoods in the tropical dry deciduous forest of Sheopur, Madhya Pradesh, India. Int For Rev 25(4):435–448

Rao BR (2015) Genetic diversity, genetic erosion, conservation of genetic resources, and cultivation of medicinal plants. Genet Divers Erosion Plants: Case Hist 8:357

Samant SS, Vidyarthi S, Pant S, Sharma P, Marpa S, Sharma P (2011) Diversity, distribution, indigenous uses and conservation of the medicinal plants of Indian Himalayan region used in cancer. J Biodivers 2(2):117–125

Scheffran J (2011) Security risks of climate change: vulnerabilities, threats, conflicts and strategies. In: Coping with global environmental change, disasters and security: threats, challenges, vulnerabilities and risks. Springer Berlin Heidelberg, Berlin, Heidelberg, pp 735–756

Seddon N, Chausson A, Berry P, Girardin CA, Smith A, Turner B (2020) Understanding the value and limits of nature-based solutions to climate change and other global challenges. Philos Trans R Soc B 375(1794):20190120

Sen T, Samanta SK (2015) Medicinal plants, human health and biodiversity: a broad review. Biotechnol Appl Biodivers 59–110

Sheldon JW, Balick MJ, Laird SA, Milne GM (1997) Medicinal plants: can utilization and conservation coexist? Adv Econ Bot 12:93–104

Srivastava S, Pandey BW, Negi VS (2024) Community-based approaches to climate change adaptation and livelihood security in the high-altitude Bhagirathi Valley of Garhwal Himalaya. J Resourc Ecol 15(1):197–203

Trenberth KE, Dai A, Rasmussen RM, Parsons DB (2003) The changing character of precipitation. Bull Am Meteor Soc 84(9):1205–1218

Upadhyay RK (2020) Markers for global climate change and its impact on social, biological and ecological systems: a review. Am J Clim Chang 9(03):159

Zimmerman EM (2005) Valuing traditional ecological knowledge: incorporating the experience of indigenous people into global climate change policies. NYU Envtl. LJ 13:803

Chapter 5
Scoping of Forest Resources in Light of Climate Change Implications in Western Himalayan Region

Sugandha, Mahika Phartiyal, and Harshit Pant Jugran

Abstract The Himalayan ecosystem, particularly the western Himalayan region of the Indian Himalayan region (IHR) forms an integral part of the global ecosystem which influences the entire Indian peninsular landmass by providing irrigational facilities to the plain regions along with high content of alluvial soil deposits. More importantly, it is a kernel of natural resources and has a rich forest resource base with great contrast in the vegetation pattern changing from Temperate to Alpine with the elevation change. The rural Himalayan communities are very close to nature and highly depend on forest resources. However, the ever-increasing pressure on these ecosystems results in further ecological damage to an already fragile landscape. The huge natural wealth of the region necessitates proper governance of the natural resource base. Economical, indirect and quick benefits from forests are much higher than the direct and sustained benefits but the continuous anthropogenic activities affect the resources at the higher side. Anthropogenic activities such as grazing, ruthless lopping, land use changes, forest fire, indiscriminate non-wood forest products (NWFP) collection, exploitation, excessive resource extraction using faulty methods, and ethical and moral changes towards sustainable management are causing disturbances in the Himalayan forests by leading to the change in the structure of forest and altering the species composition which further can accelerate succession. Though concerns are being raised about the loss of native knowledge and the possible extinction of plant resources, hence conservation and sustainable management of key habitats could contribute to safeguarding the heritage of this economically and socio-ecologically rich region.

Sugandha
Silviculture and Forest Management Division, Forest Research Institute, Deemed to Be University, Dehradun 248195, India

M. Phartiyal
Jawaharlal Nehru University, New Delhi 110067, India

H. P. Jugran (✉)
Centre for Socio-Economic Development, G.B.Pant National Institute of Himalayan Environment, Kosi-Katarmal, Almora, Uttarakhand 263643, India
e-mail: harshit.pantj@gbpihed.nic.in

D. Arya et al. (eds.), *Climate Change Impact on Himalayan Biodiversity*,
Environmental Science and Engineering, https://doi.org/10.1007/978-3-031-77149-1_5

Keywords Biodiversity · Conservation · Disturbance · Community dependence · Governance · Resources

5.1 Introduction

Himalayas are one of the youngest mountain ranges on the planet Earth and consist mostly of uplifted sedimentary and metamorphic rock. These young mountains are formed by the continuous continental plate collision of India and Eurasia. The continuous shifting of the Indian plate towards the north and collision with the Eurasian plate leads to the upliftment of the Himalayas and geologically divides it into three parts i.e. The Great Himalayas, The Middle Himalayas and The Outer Himalayas. The Himalayas form the world's largest mountain chain (Mani 1974) and spread across five countries: Bhutan, India, Nepal, China, and Pakistan. The Indian Himalayan Region (IHR) covers approximately 18% of India's land surface, which cover over 5.3 lakh km^2 area (Fig. 5.1).

Rodgers and Panwar (1988) have given the biogeographic classification of IHR which consists of three Biogeographic Zones (i.e., Trans Himalaya, Himalaya and North-east India) and nine provinces viz., Ladakh Mountain, Tibetan plateau, Trans Himalaya of Sikkim, North-West Himalaya, West Himalaya, Central Himalaya, East Himalaya, Brahmaputra Valley and Assam Hills. The Himalayan range is bordered on the northwest by the Karakoram and Hindu Kush ranges, on the north by the Tibetan Plateau, and on the south by the Indo-Gangetic Plain. Some of the world's major rivers, the Indus, the Ganges, and the Tsangpo-Brahmaputra, rise in the Himalayas,

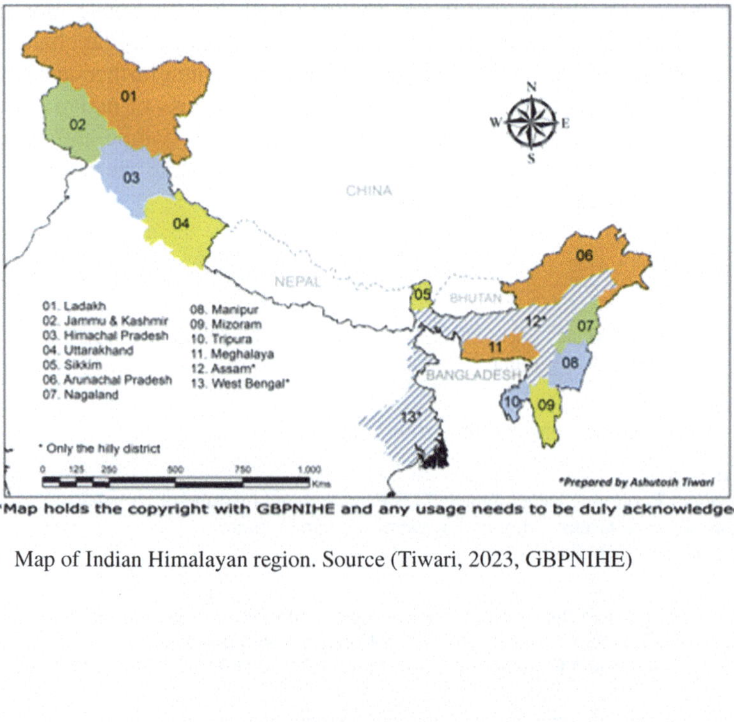

Fig. 5.1 Map of Indian Himalayan region. Source (Tiwari, 2023, GBPNIHE)

and their combined drainage basin is home to roughly 600 million people. Its western anchor, Nanga Parbat, lies south of the northernmost bend of the Indus river, while its eastern anchor, NamchaBarwa, lies west of the great bend of the Tsangpo river. The Himalayan Mountain ranging from 27 ° 38' north latitude to 72 ° 98' east longitude embody a diverse and characteristics vegetation distributed over a wide range of topographical variation (Gupta 1963).

5.1.1 State of Indian Himalayan Region

The Himalayas encompass diverse and characteristic vegetation distributed over a wide range of climatic and topographical variation (Dhaulkhandi et al. 2008). It ranges from tropical dry deciduous forests in the foothills to alpine meadows above timberline (Singh and Singh 1992). Both structure and diversity of vegetation have strong functional role in controlling ecosystem process like biomass production, cycling of water and nutrient (Gower et al. 1992). The forest type in Indian Himalayan region varies from moist temperate to tropical and dry deciduous forest (Singh and Singh 1987). The Himalayan moist temperate forest extends from 1500 to 3000 m asl in Western Himalaya which covers the area of Himachal Pradesh and Uttarakhand. The variable topography of this area supports luxuriant vegetation between 2200 and 2800 m which exhibits a dense canopy of *Quercus* species at moist situations (*Quercus leucotrichophora, Quercus floribunda, Quercus semecarpifolia, Quercus lenuginosa* and *Quercus glauca*). Above the 2800 m altitude Oak-conifer association occurs where, *Quercus semecarpifolia, Abies pindrow, Picea smithiana, Pinus wallichiana* and *Taxus baccata* Var. *wallichiana* are in the dominant form while the foothill occupies *Shorea robusta, Tectona grandis, Dalbergia sissoo* as dominating species.

Himalayas are the most vulnerable environment, which harbor rich repositories of biodiversity as they are considered as hotspots of biodiversity due to the wide variation in geological and geographic attributes, which cause higher diversity in particular habitats and microclimatic conditions which can be site-specific. The distribution and dynamics of biological resources must be understood to provide a rational basis for planning and management decisions, without which conservation of these resources in the natural habitats would be impossible (Keith and Sanders 1990; Khoshoo 1992).

The geographical position of the Indian Himalayan region manifests rich plant diversity which includes more than 8000 angiosperms, 44 gymnosperms, 600 pteridophytes, 1737 bryophytes, 1159 lichens, etc. These biological resources have been a source of medicine for millions of people in the country as well as among the different communities of world while the IHR supports over 1748 (32.2% of India) plant species of known medicinal value (Samant et al. 1997). The unique diversity of MPs in the region is manifested by the presence of a number of native (31%), endemic (15.5%) and threatened elements. Approximately 14% of total Red Data plant species occurs in IHR and 3.5% of total MPs in different threat categories of Conservation Assessment and Management Plan (CAMP) occurs in this region (Samant et al. 1997).

5.1.2 Rationale

Disturbance has become a widespread feature in most of the forests all over the Himalayan region (Singh and Singh 1992) and the vast reservoir of plant resources in India is in peril due to increasing anthropogenic pressure. As anthropogenic pressure can cause uncontrolled influences that bring changes in floral and faunal diversity, habitat, landscape, and soil degradation and may also lead to considerable alterations in the environmental conditions therefore, knowledge of ecological processes and biotic pressure can help in understanding the persistence of long-lived plant communities.

As the livelihood in the rural areas of the mountainous regions depends to a large extent on subsistence farming, the possibility of disturbance towards resource management also increases. The forest provides a large number of natural resources and a number of people in India who rely on forest products is assumed to be 50 million (Shiva 1993) and the disturbance caused to these resources affects the forest product quality and quantity in long run. Many traditional societies utilize their natural resource based on their traditional knowledge, their perceptions, experience and response to patterns of resource uses by others communities (Chettri et al. 2006).

Economical, indirect, and quick benefits from forests are much higher than the direct and sustained benefits but the continuous anthropogenic activities affect the resources on the higher side. Timber extraction itself to a large extent contributes to the shrinkage in forest cover. Similarly other anthropogenic activities such as grazing, ruthless lopping, land use changes, forest fire, indiscriminate NWFP collection, exploitation, excessive resource extraction using faulty methods, ethical and moral changes towards sustainable management are causing disturbances in the community forests by leading to the change in the structure of forest and altering the species composition which further can accelerate succession. Though concerns are being raised about the loss of native knowledge and the possible extinction of plant resources, hence proper documentation of traditional knowledge regarding plant use, along with conservation and sustainable management of key habitats could contribute to safeguarding this heritage (Bannister 2006). Information on forest composition and association, biotic pressure and type of species surviving, and the extent of biomass removal can help to rejuvenate depleting forests through silvicultural practices and community involvement (Ramakrishnan and Toky 1981; Singh and Singh 1987; Sundriyal et al. 1994a).

To ensure the stability of healthy forest growth and its beneficial functions, the risks of catastrophic removal of resources need to be reduced, and greater diversity should be maintained in the forest, primarily by providing opportunities for new seedlings to become established (Singh and Singh 1992). For sustained regeneration and growth of all species in the presence of older plants is required for better growth of any plant community (Ramakrishnan et al. 1981; Taylor and Zisheng 1988) which leads to the concept of sustainable resource utilization and conservation of biological diversity. The conservation of biological diversity is vital to the sustainability of many sectors including forestry, agriculture, fisheries, healthcare, science, tourism and industry (MacKinnon 1998). International concern over the loss of biological diversity and the direct and indirect services it provides culminated in the signing of the Convention on Biological Diversity following the United Nations Conference on Environment and Development in Rio de Janeiro, 1992. This advocated the protection of natural biological capital and alleviation of poverty in a sustainable way (Primack 1995) and continued efforts to address these issues were reiterated at Rio^{+10}, the 2002 World Summit on Sustainable Development in Johannesburg (Lee 2002).

Indian societies at all stages of socioeconomic and cultural evolution exhibited restrained resource use that may have consequences for prudent use of natural resources (Gadgil and Guha 1992). The intervention of local communities to preserve and sustainably utilize and manage natural resources has a high success rate and it directly helps them to overcome ecological disasters and threatened livelihoods. Yet the government has been promoting centralized management of resources by curtailing the functions of locally formed institutions by forgoing the traditional wisdom and cultural practices of locals that qualify them as the best decision-makers. However, in recent years with increased migration and the low interest of younger generations, this information regarding traditional knowledge is fading with the end of the older generation. To overcome the issues of out-migration, to diminish economic disparity and to increase the per capita income of the people living amidst diverse ecology in this region, certain sustainable measures need to be taken. Bio-ecological inputs and social inputs and can effectively minimize depletive forests which can be further catalyzed by simple technology and physical processes with the help of external agents. In the broader picture, this will intensify renewable intrusion in global resource management through decentralized resource management plans executed in different regions of the world.

5.1.3 Climatic and Anthropogenic Impacts on Forest Resources

The increase in human population and its continuous shifting near to or within forested areas have severely increased the degradation of floral and faunal diversity and loss of biodiversity as well as forest cover (Anitha et al. 2009; Davidar et al. 2010). Diverse anthropogenic disturbance factors including harvesting of fuel wood and timber have been reported to greatly influence the diversity of forest communities (Sayer & Whitmore 1991) leading to change in species composition and vegetation structure (Berkmuller et al. 1990; Kouki 1994). Uncontrolled grazing by domestic livestock also directly impacts the regeneration processes of forests (Kelt & Valone 1995; Saberwal 1995). The increasing human influence over the forest resources leads to disturbances, naturally or due to anthropogenic activities (Gairola et al. 2009, 2015; Rawal et al. 2012).

Disturbance could be divided into various categories such as geological disturbances which are landslide, soil erosion and earthquakes while the anthropogenic disturbances include deforestation, grazing, lopping of tree branches for fodder and fuel wood, removal of leaf and wood litter from the forest floor and frequent fire incidents. Climatic anomalies form another part of disturbance which is neither anthropogenic nor entirely natural. These disturbances effect ecosystem stability and hamper the successional process of forest ecosystems. Anthropogenic disturbances are usually chronic in nature and involve slow but steady decline in resource status such as removal of biomass at a given time which keeps on going throughout the year, leaving behind minimal chance of recovery (Singh 1998).

Disturbance has become a widespread feature in most of the forests all over the Himalaya (Singh and Singh 1992), therefore, knowledge on ecological processes and biotic pressure can help in understanding the persistence of long-lived plant communities. Saberwal (1996) have reported that grazing and migration of the Gaddi tribe of Gujjars in Himachal Pradesh had destroyed the land and led to the introduction of the weed *Rumex nepalensis*. Present census of department of livestock, Uttrakhand (2019) have reported that there is a significant decline of cattle population in the state.

Activities like deforestation, grazing, mining and road construction cause degradation of vegetation in the Himalaya and it is reported that there is a loss of 25.6% in the forest cover in 22 years (Jha et al. 2022). Human activities as well as uncontrolled livestock grazing are the major source of anthropogenic pressure (Silori 2001). Kumar and Ram (2005) have reported that human made disturbance cause decrease in dominance of single species and increase in plant biodiversity by mixing of different successional status. Joshi and Yadav (2015) has reported that increased disturbance activities in *Q. leucotrichophora* forest resulted in decline of species richness and the canopy cover was more than 65–70% in un-disturbed sites in comparison to 45–50% in moderately disturbed stand and 35% in highly disturbed stand.

In the North-eastern Himalayan region of Tripura, an increasing population over recent decades and their consequent dependence on forest products has led to both over-exploitation and deforestation of forests for fuelwood and timber collection and commercial plantation (Majumdar et al. 2012). Similarly in the Indian Himalayan state of Uttarakhand, Munsi et al. (2009) have reported that anthropogenic disturbances are the main drivers of land use change in the form of increased agricultural activities and human settlements.

In the Western Himalayan region, broadleaf forests between 1400 and 1800 m asl, mainly dominated by Oaks and associated species, are gradually shifting to Chir pine (*P. roxburghii*) dominated forests due to heavy human pressure such as lopping, grazing and leaf litter collection as well as increasing forest fires (Bhandari et al. 2012). Fire assisted invasion by Pine in the Oak dominated forests completely changes the microclimatic conditions of the forest ecosystem which makes them more suitable for invasion of *P. roxburghii* which expand its territory year after year due to higher regeneration capacity. The consequences of degradation in the Himalayan ecosystems have been more severe since they are considered as ecologically fragile

(Mani 1974) and the regeneration of such degraded ecosystems is extremely difficult due to the physical instability and environmental characteristics of the area.

To address these issues the present study was focused on following research objectives:

1. To assess the forest resources across the study area in light of climate change.
2. To suggest conservation-based scoping of the assessed forest resources.

5.2 Study Area

The study area was in the state of Uttarakhand which forms the 27th district of the Indian constitution and is a part of North-western part of IHR (Indian Himalayan region). It covers an area of 53,483 sq. km (Nandy et al. 2006) and is located in the foothills of the Himalayas between 28⁰43'N to 31⁰28'N latitude and 77⁰34'E to 81⁰3'E longitude. The climate of Uttarakhand is temperate except in the plain areas where the climate is tropical with temperatures ranging from sub-zero to 43⁰C with average annual rainfall of about 1,550 mm. The cold weather conditions start at the end of November, with a regular drop in temperature. The soil type of the study area is mainly brown forest soil.

5.2.1 *Description of Study Site*

The present study sites were located in Almora district of Uttarakhand at an altitude ranging between 1460–1655 m asl.

S.N	Site	Longitude	Latitude	Altitude (m)	Forest type/Dominant species
1	Katarmal	29⁰ 37′23"	79⁰36′779"	1465	*Pinus roxburghii*
2	Mehatgaon	29⁰39′20"	79⁰37′442"	1326	*Pinus roxburghii*
3	Chaani-Lweshal	29⁰50′13"	79⁰35′04"	1655	*Quercus leucotrichophora*
4	Jyoli	29°38′23.40"	79°37′23.37"	1388	*Pinus roxburghii*
5	Dhamas	29°35′54.69"	79°39′24.89"	1460	*Pinus roxburghii*

5.2.2 *Materials and Methods*

For addressing the selected objectives specific study design was enacted and site selection and methodological proposition was conducted.

5.2.2.1 Assessment of Resources and Disturbance Status

For the resource assessment the questionnaire-based survey was conducted across the study sites. The following factors were taken into consideration for the data collection:

i. Questionnaire assessment and PRA including focused group discussions and participant observation.
ii. Field survey and documentation
iii. Validation and scope assessment

Disturbance status of the study sites was calculated for the diverse form of disturbances. It was recorded by conducting rapid questionnaire survey and documentation at selected sites and the information was generated on different types of disturbance present at the study sites.

5.2.3 *Lopping Intensity*

The status of lopping intensity was calculated following the standard methodology of Rawal et al. (2012) by counting lopped tree individuals at 10×10 m^2 area of the study site and lopping percentage was estimated using the following formula-

$$Lopping(\%) = \frac{Totalnumberofloppedtreesinaplot}{Totalnumberoftreesinaplot} \times 100$$

The lopping intensity was classified as

$$Lowlopping =< 30\%,$$

$$Moderatelopping = 30 - 60\%$$

$$Highlopping => 60\%$$

5.2.4 *Grazing Intensity*

Grazing intensity was estimated using the standard methodology of (Cierjacks and Hensen 2004) and Singh et al. (2015) by calculating dung pellets present at the study sites. The higher number of dung pellets indicated the presence of cattle grazing and the grazing pressure in the forest area and the variety of dung or animal droppings

Fig. 5.2 Disturbance analysis across the study sites. Grazing observed at one of the circular plot laid at Katarmal village

present at the study site indicated the variety of cattle grazing in the study area (Figs. 5.2 and 5.3).

5.3 Results

The results were described under three parts which are, a. disturbance assessment, b. people's perception and c. observation based documentation.

5.3.1 Disturbance assessment

The assessment of disturbance to the forest resources was conducted across the study sites and it was reported that the lopping intensity ranged between 45.63% and 86.77% across the study sites (Fig. 5.2) and it was maximum at Pine dominated Mehatgaon site (86.77%), whereas Oak dominated Chaani-Lweshal site showed moderate lopping signs (45.63%), while the average grazing intensity ranged between 0.6 and 6.6 across the study sites and it was maximum at Pine dominated Katarmal site (6.6) and minimum at Oak dominated Chaani-Lweshal site (0.6). The regeneration

Fig. 5.3 Disturbance analysis across the study sites. 1 X 1 m quadrat laid for counting number of dung pellets in Mehatgaon village

status relates with the lopping intensity as it was comparatively low in heavily lopped stands (Table 5.1).

a. People's perception for climate change and resource management

As the religion is an integral part of Hill culture and values, where the sacred groves and religious trees occupy an important part of daily rituals of villagers, the resource management across the study sites was mostly focused on the sacred values. Medicinal species like *Ocimum sanctum* (Tulsi), *Ficus* (pipal) and *Cedrus* (Deodar) etc. has religious value which is present at most of villages. Majority of the people are aware about the religious plants around them.

The questionnaire based analysis showed that 40% of the respondents of Katarmal have reported that the lack of information regarding conservation status and trainings for resource extraction can be beneficial. Similar responses were obtained from the

Table 5.1 Lopping intensity and grazing intensity across the sites

Site	Lopping intensity (%)	Grazing intensity
Katarmal	77.8	6.6
Mehatgaon	86.77	3.1
Chaani-Lweshal	45.630	0.6
Jyoli	65.28	4.9
Dhamas	73.61	5.3

VPs of Mehatgaon, Chaani-Lweshal, Jyoli and Dhamas where 53% people of Mehatgaon and 27% people of Chaani-Lweshal, 39% people of of Jyoli and 51% people of Dhamas also have the response that proper awareness among the stakeholders can lead to successful preservation of the collected resource.

Regarding the natural resources status the water resources have been reported as dried up ones due to temperature and rainfall anomalies. Irrespective of the fact that the water is supplied through pipelines too in the villages, the people frequently use water for domestic and irrigation purpose from *naulas* and *srot* present in the village area, which have the dwindling water level due to climate change impacts. 53% respondents of Chaani-Lweshal, 45% of Katarmal and 30% of Mehatgaon along with 43% respondents of Jyoli and 39% respondents of Dhamas has reported the decline in water resources since last decade and insisted on water conservation practices as in one case the respondents from Chaani-Lweshal has reported that 80% of their time usually spent on the collection of water from nearby water resources.

There is no reported change in agricultural practices in three study sites of Katarmal, Mehatgaon and Chaani-Lweshal as the villagers are mostly dependent upon the marked based economy. 100% of the people of all these three sites have reported that there is no change in any agricultural practices while 80% of the stakeholders from Jyoli and Dhamas have reported change in certain agricultural practices. The farmers have not noticed changes in their traditional practice of agriculture as they still use the same methods for sowing, harvesting, irrigation which was used earlier too. The seeds sometimes supplied by the government agencies are not trusted upon; hence home grown or locally barter seeds are preferred by them. 33.33% respondents from Katarmal, 20% respondents from Mehatgaon and 26.66% respondents from Chaani-Lweshal have reported that there is a change in seed sowing timing and crop ripening and usually the seeds are being sown 2 to 3 months prior the earlier sowing season while 29% respondents in Jyoli and 38% respondents in Dhamas have reportedly stopped sowing the wheat due to climatic anomalies.

The survey also revealed that across the study sites the villagers expect initiatives from government side such as soil testing which would help them to use their agricultural land to sow and harvest an optimum amount of crop. Similarly land use patterns as reported by 46% respondents from Katarmal, 60% respondents from Mehatgaon and 33.33% respondents from Chaani-Lweshal, 45% respondents from Dhamas and 57% respondents from Jyoli showed changes over the last decade and the main reason behind this was increasing deforestation for construction purpose as well as for agriculture. The respondents across the study sites were of common perception that this increase in deforestation has led to the change in climatic conditions.

The study also showed that 40% respondents from Katarmal, 53.33% respondents from Mehatgaon, 13.33% respondents from Chaani-Lweshal, 45% respondents from Jyoli and 34% respondents from Dhamas have reported that there is decline in species composition. The respondents from Mehatgaon have reported that species specially like *Citrus* (lemon, Malta) which earlier used to be fruiting every year whereas now the tree bears no flowering and there is hardly any fruiting. Similarly 40% respondents from Katarmal, 50% respondents from Mehatgaon and 55% respondents from Chaani-Lweshal, 45% respondents from Jyoli and 51% respondents from

Dhamas have reported that growth of some new species such as guava, orange, plum, pomegranate etc. which earlier were never seen in their area, have been increasing since last two decades. It is also reported by 60% respondents from Mehatgaon, 68% respondents from Katarmal and 80% respondents from Chaani-Lweshal, 76% respondents from Jyoil and 69% respondents from Dhamas had reported that there is a decline of native plant growth as the elevation increases, which is replaced by other invasive species such as *Eupatorium odoratum* and *Lantana camara* as well as instances of *Pine* replacing *Oaks* have been reported in Chaani-Lweshal.

The study also shows that 53% respondent from Katarmal, 40% respondent from Mehatgaon and 60% respondent from Chaani-Lweshal, 54% respondents from Jyoli and 49% respondents from Dhamas have reported the presence of invasive species in their area. They also believe introduction of pine in their forest by the Britishers was one of the biggest blunder ever made. Pine is referred to as invasive by a majority of the people as they consider that pine is slowly covering up the complete forest area and acting as a hindering factor in the growth of other fruitful species (Fig. 5.4). They consider pine forest as a source of income for government, which is also frequently under fire and reduces the moisture of the soil and the common response across the study sites was that the Pine does not help much as it damages. Pine is only beneficial in some way that it provides a minimal amount of money from resin tapping. Usually there is No private resin tapping reported from the study area as it is tapped by private or government contractors which is taken up for governmental selling and 30–40% percent of money is provided to the gram sabha which is envisaged by the VP rules.

Fig. 5.4 Spread of Pine in the forest region and the adversities (credit- Fig. 5.4). (Harshit Pant Jugran)

5.3.2 Observation based documentation

Across the study sites it was also reported that 80%–95% of the people have the knowledge of common wild plants uses, though they don't use it. They have the common perception that earlier the vegetation was quite good as compared to the present situation and the rising temperature, presence of pine and less water is the main reason of vanishing medicinal plants and wild edible plants of the area. Passing of the information by the aged lot of the village to the young generation was rarely experienced which lead to very less knowledge of wild plants and their use as medicines as the young mass doesn't identify the medicinal plants. It was observed that 80% respondent from Katarmal, 78% respondent from Mehatgaon and 88% respondent from Chaani-Lweshal, 73% respondents from Jyoli and 69% respondents from Dhamas have reported that they usefully do not identify the medicinal plants which grow in their nearby forest area.

It was also observed that the abundance of pine needles in the litter is diminishing the nutrients in the soil as there is a leach out of its cation during decomposition (Schreier et al. 1995). The fires cause incomprehensible damage and loss both to the forest and the soil (Fig. 5.5), which is deprived of it ability of infiltration of rainwater and recharges springs (Mehta 1996). Since the forest is not very resourceful the Van Panchayat is not is a very active state at Jyoli although the micro plan has been proposed by the Van Panchayat members. In Dhamas the forest type is mixed Chir Pine (*Pinus roxburghii*) is one of the predominant species of this forest followed by Oak (*Quercus floribunda* and *Quercus leucotrichophora*). The forest area is also immense in Dhamas which clearly defines the dependence on forest is higher than Jyoli and the Van Panchayat is also stronger as it has a richer resource base. The villagers are charged 600/Rs and 500/Rs for different sizes of logs that they collect from the forest village. Therefore in most of the cases they bring woods from the fallen trees covertly from the forest guard which is assigned by the forest department. The forest department yearly provides the village their 'rightful wood' in specific quantity.

Forest resource are mainly use for fuel wood, fodder, edible fruits, household and agricultural purposes, pine needles are used as bedding for animals which is later used as manure. Locally known medicinal plants are found in the forest. Pine forest was planted by the British to fulfil their commercial plan of resin collection. Before the British era the lesser Himalayan region was known for its broad leaved forest where the human dependence was also high (Guha 2000). The commercial needs surpassed the local subsistence needs it not only disrupted the local relationship with the forest but it changed the entire forest ecosystem. In today's scenario the pine forest has become a big predicament for the locals as the large scale forest not only destroy the forest but at times enter their agricultural fields and even houses.

Due to the changing weather patterns (temperature and precipitation) the traditional framing pattern is being replaced by more sturdy crops with the use of fertilizers. Various schemes and projects under Central organizations of ICAR (Indian Council of Agriculture Research) such as VPKAS (Vivekanand Krishi avem

Fig. 5.5 Incidents of forest fire in Pine Forests (credit- Fig. 5.5). (Behera 2023)

Anusandhan Kendra), Hawalbagh, Almora and various KVK (Krishi Vigyan Kendra) of GBPUAT (G.B. Pant University of Agriculture and Technology) have intervened to solve this issue and are working towards it in various ways.

In both the villages of Jyoli and Dhamas wheat has not been cultivated since last two years due to scanty rainfall in winter season. The local people complained that, 'the amount of wheat we ingest to work in the field even that amount is not cultivating.' Wheat requires at least 12to 15 inches of water for proper cultivation with the decrease in rainfall and the rise in temperature wheat cultivation is becoming impossible. The locals even complained, 'not even a sack of wheat was cultivated in the whole village.' Four to five years back these villages sold surplus crops of lentils, grains and vegetables but in the present scenario the crops are barely available for self-consumption lasting 2–6 months depending on the land area cultivated.

5.4 Discussion

Mountain communities are known for women dominating societies. As the man of the family migrates out of the village for work and the women becomes the head of the family. Women are a subject to drudgery in the hills as she has role to play in all the physical activities that takes place around her. The collection of fuel wood, water household work and in the rural mountain communities practicing agricultural are the duty of women, these are highly time consuming physical activities. Burning wood is an unsustainable fuel as it causes indoor pollution which has adverse health implications causing cardiac and respiratory diseases and it also lowers the life expectancy of the people; it also inflicts obstacles in education of children. The village communities are dependent on agriculture as the basis of primary subsistence. Almost 90% of the people use compost and practice organic farming. However the region is rain fed and the only source of irrigation is rainfall. The change in climate has resulted in decrease in rainfall over the years which have adversely affected the crop yield. In both the villages wheat yield has been negative since two years and the villagers have stopped cultivating wheat. Adduced to this the land reforms have not been implemented therefore there is no provision for Gherabandi/Chakbandi or land consolidation. Due to which landholdings are dispersed this makes the labour intensive farming in such undulating topography all the more difficult and un-remunerative. The slope of the terraced hills usually extends over 45°. The crops grown are at times not sufficient for self consumption, crops usually last 4–5 months on average. No surplus is generated and hence there is no income generation which has compelled the men of the villages to migrate. This pattern is witnessed across central mountain community belt. High male out-migration has not only lead to an additional burden on women in terms of physical work but has also implied their active involvement in the village household economy and common life. About 21% of the households in the state are estimated to be headed by women which has been declined since last decade (IIPS 2021). These activities preoccupy women in drudgery constraining their leisure time, education and health. The sex ratio clarifies that female are more in number yet they have a low literacy rate. Despite of the facts the community rests on the shoulders of vulnerable women for maintaining the socio-economic status of the family.

The changing climate has lead to the decline in the crop productivity which has further lowered the per capita income of the villagers. There is no scope for employment rather than agriculture on which every family is decedent and bamboo crafting which is only limited to scheduled caste. The employment dearth has compelled the male to temporarily migrate out in search of job this is also leading to the expansion of the towns and cities and thereby is exerting immense pressure on urban resources. The crop yield is also adversely affected due to the wildlife conflict which has increased in the village, the monkey and wild boar population is rising with every passing day. Earlier the village communities used to pool out money and hired a guard who would look after their crops during night. However, in the present there is limited profit which is being incurred form agriculture, the families are resistant in hiring a guard because as such the crop yield is low.

The interior villages still do not have the basic infrastructure which is a major problem. Inaccessibility is negatively correlated to development. The only mode to reach the interior villages is by hours of road journey. High education institutes and government hospitals are also at a distance. There is less exposure and opportunity on one hand and erosion of traditional knowledge on the other hand. Increased population has increased food security and it in addition exerts exploitative pressure on the natural resources.

5.5 Conclusion

The study emphasizes the need of awareness among villagers needs to be enhanced for the proper conservation of resources so that they can obtain the benefits regarding utilization as well as possible market linkage. This can only be obtained after the information dissemination regarding natural resource conservation and community based natural resource management. India being one of the signatory in climate change adaptation and mitigation scenario of UNFCCC, has to incorporate the awareness regarding resource conservation which is directly linked to convention of natural diversity. Furthermore the awareness at grassroots level regarding the market based intervention such as REDD and REDD plus has to be incorporated at village and district level so that the stakeholders enthusiastically cultivate medicinal plants to develop market and obtain benefit out of it. The inclusion of ABS and utilization of forest resources in the form of FES (Forest ecosystem services) is a must for the village development plans in the region for the fruitful outcomes.

References

Anitha K, Joseph S, Ramasamy EV, Prasad SN (2009) Changes in structural attributes of plant communities along disturbance gradients in a dry deciduous forest of Western Ghats India. Environ Monit Assess 155:393–405

Bannister K (2006) Prophet River Ethnobotany: A report on traditional plant knowledge and contemporary concerns of the Prophet River First Nation. British Columbia: Prophet River First Nation 54

Behera RS (2023) Unraveling the manmade catastrophe of forest fires. https://socialdhara.com/unraveling-the-manmade-catastrophe-of-forest-fires/

Berkmuller K, Mukherjee SK, Mishra BK (1990) Grazing and cutting pressure on Ranthambhore National Park, India. Environ Conserv 17:53–58

Bhandari BS, Mehta JP, Semwal RL (2012) Forest fire in Uttarakhand Himalaya: an overview.In: Glimpses of Forestry Research in the Indian Himalayan Region. GB Pant Institute of Himalayan Environment & Development, Almora, 167–176

Chettri SK, Singh KK, Krishna AP (2006) Anthropogenic pressures on the natural resources in fringe areas of Khangchendzonga Biosphere reserve. Int J Eco Envir Sci 32(3):229–240

Cierjacks A, Hensen I (2004) Variation of stand structure and regeneration of Mediterranean holm oak along a grazing intensity gradient. Plant Ecol 173:215–223

Davidar P, Sahoo S, Mammen PC, Acharya P, Puyravaud JP, Arjunan M, Garrigues JP, Roessingh K (2010) Assessing the extent and causes of forest degradation in India: where do we stand? BiolConserv. 143:2937–2944

Dhaulkhandi M, Dobhal A, Bhatt S, Kumar M (2008) Community structure and regeneration potential of natural forest site in Gangotri India. J Bas Appl Sci 4(1):49–52

Gadgil M, Guha R (1994) Ecological conflicts and the environmental movement in India. Develop and Change 25(1):101–136

Gairola S, Rawal RS, Dhar U (2009) Patterns of litterfall and return of nutrients across anthropogenic disturbance gradients in three subalpine forests of west Himalaya, India. J Res 14:73–80

Gower ST, Vogt KT, Grier CC (1992) Carbon dynamics of rocky mountain Douglas-Fir: Influence of water and nutrient availability. Eco Monographs 62(1):43–65

Guha R (2000) The unquiet woods: ecological change and peasant resistance in the Himalaya. Univ of California Press

Gupta AL (1963) Annual precipitation and vegetation of de dry temperate coniferous region of North West Himalay. Jou Int Bot Soc 42(2)

International Institute for Population Sciences (IIPS) and ICF. 2021. National Family Health Survey (NFHS-5), India, 2019–21: Uttarakhand Mumbai: IIPS

Jha SK, Negi AK, Alatalo JM, Prabhu V, Jha MB, Kumar H (2022) Forest degradation index: A tool for forest vulnerability assessment in Indian Western Himalaya. Sustainability 14(23):15655

Joshi A, Yadav AK (2015) Effect of anthropogenic disturbances on plant diversity in oak dominated forest of Nainital, Kumaon Himalaya, India. New York Sci J 8:22–27

Keith DA, Sanders JM (1990) Vegetation of Eden region, south-eastern Australia: Species composition, diversity and structure. J Veg Sci 1:203–232

Kelt DA, ValoneTJ (1995) Effects of grazing in the abundance and diversity of annual plants in Chinuahaun desert scrub habitat. Oecologia 103:191–195

Khoshoo TN (1992) Plant diversity in the Himalaya: Conservation and Utilization; Govind Ballabh Pant Institute of Himalayan Environment and Development: Almora, India

Kouki J (ed) (1994) Biodiversity in the Fennoscandian boreal forests: natural variation and its management. Annales Zoologici Fennici 31:217

Kumar A, Ram J (2005) Anthropogenic disturbance and plant diversity in forest of Uttaranchal, Central Himalayan. Bio Conser 14:309–331

Lee M (2002) The state of the planet. Ecologist 32(7):6–11

MacKinnon K (1998) Sustainable use as a conservation tool in the forests of South East Asia. In: Chap. 5 in Milner-Gulland, E. and Mace, R. Conservation of Biological Resources London: Blackwell Science Ltd

Majumdar K, Datta BK, Shankar U (2012) Ten new additions of tree species to the Tripura state, North East India: Distribution of extension and geographic map. Nebio 3(1):17–24

Mani MS (1974) Ecology and biogeography in India. The Hague. Dr. Junk Publisher. Monographiae Biologicae 23:727

Munsi M, Malaviya S, Oinam G, Joshi PK (2009) A landscape approach for quantifying land-use and land- cover change in middle Himalaya. Reg Enviro Change 10:145–155

Nandy SN, Dhyani PP, Samal PK (2006) Resource information database of Indian Himalaya, ENVIS Monograph 3, G.B. Pant Institute of Himalayan Environment and Development

Primack R (1998) Essentials of conservation biology Sinaur Publishers

Ramakrishnan PS, Toky OP, Mishra BK, Saxena KG (1981) Slash and bum agriculture in north east India. In: HA

Ramakrishnan PS, Toky OP (1981) Soil nutrient status of hill agro-ecosystems and recovery pattern after slash and burn agriculture (jhum) in North-eastern India. Plant Soil 60:41–64

Rawal RS, Gairola S, Dhar U (2012) Effects of disturbance intensities on vegetation patterns in Oak forests of Kumaun, West Himalaya. J Mou Sci 9:157–165

Rodgers WA, Panwar HS (1988) Planning wildlife protected area network in India, WII, Dehradun, 12

Saberwal VK(1995) Pastoral politics: Gaddi grazing, degradation and biodiversity conservation in Himachal Pradesh, India. Conser Biol 10:741–749

Samant SS, Rawal RS, Dhar U (1997) Diversity, endemism and economic potential of wild edible plants of Indian Himalaya. I J Sus Devel World Ecol 4:179–191

Sayer JA, Whitmore TC (1991) Tropical moist forests: destruction and species extinction. Bio Conser 55:119–214

Shiva MP (1993) Solutions to overcome impediments in forest development through MFP based management. In: Proceedings of the International Seminar on Minor Forest Products in Forestry, 17–18 April. Dehra Dun

Silori CS (2001) Status and distribution of anthropogenic pressure in the buffer zone of Nanda Devi Biosphere Reserve in Western Himalaya, India. Bio Conser 10:1113–1130

Singh JS, Singh SP (1987) Forest vegetation of the Himalaya. Botl Review 53:80–192

Singh SP (1998) Chronic disturbance, a principal cause of environmental degradation in developing countries (Editorial). Environ Conserv 25:1–2
Singh R, Sharma RK, Babu S (2015) Pastoralism in transition: Livestock abundance and herd composition in Spiti Trans-Himalaya. Human Eco 43(6):799–810
Singh JS, Singh SP (1992) Forest of Himalaya, structure, functioning and impact of man, Gyanodaya Parkashan, Nainital, India
Sundriyal RC, Sharma E, Rai LK, Rai SC (1994a) Tree structure, regeneration and woody biomass removal in a subtropical forest of Mamaly watershed in the Sikkim Himalayan Vegetation 113:53–63
Taylor AH, Zisheng Q (1988) Regeneration patterns in old growth Abies-Betula forests in the Wolong natural reserve, Sichuan China. J Ecol 76:1204–1218

Chapter 6
Keystone Insects: Influences on Ecosystem Structure and Their Function in the Himalayan Region

Suman Upadhyay, Sunaullah Bhat, Sandeep Kumar, and Sangeeta Rawat

Abstract The Himalayan region, renowned for its rich biodiversity and stunning landscapes, harbors a delicate ecosystem where certain insects act as keystone species, exerting a disproportionate influence on the entire system. This chapter explores the significance of keystone insects, particularly ecosystem engineers like ants and termites in the Himalayas, and aquatic insects in stream and river ecosystems. Keystone insects play crucial roles in nutrient cycling, habitat modification, and the regulation of other species, contributing to the resilience of these ecosystems. This review study also addresses the impact of human activities on keystone insects and delves into the effects of climate change, emphasizing the vulnerability of these species to environmental shifts. Additionally, the role of predatory insects in herbivore regulation is discussed, showcasing their importance in maintaining ecological balance. The chapter underscores the need for research and conservation efforts to safeguard keystone insects and preserve the intricate relationships within these ecosystems.

Keywords Keystone insects · Ecosystem engineers · Impact of climate change on Keystone insect · Ecological resilience · Himalayan region

6.1 Introduction

The Himalayan region is known for its rich biodiversity as well as stunning landscapes, serving as the habitat for a complex and intricate network of life. Within this intricate ecosystem, certain insects play a crucial role as "keystone species," influencing the structure and function of the entire system. Keystone species are

S. Upadhyay · S. Bhat · S. Kumar (✉)
Department of Zoology, Soban Singh Jeena University Campus, Almora 263601, Uttarakhand, India
e-mail: sandeepento@gmail.com

S. Rawat
Department of Zoology, Insect Systematics Laboratory, KU SSJ, Almora, India

D. Arya et al. (eds.), *Climate Change Impact on Himalayan Biodiversity*, Environmental Science and Engineering, https://doi.org/10.1007/978-3-031-77149-1_6

characterized by their disproportionately high value, influence and significance, in comparison to their numerical abundance within an ecosystem. While these species can take various forms across the spectrum of living organisms, their shared trait is their profound impact on their habitat, exceeding what would be expected based on their biomass (Budnukaeku et al. 2021). The absence of these species leads to significant alterations in the composition of species and other key attributes within the ecosystem.

This removal triggers notable shifts, impacting the intricate balance and various relationships among different organisms and elements in the ecosystem. Keystone species may be rare or common, and they can exhibit either generalist or specialized dietary behaviors (Power et al. 1996). These species frequently affect nutrient cycles, limit the expansion of invasive species, and control the populations of other creatures. Keystone insects are essential for preserving ecological resilience and supporting a wide variety of species in the harsh environmental conditions of the Himalayan region. The main objective of this chapter is to highlight the important role played by keystone insects in influencing the structure and function of Himalayan ecosystems. Through a thorough exploration of various insect species and their ecological contributions, the goal is to deepen our understanding of the intricate relationships within these ecosystems. Moreover, the chapter delves into the contemporary challenges arising due to climate change, emphasizing its potential impact on the distribution and behavior of key insect species and the resulting implications for the stability of Himalayan ecosystems. Serving as a comprehensive resource, the chapter compiles current knowledge on keystone insects in the region, shedding light on their ecological significance, the challenges posed by a changing climate and human activities, and the imperative for research and conservation efforts to safeguard the ongoing health and functionality of these unique keystone species.

6.2 Ecosystem Engineers in Himalayan Region: Termites and Ants

Ecosystem engineers refer to the organisms having ability to modify and shape their environment, there by impacting the physical and ecological structure of ecosystems. In the Himalayan region, ants and termites are a notable example of ecosystem engineers (Nithyatharani et al. 2018; Saha et al. 2018). These insects play a crucial role in maintaining the balance and functioning of the ecosystem through their activities such as burrowing, tunneling, and creating habitats. They create intricate tunnel systems that help aerate the soil, improve water infiltration, and influence the distribution of seeds and nutrients. This activity has a significant impact on the ecosystem's overall health and functioning. Ants, as powerful ecosystem engineers, have the capability to actively manage their own ecosystems by influencing the physical and chemical characteristics of the soil, thereby controlling the soil microbial community. The foods, food residues, and excreta collected by ants serve as important sources of nutrients for the microorganisms within their nest area. This interaction between ants and soil microorganisms plays a significant role in shaping and maintaining the ecological balance within their habitat (Saha et al. 2018).

Similarly, termites or white ants in the Himalayan region are also considered ecosystem engineers. They build extensive underground nests that contribute to soil fertility and aid in the decomposition of organic matter. This, in turn, influences nutrient cycling and overall ecosystem dynamics in the region. Termites are a group of social insects belonging to the Order Isoptera. There are approximately 3000 species of termites around the world, with approximately 75% of them being soil feeders (Nithyatharani et al. 2018). Termites are frequently used as a sign of soil fertility because they have a significant impact on processes such as nitrogen fixation, acetogenesis, methanogenesis, soil transportation, and nutrient movement in the soil. They have a significant role in nutrient recycling, its movement and their transportation to soil material. Termites are considered as ecological engineers in mounds building which improves organic carbon content, clay and nutrient of the soil. The soil in mound is redistributed by erosion, affecting the microstructure as well as fertility of the soil (Nithyatharani et al. 2018). Termites feed mostly on the plant debris therefore acts as decomposers. Their gut is modified and adapted for rising of the pH, hydrogen and oxygen which are essential for physical and soil and chemical modifications of the soil (Léonard and Rajot 2001).

6.3 Aquatic Insects in Stream and River Ecosystems

Aquatic insects stand out as a remarkably diverse group thriving in various freshwater environments, encompassing stagnant or lentic water bodies (like lakes and ponds) as well as flowing or lotic water bodies (such as rivers and streams). Their significance in freshwater ecosystems extends beyond mere diversity, as they play pivotal roles in the food chain, energy flow pathways, purifying water and contribute substantially to biomass (Kulkarni and Zade 2020). Notably, aquatic insects serve as a primary food source for fish and amphibians, shaping the dynamics of these ecosystems (Suter and Cormier 2015). These insects are integral to the transfer of energy across different populations, bridging the gap between microbial and vertebrate communities and connecting aquatic and terrestrial ecosystems. Their adaptability to a wide range of environmental conditions, including dissolved oxygen concentration, pH, temperature, and nutrient levels (such as phosphorus and nitrogen), makes them valuable indicators of the physical, chemical, and biological nature of water bodies (Arumugam and Athikesavan 2021). Table 6.1 represents some of insect orders along with examples and their role in the ecosystem.

Table 6.1 Lists of various entomofauna present in water bodies and their role in the aquatic ecosystem

Name of order	Examples	Role in ecosystem
Collembola	Springtails	• Small arthropods with specialized tails for jumping; Common around vegetation and quiet edge waters
		• Play a role in nutrient cycling and decomposition
Ephemeroptera	Mayflies	• Best-known aquatic insects; Essential for ecosystems
		• Sensitive to water quality; Indicators of environmental health
Odonata	Dragonflies and Damselflies	• Predators of insects and small fishes, Associated with wetlands, backwaters, and vegetation
		• Control mosquito populations; Key players in aquatic food webs
Plecoptera	Stoneflies	• Primarily found in streams; Oxygen-sensitive; Perform "pushups" to stimulate oxygen flow
		• Indicators of clean, well-oxygenated water; Sensitive to pollution
Hemiptera	Water striders, Giant water bugs	• Predacious on insects, including mosquito larvae; some species are fish killers
		• Play a role in insect population control; Indicators of water quality
Megaloptera	Hellgrammites, Alderflies	• Hellgrammites common in stream riffles; Alderflies found in backwaters
		• Contribute to nutrient cycling; Serve as food for fish and other aquatic organisms
Neuroptera	Spongillaflies	• Mainly terrestrial; a small group of aquatic species associated with freshwater sponges
		• Minor order with limited aquatic representation
Trichoptera	Caddisflies	• Build casings for hiding; some use webs to capture food
		• Sensitive to water quality; Indicators of stream health; Provide a food source for aquatic animals

(continued)

Table 6.1 (continued)

Name of order	Examples	Role in ecosystem
Lepidoptera	Butterfly and moth larvae	• The majority of butterfly and moth larvae are land-based, but several have aquatic phases that feed on water vegetation. While some are harmful to crops, others have the potential to be used as biological controls for invasive aquatic plants
Coleoptera	Water Beetles	• Diverse aquatic insects; some have larval stages in water, while others live in water as adults
		• Varied roles including predation, scavenging, and herbivory
Hymenoptera	Parasitic wasps	• Parasitic to aquatic species; Dive underwater to lay eggs; Larvae develop by consuming the parasitized host
		• Play a role in population control of aquatic organisms
Diptera	Mosquito larvae, Midge larvae	• Found in diverse aquatic habitats; Mosquitoes are well-known
		• Midges are important food for fish
		• Serve as indicators of water quality; Essential for nutrient cycling

Source Elango et al. (2021)

6.3.1 Role of Aquatic Insects in Maintaining Water Quality

Aquatic insects are employed to assess the well-being of aquatic environments due to their distinct reactions to stimuli within their water habitats. Their varied responses serve as indicators, aiding in the evaluation of the environmental quality in those aquatic bodies.

6.3.1.1 Bioindicators of Water Quality

Aquatic insects serve as reliable indicators of water quality due to their susceptibility to the physical, chemical, and biological factors present in the water body. They are unable to evade the impacts of pollution, making them reflective of both immediate and prolonged pollution occurrences (Elango et al. 2021).

Various levels of disturbance in a water environment are assessed by observing certain aquatic insect families. In the context of water pollution, the families Lestidae, and Aeschnidae, Libellulidae, (Odonata), Chironomidae (Diptera), Gyrinidae (Coleoptera), and Belostomatidae and Gerridae (Hemiptera) are considered as pollution indicators. The existence of water boatman (*Hesperocorixa* sp.), backswimmers

(*Notonecta* sp.), and midge larvae in aquatic habitats suggests water contamination, as these organisms are tolerant to pollution. Conversely, the presence of damselfly, dragonfly and dobsonfly nymphs indicates mild or no water contamination (Thaware 2023).

Mayflies (Ephemeroptera) are recognized as crucial "keystone" species, and their presence is seen as a significant sign of low to moderately productive i.e., (Oligotrophic Mesotrophic) conditions in flowing waters. The elevated presence of *Chironomus* sp. in aquatic environments signifies the eutrophic nature of the water body (Papáček 2013).

6.3.1.2 Nutrient Cycling and Litter Decomposition

Stream insects play a crucial role beyond nutrient retention, actively contributing to the alteration of nutrient elements through mineralization and assimilation. They also exert influence on microbes involved in oxidation or reduction processes, such as nitrification or denitrification. The key to leaf decomposition in streams lies in the activities of two important feeding guilds: the shredders and collectors. These guilds play a crucial role in breaking down and processing leaf litter, contributing to the nutrient cycling and overall ecosystem functioning in aquatic environments. Detrimentally, in the absence of detritivorous insects, the gravel substrates in forest streams might become clogged with incompletely decomposed leaves. This accumulation poses a threat to the habitat of salamanders, as well as the well-being of fish eggs and larvae, leading to a reduction in interstitial dissolved oxygen (Suter and Cormier 2015). Table 6.2 represents a list of aquatic insects trophic categories and their role in decomposition process.

6.4 Human Impacts on Keystone Insects

Keystone species play a major role within their community, just as the "keystone" in the arch holds the arch together, its removal will result in total collapse of the arch, so as the ecological keystone species holds the ecosystem together (Davic 2003). The Keystone species is a very important part of the ecosystem, directly or indirectly their existence affects man in his day -to -day live, but anthropogenic factors (Fig. 6.1) often drive declines in keystone species populations.

6.5 Climate Change and Keystone Insects

Climate change is a matter of significant global concern, standing out as a leading force behind worldwide environmental transformations that affect biodiversity and ecosystem operations (Olsen et al. 2022). Insects, being particularly susceptible to

Table 6.2 List of aquatic insect's trophic categories and their role in decomposition

Trophic category	Feeding strategy	Examples of organic matter consumed	Ecological significance
Collectors	Feed on fine particulate organic matter (FPOM)	Small organic matter particles (<1 mm), often plant debris	Important contributors to decomposition, breaking down small particles
Shredders	Consume coarse particulate organic matter (CPOM)	Larger plant particles (>1 mm)	Key players in breaking down larger plant debris, aiding in decomposition
Scrapers	Ingest periphyton	Feed on attached algae and other microorganisms	Contribute to the breakdown of periphyton and associated organic matter
Piercers	Feed on dissolved organic matter	Consume dissolved organic matter, influencing nutrient dynamics	Play a role in the cycling of nutrients through decomposition
Predators	Hunt other macro invertebrates	Consume other aquatic insect larvae and macro invertebrates	Participate in the breakdown of other organisms, contributing to decomposition

Source Cummins and Klug (1979)

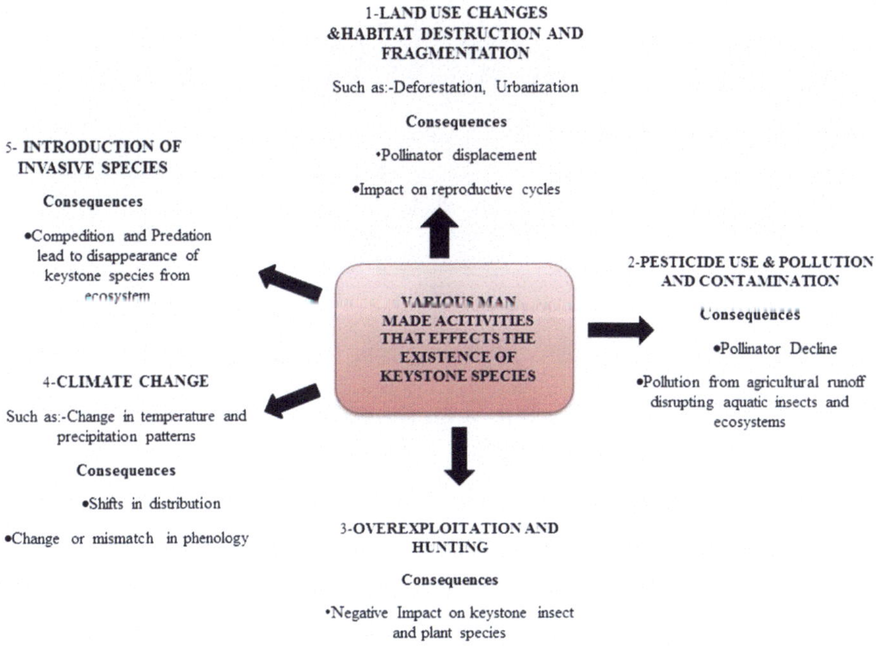

Fig. 6.1 The impact of human activities on the vitality of keystone insect populations. *Source* Hegland et al. (2009), New et al. (2009)

climate variations, are influenced by factors like temperature fluctuations, relative humidity, concentrations of CO_2 and O_3, wind patterns, and average annual precipitation (Fig. 6.2).These environmental dynamics can lead to abnormal behaviors in insects, thereby impacting the distribution, abundance, ecology, and behaviour of various species, especially considering the poikilothermic nature of insects. climate change is primarily driven by both natural and human factors.

Climate change has both direct and indirect repercussions on insect populations. The direct effects involve fundamental processes such as development, reproduction, and survival rates. Conversely, climate change indirectly influences insects by disrupting their interactions with the environment, encompassing natural enemies, competitors, vectors, and mutualists (Fig. 6.3). This interplay assumes a pivotal role in the realms of global agriculture, health, and food security (Anand et al. 2014).

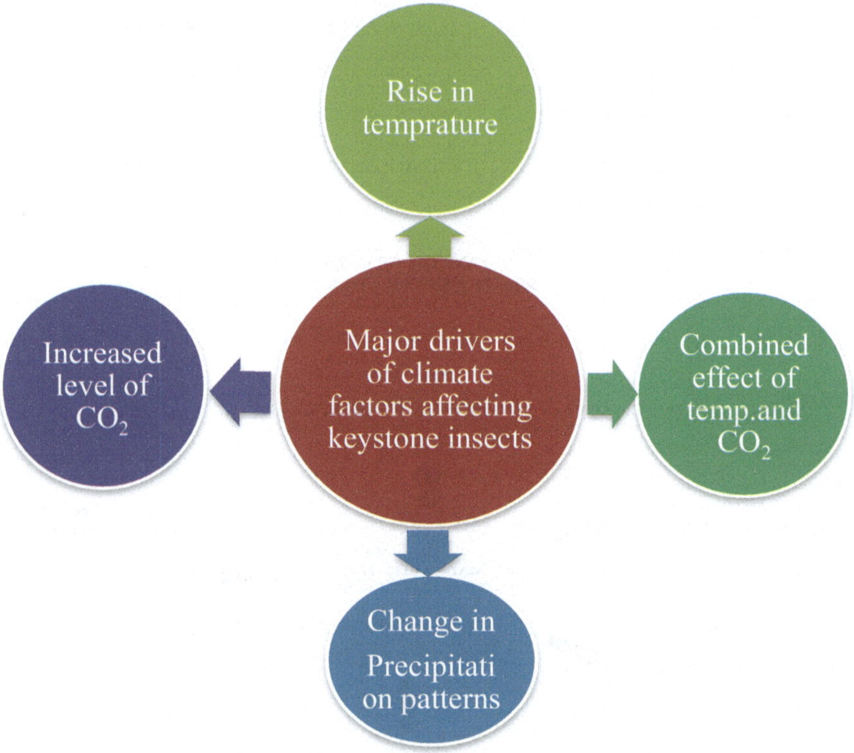

Fig. 6.2 Major climatic factors affecting insects. *Source* Bhagarathi and Maharaj (2023)

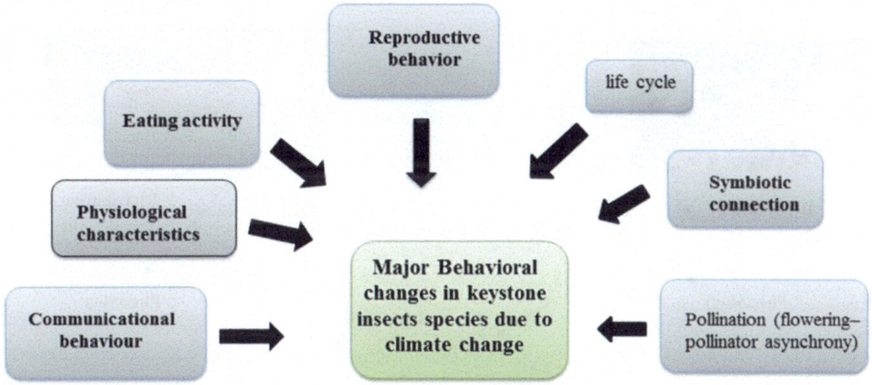

Fig. 6.3 Major behavioral changes in key insects due to climate change. *Source* Anand et al. (2014)

6.6 Effects of Climate Shift on Pollinators (Key Insect Species in Ecosystem)

Over 33% of global food crops depend on commercial bees for pollination (Sharma and Singla 2021). However, alterations in climate directly impact the phenology and distribution of these pollinator species, causing disruptions in the temporal and spatial alignment of plant-pollinator ecological interactions. If the response to climate change results in changes to the distribution and blooming periods of flowering plants, the spatial arrangement of pollinators and the timing of their nectar requirements must also adapt. This adjustment can result in a deficiency of nectar resources for pollinators during crucial periods. The consequent mismatch holds the potential for cascading effects on ecosystem functions (Fig. 6.4), particularly for plants and pollinators with specialized relationships (Hegland et al. 2009).

6.7 Effect of Climate Change on Distribution of Species

Foreseen shifts in temperature, CO_2 levels, and precipitation patterns resulting from climate change are likely to have repercussions on the distribution, geographic spans, and migratory behaviors of insect species. The geographical range of insects is intricately tied to local climate conditions, with unfavorable environments acting as defining boundaries. Remarkably, low temperatures, often exerting a more significant influence than high temperatures, play a pivotal role in determining the geographical range of a species. The predicted climate changes have the potential to reshape the distribution, migration patterns, and overall ranges of insect species, allowing them to traverse boundaries and eventually posing potential threats to global food security. In response, insects may undergo genetic, physiological, or behavioral adaptations to

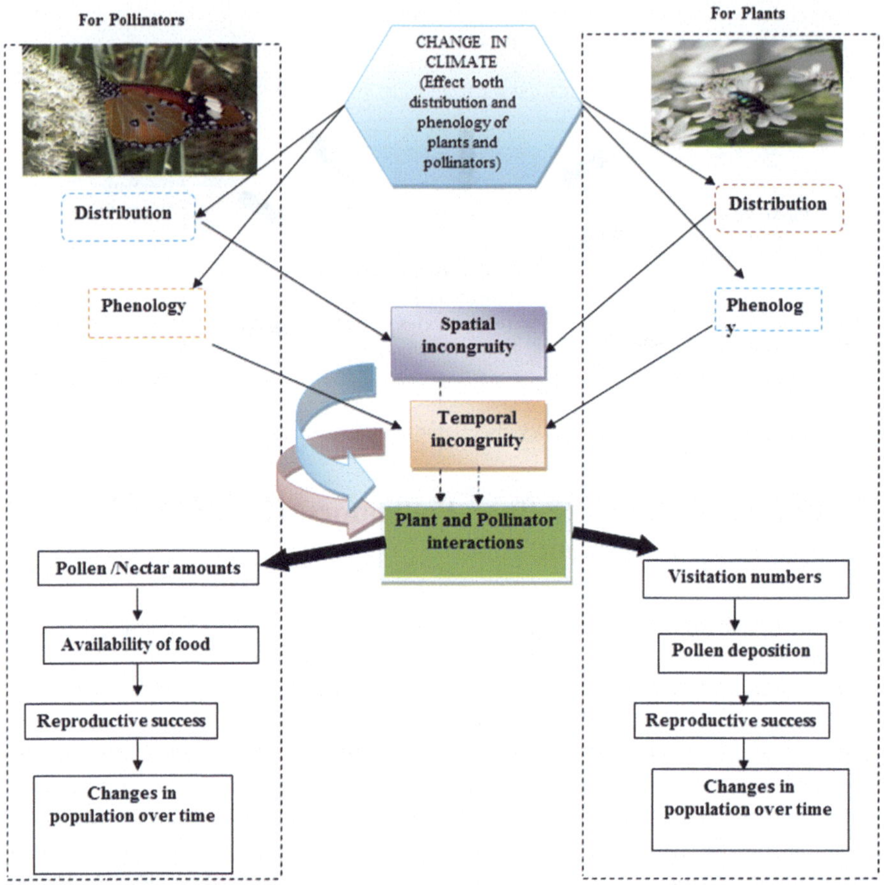

Fig. 6.4 Impact of climate change on plant-pollinator interactions. *source* Hegland et al. (2009)

cope with new climates, expand their current range, or face extinction if adaptation is beyond their capacity. The ability to adapt to rising temperatures and expand ranges is expected to vary based on the specific life history traits of each insect species (Chandrakumara et al. 2023).

6.8 Predatory Insects and Herbivore Regulation

The intricate balance of ecosystems depends on a delicate interplay between numerous creatures, each of which plays a specific function in maintaining equilibrium. One critical part of this equilibrium is the regulation of herbivore populations by predatory insects. This delicate equilibrium between predator and prey is a remarkable illustration of nature's long-term pest management tactics.

Predatory insects play a crucial role in the insect food chain by consuming other insects, including herbivores and small invertebrates (Table 6.3). Their presence is vital for controlling herbivorous insect populations, preventing them from becoming excessively destructive. The act of predation exerts a significant impact on the behavior, distribution, and population dynamics of both the predators and their prey (Finke and Snyder 2008).

Herbivores, which include insects and mammals, are a persistent threat to plant life. If left unchecked, they have the potential to devastate crops, damage ecosystems,

Table 6.3 List of various predatory insects controlling herbivore population in the ecosystem

Natural enemies	Herbivores controlled	Citation
1. Ladybird Beetles (Coleoptera: Coccinellidae)	Aphids, mites, scale insects	Omkar and Pervez (2016)
2. Praying mantis (Dictyoptera: Mantidae)	Various insects, including caterpillars	Mebs et al. (2017)
3. Ground Beetle (Coleoptera: Carabidae)	Snails Slugs,	Holland and Luff (2000)
4. Hoverflies (Diptera: Syrphidae)	Aphids, thrips, scale insects	Rotheray and Gilbert (2011)
5. Dragonflies (Odonata: Libellulidae)	Mosquito larvae, small aquatic insects	Vatandoost (2021)
6. Damselflies (Odonata: Nabidae)	Aphids, mites, chironomids, flies, horseflies, and mosquitoes	Pfitzner et al. (2015)
7. Spiders (Araneae: Araneidae)	Various insects, including flies and beetles	Nyffeler and Birkhofer (2017)
8. Hymenopteran Wasps		Rosenheim (1998)
Braconidae	Caterpillars, grubs, aphids and other soft bodied insects	
Ichneumonidae	Caterpillars, grubs, sawflies, true bugs, Scales, Aphids and other soft bodied insects	
Chalcididae	Caterpillars, spiders, ticks and mites, pseudoscorpions and even gall-forming nematodes	
Eulophidae	Aphids, caterpillars, leaf miners, scales	
Pteromalidae	Aphids, flies, leaf miners, beetles	
	Mealybugs, pupa of Lepidopteran	
Encyrtidae	Scale insects and other soft bodied insect	
9. Predatory Mites (Phytoseiidae)	Spider mites small arthropods, thrips	McMurtry and Croft (1997)

and cause cascading impacts across the food web. Nature has produced a vast assortment of predatory insects that serve as critical regulators, safeguarding the health and stability of ecosystems.

6.8.1 Implications Natural Enemies for Plant Growth and Diversity

6.8.1.1 Increased Plant Diversity

Natural enemies of insects contribute to enhanced plant diversity by controlling herbivore populations. Reduced herbivore pressure reduces the dominance of certain plant species, resulting in a more diverse and resilient environment (Letourneau et al. 2011).

6.8.1.2 Enhanced Plant Health

The presence of natural enemies insects prevents herbivores from reaching damaging population levels. This protection allows plants to allocate more resources to growth and reproduction, promoting overall plant health (Straub et al. 2008).

6.8.1.3 Sustainable Agriculture

In agricultural contexts, the use of natural enemies/biotic factors for pest control offers a sustainable alternative to chemical pesticides. This approach minimizes environmental impact, preserving biodiversity and ecosystem health (Landis et al. 2000).

6.9 Conclusion and Future Prospects

In conclusion, keystone insects play pivotal roles in shaping the structure and function of ecosystems in the Himalayan region and aquatic environments. The ecosystem engineers, ants, and termites contribute significantly to soil health and nutrient cycling in the Himalayas, while aquatic insects act as bioindicators and key players in nutrient cycling in freshwater ecosystems. The interplay between keystone insects and their environment is intricate and sensitive, making them susceptible to human impacts and climate change. As climate change poses challenges to the distribution and behavior of key insect species, there is a growing urgency for conservation efforts. The delicate balance maintained by predatory insects in herbivore regulation further emphasizes the interconnectedness of species within ecosystems. To ensure the continued health and functionality of these ecosystems, concerted efforts are required, including research, conservation, and sustainable practices that account for the vulnerabilities of keystone insect species. This will open vistas for sustainable development in terms of global food security, healthy ecosystem and healthy environment.

References

Anand P, Jagadiswari R, Mukherjee AKJB, Somnath SP, Totan A, Sushmita M, PRS (2014) Climate Change: Impact on Crop Pests Res Today 2 5 327 329

Arumugam S, Athikesavan S (2021) Diversity and distribution of aquatic insects in pond ecosystem in Cheyyar, Thiruvannanalai District of Tamil Nadu, Indiaa. Uttar Pradesh J Zool 10–15

Bhagarathi LK, Maharaj G (2023) Impact of climate change on insect biology, ecology, population dynamics, and pest management: a critical review. World J Adv Res Rev 19(3):541–568

Budnukaeku AC, Ichendu CB, Irene D (2021) Human Impacts on keystone species within the ecological food web. MOJ Ecol Environ Sci 6(5):182–186

Chandrakumara K, Sau AK, Tanwar AK, Hadimani BN (2024) Variations in the biological and ecological attributes of insects due to climate change: a review

Cummins KW, Klug MJ (1979) Feeding ecology of stream invertebrates. Annu Rev Ecol Syst 10(1):147–172

Davic RD (2003) Linking keystone species and functional groups: a new operational definition of the keystone species concept. Ecol Soc 7(1)

Elango K, Vijayalakshmi G, Arunkumar P, Sobhana E, Sujithra P (2021) Aquatic insect's biodiversity: importance and their conservation. Biol Divers Curr Status Conserv Policies 289–303

Finke DL, Snyder WE (2008) Niche partitioning increases resource exploitation by diverse communities. Science 321(5895):1488–1490

Hegland SJ, Nielsen A, Lázaro A, Bjerknes AL, Totland Ø (2009) How does climate warming affect plant-pollinator interactions? Ecol Lett 12(2):184–195

Holland JM, Luff ML (2000) The effects of agricultural practices on Carabidae in temperate agroecosystems. Integr Pest Manag Rev 5(2):109–129

Kulkarni RR, Zade SB (2020) Role of aquatic insects in the enhancement of biodiversity of a freshwater reservoir, Ramala, Chandrapur, Maharashtra. Environ Conser J 21(3):89–92

Landis DA, Wratten SD, Gurr GM (2000) Habitat management to conserve natural enemies of arthropod pests in agriculture. Annu Rev Entomol 45(1):175–201

Léonard J, Rajot JL (2001) Influence of termites on runoff and infiltration: quantification and analysis. Geoderma 104(1–2):17–40

Letourneau DK, Jedlicka JA, Bothwell SG, Moreno CR (2011) Effects of natural enemy biodiversity on the suppression of arthropod herbivores in terrestrial ecosystems. Annu Rev Ecol Evol Syst 40:573–592

McMurtry JA, Croft BA (1997) Life-styles of phytoseiid mites and their roles in biological control. Annu Rev Entomol 42(1):291–321

Mebs D, Wunder C, Pogoda W, Toennes SW (2017) Feeding on toxic prey. The praying mantis (Mantodea) as predator of poisonous butterfly and moth (Lepidoptera) caterpillars. Toxicon 131:16–19

New E, To I, Ecology IN (2009) Concepts and synthesis 90(8):2068–2076

Nithyatharani R, Kavitha US (2018) Termite soil as a bio-indicator of soil fertility. IJRASET 6(1):659–661

Nyffeler M, Birkhofer K (2017) An estimated 400–800 million tons of prey are annually killed by the global spider community. Sci Nat 104(3):1–12

Olsen SL, Evju M, Åström J, Løkken JO, Dahle S, Andresen JL, Eide NE (2022) Climate influence on plant–pollinator interactions in the keystone species Vaccinium myrtillus. Ecol Evol 12(5)

Papáček M (2013) Small aquatic and ripicolous bugs (Heteroptera: Nepomorpha) as predators and prey: the question of economic importance. EJE 98(1):1–12

Power ME, Tilman D, Estes JA, Menge BA, Bond WJ, Mills LS, Paine RT (1996) Challenges in the quest for keystones: identifying keystone species is difficult—but essential to understanding how loss of species will affect ecosystems. Bioscience 46(8):609–620

Pfitzner WP, Beck M, Weitzel T, Becker N (2015) The role of mosquitoes in the diet of adult dragon and damselflies (Odonata). J Am Mosq Control Assoc 31(2):187–189

Pervez A (2016) Ladybird beetles. In: Ecofriendly pest management for food security. Academic Press, pp 281–310

Rosenheim JA (1998) Higher-order predators and the regulation of insect herbivore populations. Annu Rev Entomol 43:421–447

Rotheray GE, Gilbert FS (2011) The Natural Enemy Impact Index (NEII): a novel measure of biocontrol agent efficacy in arable crops. Ecol Entomol 36(6):724–731

Saha SP, Chhetri S, Rai S, Chakraborty H (2018) Evaluation of ant nest microenvironment from Darjeeling Himalaya. J Mycopathol Res 56(1):21–27

Sharma T, Singla A (2021) Potential impacts of climate change on insect behavior. J Life Sci 11(3):29–35p

Straub CS, Ives AR, Gratton C, Mueller E (2008) Invasive species response to a multiple predator approach. Ecology 89(7):1839–1846

Suter GW, Cormier SM (2015) Why care about aquatic insects: uses, benefits, and services. Integr Environ Assess Manag 11(2):188–194

Thaware VH (2023) Study of aquatic insects, their role in improving biodiversity and balancing the food web of a freshwater ecosystem at Karadkhed Dam in Nanded, Maharashtra. J Entomol Zool Stud 11(1):173–176

Vatandoost H (2021) Dragonflies as an important aquatic predator insect and their potential for control of vectors of different diseases. J Mar Sci 3(3):13–20

Chapter 7
Preliminary Field Observation on Spatio-Temporal Changes on Rajma Crop and Citrus Fruit Cultivation Areas in the Himalayan Watershed Using Interactive Geo-Informatics Technique

D. S. Parihar, Mahendra Singh, and Deepak

Abstract Global warming and climate change are the most important environmental challenges that affect the entire natural ecosystem including agriculture and horticulture of the globe. Different villages of the study area in different aspects, i.e., agriculture, horticulture, economy and development etc. had found large implication of global warming. The state of the farmers and farming crops is a on the state of affairs about farmers' livelihoods and economic wellbeing in the Himalayan watershed, i.e., Gori Ganga. The data and information presented here, has been gathered using different means. Primary data was sourced through interview, group discussions in sample villages. Each village selected for field surveying represented a major farming culture of related crop and care was taken that these are picked up from across the watershed area. The present study is based on exploratory and analytical research design As a result, it was observed that total sample sites are 55 (32.74%) village out of 168 villages where 28 (16.67%) villages were belongs to Rajma cultivation and 27 (16.07%) villages citrus fruit cultivation. The highest affected villages by temporal changes on Rajma crop and Citrus fruit cultivation areas were observed after 2010. Where the crop quality, size and real local taste were disappear.

Keywords Crops · Temporal changes · Field observation · Himalayan watershed and geo-informatics

D. S. Parihar (✉)
Department of Geography, Kumaun University, D.S.B. Campus, Nainital, Uttarakhand, India
e-mail: devendraparihar@kunainital.ac.in

M. Singh
Department of Geography, Hemwati Nandan Bahuguna Garhwal Central University, S.R.T. Campus, Tihari, Uttarakhand, India

Deepak
Department of Geography, S.S.J. University and Campus, Almora, Uttarakhand, India

D. Arya et al. (eds.), *Climate Change Impact on Himalayan Biodiversity*,
Environmental Science and Engineering, https://doi.org/10.1007/978-3-031-77149-1_7

7.1 Introduction and Review Literature

Climate change is the dynamics of average atmospheric conditions over a long time and has been chiefly caused by anthropogenic activity. Every year increasing green house gases by consumption of fossil fuels, i.e., Natural Oil, Coal etc. helped for more of the sun's rays and heat to absorb through in the atmosphere. The result of these changes in the atmosphere can be severe or extreme weather events, i.e., more frequent and intense storms, flooding, scorching heat waves, freezing cold snaps, wildfires, and water shortages. Growers are witnessing their entire agricultural fields of crops damaged in one swoop (Owens 2021). Climate change will alter pest, plant disease and weed distributions, with potential to reduce crop yields, including of staple crops like wheat, soybeans, and corn. Warmer temperatures can increase the metabolic rate and number of breeding cycles of insect populations (Malhi et al. 2021). The regions facing the most issues right now are those that are already battling heat and drought.

Some regions in Europe have started shifting upwards wine grape vineyards similarly north or to a suitable elevation, adjusting to the planet's changing climate. Table grapes themselves are extra warmth tolerant, and IFG's breeding program has used the maximum current heat wave to screen our varieties for heat tolerance. However, there may be nonetheless a most quantity of warmth that fruit can tolerate, and as climate change continues to alter the environment, it will affect global agriculture and food production (Owens 2021). The parts of the world suitable for growing coffee, cashews and avocados will change dramatically as the world heats up, according to a new study. Key coffee regions in Brazil, Indonesia, Vietnam and Colombia will all "***Drastically Decrease***" by around 50% by 2050. Suitable areas for cashews and avocados will increase but most will be far from current sites of production (McGrath 2022). In Himachal Pradesh, apples were known to be cultivated at altitudes between 1200–1500 m in the early 1980s (NHB 2012). Orchards have shifted to 1500–2500 m in the 2000s. More recently apples are now being cultivated at more than 3500 m elevation in the Himachal Pradesh, owing to warming of the surface temperature since the 1980s (Kothawale and Kumar 2005). Unable to adapt to effects of climate change such as forest fires and scarce rainfalls, smallholder farmers in the Himalayan state risk losing their livelihoods. The Himalayas are warming faster than the global average, which has immediate repercussions on agriculture, altering crop cycles, growing seasons and soil moisture (Thapa 2023).

Due to population growth, rapidly increased in agricultural and horticulture productivity demands optimum utilization of natural resources like land and water. The surface mulch favorably influences the soil moisture regime by controlling evaporation (Ramakrishna et al. 2006; Montenegro et al. 2013). The temperature of globe is basically determined by the amount of energy projected at it by the sun and how much of that energy is trapped by greenhouse gasses in the global atmosphere, including water vapor, carbon dioxide, methane, nitrous oxide, sulfur-di-oxide, ozone and some artificial chemicals such as Chloro Fluoro Carbons (CFC) etc. The global warming substantially accelerated in the 1950 due to post-war boom in manufacturing, land use and transportation along with other socio-economic activities and rapidly population growth triggered greater emissions of greenhouse gasses. Overall, worldwide temperatures were estimated to be 1 °C greater than pre-industrial levels (the average from 1850 to 1900) and to be increasing by 0.2 °C per decade (IPCC 2018). Anthropogenic global warming and climate change, driven by the exponential increase in Greenhouse Gas Emissions (GHGs) since the industrial revolution, will continue to impact the ocean through a variety of channels and severity of effects will depend greatly on the extent of warming reached through GHG emissions (IPCC 2019). Mountain ecosystems are sensitive to global warming and show signs of fragmentation and degradation (Xu and Wilkes 2004; Korner 2004). Species in high-elevation ecosystems are projected to shift to higher altitudes, although alpine plant species with restricted habitat availability above the tree line are projected to experience severe fragmentation, habitat loss, or even extinction if they cannot move to higher elevations (Dirnbock et al. 2003). While being the result of social relations and a driver of contemporary social actions, sociologists have always been reluctant to address the issue of global warming (Lever-Tracy 2008). Research, for example, has linked increased population density, age and well-being to increased levels of greenhouse gas emissions (Zehr 2015). The analyzes also suggested that global warming is produced by specific state characteristics, such as levels of land inequality, the penetration of foreign capital and a state's dependence on exports (Crenshaw and Jenkins 1996).

Climate change has been reported to affect lawn productivity, ecosystems, and the distribution and composition of plant communities (Wilkes 2008). Some rangelands might suffer from degradation due to the warmer and drier climate (Dirnbock et al. 2003). Upward movement of tree line and encroachment of woody vegetation on alpine meadows are widely reported. In the eastern Himalayas, the tree line is increasing at a rate of between 5 and 10 m per decade (Baker and Moseley 2007). As temperature rises, species shift their ranges to follow their principal habitats and climatic optima. Increasing temperatures and water stress are expected to lead to a 30% crease in crop yields in Central and South Asia by the mid-twenty-first century (UNDP 2006). The fundamental objectives of the present investigation incorporate the follows: geographical study of temporal changes, causes and mapping by using geo-informatics tool of historical growing sites and current status of agriculture and horticulture crops in the study area.

Due to global warming Rajma crop and citrus fruit production is affected. During field surveying in 2019 there are so many Rajma and citrus fruit cultivated villagers suffering with the less productivity, small size and low quality of crops. Near about one decade some upward situated villages in the valley were not grown those crops, at present (2023) they did produce good quality Rajma and citrus fruits in their villages. This preliminary field observation triggered my interest to the studying the present situation and temporal changes of Rajma crop and citrus fruits in different villages.

7.2 Material and Methods

For conducting the present study, three different types of data were used. These are satellite data, primary data and secondary data. To carry out present research work various kinds of data and maps are required from different sources. Figure 7.1 depicts the diagrammatic illustration of the research methodology flow chart developed for the present study which reveals that the present study is based on various primary and secondary data using latest technologies of geo-informatics. A brief account of data, software and various techniques and methods used in the present study is presented in the following paragraphs.

7.2.1 Data Collection and Uses

Data collection (Images P, GPS Coordinates, Personal Interview related growing fruits and crops in old and new cultivation villages) and mapping for the present study is based on primary source which is collected during field surveying. This was conducted in the month of May 2019 and May 2023. Present study is broadly divided into two parts: Firstly, land resources analysis which contains Ground Control Points (GCPs), location of cultivated villages, field photographs, historical rainfall records, DEM, surface relief, slope and slope aspects data. For this purpose Cartosat-1 satellite data were used. Secondly, a demographic aspect based on censes 2011 data. For the literature and manuscript prepare was used research papers, journal and books etc.

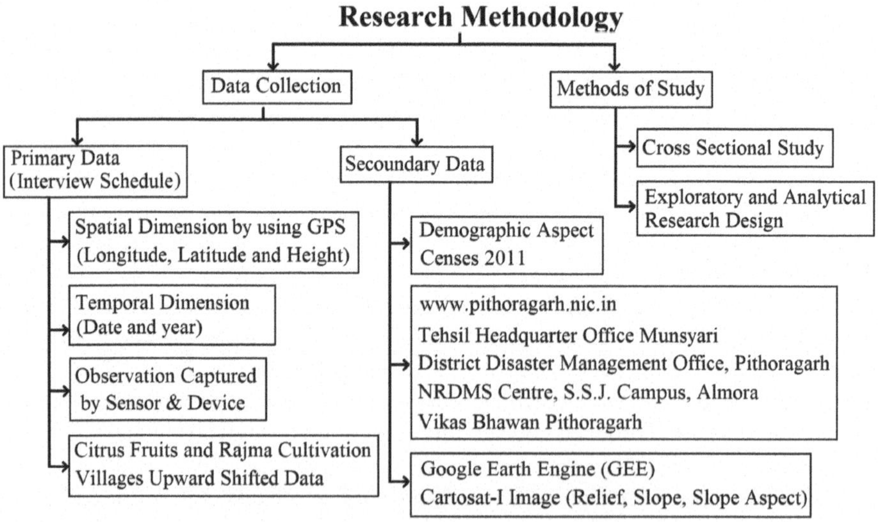

Fig. 7.1 Methodological flow chart of the present study (*after authors*)

The secondary data were collected from many government offices such as Vikas Bhawan Pithoragarh, Tehsil/Block/Forest headquarter Office Munsyari, Dharchula and Didihat, and District Disaster Management Office (DDMO) Pithoragarh, District Development Office Pithoragarh, NRDMS Centre, S.S.J. University and Campus Almora etc. Census data for the year 2011 of Gori Ganga watershed were acquired from Census of India.

7.2.2 Instruments and Software Uses

Garmin Digital Global Position System (extrex-20) for GCPs and DSLR-200D for field photograph capturing. In order to present satellite data and to develop different maps of the study area, the GIS and Remote Sensing software such as Q-GIS-2.18, Arc GIS-10.8 (Crack Version), Google Earth Engine (GEE) Pro, Microsoft Word and Microsoft Excel were used.

7.2.3 Data Sampling

Stratified random sampling was used for selecting sample sites. In this sampling method, a population is divided into subgroups to obtain a simple random sample from each group and complete the sampling process. These small groups are called strata. The small group is created based on a few features in the population. After dividing the population into smaller groups, the researcher randomly selects the sample. Total sample sites are 55 (32.74%) village out of 168 villages where 28 (16.67%) villages were belongs to Rajma cultivation and 27 (16.07%) villages citrus fruit cultivation. Sample crops are listed in Table 7.1 and ground photograph with their elevation from mean sea level are presented in Plates 7.1 and 7.2.

Table 7.1 Sample crops with scientific name of the study

S.N	Variable	Crop name	Botanical name	Family name	Common name
1	Agriculture	Rajma	*Phaselous Vulgaris*	*Leguminancea*	*White kidney beans*
2	Horticulture	Mosambi	*Citrus limon*	*Rutaceae*	*Lemon*
		Orange	*Citrus sinensis*	*Rutaceae*	*Navel orange*
		Lime	*Citrus aurantifolia*	*Rutaceae*	*Sour lime*
		Citron	*Citrus medica*	*Rutaceae*	*Fingered Citron*

7.3 Geographical Extent

The study area, viz., the Gori Ganga watershed, Kumaun Himalaya, Uttarakhand (Fig. 7.2) extends between 29° 45′ 0″ N to 30° 35′ 47″ N latitudes and 79° 59′ 33″ E to 80° 29′ 25″ E longitude, and encompasses an area of about 2191.63 km^2. The Gori Ganga watershed lies in the Great and Lesser Himalayan ranges of the Kumaun Himalaya in the Uttarakhand state. The relief of this watershed varies in between 626 m at Jauljibi to 6639 m in the snow capped mountains. The Gori Ganga watershed extends into three blocks, namely Dharchula, Munsyari, and Didihat, in three Tehsils, i.e., Munsyari, Dharchula and Didihat, and in one Sub-Tehsil known as Bangapani. Munsyari remains one of the last highway-accessible hill stations in the region. There are many tribes in this watershed which are Bhotiya, Barpatiya and Anuwal Samuday. Three main valleys i.e., Gori Ganga valley, Ralam valley and Johar valley are famous valleys in the study area.

Relief: About 34.52 km^2 (1.57%) area (smallest area of relief zone below 1000 m) falls under the low relief zone where 13 villages (7.74%) total and this relief zone between 3000–5000 m (high relief zone) is the largest relief zone of the study area. About 1133.73 km^2 (51.73%) area and 13 villages (7.74%) are situated in the Gori Ganga watershed (Fig. 7.3a, Table 7.2). Highest villages 142 (84.54%) are situated in the medium relief zone varies between 1000–3000 m.

Slope: The smallest slope zone is below 10° (gentle slope) encompasses about 180.99 km^2 (8.26%) area where 21 villages (12.5%) are situated and steep slope zone (20°–30°) is the largest slope zone which encompasses about 406.59 km^2 (31.17%) where highest villages 58 (34.52%) are situated in the Gori Ganga watershed (Fig. 7.3b, Table 7.2).

Slope Aspect: The level land about 3.33 km^2 (0.15%) area shows flat surface of the Gori Ganga watershed where villages are not situated because of this level terrain is composed of close river terrace deposits. The smallest aspect zone is 337.5°–22.5°

Plate 7.1 Rajma cultivation villages in the Gori Ganga watershed: **a** at Bona village—2077 m and **b** Golpha village—1804 m (*based on field surveying*)

Plate 7.2 Citrus fruit cultivation villages in the Gori Ganga watershed: **a** at Lodi village—1454 m, **b** at Golpha village—1804 m, **c** at Chulkote village—1983 m and **d** at Okhali village—1895 m (*based on field surveying*)

(Northern Aspect) encompasses 221.27 km^2 (10.1%) area where 7 villages (4.17%) are situated and south-eastern aspect (112.5°–157.5°) is the biggest aspect which includes 361.89 km^2 (16.51%) area where 24 villages (14.29% villages) are situated of the Gori Ganga watershed (Fig. 7.3c, Table 7.2). Highest 31 villages (18.45%) of the study area are situated in the south-western aspect of the Gori Ganga watershed.

7.3.1 *Rainfall*

Rain is a critical component of agriculture and is essential for crop growth and development timely in the study area. Without rain, plants would not be able to survive, and crops would suffer because agriculture and horticulture land are not connected by irrigation canals in the study area. Figure 7.4a, b presents graphically 13 year monthly and yearly rainfall data from January 2011 to July 2023 obtained from the Tehsil Headquarter Office, Munsyari (meteorological station). The summary of these data is presented in the hyetograph (Fig. 7.4a) which reveals that the maximum rainfall occurred in the month of July 2020 (889 mm) which accounts for 39.18% of total annual rainfall.

Table 7.2 Distribution of villages under different relief, slope aspects and slopes of the Gori Ganga watershed (*based on Censes of India-2011 and Cartosat-1 satellite data*)

S.N	Variables	Zones	Groups	Area		Villages	
				in km^2	in %	in no	in %
(A)	Relief (in meter)	Low	<1000	34.52	1.57	13	7.73
		Medium	1000–3000	636.17	29.03	142	84.54
		High	3000–5000	1133.73	51.73	13	7.73
		Very high	>5000	387.21	17.67	00	0.00
(B)	Slope (in Degree)	Gentle	<10	180.99	8.26	21	12.5
		Moderate	10–20	406.59	18.55	56	33.33
		Steep	20–30	683.09	31.17	58	34.52
		Very steep	30–40	600.69	27.41	25	14.88
		Very high steep	>40	320.27	14.61	8	4.77
(C)	Slope aspect (in Degree)	Level land	0	3.33	0.15	0	0
		Northern	337.5–22.5	221.27	10.1	7	4.17
		North-Eastern	22.5–67.5	252.39	11.51	28	16.67
		Eastern	67.5–112.5	233.59	10.66	21	12.5
		South-Eastern	112.5–157.5	361.89	16.51	24	14.29
		Southern	157.5–202.5	238.39	10.88	16	9.52
		South-Western	202.5–247.5	310.99	14.19	31	18.45
		Western	247.5–292.5	265.69	12.12	20	11.9
		North-Western	292.5–337.5	304.09	13.88	21	12.5

7.3.2 Demography

Gori Ganga watershed is the part of Kumaun Himalayan region, where Gori Ganga and its tributaries flow from north to south direction and develop narrow valleys. Distributions of 168 villages are under three Tehsils, i.e., Muansyari-120, Dharchula-08, Didihat-40. Gori Ganga watershed has total population 40,616 where 20,249 are male which accounts 49.85% and 20,367 are female which accounts 50.15% of the total population. The total population of Scheduled Tribes in the study area is 9840 where 4933 are male which accounts 50.13% and 4907 are female which accounts 49.87% of the total schedule tribe population. The Scheduled Cast total population is 6634 where 3199 (48.22%) are male and 3435 (51.78%) are female. Female population in the study area is higher than the male population and the sex ratio of male to female stands at 1000:1006. The population density in the study area stands at 18.53 people per km^2. The total literate people are 27,271 which accounts 67.14% in total population of the study area where 15,596 (57.19%) are male and 11,677 (42.81%) are female, and the illiterate are 13,345 (32.85%) where 4654 (34.87%) are male and 8690 (65.13%) are female.

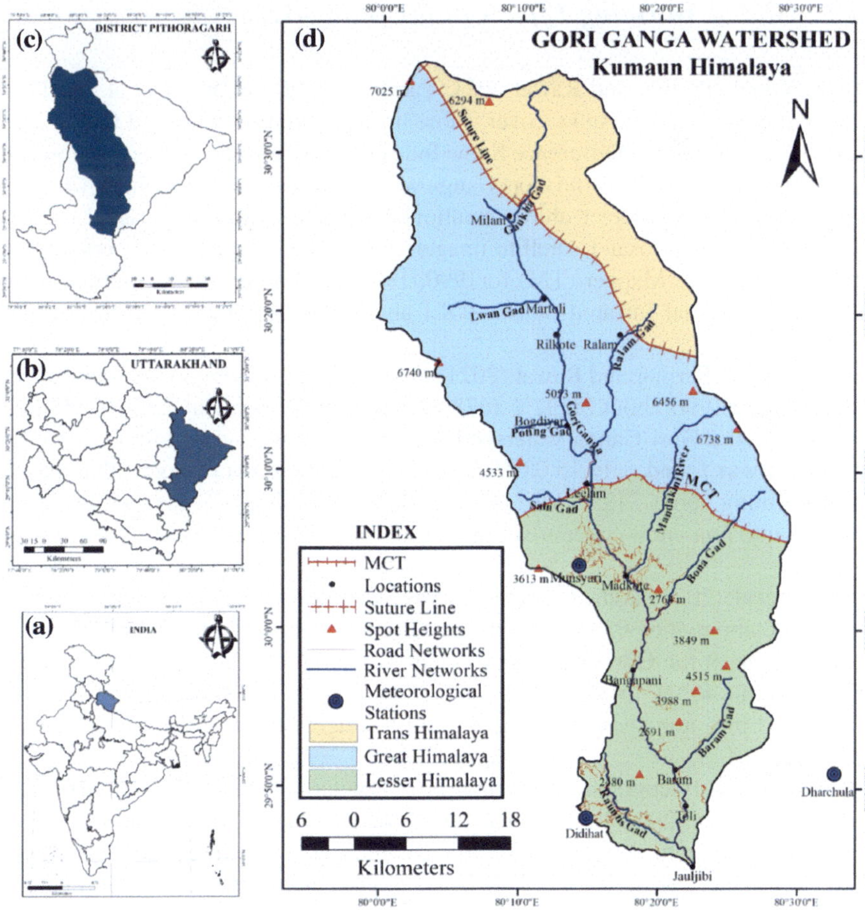

Fig. 7.2 Location and extent of the study area: **a** India, **b** Uttarakhand, **c** Pithoragarh District and **d** Gori Ganga watershed, Kumaun Himalaya region (Uttarakhand)

7.3.3 Occupational Structure

Total working population of the study area is about 20,997 where 10,886 are male which accounts 51.85% and 10,111 are female which accounts 48.15%. The total population of cultivator in the study area is 9179 in which 3901 (42.50%) are male and 5278 (57.50%) are female. The total agriculture labor population is 122 where 75 (61.48%) are male and 47 (38.52%) are female. The total population of main workers is 14091 in which 7504 are male which accounts 53.25% and 6587 are female which accounts 46.75%. The total population of non-workers is 19619 where 9363 (47.72%) are male and 10,256 (52.28%) are female.

7.3.4 Global Warming Effects in the Study Area

Earlier studied (Parihar and Rawat 2021; Parihar 2021a, b; Parihar et al. 2021) to examine the dynamics of snow cover, snow lines, vegetation lines and timberlines by using the Normalized Difference Snow Index (NDSI) and Normalized Difference Vegetation Index (NDVI) in the Gori Ganga watershed, Kumaun Himalaya, Uttarakhand (India). For the detect above mentioned variables dynamics used Landsat-5, Landsat-8 and Cartosat-1 satellite imageries of three different time periods like Landsat-5 Thematic Mapper (TM) for1990, 1999 and Landsat-8 Operational Land Imager and Thermal Infrared Sensor (OLI and TIRS) for 2016 and Cartosat-1of 2008.

According to Parihar and Rawat (2021) Geographical distribution of snow cover reveals that in 1990 about 30.97% (678.87 km^2), in 1999 about 25.77% (564.92 km^2) area of the Gori Ganga watershed was under snow cover while in 2016 the snow cover was found only 15.08% (330.44 km^2). These data suggest that due to global warming about 348.43 km^2 snow cover of Gori Ganga watershed has been converted into non-snow cover area at an average rate 13.40 km^2/year during the last 26 years.

Parihar et al. (2021) studied to detect average height of snow line dynamics reveals that in 1990 about 4665.02 m, 1999 about 4764.97 m and in 2016 about 5068.95 m average height of the Gori Ganga watershed. These data suggest that due to global warming about 403.93 m snow line average height of Gori Ganga watershed has been shifted into non-snow cover area at an average rate of 15.53 m/year during 1990–2016.

Parihar (2021a) Geographical distribution of vegetation line average height reveals that in 1990 about 4294.97 m, 1999 about 4394.51 m and in 2016 about 4758.55 m which is change from non vegetation to vegetation area. During 1990–1999 about 150.67 km^2 areas and during 1999–2016 about 434.03 km^2 area of the Gori Ganga watershed was converted from non-vegetation cover to vegetation cover area. These data suggest that due to global warming about 463.58 m vegetation line average height and 584.70 km^2 vegetation area of Gori Ganga watershed has been shifted into non-vegetation cover to vegetation cover area at an average rate of 17.83 m/year and 22.49 km^2/year during 1990–2016.

Parihar (2021b) Geographical distribution of average timberline height reveals that in 1990 about 3516.11 m, in 1999 about 3680.69 m and in 2016 about 4060.58 m which is change from vegetation to timber area and during 1990–1999 about 197.58 km^2 areas and during 1999–2016 about 463.95 km^2 area of the Gori Ganga watershed was converted from non-timber to timber area. These data suggest that due to global warming, about 544.47 m timberline average height and 661.53 km^2 timber area of Gori Ganga watershed has been shifted into non-timber cover to timber cover area at an average rate of 20.94 m/year and 25.44 km^2/year from 1990 to 2016.

7.4 Results and Discussion

Due to global warming, the atmosphere and land both have a large impact on the overall economic resources i.e., agriculture, horticulture, herbal cultivation and small scale industries in the Gori Ganga watershed. Agriculture and horticulture are strongly influenced by changing weather due to global warming. Different crops show different responses to warming. Many crops of agriculture and horticulture are shifting upward in the study area. Rajma cultivation and citrus fruit cultivation have maximum upward shifting. For completion of present study Fig. 7.7 depicts the spatial distribution of Rajma and citrus fruit cultivation old and new villages in the study area which is registered in Table 7.3.

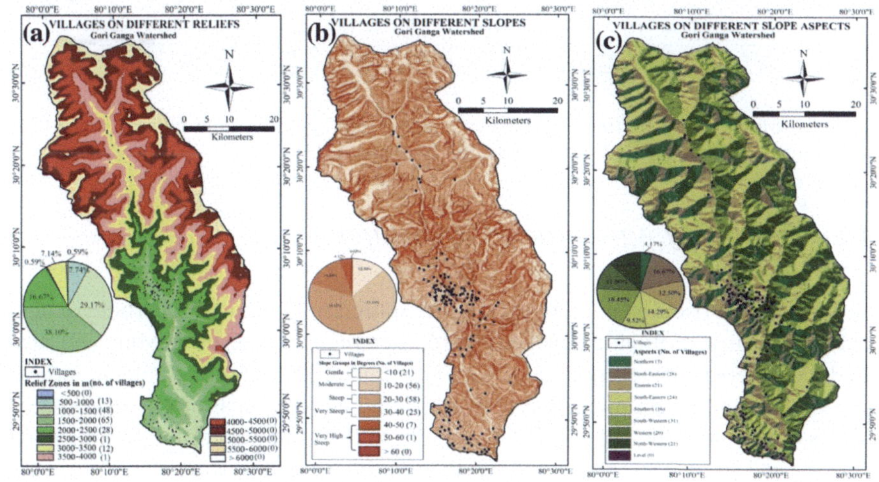

Fig. 7.3 Distribution of villages on different topographic aspects **a** relief, **b** slope and **c** slope aspects of the Gori Ganga watershed (*based on Censes of India and Cartosat-1 satellite data*)

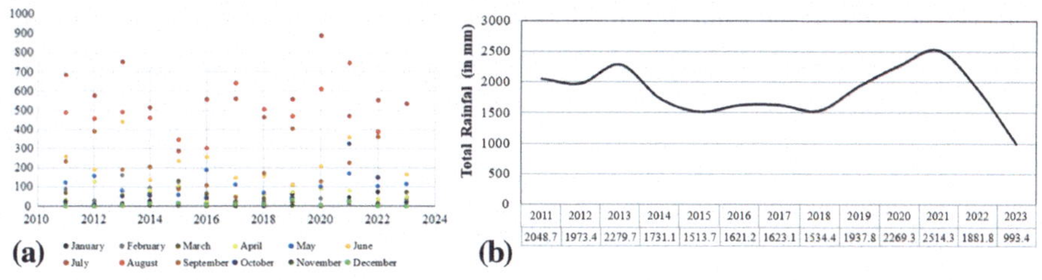

Fig. 7.4 Rainfall (in mm) records of 13 years (January 2011–July 2023) at Munsyari metrological stations of Gori Ganga watershed: **a** Monthly Rainfall and **b** Annual Rainfall *Source DDMO* (2023)

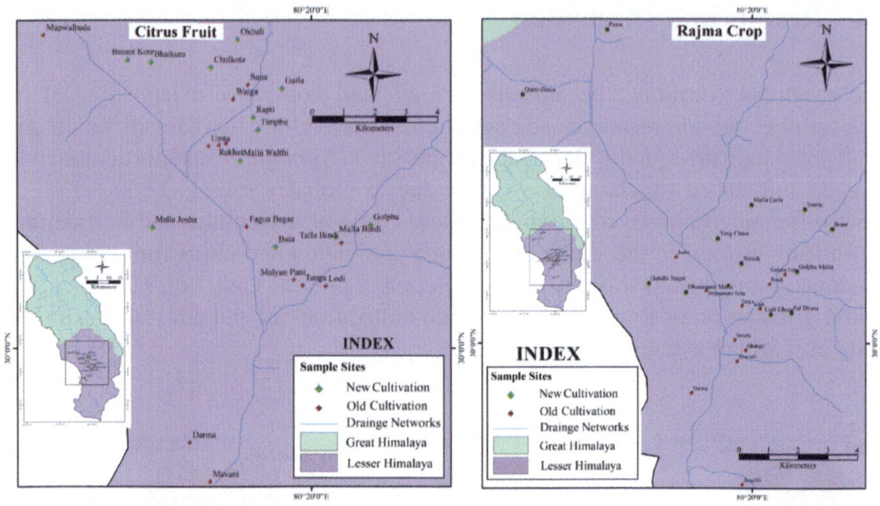

Fig. 7.5 Distribution and shifting of Rajma crop and citrus fruit cultivation villages in the Gori Ganga watershed (based on field survey and GPS, 2019 and 2023)

7.4.1 Shifting of Agriculture Crop (Rajma)

Munsyari Rajma (rich in vitamins and good source of protein and carbohydrates) is one of the most profitable agriculture crop derives its name from Munsyari also got GI Tag (Geographical Indication Tag) in October 2021. A geographical indication is a sign used on products that have a specific geographical origin and possess qualities or a reputation that are due to that origin. It is famous for totally free from chemicals and pesticides. Due to global warming the existing Rajma crop growing areas have becoming unsuitable for crop from the last one decade. Therefore, the Rajma crop cultivation has started shifting upward in mountains in search of suitable temperatures. Table 7.3 reveals that for cultivation of Rajma crop, the suitable height in the study area was 1450–1800 m but now due to global warming the cultivation of Rajma has shifting in the elevation of 1800–2400 m. the production of Rajma is more healthy in new altitude area (1800–2400 m) compare to old sites of Rajma growing area in between 1450–1800 m. this change has been observed after 2010.

7.4.2 Shifting of Horticulture (Citrus Fruit)

During this study, it was observed that before shifting of vegetation and tree line, the fruit belts are also shifting upward gradually due to global warming in the watershed. The significant shifting was found in cultivation of citrus fruit. Normally the citrus belt in the study area is found in between 1000–1590 m but now this citrus belt

Table 7.3 List of villages Rajma and citrus fruit production which has been shifted after 2010 and old growing villages with their height (based on field survey and GPS, 2019 and 2023)

	Height group (m)	Sample village name and height (in m)	Status
Rajma Cultivation	1450–1800	Lodi (1454), Talla Bindi (1454), Sirtola (1456), Josha (1467), Jarajibli (1470), Ghangali (1532), Tanga (1537), Darma (1565), Khartoli (1592), Bata Dhura (1715), Dhunamani Talla (1707), Umdada (1709), Dolma (1756) and Golpha Talla (1797)	Old Rajma cultivation villages—14 (Before 2010)
	1800–2400	Dhunamani Malla (1910), Lodi Dhura (1933), Chulkote (1983), Golpha Malli (1985), Yangchaur (2052), Bona (2077), Gandhi Nagar (2115), Quiry Zimiya (2137), Paton (2159), Bothi (2167), Pal Dhura (2211), Nirtoli (2272), Malla Gaila (2332) and Tomic (2391)	New Rajma cultivation villages—14 (After 2010)
Citrus Fruit Cultivation	1000–1590	Mavani (1050), Fagua Bagar (1175), Timulkhet (1421), Bindi Talla (1454), Umta (1456), Lodi (1454), Rakhet (1507), Sana (1511), Barkhana (1529), Tanga (1537), Mulyanpani (1542), Waiga (1571), Rapti (1572) and Mapwalbada (1574)	Old citrus fruit cultivation villages—14 (Before 2012)
	1590–2000	Basant Kote (1590), Malla Bindi (1593), Umdada (1709), Bata Dhura (1715), Timphu (1741), Darma Malla (1751), Golfa (1804), Bhatkura (1821), Gaila Talla (1881), Malla Walthi (1885), Malla Josha (1921), Okhali (1895) and Chulkote (1983)	New citrus fruit cultivation villages—13 (After 2012)

has been shifted upwards up to 2000 m in the study area. Figure 7.7 depicts the distribution of old and new villages which has come after 2012. Table 7.3 presents the details of new villages which have started production of citrus after 2012 and also contains the list of villages producing citrus before 2012. These data reveals that the citrus fruit cultivation has been shifted about 100–400 m upward in the study area due to global warming. Due to increase temperature, the quality of citrus in term of its size, quality and taste has been gone down in old citrus growing areas while the quality of citrus fruit in term of its size, quality and taste is very good in the new citrus fruit growing areas due to ideal temperature in the higher elevations.

7.4.3 Causes of Shifting

As the pressure of population rapidly growing, plastic uses in large amount, more and more development of human concrete settlements, bitumen roads (Damar road), dams, tunnels, towers and other public utilities are increasing in the Himalayan watershed. Beside lopping of trees for fuel or fodder, overgrazing, increasing domestic and industrial consumptions of timber, increasing population in bugyal for herbal collection felling and burning of trees and shrubs for cooking and heating including the open burning of biomass, forest open area burning and firing in the alpine region during herbs collecting period are adversely affecting the natural environment. The Gori Ganga watershed has subsistence agriculturally based economy with high cropping cost and low productivity due to uncertainty of weather condition caused by climate change. Due to frequently occurring hydrological and climatic disasters increasing tendency of out-migration for searching better livelihoods and employment is increasing in the study area.

Plains associated with relief are more suitable for agriculture than the highlands. Moreover, it turns into less difficult to apply current device on flat agriculture fields, while it's miles tough to apply machines on rugged topography (Shah and Vijayshankar 2022). Gori Ganga watershed has more than 80% area covered by above 2000 m topography which is not very suitable situation for crop cultivation. One of the main challenges of mountain agriculture is the steep slopes, which can make it difficult to cultivate crops and raise livestock. Mountainous terrain often presents challenges such as steep slopes, which can affect irrigation, soil erosion, and accessibility. Terracing, contour farming, and other soil conservation measures can be used to mitigate these challenges and create suitable areas for agriculture (Chapagain and Raizada 2017). It is apparent from the experimental results that the slope steepness significantly affected grain yields of all the three crops, Mung, Mash and Millet (Ahmad et al. 2022). Yields were declined with increasing steepness. Gori Ganga watershed has more than 75% area under moderate to very steep slope conditions which is not so much favorable for good cultivation. Pook and Moore (1966), Cantlon (1953) said, in general, the polar-facing slopes (Northern Slope Aspect) are wet and cool, with rich organic matter and deeper soil, and associated with mesic vegetation (*low tolerance for drought, instead they prefer the continuity of constant irrigation*), while the equator-facing slopes (Southern Slope Aspect) are hot and dry, with low soil nutrient and severe soil erosion, and are occupied by xeric vegetation (*all sorts of varied trees, shrubs, perennials, and even many bulbs, which need a dormant, dry season each summer to recharge*). Gori Ganga watershed has more than 30% area under (Northern, North-eastern and North-western) and more than 40% area under (Southern, South-Eastern and South western). Where find similar climatic conditions for cultivations. According to i-kisan Agri Infomatics and Services, "Citrus species do best in regions with an annual rainfall of 1250–1850 mm. As much, in arid and semi-arid areas, an adequate irrigation water supply is required for maintaining desirable soil moisture conditions." On the basis of above-mentioned rainfall condition for citrus fruit cultivation study area had received more than rainfall in the year 2011, 2012, 2013, 2019, 2020, 2021 and 2022 (Fig. 7.6).

7.5 Conclusion

Global warming is causing the melting of snow and glaciers and upward shifting vegetation/timber line in the Himalaya region. Therefore, the qualitative analysis reveals the urgency to strengthen climate data collection network to meet research needs on various aspects of climate change impacts, mitigation and adaptation. Application of remotely sensed data made possible to study the impact of global warming and climate change in less time, at low cost and with better accuracy in combination with Geo-informatics which provides a suitable platform for data analysis, updating and retrieval. Geo-informatics-based studies help in formulate of policies for better decision making, planning, development and management future planning of natural as well as of human resources.

Acknowledgements We are highly thankful to acknowledge the help of numerous researchers, institutes, organizations especially NRDMS Almora for providing Cartosat-1 image of the study area and information officers who provided valuable publications.

Funding Sources The author(s) received no financial support for the research, authorship or publication of this article.

Conflict of Interest The author(s) declares no conflict of interest.

Data Availability Statement The datasets analyzed during the current study are available from the corresponding author on reasonable request.

References

Ahmed S, Umair A, Rahim M, Yunas M, Waheed A, Ali S, Farooqi MH (2022) Slope steepness and crop vegetative cover impact on soil and water losses under high rainfall zone of Pothowar. Nat Volatiles Essent Oils 9(2):871–878. https://www.nveo.org/index.php/journal/article/view/5282/4171

Baker BB, Moseley RK (2007) Advancing tree line and retreating glaciers: implications for conservation in Yunnan, Province Reason. China. Arct, Antarct Alp Res 39(2): 200–209. https://doi.org/10.1657/1523-0430(2007)39[200:ATARGI]2.0.CO;2

Cantlon JE (1953) Vegetation and microclimates on north and south slopes of Cushetunk Mountain, New Jersey. Ecol Monogr 23(3):241–270. https://doi.org/10.2307/1943593

Chapagain T, Raizada M (2017) Agronomic challenges and opportunities for smallholder terrace agriculture in developing countries. Front Plant Sci 8, Article 331. https://doi.org/10.3389/fpls.2017.00331

Crenshaw EM, Jenkins JC (1996) Social structure and global climate change: sociological propositions concerning the greenhouse effect. Sociol Focus 29(4):341–358. https://doi.org/10.1080/00380237.1996.10570650

Dirnbock T, Dullinger S, Grabherr G (2003) A regional impact assessment of climate and land-use change on alpine vegetation. J Biogeogr 30:401–417. https://doi.org/10.1046/j.1365-2699.2003.00839.x

IPCC (2018) Global warming of 1.5 °C: an IPCC special report on the impacts of global warming of 1.5 °C above pre-industrial levels and related global greenhouse gas emission pathways, in

the context of strengthening the global response to the threat of climate change, sustainable development, and efforts to eradicate poverty. Intergovernmental Panel on Climate Change.
IPCC (2019) IPCC Special Report on the Ocean and Cryosphere in a Changing Climate (SROCC). In: Portner HO, Roberts DC, Masson-Delmotte V, Zhai P, Tignor M, Poloczanska E, Mintenbeck K (eds) Intergovernmental Panel on Climate Change, Geneva
Korner C (2004) Mountain biodiversity, its causes and function. AMBIO: A J Hum Environ 33(sp13):11–17. https://doi.org/10.1007/0044-7447-33.sp13.11
Kothawale DR, Kumar KR (2005) On the recent changes in temperature trends over India. Geophys Res Lett 32:1–4
Lever-Tracy C (2008) Global warming and sociology. Curr Sociol 56(3):445–466. https://doi.org/10.1177/0011392107088231
Malhi GS, Kaur M, Kaushik P (2021) Impact of climate change on agriculture and its mitigation strategies: a review. Sustainability 13(3):1318. https://doi.org/10.3390/su13031318
McGrath M (2022) Climate change: key crops face major shifts as world warms. Access 11 October 2023
Montenegro AAA, De Lima JLMP, Abrantes JRCBO, Santos TEM (2013) Impact of mulching on soil and water conservation in semiarid catchment. Die Bodenkultur 64(3–4):79–85
NHB-National Horticulture Board, Apple (2012) Access 11 October 2023
Owens C (2021) How changing climates are affecting global fruit production areas that are being impacted most severely have historically seen significant effects from climate change. Access 11 October 2023
Parihar DS (2021a) Due to global warming: snow line dynamics in the Gori Ganga watershed, Kumaun Himalaya by using RS & GIS. Int J Ecol Environ Sci 3(1):226–233
Parihar DS (2021) Dynamics of timberline due to Spatio-temporal changes using GIS and RS in the Gori Ganga watershed, Kumaun Himalaya, Uttarakhand. SSRG Int J Geo-Inf Geol Sci 8(1):42–47. https://doi.org/10.14445/23939206/IJGGS-V8I1P105
Parihar DS, Rawat JS (2021) Spatio-temporal change in snow cover area using RS & GIS in the Gori Ganga watershed, Kumaun Himalaya. Int J Adv Res 9(3):30–34. https://doi.org/10.21474/IJAR01/12550
Parihar DS, Rawat JS, Singh Mahendra, Pant NC (2021) Spatio-temporal change of vegetation line in the Gori Ganga watershed, Kumaun Himalaya by using Remote Sensing and GIS techniques. Int J Trend Sci Res Dev 6(1):409–416. https://www.ijtsrd.com/papers/ijtsrd47806.pdf
Parihar DS, Tamta MK, Deepak (2020) Detailed study of agriculture and types of land in higher Himalayan Region: a case study of Tehsil Munsyari. Int Res J Manag Sociol Hum 11(1):127–138
Pook E, Moore C (1966) The influence of aspect on the composition and structure of dry sclerophyll forest on Black Mountain, Canberra. ACT. Aust J Bot 14(2):223–242. https://doi.org/10.1071/BT9660223
Ramakrishna A, Tam HM, Wani SP, Long TD (2006) Effect of mulch on soil temperature, moisture, weeds infestation and yield of groundnut in Vietnam. Field Crop Res 95(2/3):115–125. https://doi.org/10.1016/j.fcr.2005.01.030
Shah M, Vijayshankar PS (2022) Symbiosis of water and agricultural transformation in India. In: Chand R, Joshi P, Khadka S (eds) Indian agriculture towards 2030. India studies in business and economics. Springer, Singapore, pp 109–152. https://doi.org/10.1007/978-981-19-0763-0_5
Thapa S (2023) Caste, gender, climate change: farmers in Uttarakhand struggle for survival. Climate Emergency, Indian Development Review (IDR). Access 11 October 2023
UNDP (2006) Human development report. Beyond scarcity: power, poverty and the global water crisis. United Nations Development Programme, New York
Wilkes A (2008) Towards mainstreaming climate change in grassland management policies and practices on the Tibetan Plateau. Working Paper No. 67. World Agro-forestry Centre, ICRAF-China, Beijing

Xu JC, Wilkes A (2004) Biodiversity impact analysis in Northwest Yunnan, Southwest China. Biodivers Conserv 13(5):959–983
Zehr S (2015) The sociology of global climate change. Wiley Interdisc Rev Clim Change 6(2):129–150. https://doi.org/10.1002/wcc.328

Chapter 8
Conservation Challenges and Adaptation Strategies for Indian Himalayan Biodiversity in a Changing Climate

Tridipa Biswas, Sivaranjani Subramanian, Rajashekhar Niyogi, Wishfully Mylliemngap, Sohini Mitra, Dhiraj Kumar Das, and Mriganka Shekhar Sarkar

Abstract The Indian Himalayan Region (IHR), renowned for its rich biodiversity and cultural diversity, confronts formidable challenges from ongoing climate change, habitat degradation, and human activities. This review-based study reveals the profound vulnerability of the IHR to climate change impacts, including rising temperatures, glacial retreats, and altered precipitation patterns, threatening the region's unique flora, fauna, and ecosystem services. However, amidst these challenges, we identify key adaptation strategies crucial for conserving Himalayan biodiversity. Strengthening protected areas and corridors, empowering local communities through community-based conservation initiatives, and implementing sustainable land-use planning emerge as pivotal measures to mitigate threats and foster resilience. Additionally, leveraging technological innovations such as remote sensing and conservation genetics offers promising avenues for monitoring biodiversity changes and implementing proactive conservation measures. We emphasize the imperative of collaborative action, robust policy frameworks, and interdisciplinary approaches to safeguard the invaluable natural and cultural heritage of the IHR and build resilience against ongoing environmental changes.

T. Biswas · S. Subramanian · W. Mylliemngap · M. S. Sarkar (✉)
North-East Regional Centre, G.B. Pant National Institute of Himalayan Environment, Chandranagar, Itanagar 791113, Arunachal Pradesh, India
e-mail: mriganka.bio@gmail.com

R. Niyogi
Department of Biological Sciences, Indian Institute of Science Education and Research, Kolkata 741246, Mohanpur, India

S. Mitra
Adamas University, Barasat, Jagannathpur, Kolkata 700126, India

D. K. Das
Department of Zoology, Assam Don Bosco University, Tepesia, Sonapur 782402, India

D. Arya et al. (eds.), *Climate Change Impact on Himalayan Biodiversity*, Environmental Science and Engineering, https://doi.org/10.1007/978-3-031-77149-1_8

Keywords Himalayan biodiversity · Climate change · Adaptation strategies · Conservation challenges · Community-based conservation · Sustainable land-use planning · Technological innovations

8.1 Introduction

The Himalayas stand as the loftiest and most expansive mountain range globally. Boasting 15,000 glaciers, the world's highest concentration, they serve as the wellspring for Asia's eight largest rivers, supplying freshwater to plains and arid regions (Xu et al. 2009; Shrestha et al. 2012). The Indian Himalayan Region (IHR) encompasses nine states, two Union Territories, and two partially covered hill areas within plain-dominating states. These nine states, from northwest to northeast, include Himachal Pradesh in the northwest; Uttarakhand in the Central Himalayas; and Sikkim, Arunachal Pradesh, Nagaland, Manipur, Mizoram, Tripura, and Meghalaya in the northeast Himalayas. Additionally, there are two hilly districts in West Bengal and Assam in the northeastern part of the Himalayas, as well as the recently formed Union Territories of Jammu & Kashmir and Ladakh in the western part of the Indian Himalayas (Fig. 8.1). The IHR is a global biodiversity hotspot, harboring a staggering array of flora and fauna (Chitale et al. 2015). The region is home to numerous endemic species, rare and endangered animals like the snow leopard, bengal tiger, red panda, and a wide variety of plant life. Its diverse ecosystems, ranging from alpine meadows to temperate forests and subtropical jungles, contribute to its ecological significance. IHR is also inhabited by diverse ethnic groups, each with its own language, culture, and traditions (Berreman 1963). Communities have adapted to the challenging mountain environment over centuries, relying on agriculture, pastoralism, and trade for sustenance. Residents of environmentally delicate areas, such as the IHR, bear a greater brunt of the escalating vulnerabilities induced by climate change (Rattani 2018; Yadav and Singh 2024). Variations in the yearly rainfall cause the higher out-migration rate in their study conducted in the Hindu Kush Himalayan (HKH) region. Furthermore, this region's agricultural output is declining due to extreme temperatures, directly impacting household food security (Hussain et al. 2016; Ghosh et al. 2021). Agriculture, tourism, and traditional practices, fuelled by the region's breathtaking scenery and rich cultural heritage, are pivotal to local economies, offering employment and income prospects.

However, as per the available reports, IHR harbors about 8,000 species of Angiosperms (40% endemics), 44 species of Gymnosperms (15.9% endemics), 600 species of Pteridophytes (25% endemics), 1,737 species of Bryophytes (32.5% endemics), 1,159 species of Lichens (11.2% endemics) and 6,900 species of Fungi (27.4% endemics) (Singh and Hajra 1996). The IHR is geographically divided into Eastern and Western Himalayas, each boasting its unique biodiversity significance. The Eastern Himalaya stands out as a critical area within various ecological classifications, including 'crisis ecoregions,' 'biodiversity hotspots,' 'endemic bird areas,' 'mega diversity countries,' and 'global 200 ecoregions' (Brooks et al. 2006). This

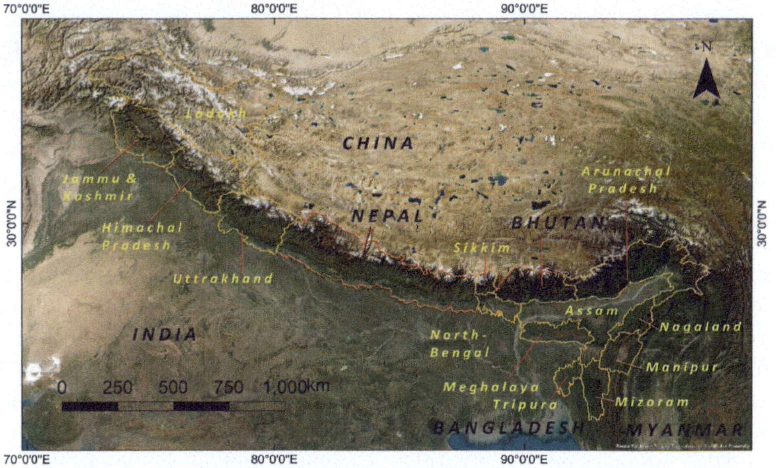

Fig. 8.1 The study area map of the Indian Himalayan Region

region serves as a convergence point for the Indo-Malayan, Palaearctic, and Sino-Japanese biogeographical realms, characterized by diverse ecological and altitudinal gradients, fostering a rich diversity of flora and fauna (CEPF 2005, 2007). Moreover, the Eastern and Western Himalaya forests play a crucial role in sustaining numerous rivers, which serve as lifelines for downstream provinces and countries. Beyond supplying water and supporting biodiversity, these rivers and landscapes offer invaluable ecosystem services such as soil retention, climate regulation, carbon sequestration, and acting as reservoirs of pollinators and natural predators. The well-being of millions of people downstream is intimately connected with the natural resources of IHR.

In the past few decades, the Himalayan ecosystems have been exceptionally vulnerable to the impacts of climate change (Chaturvedi et al. 2011). Recent studies have indicated a notable increase in the rate of warming in the Himalayas over the past decade, ranging from 0.3 to 0.5 °C (Diodato et al. 2012). The climate projections suggest that at higher altitudes and latitudes within the Himalayan region, temperatures could rise by 5–6 °C, accompanied by a 20–30% increase in rainfall by the end of the twenty-first century (Solomon 2007; Kohler and Maselli 2009). Notably, the temperature rise is more pronounced in the upland areas than in the lowlands of the Himalayas (Shrestha et al. 1999). Additionally, the Himalayas represent a significant global hotspot for black carbon emissions, ranking as the region's second-largest contributor to atmospheric warming following carbon dioxide. The combined effects of black carbon and greenhouse gases contribute to a warming trend of 0.25 °C per decade in the Himalayas. This heightened warming has led to an increase in the frequency of flooding events from glacial lakes. Records indicate thirty-four instances of glacial lake outburst floods on the northern slopes of the Himalayas in the past 50 years (Kohler and Maselli 2009; Chaturvedi et al. 2011; Padma 2014; Salick et al. 2014; Tewari et al. 2017). The Himalayan region's vulnerability to climate change is further compounded by stresses such as deforestation, anthropogenic pressures, exploitation of natural resources, and infrastructural developments (Ravindranath et al. 2006). In the western Himalayas, (Dimri and Dash 2012) have observed several changes in surface air temperature, including an increase overall, a rise in the number of warm days, and a decrease in cold days, alongside a slight uptick in precipitation with inconsistent patterns between 1975 and 2006. Additionally, there was a cooling trend before the monsoon in the latter part of the twentieth century (Yadav et al. 2004). From 1986 to 2000, there was a decline in spring snow cover and a faster rate of snow melting from winter to spring post-mid-1990s in the western Himalayas (Kripalani et al. 2003). Another study examined the impacts of climate change on plant pheno-phases and their influence on the traditional healthcare system in Uttarakhand Himalaya. Their study, based on perceptions of Vaidyas (traditional healers), indicates a slight shift in the collection period of medicinal plants, contrary to tradition, leading healers to identify substitute plant species for treating ailments. Kumar (2012) emphasized that 13–16% of medicinal plant species in the Sikkim Himalaya may lose their habitats by 2050 and 2070. Species with specific localities and narrow elevational distributions are particularly vulnerable and face extinction due to climate change in the Himalayas. Climate change has recently impacted various aspects of plant life, including phenology, life cycle, reproductive success, and distribution patterns. High-temperature stress has further affected the survival, phytochemical composition, and production of secondary metabolites and volatile organic compounds in valued medicinal plants across the Indian Himalayan Region.

With the above context, the study encompasses a comprehensive review and documentation of the major conservation challenges facing the IHR, including loss of biodiversity and habitat degradation. The current overview of the threatened species and ecosystems, coupled with the repercussions on ecosystem services, underscores the urgent need for intervention. Climate change exacerbates these challenges, with rising temperatures leading to shifts in species distribution, glacial retreat impacting freshwater ecosystems and dependent species, and altered precipitation patterns affecting vegetation and wildlife while increasing vulnerability to extreme weather events. We documented major adaptation strategies to address these issues, focusing on existing protected areas and corridors, community-based conservation initiatives, and sustainable land-use planning to balance development and conservation efforts while mitigating habitat fragmentation, documenting the effective policy frameworks and governance, both nationally and internationally, alongside ongoing cross-border collaborations to achieve conservation goals. Furthermore, technological innovations such as remote sensing, conservation genetics, and early warning systems offer valuable tools for monitoring current day biodiversity changes and implementing proactive conservation measures and adaptation strategies that safeguard the IHR biodiversity in the future.

8.2 Conservation Challenges in the Himalayan Region

8.2.1 Loss of Biodiversity

8.2.1.1 Threatened Species and Ecosystems

As per a survey conducted by UNESCO in 2021, climate change and biodiversity depletion stand out as the predominant global challenges of the decade. The diminishment of biodiversity results from a confluence of natural and anthropogenic influences, encompassing the proliferation of invasive species, shifts in climate patterns, overexploitation of species, degradation of forests, and the obliteration of unique habitats (Butchart et al. 2010). Notably, the latter half of the previous century witnessed an alarming pace of plant extinction, marked by the loss of 137 species daily (Mora et al. 2011), a rate estimated to be 1000–10,000 times swifter than natural occurrences (Hilton-Taylor and Brackett 2000).

The IHR is well-known for its rich floral and faunal diversity, supporting nearly 50% of the total flowering plants (Singh 2006) and 31% of faunal species in India (Chandra et al. 2019). Despite this rich diversity, this Himalayan biodiversity hotspot is also highly vulnerable, situated in a tectonically active zone and due to anthropogenic disturbances. In the IHR, wildlife species face threats from illegal wildlife trafficking, poaching, human-wildlife conflicts, habitat loss, fragmentation, and degradation from human development and extraction of natural resources (Gaston et al. 1983; Sathyakumar and Bashir 2010). These have led to a decline in wildlife population, reduction in distribution range, and local extinction of species in some cases. Several studies have been conducted to gather information on the threatened plants of IHR (Kala 2000; Aldam 2002; Badola and Mohinder Pal 2003; Jishtu and Goraya 2008; Prabhu 2008; Rawat et al. 2016; Barik et al. 2018a, b). A list of 175 mammal species and more than 500 bird species are known from the Himalayan biodiversity hotspot (WWF 2001). Of the recorded mammal species, 45 were listed under various threat categories (14 endangered, 29 vulnerable, and two critically endangered) (CEPF 2005; Chettri et al. 2010).

Barik et al. (2018b) documented a comprehensive inventory comprising 2,704 plant species facing threats within India, distributed across 1,031 genera and 217 families. Mehta et al. (2020) compiled information regarding endangered species in the IHR by synthesizing data from three main sources: IUCN (2016), CAMP (Ved et al. 2003a, b) and the Red Data Book (Nayar and Sastry 1987, 1988, 1990). Their compilation identified 713 threatened taxa, encompassing 383 genera and 126 families. This assemblage comprised 663 angiosperms, 24 gymnosperms, 22 pteridophytes, and 4 bryophytes. Notably, 64 species were classified under the higher risk of extinction categories within these threatened taxa according to IUCN standards, comprising 13 as critically endangered, 28 as endangered, and 23 as vulnerable.

8.2.1.2 Impact on Ecosystem Services

Ecosystem services (ES) denote the broad spectrum of advantages humanity gains from the natural environment, including tangible resources like food and water and intangible values such as air purification and cultural significance. Nonetheless, in recent years, the deterioration of ecosystem services has arisen as a notable worldwide apprehension (Singh et al. 2021a). The global devaluation of ecosystem services resulting from land-use alterations has been approximately $20 trillion annually from 1997 to 2011 (Mondal and Zhang 2018). The Hindu-Kush Himalaya (HKH) region, spanning eight countries including Afghanistan, Bangladesh, Bhutan, China, India, Myanmar, Nepal, and Pakistan, earns its name as "The Third Pole" due to its vast reserves of frozen water, after Antarctica and the Arctic regions. This region's rich biodiversity gives rise to numerous ESs (Xu et al. 2019), benefiting mountain communities and those downstream or in lowland areas. Forest ecosystems play a crucial global role, contributing to national economies through tangible services like employment generation and increased local incomes (Patterson et al. 2017; Shekhar 2021). Yet, the IHR faces threats from anthropogenic pressures such as unsustainable resource extraction, urbanization, and inadequate waste management, resulting in biodiversity loss, reduced productivity, and disrupted service provision (Kumar et al. 2019; Rani et al. 2021). In the HKH region, preserving the flow of ecosystem services presents a critical challenge amidst rapid habitat degradation and insufficient biodiversity protection measures (Xu et al. 2019). Over the past century, warming climates have severely impacted vital river basins in the region, such as the Gangetic Plain, the Indus Delta, the Yellow, and the Yangtze, all of which are crucial for regional and global food and water security (Zomer and Sharma 2009). The swift melting of glaciers and diminished ice storage capacities have resulted in heightened meltwater runoff during summer monsoons and reduced runoff during dry seasons. This phenomenon has contributed to water-related hazards and alterations in streamflow patterns (Zomer and Sharma 2009; Xu et al. 2009). These changes pose uncertainties in future water resource management due to precipitation variations (Immerzeel et al. 2020; Nie et al. 2021).

Functional plant traits, such as morphology, phenology, and physiology, directly influence ecosystem functioning by mediating biotic interactions and responding to environmental changes (Wood et al. 2015). However, the ecosystem functioning in the Himalayas is constantly threatened by climate change and other human disturbances, which are causing the alteration of ES in IHR. Given the challenges posed by climate warming and their profound impacts on food and water security in the Himalayan region, it becomes crucial to manage key ES on time, considering their functional characteristics. Moreover, the interdependent relationship between tourism and the environment in the IHR highlights the necessity of conserving cultural ES, which has historically played a vital role in socio-economic development (Goodwin 1996; Waitt et al. 2003).

8.2.2 *Habitat Degradation*

8.2.2.1 Deforestation and Land-Use Changes

The IHR faces significant conservation challenges due to habitat degradation, another prominent cause driven by deforestation and land-use change (Tiwari 2000; Pandit et al. 2014; Mondal and Zhang 2018; Rana et al. 2021). The fragile ecosystem of the Himalayan mountains exhibits forest fragmentation due to natural and anthropogenic disturbances (Chakraborty et al. 2017). Forest fragmentation is a major conservation issue, contributing to a general decrease in forest cover and an increase in agriculture expansion and urbanization (Roy et al. 2013; Reddy et al. 2013; Chakraborty et al. 2017). Over 75% of the original Himalayan habitat has been destroyed or degraded, with causes including fuelwood and fodder collection, extensive livestock grazing, and rapid development, leading to habitat fragmentation and population loss (WWF 2024). Deforestation in the Himalayas started during British rule in the nineteenth century, and commercial exploitation of forests has continued unabated across the Himalayan nations in the last three to four decades. A study has projected that deforestation will reduce the forest cover, resulting in widespread species extinctions in the Himalayan region, where plant species endemism is high (Pandit et al. 2014). Deforestation in IHR is primarily due to anthropogenic activities such as agriculture expansion, infrastructure development, and logging, which not only directly diminish forest cover but also disrupt crucial habitats for numerous species, including endemic and endangered ones like the Himalayan musk deer and the snow leopard (Pandit et al. 2014; Rana et al. 2021). Deforestation also leads to soil erosion, loss of biodiversity hotspots, and disruption of critical ecosystem services such as water regulation and carbon sequestration. Land-use changes, particularly the conversion of forests into agricultural lands or urban areas, further exacerbate habitat fragmentation and loss, limiting the movement and dispersal of wildlife populations (Watt 2020). According to the ICIMOD assessment report, 70–80% of the region's original habitat has already been lost, and this may increase to 80–87% by 2,100. Additionally, a quarter of endemic species in the Indian Himalayas alone could be wiped out by 2,100. The report also emphasizes that climate change, along with new infrastructure development, trade routes, and hydropower dams, is set to exacerbate the loss of habitat and biodiversity in the fragile region (Wester et al. 2023; The Third Pole 2019).

8.2.2.2 Infrastructure Development and Its Implications

The IHR has been undergoing rapid infrastructure development. While infrastructure projects promise economic growth and improved connectivity, they often come at a significant environmental cost. Habitat degradation resulting from infrastructure development poses a grave threat to the delicate ecosystems of the Himalayas, with far-reaching implications for both the environment and human communities dependent on its resources (Stanturf 2015).

Despite international initiatives such as the United Nations Framework Convention on Climate Change (UNFCCC), the United Nations Convention to Combat Desertification (UNCCD), the Convention on Biological Diversity (CBD), and Disaster Risk Reduction (DRR), aiming to tackle climate change, combat landscape degradation, and conserve biodiversity, degradation persists, leading to severe consequences like land use changes and habitat destruction (Parrotta et al. 2012; Mansourian and Parrotta 2019). habitat degradation exacerbates the vulnerability of local communities to natural disasters such as landslides, flash floods, and avalanches (Nyaupane and Timothy 2022). Deforestation destabilizes slopes, increasing the risk of landslides during heavy rainfall or seismic activity. Similarly, the alteration of river courses and the construction of dams can amplify the impact of floods downstream, jeopardizing lives and livelihoods. Figure 8.2 provides a glimpse of the human footprint that disrupts Himalayan ecosystems.

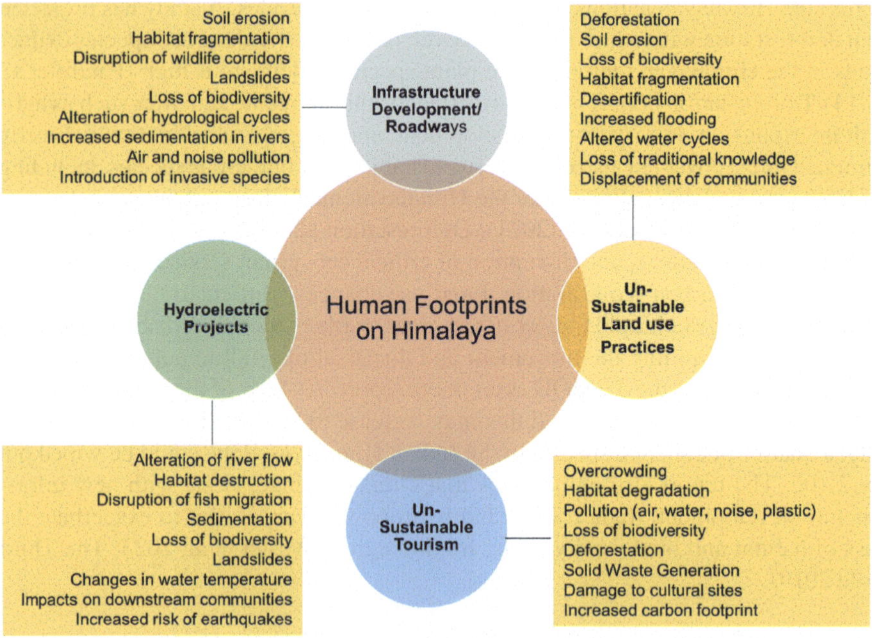

Fig. 8.2 Infrastructure developmental projects, hydroelectric projects, unsustainable tourism and unsustainable land use practices causing damage to Himalayan ecosystems (Chandy et al. 2012; Diduck et al. 2013; Batar et al. 2017; Pathak et al. 2019)

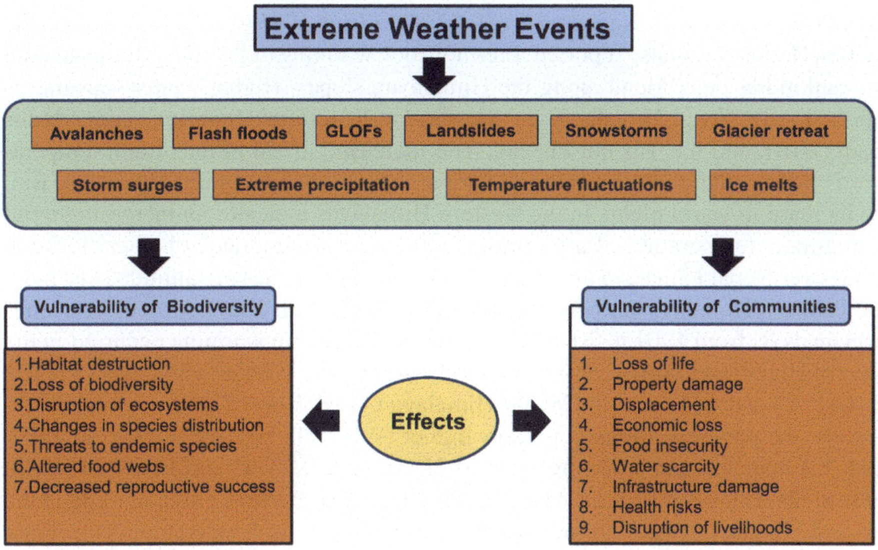

Fig. 8.3 The effects of extreme weather events on Himalayan biodiversity and indigenous communities

8.3 Climate Change Impacts on Himalayan Biodiversity

8.3.1 Temperature and Precipitation Patterns Influence Life in Himalaya

8.3.1.1 The Effects of Rising Temperatures and Altered Precipitations on Species Distribution

The IHR is experiencing substantial challenges due to climate change, notably with rising temperatures that are affecting the environment, ecosystems and local communities (Eriksson et al. 2009; Shrestha et al. 2012; Tewari et al. 2017; Dimri et al. 2021; Banerjee et al. 2022). Scientific studies have revealed a concerning warming trend in the Himalayas, with temperature increases observed at varying rates across different altitudes and regions (Shrestha et al. 2012; Tewari et al. 2017; Dimri et al. 2021). Also, some places in the IHR recorded temperatures higher than the global average (Dimri et al. 2021; Banerjee et al. 2021). For instance, a study reports that the temperature in the Himalayas rose by 1.5 °C between 1982 and 2006, with an average rate of 0.06 °C per year, which is considerably higher than the global average increase during a similar period. This warming trend is consistent with previous localized studies across different parts of the Himalayas, indicating warming trends ranging from 0.03 to 0.06 °C per year (Shrestha et al. 2012). Another study showed that increasing temperature trends of 0.12 and 0.4 °C/yr are reported over the middle Mountains and

Himalaya, respectively. An increase of 0.03 °C/yr from 1985 to 2002 in Bhutan in the Eastern Himalayas is also reported. Pre-monsoon warming of 2.7 °C/yr due to aerosol interaction has been found along the Himalayan slopes. Higher winter warming is reported over Nepal Himalaya, the western/northwestern Himalaya, upper Indus basin (UIB), and the Tibetan Plateau. The increasing trend in maximum temperature is higher than that of minimum temperature over the western Himalaya. But, the increase in warm nights in the western Himalayas indicates an increasing trend in minimum temperatures. Furthermore, a greater warming rate at higher elevation zones over Nepal Himalaya and greater warming with increasing altitude over India and Pakistan Himalaya has been documented (Dimri et al. 2021). Moreover, a long-term analysis from 1901 to 2016 shows that the maximum warming occurred in the Eastern Himalaya (74.67% or 0.93 °C), followed by the Western Himalaya (52.56% or 0.64 °C) and minimum in Central Himalaya (44.31% or 0.73 °C) regions. Additionally, the winter warming is notably higher (WH: 1.11 °C, CH: 1.19 °C and EH: 1.41 °C) than summer warming (WH: 0.31 °C, CH: 0.26 °C and EH: 0.54 °C) in these areas (Garg et al. 2022). Also, a study found a wintertime climatic trend in the western Himalayas, with an increasing trend in minimum temperature (Dimri and Dash 2012).

Research indicates a complex scenario with varying trends, including both increasing and decreasing precipitation in different parts of the Himalayas (Sabin et al. 2020). The IPCC 2007 describes increasing trends of precipitation in northern and central Asia and predicts that average annual precipitation will increase by 10–30% on the Tibetan Plateau as a whole by 2080. Similarly, a net 60–206 mm (5–13%) increase in precipitation over the simulated 1970s rainfall in the Indian Himalayas by 2030s is expected (Shrestha et al. 2012). The Himalayan region is also experiencing unpredictable rain patterns, with the amount of rain remaining the same, but the number of days when it rains decreases, leading to more intense rainfall events and flooding (Saini et al. 2023). Rapid glacier melting, glacial lake outburst flooding, and changing rain patterns are causing significant challenges for local communities, farmers, and ecosystems.

The rising temperatures in the Indian Himalayan region are posing a severe threat to the biodiversity of this unique ecosystem (Shrestha et al. 2012). The intricate web of plant and animal life in the Himalayas is being disrupted by climate change, leading to habitat loss, population declines, and altered ecological dynamics. Various studies have highlighted the cascading effects of temperature rise on biodiversity, including changes in species composition, reduced genetic diversity, and increased vulnerability to invasive species (Sharma et al. 2009; Tewari et al. 2017). The warming temperatures are causing a series of ecological changes threatening the delicate balance of biodiversity in this sensitive ecosystem (Sharma et al. 2008). This increase in temperature has triggered changes in the distribution and behavior of various species, affecting their survival and reproductive patterns (Tewari et al. 2017). The latest report from the International Centre for Integrated Mountain Development (ICIMOD) highlights that rising temperatures, even if limited to a 1.5 ℃ increase, significantly impact biodiversity and ecosystem services in the HKH region (Down to Earth 2023). While some species may benefit from the rising temperatures, the majority are experiencing negative effects. The changing climate in the Himalayas is altering the phenology of plant species, affecting their flowering and fruiting patterns and timing of leaf-fall, leading to a decrease in plant survival and threatening species' vulnerability. For instance, alpine ginger Roscoea species in the western Himalayas are experiencing early flowering, while the Himalayan rhododendron is showing advanced and delayed flowering in different regions (Down to Earth 2023; Singh et al. 2011).

The warming temperatures also affect the fauna of the Himalayan region. The declining fauna in the region is also a cause for concern, with mammals, insects, microbes, birds, amphibians, and fishes becoming extinct or experiencing genetic and behavioural changes. Species like the snow leopard, Himalayan musk deer, golden snub-nosed monkeys and Himalayan grey langurs have already experienced range shifts with declining populations due to the changing climate. The upward movement of the snowline is particularly concerning for snow leopard habitats, with predictions indicating potential habitat loss by 2080 (Chaudhary et al. 2023; Down to Earth 2023). The loss of glacier-fed streams and rivers threatens the survival of cold-water species like trout, adapted to specific temperature ranges (Kattel 2022). Research highlights that changes in the cryosphere have already impacted ecosystems and biodiversity. Future estimates predict a significant decline in habitat for various species in the Himalayas. The habitat for the Himalayan grey langur is forecasted to decrease by over 60% by 2050. Similarly, the Himalayan and Tibetan brown bears are expected to lose habitats by 34 and 32%, respectively. The blue sheep may experience a decline of 56–58%, while the habitat for the Himalayan ibex is likely to reduce by 33.7 to 64.8% in the future. These projections indicate a concerning trend of habitat loss for several species in the Himalayan region (Chaudhary et al. 2023). Birds like the satyr tragopans have shifted their habitat to higher elevations, shrinking their range. Black-necked cranes have been observed to adjust their behavior in terms of incubation based on weather conditions and the thermal needs of their eggs (Down to Earth 2023). Climate change poses a significant threat to amphibians, with numerous studies highlighting their vulnerability (Blaustein et al. 2010; Steigerwald 2021). Among the most affected are the Kashmir paa frog and the Himalaya paa frog, both endemic to Pakistan. These species are experiencing deformities, reduction in body size, altered metamorphosis, and increased mortality due to climate change (Chaudhary et al. 2023). The Sikkim paa frog has also shown an advancement in breeding by three months (Acharya and Chettri 2012). Insects and other invertebrates are not immune to the effects of climate change in the Himalayas. The monotonic ants in the Eastern Himalayas, as well as caterpillar fungus across the region, are facing population declines (Down to Earth 2023; Hopping et al. 2018). Moreover, new species are emerging, and altitudinal changes are being observed in various insect populations, indicating a significant ecological shift. Furthermore, the introduction and spread of invasive species are also threatening the health, sustainability, and productivity of native ecosystems in the region (Thapa et al. 2018). Invasive species like *Ageratina adenophora* and *Lantana camara* are projected to increase their spread by 45.3 and 29.8%, respectively, under the Representative Concentration Pathway (RCP) 8.5 scenario, particularly in the Kailash Sacred Landscape of India, which is above 5,000 m asl. This increase in the spread of invasive species is expected to have detrimental effects on biodiversity, food security, and economic losses. In fact, 75% of 26 invasive plant species are predicted to expand their range, while 25% will contract, further threatening the delicate balance of native ecosystems (Gupta et al. 2021; Chaudhary et al. 2023). As well, local communities in the Indian Himalayan region are particularly vulnerable to these changes. The impacts on agriculture are substantial, with heat stress affecting crop yields and food production declining in some areas (Dahal et al. 2022). Indigenous communities reliant on traditional farming practices face challenges adapting to these changing conditions, jeopardizing food security and livelihoods.

8.3.1.2 Shifts in Species Distribution

During the last century, the global average temperature increase has initiated shifts in species geographic range and extinctions. Montane flora and fauna are highly sensitive to the changing climate (Telwala et al. 2013). Climate change will likely affect the persistence of large, space-requiring species through habitat shifts, loss, and fragmentation. Climate-induced range shift is the most documented fingerprint of climate change, and such shifts in forest ranges have probably already started the range shift processes among birds and large mammals associated with these forest types (Forrest et al. 2012). Elevation does not directly affect plant physiology but indicates that its association with climatic variables influences it (Bobrowski et al. 2021).

Increasing temperature and changes in hydrological regimes will significantly affect the habitat suitability for many species, like Himalayan dragonfly (*Epiophlebia laidlawi*). According to predictions, the suitable habitat area of the Himalayan dragonfly will shift towards uphill by 2050 and 2080 due to climate change (Shah et al. 2012). It has been predicted that about 30% of snow leopard habitat in the Himalayas may be lost due to a shifting treeline and consequent shrinking of the alpine zone, mostly along the southern edge of the range and in river valleys. The habitat of snow leopards in Bhutan, Myanmar, and Nepal is likely to be pushed northward toward the China border, while local populations in eastern China and northern India may become completely isolated. As a result of the shift in the forest, the habitat of snow leopards will likely be occupied by common leopards (*Panthera pardus*), wild dogs (*Cuon alpinus*), and in Bhutan, tigers (*Panthera tigris*), which will lead to resource competition among these species (Forrest et al. 2012). The Monocled and King cobras in the Sikkim Himalayas have shifted to higher elevations, moving from 1,000 to 1,700 m asl (Bashir et al. 2010). According to another research report, due to climate warming, Satyr Tragopan (*Tragopan satyra*) in the Himalayas are gradually shifting their habitat towards higher altitudes and latitudes. It also predicted that under different concentrations of greenhouse gas emission, the suitable habitat range of Satyr Tragopan will shrink, and the most suitable habitat would be lost in low altitude and latitude landscapes under future climate conditions. The maximum total preferable habitat of Satyr Tragopan (*Tragopan satyra*) will also decrease by

11,446.96 km^2 or 1.56% of the present suitable habitat area under the RCP6.0 climate projection model. This study also exhibits that future global climate change will significantly impact pheasant prevalence and distribution in the Himalayas (Chhetri et al. 2021). The rates and patterns of interim treeline dynamics that are driven by climate change may be intrinsic (Dullinger et al. 2004).

Many plant species such as the Himalayan pine have shifted upwards at the rate of 11–54 m/decade in the western Himalayan regions of India (Yadava et al. 2017). According to reports, treeline is reported to have moved 388 ± 80 m asl upslope in the IHR between 1970 and 2006 (Dahal et al. 2021). It was found that endemic plant species of alpine Sikkim Himalaya have shifted their elevational range margins by nearly 90%. More than 50% of range contraction than their recorded historical range extents were shown by six species (*Rheum nobile* Hook. f. and Thomson, *Saussurea stella* Maxiim., *Rhodiola bupleuroides* Wallich exi Hook. f. and Thomson, *Saussurea uniflora* Wall., *Saussurea leontodontoides* (DC.), *Ponerorchis chusua* (D. Don) Soo.), while on the other hand, four species (*Microgynaecium tibeticum* Hook. f., *Meconopsis simplicifolia* Walp., *Pedicularis trichoglossa* Hook. f. and *Lancea tibetica* Hook. f. and Thomson) showed range expansion by more than 100% of their historical range border (Telwala et al. 2013). The distribution of the three treeline species (*Abies spectabilis*, *Betula utilis*, and *Pinus wallichiana*) in the Nepalese Himalaya indicates that climatic shifts will likely influence future habitat suitability. It is predicted that under future climatic conditions, the suitable habitat for the three treeline species will shift toward higher elevations and may also undergo an overall range expansion (Chhetri et al. 2018). According to a study, the habitat of *B. utilis* will shift northward throughout all the Himalayas (Schickhoff et al. 2016), and over 400 masl upward shift will be shown by subalpine conifer forest zones between 2000 and 2050 under future climate scenarios (Zomer et al. 2014). A shrub species *Juniperus polycarpos* has been found at 4,000 m asl above the treeline. About 90% of the endemic species in the Sikkim Himalayas have been displaced at the rate of 27–22 m/ decade. Many species, including *Potentilla pamirica*, found in the northwest Himalayas of eastern Ladakh, have moved upwards by about 150 m asl above the plant distribution limit (Chaudhary et al. 2023). According to another study, it is also predicted that due to future climate scenarios the habitat of *Rhododendron spp.* will shrink under in the region of Sikkim, Indian Himalayas. The study also indicates that, under a wide range of climate change scenarios, the distributions of ten selected *Cyananthus* and *Primula* alpine species will shift upward in elevation and northward in latitude (Kumar 2012). Changing climates are also leading to shifts in vegetation phenology, with the start of the growing season advancing by 4.7 days per year while the end of the growing season remains unchanged (Zeng et al. 2011).

The climatic scenario significantly affects the rate of species range expansion too. With rising temperatures, the expansion rate increases, directly affecting the growth rates and fecundity levels and lowering the risk of faunal mortality, like *P. mugo*. The effect of climate change is not the same for everyone. It varies from one species to another (Dullinger et al. 2004). Finally, it can be inferred that species will vary not only in their response to increasing temperatures in terms of their growth rates but also in their dispersal capacities and competitive abilities during their recruitment phase.

8.3.1.3 Population Distribution of Himalayan Flora and Fauna

Population distribution refers to the spatial arrangement of different species across landscapes. It is influenced by factors such as habitat suitability, resource availability, climatic conditions, and interactions with other species (Krebs 1972). Assessing population distribution, researchers can identify areas crucial for biodiversity conservation, assess the impacts of habitat fragmentation or degradation, and develop effective management strategies to maintain healthy ecosystems and ensure the survival of vulnerable species (Ewel 2001).

The majority of plant population distribution studies have been conducted in the Eastern Himalayas and the Central Himalayas. The Western Himalayas have also received substantial attention; however, the Trans-Himalayas have not been extensively studied. Research on animal population distribution has become more widespread across the Himalayas. While the Central Himalayas have received relatively more attention, the Trans-Himalayas remain less studied. Overall, there is a need for more comprehensive research across all regions of the Himalayas to gain a holistic understanding of plant and animal population distribution and their responses to environmental changes.

Numerous studies conducted across the Himalayan region shed light on the complex interplay between climate change, habitat loss, and wildlife populations. Research focusing on large ungulate populations highlights the importance of effective wildlife management in maintaining prey densities crucial for large carnivore conservation (Muthamizh Selvan et al. 2014). Similarly, investigations into the distribution of Asiatic black bears highlight the need for proactive conservation strategies to mitigate the adverse impacts of climate change on vulnerable species (Zahoor et al. 2021). Studies on the Royle's pika and Himalayan pheasants emphasize the significant loss of habitat suitability under future climate scenarios, urging the implementation of conservation measures to safeguard these species (Bhattacharyya et al. 2019; Chhetri et al. 2021). Furthermore, research on snow leopards and blue sheep underscores the potential disruption of predator–prey interactions due to climate-induced habitat shifts, necessitating adaptive management approaches (Aryal et al. 2016). Additionally, studies on Kashmir musk deer and Himalayan Musk Deer emphasize the importance of transboundary conservation efforts in protecting species facing habitat fragmentation and degradation (Lamsal et al. 2018). Finally, research on Himalayan langurs highlights the urgent need for habitat management strategies to mitigate the impacts of climate change on species distribution and connectivity (Bagaria et al. 2020). These studies underscore the critical importance of proactive conservation measures in safeguarding Himalayan biodiversity amidst ongoing environmental changes.

There are several studies on vegetation distribution and responses of treelines and associated vegetation to climate change in the Himalayan region. While Paeonia species in the Himalayan-Hengduan Mountains exhibit complex genetic differentiation, with *P. ludlowii* possibly evolving from *P. delavayi* due to niche shifting (Zhang et al. 2018), treeline distribution in Langtang National Park, Nepal, reveal upward shifts in *Rhododendron campanulatum* and relatively stable conditions for *Abies spectabilis*, highlighting the importance of pristine environments for assessing climate impacts (Mainali et al. 2020). Dendrochronological analysis in central Nepal demonstrates an upward shifting trend in *Abies spectabilis*, driven by temperature changes (Gaire et al. 2014), while studies in the Western Himalaya emphasize the importance of monitoring forest composition shifts and promoting afforestation efforts in response to changing conditions (Tiwari et al. 2018). Further, population structure analyses across different altitudinal strata in the Indian western Himalaya reveal varied regeneration potential and the possibility of long-term population decline (Gairola et al. 2014). Additionally, projections for future plant distribution in Sikkim Himalaya suggest significant habitat loss for endemic angiosperm species, potentially altering ecosystem structure and function (Manish et al. 2016). Finally, investigations into treeline distribution in environmentally contrasting areas of the Nepalese Himalaya underscore the influence of both climatic and non-climatic factors on treeline stability and highlight the need for region-specific analyses to understand future trends (Shrestha et al. 2015; Schickhoff et al. 2016). Overall, these studies emphasize the complexity of Himalayan ecosystems' responses to climate change and underscore the importance of interdisciplinary research for effective conservation and management strategies.

8.3.2 Glacial Retreat

8.3.2.1 Impact on Freshwater Ecosystems

The retreat of glaciers in the IHR has profound implications for freshwater ecosystems in the region (Gosavi et al. 2020). As glaciers shrink due to rising temperatures, they release large volumes of meltwater, altering river and stream flow patterns and hydrology. This influx of water initially leads to increased flow rates, posing risks of flash floods and erosion. However, as glaciers continue to recede, the reduced meltwater input can decrease water availability during dry seasons, exacerbating water scarcity issues for downstream humans and wildlife (Sharma et al. 2022a). Furthermore, the loss of glacial meltwater affects freshwater bodies' temperature and nutrient levels, impacting the survival and distribution of aquatic species, especially fishes, amphibians, and insects adapted to specific environmental conditions in hill streams (Robinson et al. 2010). Ultimately, the glacial retreat in the Himalayas threatens the stability and resilience of freshwater ecosystems and jeopardizes the essential services they provide to local communities, including water supply, fisheries, and agriculture.

8.3.2.2 Consequences for Dependent Species

Throughout the twentieth century, the Himalayas and Tibetan plateau have undergone significant climate warming, particularly felt in the Hindu-Kush Himalayas, where dwindling snowfall and receding glaciers paint a vivid picture of environmental change (Sabin et al. 2020). This glacier retreat isn't just a local issue—it's a global trend with far-reaching consequences for water resources, biodiversity, and the intricate web of ecosystem services. Delving into this complex tapestry of nature's response to glacial retreat, Cauvy-Fraunié and Dangles (2019) examined global literature, uncovering diverse species reactions. While specialist species endemic to glacier environments face threats, adaptable generalist species downstream may find new opportunities for colonization (Cauvy-Fraunié and Dangles 2019; Bosson et al. 2023).

The cascading effects of glacier melting extend beyond species survival. As permafrost inches upwards due to glacial thawing, Himalayan ecosystems stand to lose their endemic species, sending ripples through biodiversity and the vital services these ecosystems provide (Xu et al. 2009). Rising temperatures promise to reshape the landscape, triggering accelerated glacier melt and prompting forests to shift their bounds, even reclaiming newly exposed lands once ice-dominated (Burga et al. 2010). One striking example of this ecological metamorphosis is the Tipra Valley glacier nestled in the heart of India's Central Himalayas. Over the past 46 years, it has receded a staggering 535 m, opening up fresh spaces for vegetation to take root (Singh et al. 2016). Witnessing this shift firsthand, three montane species—*Pinus wallichiana* A.B. Jackson, *Picea smithiana* (Wall.) Boiss., and *Populus ciliata* Wall. ex Royle—have embarked on a vertical journey, migrating upwards to altitudes previously beyond their reach (Singh et al. 2016).

The interplay between climate change and glacial retreat orchestrates a symphony of ecological adjustments, from elevational species migrations to reforestation of newly exposed landscapes. These transformations reverberate throughout the intricate fabric of Himalayan ecosystems, reshaping their very essence and challenging the resilience of life within (Schickhoff et al. 2016).

8.3.3 Increased Vulnerability Due to Extreme Weather Events

In recent times, the Himalayan biodiversity and the welfare of local communities have grown more fragile, presenting considerable obstacles to conservation endeavors and the sustenance of the millions reliant on these environments for their livelihoods. Due to highly variant topography and altitude-dependent climate, the Himalayan region is vulnerable to extreme rainfall events and cloudbursts (Nandargi and Dhar 2011; Thayyen et al. 2013). For instance, the Kedarnath tragedy, triggered by intense rainfall measuring 325 mm within 48 h, resulted in severe floods, landslides, and significant ecosystem damage, inflicting a heavy toll on human life (Dobhal et al. 2013; Rawat et al. 2016). This calamity wreaked havoc on vegetation, with forests severely fragmented in numerous areas, posing a serious threat to the local extinction of many plant species (Rawat et al. 2016). Moreover, Arunachal Pradesh witnessed two unprecedented cloudbursts in 2008 and 2010, causing devastating flash floods and extensive damage to both forested and agricultural lands (Bharali and Khan 2011).

The projected impacts of climate change in the Eastern Himalayas were also identified in various studies. One of the stresses caused is through more precipitation. It leads to multiple impacts on the communities, i.e., a Decrease in controlled and prescribed burning under shifting cultivation and other forest use areas (North East India), Wetland degradation (Umiam Lake, Barapani in Meghalaya), degradation of riverine island ecosystems (Majuli) and associated aquatic biodiversity, Widening of river basins (Brahmaputra breadth widened from 1.8 to 20 km near Kamakhya Temple) eroding agricultural land Increase in water-borne diseases (kala-azar, cholera, diarrhoea), Riverbank erosion, Flood-water related sedimentation, Increase in landslides and landslide dam outburst floods, Flood-water inundation (Dibrugarh and Dhubri) and Shallow water table pollution (Tse-ring et al. 2010). Heavy rainfall can trigger flash floods and landslides, destroying habitats, displacing wildlife, and causing soil erosion. Conversely, droughts can lead to water scarcity, habitat degradation, and loss of vegetation cover, exacerbating the vulnerability of species struggling to cope with environmental stressors.

8.4 Adaptation Strategies for Himalayan Biodiversity

8.4.1 Strengthening Protected Area Network and Corridors

Area-based conservation, notably through protected areas, serves as the leading strategy worldwide to combat biodiversity decline, encompassing 8% of the world's oceans and 17% of its lands. Following the adoption of the Kunming-Montreal Global Biodiversity Framework within the Convention on Biological Diversity, area-based conservation, including protected areas and other effective conservation measures (OECMs), is poised to broaden and swiftly grow. This expansion aligns with the 30 × 30 target, aiming to safeguard 30% of the planet's ecosystems by 2030 (Gurney et al. 2023). However, in the Himalayan context, currently, all the Himalayan states, including Assam and the hilly districts of West Bengal, have cumulative land of different statuses of protection of 50,805 km^2 area, which is only 8.24% of the total administrative land mass of the IHR (Table 8.1). Hence, safeguarding additional undisturbed land through OECMs, Community Reserves, and Conservation Reserves could bolster the creation of corridors for both plant and animal dispersal. These areas could also function as satellite core zones, facilitating the movement of large carnivores. Strengthening protection beyond current PAs will further aid in reducing human-wildlife conflicts across the IHR.

Table 8.1 Various conservation designations of PAS across Himalayan States, including Assam and North Bengal (NP = National Park; WLS = Wildlife Sanctuaries; ComR = Community Reserve; ConR = Conservation Reserve; PAs = Protected Areas)

States	NP (no.)	NP Area (Km^2)	WLS (No.)	WLS Area (Km^2)	ComR (No.)	ComR Area (Km^2)	ConR (No.)	ConR Area (Km^2)	No. of PAs	Total Area of PAs (Km^2)
Jammu and Kashmir	4	2,432.5	14	1,754.1	30	676.9	0	0	48	4,863.5
Ladakh	1	3,350.0	2	9,000.0	5	249.0	0	0	8	12,599.0
Uttarakhand	1	520.8	7	2,690.1	1	58.2	0	0	9	3,269.2
Himachal Pradesh	5	2,256.3	26	5,965.0	4	21.4	0	0	35	8,242.7
Arunachal Pradesh	2	2,290.8	13	7,614.6	0	0	9	131.6	24	10,037.0
Manipur	2	140.0	7	708.1	0	0	11	112.6	20	960.7
Meghalaya	2	267.5	4	94.1	0	0	74	141.7	80	503.3
Mizoram	2	150.0	7	720.7	0	0	0	0	9	870.7
Nagaland	1	202.0	4	43.9	0	0	114	851.8	119	1,097.7
Sikkim	1	1,784.0	7	399.1	1	0.06	0	0	9	2,183.2
Tripura	2	36.7	4	603.6	1	12.9	0	0	7	653.3
Assam	7	2,664.6	17	1,729.0	0	0	0	0	24	4,393.5
North Bengal	5	651.4	7	473.8	1	5.8	0	0	13	1,131.1
Total	35	16,747.0	119	31,796.0	43	1,024.5	208	1,237.7	405	**50,805.0**

8.4.2 Community-Based Conservation Initiatives

As the human population expands, urbanization steadily encroaches upon habitats like forests, grasslands, and wetlands, posing a significant threat to biodiversity. Hence, the active involvement of local communities is essential for conservation efforts at the grassroots level (Lamichhane et al. 2020). In the Himalayan region, researchers have reported several participatory conservation activities and pilot projects that have proven successful outcomes. In the Ladakh region of the IHR, Sonam et al. (2022) initiated a pilot community-based conservation for wolves *Canis lupus,* which has long conflicted with the local communities living in the area. This pilot program looks promising for replication in culturally similar areas of the Trans-Himalayan and Tibetan regions. In the Hindu Kush Karakoram Pamir Landscape (HKKPL), a similar environment to trans-IHR, Adhikari et al. (2021) described a community-based trophy hunting program implemented in Pakistan and Tajikistan which has contributed to increased populations of key wildlife species and improved the socio-economic condition of the local community wellbeing. On the contrary, Rashid et al. (2020) observed that trophy hunting and mass tourism introduced in the

Khunjerab National Park, northern Pakistan, have provided economic benefits but have resulted in various ecological issues such as pollution and impact on wildlife.

The strong commitment by the community to the conservation of their wildlife can be seen in two examples of community-led organizations, viz., *Singchung Bugun Village Community Reserve Management Committee* in Arunachal Pradesh and the Lemsachenlok Organization in Nagaland which were honoured with the 2018 India Biodiversity Awards (the award constituted by Indian Government in partnership with UNDP initiated in 2012) in recognition for their conservation of wild species. The *Singchung Bugun Village Community Reserve Management Committee* in Arunachal Pradesh has been involved in the conservation of critically endangered and endemic bird species, Bugun Liocichla *(Liocichla bugunorum).* The Lemsachenlok Organization is involved in managing and conserving the Yaongyimchen Community Biodiversity Conservation Area (YCBCA) in the Longleng district of Nagaland by prohibiting hunting to conserve wild species. These two examples demonstrated another form of community-based conservation in partnership with the forest department (Ghosh 2018).

Badola et al. (2021) designed an incentive-based mechanism based on the principles of PES (payment for ecosystem services) to address human-wildlife conflict in the corridors connecting the Rajaji and Corbett Tiger Reserves (TRs) located in the foothills of the Western Himalayas. In order to maintain the functionality of the forest corridor, the proposed Payment to Encourage Coexistence (PEC) program would give financial support to the local communities in order to encourage coexistence with wildlife. This would reduce human-wildlife conflict by encouraging behavior changes and giving opportunities to explore alternative livelihoods. This strategy may be replicated and scaled up to other protected areas and wildlife corridors to protect biodiversity at the landscape level. Similarly, Dutta and Dutta (2023) in their study of western Arunachal Pradesh of Indian Eastern Himalaya, observed that community-based conservation initiatives should also be supported with arrangements for alternative livelihood options for the forest-dependent local communities in the vicinity of the Community Conserved Areas (CCA).

Thus, community-based conservation initiatives in the IHR promise to protect the region's native biodiversity. Through collaborative efforts, local communities can preserve fragile ecosystems and safeguard endangered species. By empowering inhabitants to engage in conservation actively, these initiatives ensure the sustainable coexistence of humans and wildlife in the IHR region. Such grassroots approaches also foster a sense of ownership and responsibility, which is crucial for preserving this unique biodiversity hotspot in India.

8.4.3 *Sustainable Land-Use Planning*

8.4.3.1 Balancing Development and Conservation

Balancing development and conservation in sustainable land use planning is crucial for the overall biodiversity conservation in the IHR region (Laladhas et al. 2017; Dhiman and Muthanarasimha 2022). The Himalayan ecosystem sustains around 1.3 billion people and is also rich in biodiversity, making it a global hotspot (Singh et al. 2021a). Sustainable management is essential to protect endemic species vulnerable to extinction due to human interference (Pathak et al. 2023). Utilizing wild edible plants can address food security issues while ensuring sustainable harvesting practices (Iannone et al. 2022). Incorporating sustainable urban design principles can further enhance conservation efforts by considering spatial aspects for native plant and animal welfare. Embracing sustainable use of wild species as a conservation tool is vital, especially in the Indian context, to bridge the gap between development and biodiversity conservation. Implementing transdisciplinary approaches that integrate conservation, livelihoods, and development is vital to achieving a balance in sustainable land use planning in the IHR. A landscape approach to biodiversity management is recognized for its ability to reconcile objectives of biodiversity conservation and sustainable development (Kotru et al. 2020).

Despite the region's wealth of natural resources, many of its inhabitants remain marginalized, living at subsistence levels. Unfortunately, the unsustainable exploitation of these resources is causing escalating environmental degradation and exacerbating the impact of natural disasters. It is imperative to develop new approaches that prioritize restoring equilibrium between economic interests and ecological needs while also respecting socio-cultural values (Singh 2006).

The Intergovernmental Panel on Climate Change (IPCC) and the Intergovernmental Science and Policy Platform on Biodiversity and Ecosystem Services (IPBES) highlight the risks to humanity from unsustainable use of natural resources and loss of biodiversity under rapid climate warming conditions. The Convention on Biological Diversity addresses the issues of biodiversity and ecosystem services in the HKH region by collaborating with other international agreements and goals. Urgent action is needed to address the challenges of biodiversity losses, poor ecosystem functioning, and reduced ecosystem services in the HKH region through a resilience management framework that includes monitoring, multi-indicator approaches, and collaboration among stakeholders (Kattel 2022). The depicted Fig. 8.4 vividly demonstrates how the synchronicity of science, policy, and practice can attain sustainable development, benefiting not just the IHR but also the entire planet.

8.4.3.2 Mitigating Habitat Fragmentation

Mitigating habitat fragmentation in the IHR is crucial for sustainable land use planning. Studies show that forest fragmentation susceptibility maps can be developed

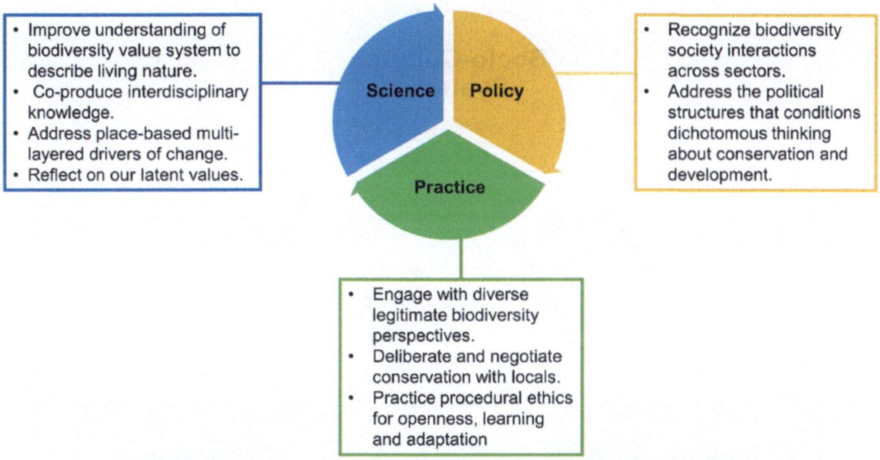

Fig. 8.4 The increasing integration of science, policy, and practice is essential for addressing the diverse aspects of the natural world, considering various worldviews, ethics, and information systems among people to enable sustainable development (Pascual et al. 2021)

using earth observation data and the weight-of-evidence method to identify causative factors like proximity to built-up areas and roads (Batar et al. 2021). Monitoring land use/land cover dynamics through remote sensing helps quantify changes like urban expansion and forest loss, driven by human activities (Parida et al. 2022). Implementing nature-based solutions like forest protection and landscape restoration can rejuvenate the land, conserve biodiversity, and support ecosystem services in the region (Duraisami et al. 2022). Efforts towards ecological restoration and rehabilitation of forests are vital to improve the environment, mitigate climate change impacts, and support local livelihoods in the IHR (Bhatt et al. 2020). Addressing urban heat islands through a better understanding of urbanization impacts on land surface temperature is also essential for ecosystem health and human wellbeing in the IHR.

Nature-based solutions (NbS) like forest protection and landscape restoration can rejuvenate the land, conserve biodiversity, ensure food security, and fulfil the critical needs of the inhabitants. The Restoration Opportunities Atlas developed by the World Resources Institute India (WRI India) provides insights into potential landscape management strategies in the IHR to contribute to India's nationally determined contribution (NDC) under the Paris Climate Agreement and the sustainable development goals (SDGs) (Duraisami et al. 2022). Figure 8.5 explains how the sociocultural system is related to ecological and economic systems in attaining sustainable development.

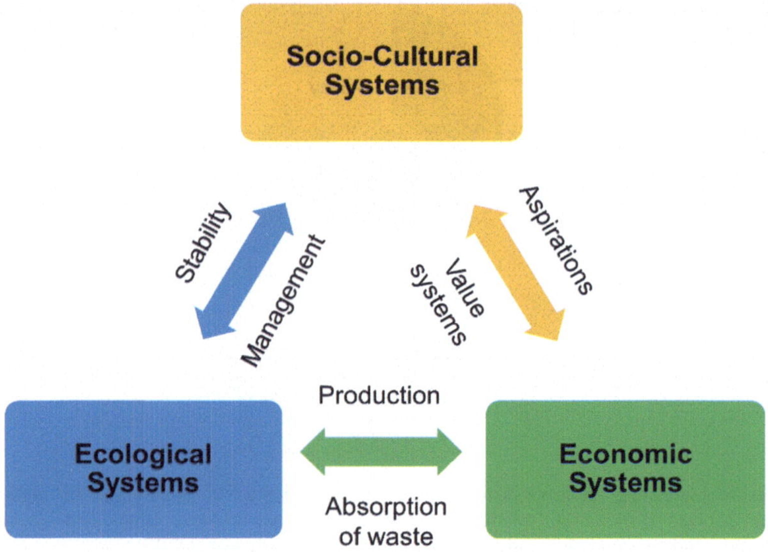

Fig. 8.5 Superimposition of the socio-cultural system on the relationship of ecological and economic systems in attaining sustainable development (Singh 1998)

8.5 Policy and Governance

8.5.1 National and International Frameworks

8.5.1.1 Existing Policies for Biodiversity Conservation

The IHR is a vital ecosystem that serves as a rich repository of plant and animal wealth in diverse ecological systems (Rawal et al. 2013). It is recognized as one of 34 global biodiversity hotspots, including twenty-five forest types and a rich assemblage of fauna species (UNESCO 2014). It is home to over 51 million people who practice hill agriculture (Himachal Tonite 2022). A complex geologic structure, snowcapped peaks, large valley glaciers, deep river gorges, and rich vegetation characterize the region. The IHR ecosystem is vital to the ecological security of the Indian landmass and occupies a strategic position along the country's northern boundary (NMSHE 2010). The conservation of biodiversity in the Indian Himalayan Region is governed by a complex interplay of national and international policies to preserve its rich ecological diversity amidst challenges like rapid biodiversity loss and habitat degradation. Addressing these challenges entails enhancing governmental ties, including those among China, India, and Nepal, to advance sustainable ecosystem management (Goodale et al. 2022). Regional cooperation and a transboundary landscape approach are highlighted as essential strategies for conserving biodiversity effectively in the Himalayan region and beyond (Xu et al. 2019; Kotru et al. 2020). The coming decade is pivotal for setting new goals and commitments that will halt the current loss of biodiversity and underscore its critical importance for sustainable development and human wellbeing (Chaudhary et al. 2022). In the Indian context, policies within the forest, wildlife, and environment sectors have been assessed over several decades to evaluate their effectiveness in sustaining and promoting biodiversity conservation. These policies are essential for safeguarding the diverse ecosystems, species, and genes that form the foundation of life in the Himalayas. At the national level, India has implemented various key policies to conserve biodiversity in the Himalayan region. The Indian Forest Act of 1927 was one of the earliest legislative measures aimed at protecting forests and wildlife. The preamble to the Indian Forest Act, 1927 (16 of 1927) states that the Act seeks to consolidate the law relating to forests, the transit of forest produce and the duty that can be levied on timber and other forest produce (NBA 1927). Subsequently, the Wildlife Protection Act of 1972 and Amendment 1991 provide for the protection of birds and animals and for all matters that are connected to it, whether it be their habitat, waterholes, or forests that sustain them. The Wildlife Protection Act of 1972 provided a comprehensive legal framework for safeguarding wildlife and their habitats. The Act was substantially amended in 2002. A major objective of this amendment was to update the law to deal with sharply escalating levels of organized poaching. The Ministry of Environment and Forest formally notified this Amendment Act on 1st April 2003 (MoEF 1972). Additionally, the Environment Protection Act of 1986 focused on environmental conservation and pollution control. These policies laid the foundation for biodiversity conservation in the IHR (MoEF 1986). The National Mission for Sustaining Himalayan Ecosystems (NMSHE) is one of India's eight missions under the National Action Plan on Climate Change (NAPCC). The mission aims to protect the interests of the people and ensure the sustainability of Himalayan ecosystems. NMSHE deliver a better understanding of the coupling between the Himalayan ecosystem and the climate factors and provides inputs for Himalayan sustainable development while addressing the protection of a fragile ecosystem (PIB 2014). This mission requires collaborative efforts from climatologists, glaciologists, and other experts and information exchange with South Asian countries and those sharing the Himalayan ecology (NMSHE 2010). Internationally, the Himalayan region is recognized as a biodiversity hotspot, and efforts are being made to protect and conserve its unique ecosystems. India is a signatory to several major conventions relating to the conservation and management of wildlife. Initiatives like the UN Convention on Biological Diversity (CBD) and the Aichi Biodiversity Targets provide a global framework for countries to commit to ambitious yet achievable goals for biodiversity conservation. India, along with seven other countries in the Hindu Kush Himalayas, is a signatory to the UN Convention on Biological Diversity (CBD) and has committed to implementing the Aichi Biodiversity Targets. These targets set ambitious yet achievable goals for biodiversity conservation on a global scale. The upcoming post-2020 Strategic Plan for Biodiversity, to be agreed upon at the 15th summit of the CBD in Kunming, China, will guide global efforts towards biodiversity conservation and restoration from 2021 to 2030. India's participation in these global initiatives reflects its commitment to preserving biodiversity within its borders and in collaboration with other nations

(Global Biodiversity Framework 2021). Efforts towards conserving biodiversity in the Himalayan region emphasize the importance of regional cooperation and a transboundary landscape approach. Transboundary landscape (TBLs) initiatives underscore the potential of collaborative approaches to facilitate effective biodiversity conservation and management amidst a backdrop of a poor rural economy and high climate change risks. The International Centre for Integrated Mountain Development (ICIMOD) has played a crucial role in partnering with member countries to conceptualize transboundary cooperation initiatives that link national commitments to global conservation goals (Kotru et al. 2020). This partnership has been particularly significant in aligning efforts with key international conventions such as the Convention on Biological Diversity (CBD), the Convention on International Trade in Endangered Species of Wild Fauna and Flora (CITES), and the Convention on the Conservation of Migratory Species of Wild Animals (CMS) (UN Biodiversity Conference 2022; IUCN WCPA 1; IUCN WCPA 2). The CBD, one of the pivotal agreements, focuses on ensuring the preservation and sustainable utilization of biodiversity resources. Through its provisions, the CBD sets the stage for countries to work together towards the common goal of biodiversity conservation and sustainable management (de Chazournes 2009; Desai et al. 2011). CITES, another critical convention, regulates the international trade of endangered species to safeguard vulnerable wildlife populations and prevent their exploitation. By overseeing and controlling the cross-border movement of endangered species, CITES plays a crucial role in global conservation efforts (CITES 1979; CITES 2022). Furthermore, the CMS brings nations together to establish legal frameworks for coordinated conservation actions for migratory species that traverse multiple countries. By facilitating international cooperation, the CMS enhances the protection of migratory animals and their habitats, fostering a harmonized approach to conservation across borders. These conventions collectively provide a robust framework for international collaboration and concerted action to safeguard biodiversity on a global scale. Through partnerships with organizations like ICIMOD, member countries can translate these conventions into tangible initiatives that promote transboundary cooperation and advance the shared goal of conserving our planet's rich biological diversity (IUCN-WCPA 2024). In addition to these policies, ongoing efforts are being made to address the challenges of biodiversity conservation and its sustainable utilization in the IHR. For instance, The Critical Ecosystem Partnership Fund (CEPF) has identified priority corridors in the Eastern Himalayas, including the Bhutan Biological Conservation Complex, Kangchenjunga-Singalila Complex, and North Bank Landscape, to maintain and restore connectivity and protect wide-ranging threatened species (CEPF 2005). Also, the Mountain Division of the National Institute of Himalayan Environment (NIHE), NMSHE, and NMHS have provided partial funding for research on Himalayan biodiversity (NMSHE 2010). Despite positive global actions towards conservation targets like the Aichi Biodiversity Target, challenges persist in biodiversity conservation in the IHR. Issues such as climate change impacts, resource degradation, and social challenges like poverty and vulnerable livelihoods complicate effective biodiversity management (Singh et al. 2021a, b). The region faces new challenges from mega-investments like the Belt and Road Initiative and hydropower development, necessitating comprehensive policies that balance economic development with biodiversity preservation (Kotru et al. 2020). These challenges require a coordinated effort to ensure the sustainability of the Himalayan ecosystem and protect its rich biodiversity (Kattel 2022). The need for decentralized natural resource policies, enhanced cross-border cooperation, and sustainable development strategies is evident (Kotru et al. 2020; Rana et al. 2021; Kattel 2022). These studies encapsulate the intricate policy landscape governing biodiversity conservation in the Indian Himalayan Region, emphasizing the importance of holistic approaches that integrate national and international strategies to safeguard this vital ecological treasure.

8.5.1.2 Assessing Their Effectiveness in the Himalayan Context

National and international policy frameworks play a crucial role in conserving Himalayan biodiversity. The Himalayan region faces challenges like biodiversity loss, climate change impacts, and ecosystem degradation. Studies emphasize the importance of integrating agrobiodiversity into national policies to enhance food security and agricultural sustainability (Baral et al. 2022). Additionally, the need for active community participation is highlighted to achieve integrated biodiversity conservation and development strategies with a more significant impact (Banerjee et al. 2022). The region's vulnerability due to climate change underscores the urgency for effective policies addressing biodiversity losses and declining ecosystem services (Kattel 2022). Furthermore, protected areas in the eastern Himalayas require specific management interventions based on vulnerabilities to ensure effective conservation (Rasul et al. 2022). A harmonized approach involving stakeholders, policy integration, and sustainable resource use is essential for successful biodiversity conservation in the Himalayan context.

A resilient ecosystem is characterized by a robust ecological boundary, where elements such as soil, vegetation, water, and sediment naturally support diverse organisms crucial for ecosystem functionality and the generation of various services. Understanding the intricacies of this resilient ecological boundary involves recognizing the evolutionary dynamics of species interactions and dispersal within conducive habitats. Losing any population from this boundary raises concerns about the ecosystem's persistence (Balvanera et al. 2014). Take, for instance, the resilient ecological boundary observed in lakes and rivers, which relies on a healthy flow regime, physical habitat structures serving as refuges for species, optimal thermal and light conditions regulating organismal metabolism, ecosystem productivity, controlled chemical and nutrient levels ensuring water quality, and the dynamics of biotic assemblages influencing ecosystem processes and community composition (Baron et al. 2002). Ultimately, the provision of ecosystem services is determined by biophysical processes, which benefit humans (Fig. 8.6).

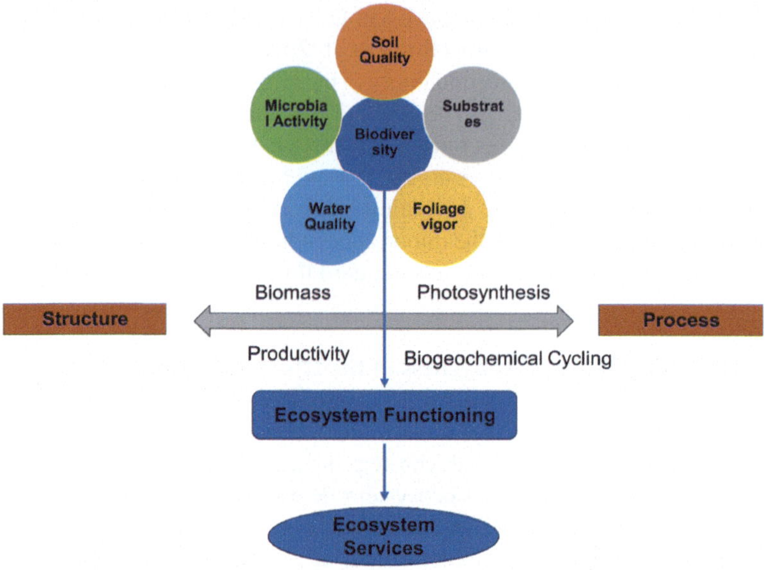

Fig. 8.6 Framework for ecological resilience boundary from which biodiversity supports ecosystem structure, processes and ecosystem services (Kattel 2022)

8.5.2 Importance of Cross-Border Collaborations

The Himalayas, often referred to as the 'Roof of the World, ' harbours a wide array of wild and domestic flora and fauna. Many of the species dwelling in these landscapes are endemic and adapted to the unique climatic conditions of this vast landscape. To survive with the changing temperature and food availability, the animals and plants of this region undergo various adaptations, including continuous seasonal migration that surpasses the political boundaries of the human world.

Transboundary cooperation entails collaboration among neighboring countries that share ecosystems or natural resources. In the context of the Himalayas, this cooperation involves nations like India, Nepal, Bhutan, China, and Pakistan, all of which have territories within the Himalayan range. Recognizing the interconnectedness of their ecosystems, these nations must work together to address common environmental concerns.

Climate change and habitat loss are gradually engulfing the Himalayan landscape. The concerned countries need a collaborative effort to tackle this problem and safeguard this fragile ecosystem. Collaborative efforts may include establishing transboundary protected areas, conducting joint research studies, and sharing knowledge.

For example, Yak rearing has traditionally been an important livelihood source of the mountain pastoral communities for thousands of years. Traditional yak migration routes across valleys in the Himalayan Hindukush region have acted as essential locations for knowledge transfer about yak rearing and the spread of different yak breeds from their origin (Sharma et al. 2016). A transboundary co-operational approach is crucial to conserve the rangelands and pasturelands and facilitate knowledge transfer through proper research. There are also many wild animals, such as the Great One-horned rhino, whose last populations are found in India and Nepal. The conservation of these animals depends largely on their ability to move freely in their rangelands, which expand between the two countries. A transboundary cooperation mechanism that exists between the two countries plays a crucial role in downlisting the species from Endangered to Vulnerable on the IUCN Red List (Talukdar and Sinha 2013).

Large predators, co-predators and their prey are also seen to have large homing ranges, often expanding beyond the political borders of a country. The availability of large, uninterrupted landscapes is crucial for these animals' genetic viability and long-term survival. The introduction of the Landscape Initiative for the Far Eastern Himalayas (HI-LIFE), which is a transboundary collaboration between India, China and Myanmar, is a step forward toward the protection of the Far Eastern Himalayan landscape, which is home to carnivores like tigers, leopards, wild dogs along with their prey (Sarkar et al. 2021).

Effective transboundary cooperation can yield numerous benefits not only for the environment but also for local communities. By conserving critical habitats and keystone species, nations can maintain ecosystem services vital for human well-being, such as water regulation, soil stabilization, and carbon sequestration. Moreover, collaborative efforts can promote ecotourism, generate economic opportunities, and foster cultural exchange among neighboring communities.

In conclusion, transboundary cooperation stands as an indispensable mechanism for safeguarding Himalayan flora and fauna in the face of mounting environmental challenges. By transcending geopolitical boundaries and working together toward shared conservation goals, nations can preserve this ecological treasure for future generations. As stewards of the Himalayas, we must prioritize cooperation, innovation, and sustainability in our collective efforts to protect and cherish this majestic mountain range.

8.6 Technological Innovations for Conservation as a New Adaptation Tool to Safeguard Himalayan Biodiversity

8.6.1 Remote Sensing and Monitoring

8.6.1.1 Tracking Changes in Biodiversity

The IHR is a biodiversity hotspot, with a rich and diverse array of species, including many endemics. However, this region is facing significant challenges due to climate change, land-use changes, and other human activities (Rana et al. 2021; Dhiman and

Muthanarasimha 2022; Wani et al. 2022). Various studies have shown that climate change is causing shifts in species distribution, phenology, and habitat suitability, leading to biodiversity losses (McElwee 2021; Shivanna 2022). The region is experiencing warming faster than the global average, with Trans-Himalaya experiencing even more rapid warming (Krishnan et al. 2019; Kattel 2022). This warming has resulted in glacial retreat, glacial lake expansion and outburst, rapid snow melt, more frequent extreme events including precipitation, drought, and desertification, a transition from a snow to rain-dominant precipitation system (Kc et al. 2022; Mongabay-India 2024). These changes have disrupted the natural cycles of species, leading to alterations in phenology, such as shifts in leaf-fall and fruiting times, impacting plant survival and species vulnerability (Shivanna 2022).

Additionally, the warming climate has resulted in the shifting of the tree line and the movement of plant species to higher elevations, affecting the distribution and habitats of various flora and fauna in the region (Antão et al. 2022). Furthermore, the introduction and spread of invasive species, habitat fragmentation, and decreased food availability have further threatened the biodiversity and ecosystems of the region (IUCN 2019; US National Park Service 2021). One of the most significant impacts of climate change on biodiversity in the IHR is the alteration of forest ecosystems. Forests in Pakistan's high-altitude and mid-altitude zones have decreased by 30.5 and 49.7%, respectively, over 39 years. A third of this loss can be attributed to agricultural expansion.

Similarly, 170,684 ha of forest was lost or degraded in Pakistan from 1990 to 2010. In Mustang, Nepal, 42% of forest was lost from 1979 to 2009 due to agricultural expansion. These forest loss and degradation trends have also been observed in other parts of the Himalayas, including the Indian mountains (Kc et al. 2022). Agricultural productivity in the IHR is also being impacted by climate change, negatively affecting staple crops at lower altitudes while potentially opening up opportunities for Climate Change Driven Agricultural Frontiers (CCDAFs) at higher elevations. Nevertheless, there is a dearth of quantitative research on the effects of climate change on local food security in the Himalayas, especially in mountainous regions (Kc et al. 2022). The utilization of remote sensing and geographical information systems (GIS) has played a pivotal role in monitoring and managing the repercussions of climate change on biodiversity in the IHR. These technologies have been instrumental in assessing land use/land cover changes, ecosystem services, and mapping bioclimatic zones and ecoregions within the region (Batar et al. 2021; Patley et al. 2024). Remote sensing has proven useful in long-term ecological and biodiversity monitoring in complex terrains such as the western Himalaya (Chawla et al. 2012). For instance, GLORIA (Global Observational Research Initiative in Alpine Environments) has established a 1,500 km trans-Himalayan transect to document the impacts of climate change on alpine plants and communities (Salick et al. 2014). Remote sensing and GIS have empowered researchers to monitor changes in vegetation cover, track glacier retreat, and assess the impacts of land-use changes on biodiversity (Prasad et al. 2015). Despite these advancements, the conservation of Himalayan biodiversity faces numerous challenges, with climate change is exacerbating existing threats, such as poaching, habitat loss, and human-wildlife conflict (Dodiya et al. 2023). Adaptation strategies are being implemented to tackle these challenges, encompassing sustainable energy transition, forest conservation, climate-smart agriculture, disaster risk reduction, climate-resilient infrastructure, and community engagement. These strategies aim to foster resilience in the face of climate change and safeguard the region's unique biodiversity (Kattel 2022).

8.6.1.2 Early Warning Systems for Extreme Events

Early warning systems for extreme events in the Himalayan region are crucial for conservation efforts. Utilizing remote sensing and monitoring technologies can enhance preparedness and mitigation strategies (Shrestha et al. 2016; Gatlin et al. 2018; Talchabhadel et al. 2022; Taloor et al. 2022). These systems can provide real-time data on precipitation variability, glacier changes, snow cover extents, and high-risk glacial lakes, aiding in detecting potential hazards like floods, landslides, and glacial lake outburst floods (Basu et al. 2022). Implementing advanced monitoring techniques, such as satellite imagery and geospatial technology, can help identify high-risk areas, monitor ecosystem health, and facilitate early warning systems for disaster management. By integrating remote sensing data into assessments of mountain ecosystems, decision-makers can better plan for conservation, improve disaster preparedness, and minimize environmental impacts in the fragile IHR.

The ongoing shifts in climatic patterns are visibly altering the landscape of glaciers, with their rapid melting accelerated by escalating global temperatures. This heightened thawing has led to a surge in glacial lake outburst floods (GLOFs), particularly in the IHR, posing a grave danger to the locals. These catastrophic GLOF events have inflicted significant casualties and property damage downstream. Hence, there's an urgent need to establish a robust early warning mechanism to pre-emptively notify communities of impending disasters (Singhal et al. 2022).

The HIWAT tool kit combines numerical weather prediction, satellite-based precipitation, and land imagery techniques to predict, observe, and effectively respond to threats and impacts posed by thunderstorms in the IHR. It aims to build the capabilities of national meteorological departments and other weather-sensitive agencies in the IHR region to assess the risks associated with high-impact convective weather events (Gatlin et al. 2018).

8.6.2 Conservation Genetics

8.6.2.1 Genetic Diversity of Himalayan Flora and Fauna

Genetic diversity refers to the variety of alleles (versions of genes) present within a population (Allendorf et al. 2012). Genetic diversity is pivotal for population resilience amidst evolving environmental conditions. Naturally, outbreeding populations generally have high genetic diversity. However, it tends to decline in endangered species, thereby compromising their ability to adapt to changing environmental conditions (Frankham et al. 2004). Genetic diversity conservation is one of the three worldwide conservation priorities recognized by the IUCN (McNeely et al. 1990). Spatial variability in genetic makeup is intricately tied to the population size, selection pressures, and rate of gene flow (Frankham et al. 2002; Niyogi et al. 2022).

Most of the genetic studies in the Himalayas have focused on assessing the genetic structure and gene flow of wild populations. These studies reveal the profound influence of landscape heterogeneity, climate fluctuations, and geographical barriers on genetic differentiation and population connectivity. Findings underscore the importance of conservation strategies prioritizing habitat protection, establishing wildlife corridors, and maintaining gene flow to ensure the long-term viability of vulnerable species in light of ongoing environmental changes and anthropogenic pressures. Recent studies on the gene flow of Himalayan fauna have mainly focused on species of large carnivores (e.g., Thapa et al. (2018), Werhahn et al. (2020), Singh et al. (2022b)), ungulates (e.g., Singh et al. (2022a), Jabin et al. (2023)), primates (e.g., Khanal et al. (2022), Arekar et al. (2022)), and birds (e.g. Kumar et al. (2020), Jiao et al. (2024)). However, very few studies have investigated the molecular adaptations (i.e. gene selection) of species to the local environment (e.g., Zou et al. (2021), Yang et al. (2022)). Similarly, many studies on the Himalayan flora have explored population genetic structure (e.g., Meena et al. (2019), Rana et al. (2021), Sharma et al. (2022a, b)), while few have looked at gene selections in response to the local environment (e.g., Zhang et al. (2019)). The reason for this distinction may be that while genetic methods, which focus on a few neutral genes (like microsatellites, mtDNA, etc.), are well-suited for assessing population genetic structure, genomic methods (like ddRAD-seq, GBS, etc.) that sequence a large portion of the genome is crucial for studying gene selection. In Himalayan genetic studies, microsatellites emerge as the most commonly employed molecular markers, followed by regions of mitochondrial DNA (Khanal et al. 2022) and chloroplast DNA (Poudel et al. 2014; Rana et al. 2021), while the use of genome-wide SNPs remains relatively infrequent (Dahal et al. 2023; Jiao et al. 2024). In many studies, a combination of two or more types of markers has been applied. Some earlier studies have also employed AFLP, SRAP, and RAPD (Mohapatra et al. 2009; Bajpai et al. 2014; Nag et al. 2015) when other markers are unavailable (i.e. not standardized) for their target species. However, the use of these markers is declining, while there is a growing preference for markers that cover larger parts of the genome.

Upon reviewing the geographical extent of the faunal genetic studies, it becomes apparent that a large number of studies have focused on the central Himalayas, a moderate number on the western and eastern Himalayas, whereas very few have explored the Trans-Himalayas. Similarly, very few floral genetic studies have focused on the Trans- Himalayas, while many more have focused on other Himalayan regions. Almost an equal number of floral studies have centered on the western, central, and eastern Himalayas. An important point to note is that- a substantial number of studies have focused on more than one part of the Himalayas.

The genetic studies on Himalayan fauna highlight the importance of genetic diversity in species' adaptability and long-term survival, particularly in the face of environmental changes and anthropogenic threats. Himalayan langurs exhibit moderate genetic diversity with structured populations corresponding to fragmented distribution, influenced by landscape features (Khanal et al. 2022; Arekar et al. 2022). Asian pikas show low genetic diversity, signaling a declining effective population size (Dahal et al. 2023). Additionally, studies found that the genetic makeup of species like Himalayan wolves, snow leopards, Bengal tigers, and Hanuman langurs are influenced by environmental factors, emphasizing the urgency of conservation measures amidst changing environmental conditions to ensure their long-term survival.

Floral studies also unveil significant insights into various species' genetic diversity and population distribution. For instance, research on *Valeriana jatamansi* revealed how habitat types and altitude influence genetic diversity, with grassland habitats harboring the highest diversity (Jugran et al. 2013). Similarly, investigations into *Taxus spp.* underscored the species' low genetic diversity and emphasized the importance of conservation efforts (Mohapatra et al. 2009; Poudel et al. 2014). Studies on *Hippophae tibetana* highlighted the role of the Himalayan mountains as a barrier to gene flow between populations (Qiong et al. 2017). Research on Podophyllum hexandrum also elucidated the intermixing of populations across the Indian Himalayas, suggesting a shared genetic heritage (Nag et al. 2015). These findings collectively underscore the importance of understanding genetic variation for effective conservation strategies to preserve the Himalayan region's unique biota. Furthermore, transboundary cooperation and community involvement are emphasized as essential components of conservation efforts to preserve genetic diversity and ensure the survival of diverse species in the Himalayan region.

8.6.2.2 Genetic Strategies for Species Conservation

Genetic strategies play a pivotal role in various facets of conservation efforts, encompassing genetic monitoring, disease surveillance, and ex-situ germplasm conservation. Genetic monitoring, often achieved through techniques like DNA barcoding and population genetics analysis, allows for the tracking of genetic diversity and population distribution within species, aiding in the assessment of their conservation status (Allendorf et al. 2012). Disease surveillance, on the other hand, employs genetic tools to monitor pathogens affecting endangered species, enabling early detection and management of potential disease outbreaks (Ng et al. 2019). Additionally, gene banks facilitate ex-situ germplasm conservation, preserving genetic diversity outside of natural habitats through cryopreservation and tissue culture techniques, safeguarding against genetic erosion and providing a resource for future restoration efforts (Mahesh et al. 2011). While we have extensively outlined genetic monitoring in the preceding section, the following paragraphs shed light on disease surveillance and ex-situ germplasm conservation.

Research conducted in the Himalayas has delved into disease surveillance using genetic tools across various species (Ng et al. 2019) assessed dog husbandry practices and Canine Distemper Virus (CDV) prevalence in Nepal's Annapurna Conservation Area, highlighting the potential spillover risk to wildlife. Khurana et al. (2020) focused on the gut microbiome of *Tor putitora*, an endangered freshwater fish, revealing both commensals and opportunistic pathogens alongside virulence factors and antibiotic resistance genes. In the study by Xu et al. (2020) *Enterocytozoon bieneusi* infection was investigated in Himalayan marmots and Alashan ground squirrels, indicating a zoonotic transmission potential and emphasizing the need for surveillance and management. The gut microbiome of Himalayan Griffons was explored by Wang et al. (2021), demonstrating a diverse microbial composition and functional structure, with potential protective roles against pathogens found in carrion. Ao et al. (2017) identified two novel Bocaparvoviruses in wild Himalayan marmots, underscoring the importance of studying potential cross-species viral transmission amid ecological and environmental changes. These studies collectively underscore the importance of genetic tools in disease surveillance efforts in the Himalayas, addressing risks to wildlife, aquatic species, and potential zoonotic threats, while emphasizing the necessity of ongoing monitoring and management strategies.

Ex-situ efforts to conserve the germplasm of Himalayan flora and fauna involve strategies to preserve genetic material outside of its natural habitat. These initiatives typically include establishing seed banks, gene banks, and living collections in botanical gardens and research institutions. Prominent research institutes and universities have focused on conserving Himalayan biodiversity through various approaches, including ex-situ conservation (Chauhan et al. 2014, 2023). Integrating genetic analysis into ex-situ conservation programs enhances their effectiveness by ensuring that the preserved germplasm represents the full spectrum of genetic variation present in Himalayan flora and fauna, thereby maximizing the potential for species survival and adaptation in changing environments. While significant progress has been made in ex-situ conservation efforts for Himalayan flora and fauna, there is still ample room for additional initiatives to enhance genetic preservation further and ensure the long-term viability of these unique species in the face of ongoing environmental challenges. Continued collaboration among research institutions, governmental bodies, and conservation organizations will be essential in expanding and strengthening these crucial conservation endeavors.

8.7 Discussion and Conclusion

Climate change poses a profound and multifaceted threat to the biodiversity of the Himalayan region, a unique ecosystem renowned for its rich array of flora and fauna. This area encompasses nearly half of India's flowering plants and a third of its faunal species, making it one of the most biodiverse regions on the planet. However, the delicate balance of this ecosystem is increasingly under strain due to a combination of climate change, habitat destruction, and human activities.

Biodiversity loss is a significant concern in the Himalayas, driven by various factors. Illegal trafficking, poaching, habitat loss, and human-wildlife conflicts are among the primary threats facing wildlife species. These pressures have led to population declines and, in some cases, local extinctions of vulnerable species. Additionally, habitat degradation, deforestation, and unsustainable land use practices further exacerbate the situation, disrupting essential ecosystem services such as water regulation and carbon sequestration. Urgent action is needed to restore degraded lands and safeguard ecosystems for the well-being of both nature and local communities.

The rising temperatures in the Himalayan region present severe threats to biodiversity. As temperatures increase, the delicate ecological balance is disrupted, leading to habitat loss and altered ecological dynamics. Phenological changes, such as shifts in flowering and fruiting patterns of plant species, are already being observed. Furthermore, genetic, behavioral, and distributional changes are occurring in various fauna, driven by changing environmental conditions. Invasive species are also spreading, posing a significant threat to native ecosystems and biodiversity.

Climate change-induced shifts in species distribution are another consequence affecting both flora and fauna across the region. Montane flora and fauna, in particular, are highly sensitive to changing climate conditions, and most species experience habitat shifts and range contractions, while projected changes indicate habitat loss for some ecologically important species. Moreover, changes in vegetation phenology and treeline dynamics are evident, signaling broader ecological shifts in the region.

Freshwater ecosystems in the Himalayas are also under threat due to glacial retreat, which alters river and stream flow patterns. Reduced meltwater input affects the temperature and nutrient levels of freshwater bodies, ultimately impacting aquatic species and the services they provide. Glacial retreat threatens the stability and resilience of freshwater ecosystems and poses risks to downstream communities reliant on these water sources.

Addressing these challenges demands urgent adaptation measures. Strengthening protected area networks and establishing corridors to facilitate the dispersal of plants and animals are pivotal steps in conserving biodiversity and mitigating habitat fragmentation. Engaging local communities through community-based conservation initiatives holds great promise in achieving conservation objectives. Those initiatives within the IHR offer a beacon of hope for preserving the region's native biodiversity. Through collaborative endeavors, local communities can safeguard fragile ecosystems and protect endangered species. By empowering residents to participate actively in conservation efforts, these initiatives foster sustainable coexistence between humans and wildlife in the IHR. Such grassroots approaches cultivate a sense of ownership and responsibility crucial for the enduring preservation of this remarkable biodiversity hotspot in India. It is imperative to integrate sustainable land-use planning, striking a balance between development and conservation efforts, to secure the long-term vitality of Himalayan biodiversity. Furthermore, implementing national and international policy frameworks is indispensable for orchestrating coordinated conservation endeavors across various scales. In the face of habitat loss, fragmentation, and climate-induced shifts, conservationists must carefully design landscapes conducive to the needs of large, space-requiring species.

Remote sensing and monitoring technologies play a vital role in assessing and managing the impacts of climate change on biodiversity in the Himalayan region. Early warning systems utilizing remote sensing data can enhance disaster preparedness and mitigation efforts, helping minimize extreme weather events' impact on ecosystems and communities. Furthermore, conservation genetics studies provide valuable insights into genetic diversity, population dynamics, and species adaptation, informing conservation strategies and management practices.

Conservation efforts in the IHR must address these intertwined challenges comprehensively, integrating scientific research, policy interventions, and community engagement. Strategies may include implementing sustainable land-use practices, such as agroforestry and community-managed conservation areas, promoting ecotourism as an alternative livelihood for local communities, strengthening protected area management, and enhancing cross-border collaboration for transboundary conservation initiatives. Additionally, fostering awareness and capacity-building among local communities, policymakers, and stakeholders is essential for fostering a holistic approach to biodiversity conservation in the Himalayan region (Dhiman and Muthanarasimha 2022). Scientific research plays a crucial role in understanding ecosystem dynamics, species distribution patterns, and the impacts of anthropogenic activities on biodiversity, providing valuable insights for evidence-based conservation planning and management (Wani et al. 2023). Collaborative research efforts involving interdisciplinary approaches are necessary to effectively address the complex challenges facing Himalayan biodiversity (Kattel 2022). By prioritizing conservation actions that safeguard the region's ecological integrity while promoting sustainable development and livelihoods, it is possible to mitigate the threats facing the biodiversity of the Indian Himalayan Region and ensure its long-term survival for future generations (Rana et al. 2021).

In conclusion, addressing the impact of climate change on biodiversity in IHR requires a comprehensive and collaborative approach involving governments, local communities, scientists, and international organizations. By implementing adaptive measures and conservation strategies, we can work towards safeguarding the unique biodiversity of this ecologically sensitive and globally significant ecosystem for future generations.

Acknowledgements We thank the Ministry of Environment, Forests and Climate Change, Government of India. We also acknowledge Dr. Sunil Nautiyal, Director, Govind Ballabh Pant 'National Institute of Himalayan Environment'(NIHE), for his continued support. This study was supported by the NATCOM, the climate change cell (Ministry of Environment, Forests and Climate Change, Government of India), and all NIHE in-house projects.

References

Acharya BK, Chettri B (2012) Effect of climate change on birds, herpetofauna and butterflies in Sikkim Himalaya: a preliminary investigation. Clim Change Sikkim: Patterns, Impacts Initiatives Gangtok: Inf Public Relat Dep 141–60

Adhikari L, Khan B, Joshi S et al (2021) Community-based trophy hunting programs secure biodiversity and livelihoods: Learnings from Asia's high mountain communities and landscapes. Env Chall 4:100175. https://doi.org/10.1016/j.envc.2021.100175

Aldam C (2002) Endangered medicinal plant species in Himachal Pradesh. Curr Sci 83:797

Allendorf FW, Luikart GH, Aitken SN (2012) Conservation and the genetics of populations. John Wiley & Sons

Antão LH, Weigel B, Strona G et al (2022) Climate change reshuffles northern species within their niches. Nat Clim Chang 12:587–592. https://doi.org/10.1038/s41558-022-01381-x

Ao Y, Li X, Li L et al (2017) Two novel bocaparvovirus species identified in wild Himalayan Marmots. Sci China Life Sci 60:1348–1356. https://doi.org/10.1007/s11427-017-9231-4

Arekar K, Tiwari N, Sathyakumar S et al (2022) Geography vs. past climate: the drivers of population genetic structure of the Himalayan langur. BMC Ecol Evo 22:100. https://doi.org/10.1186/s12862-022-02054-1

Aryal A, Shrestha UB, Ji W et al (2016) Predicting the distributions of predator (snow leopard) and prey (blue sheep) under climate change in the Himalaya. Ecol Evol 6:4065–4075. https://doi.org/10.1002/ece3.2196

Badola HK, Mohinder Pal MP (2003) Threatened medicinal plants and their conservation in Himachal Himalayas

Badola R, Ahmed T, Gill AK et al (2021) An incentive-based mitigation strategy to encourage coexistence of large mammals and humans along the foothills of Indian Western Himalayas. Sci Rep 11. https://doi.org/10.1038/s41598-021-84119-7

Bagaria P, Sharma LK, Joshi BD et al (2020) West to east shift in range predicted for Himalayan Langur in climate change scenario. Glob Ecol Conserv 22:e00926. https://doi.org/10.1016/j.gecco.2020.e00926

Bajpai PK, Warghat AR, Sharma RK et al (2014) Structure and genetic diversity of natural populations of Morus alba in the Trans-Himalayan Ladakh region. Biochem Genet 52:137–152. https://doi.org/10.1007/s10528-013-9634-5

Balvanera P, Siddique I, Dee L et al (2014) Linking biodiversity and ecosystem services: current uncertainties and the necessary next steps. Bioscience 64:49–57. https://doi.org/10.1093/biosci/bit003

Banerjee A, Dimri AP, Kumar K (2021) Temperature over the Himalayan foothill state of Uttarakhand: present and future. J Earth Syst Sci 130:33. https://doi.org/10.1007/s12040-020-01527-5

Banerjee S, Niyogi R, Sarkar MS, John R (2022) Assessing the vulnerability of protected areas in the Eastern Himalayas based on their biological, anthropogenic, and environmental aspects. Trees, For People 8:100228. https://doi.org/10.1016/j.tfp.2022.100228

Baral B, Ghimire B, Basnet DR (2022) Understanding policy coherence and interplay governing biodiversity conservation and associated livelihood practices in Karnali province, Nepal. Nepal Public Policy Rev 2:27–91

Barik S, Rao B, Haridasan K et al (2018a) Classifying threatened species of India using IUCN criteria. Curr Sci 588–595

Barik SK, Tiwari ON, Adhikari D et al (2018b) Geographic distribution pattern of threatened plants of india and steps taken for their conservation. Curr Sci 114:470. https://doi.org/10.18520/cs/v114/i03/470-503

Baron JS, Poff NL, Angermeier PL et al (2002) Meeting ecological and societal needs for freshwater. Ecol Appl 12:1247–1260. https://doi.org/10.1890/1051-0761(2002)012[1247:MEASNF]2.0.CO;2

Bashir T, Poudyal K, Bhattacharya T et al (2010) Sighting of king cobra ophiophagus hannah in Sikkim, India: a new altitude record for the northeast. J Threat Taxa 2:990–991. https://doi.org/10.11609/JoTT.o2438.990-1
Basu S, Maitra A, Halder S, et al (2022) Machine learning based Earthquake Early Warning (EEW) system: a case study of Himalayan Region. Springer, pp 275–289
Batar A, Watanabe T, Kumar A (2017) Assessment of land-use/land-cover change and forest fragmentation in the Garhwal Himalayan region of India. Environments 4:34. https://doi.org/10.3390/environments4020034
Batar AK, Shibata H, Watanabe T (2021) A novel approach for forest fragmentation susceptibility mapping and assessment: a case study from the Indian Himalayan Region. Remote Sens 13:4090. https://doi.org/10.3390/rs13204090
Berreman GD (1963) Peoples and cultures of the Himalayas. Asian Surv 289–304
Bharali S, Khan ML (2011) Climate change and its impact on biodiversity; some management options for mitigation in Arunachal Pradesh. Curr Sci 855–860
Bhatt I, Negi VS, Rawal R (2020) Promoting nature-based solution (NbS) through restoration of degraded landscapes in the Indian Himalayan Region. Nat Based Solut Resilient Ecosys Soc 197–211
Bhattacharyya S, Mungi NA, Kawamichi T et al (2019) Insights from present distribution of an alpine mammal Royle's pika (*Ochotona roylei*) to predict future climate change impacts in the Himalaya. Reg Environ Change 19:2423–2435. https://doi.org/10.1007/s10113-019-01556-x
Blaustein AR, Walls SC, Bancroft BA et al (2010) Direct and indirect effects of climate change on amphibian populations. Diversity 2:281–313. https://doi.org/10.3390/d2020281
Bobrowski M, Weidinger J, Schwab N, Schickhoff U (2021) Searching for ecology in species distribution models in the Himalayas. Ecol Model 458:109693. https://doi.org/10.1016/j.ecolmodel.2021.109693
Bosson JB, Huss M, Cauvy-Fraunié S et al (2023) Future emergence of new ecosystems caused by glacial retreat. Nature 620:562–569. https://doi.org/10.1038/s41586-023-06302-2
Brooks TM, Mittermeier RA, Da Fonseca GAB et al (2006) Global biodiversity conservation priorities. Science 313:58–61. https://doi.org/10.1126/science.1127609
Burga CA, Krüsi B, Egli M et al (2010) Plant succession and soil development on the foreland of the Morteratsch glacier (Pontresina, Switzerland): straight forward or chaotic? Flora Morphol Distrib Funct Ecol Plants 205:561–576. https://doi.org/10.1016/j.flora.2009.10.001
Butchart SHM, Walpole M, Collen B et al (2010) Global biodiversity: indicators of recent declines. Science 328:1164–1168. https://doi.org/10.1126/science.1187512
Cauvy-Fraunié S, Dangles O (2019) A global synthesis of biodiversity responses to glacier retreat. Nat Ecol Evol 3:1675–1685. https://doi.org/10.1038/s41559-019-1042-8
CEPF (2005) Ecosystem profile: Indo-Burman hotspot, Eastern Himalayan region. WWF, US-Asian Programme/ CEPF, Kathmandu
CEPF (2007) Ecosystem profile: Indo-Burma hotspot, Indo-China region. Crit Ecosyst Partnersh Fund, Cambridge
Chakraborty A, Ghosh A, Sachdeva K, Joshi PK (2017) Characterizing fragmentation trends of the Himalayan forests in the Kumaon region of Uttarakhand, India. Eco Inform 38:95–109. https://doi.org/10.1016/j.ecoinf.2016.12.006
Chandra K, Sidhu A, Raghunathan C et al (2019) Fauna of Indian Trans-Himalaya: an overview. Faunal Divers Biogeogr Zones: Ind Trans-Himalaya 1–225:1–22
Chandy T, Keenan RJ, Petheram RJ, Shepherd P (2012) Impacts of hydropower development on rural livelihood sustainability in Sikkim, India: community perceptions. Mt Res Dev 32:117–125. https://doi.org/10.1659/MRD-JOURNAL-D-11-00103.1
Chaturvedi RK, Gopalakrishnan R, Jayaraman M et al (2011) Impact of climate change on Indian forests: a dynamic vegetation modeling approach. Mitig Adapt Strateg Glob Change 16:119–142. https://doi.org/10.1007/s11027-010-9257-7
Chaudhary S, Chettri N, Adhikari B et al (2023) Chapter 4: Effects of a changing cryosphere on biodiversity and ecosystem services, and response options in the Hindu Kush Himalaya. In:

Water, ice, society, and ecosystems in the Hindu Kush Himalaya: an outlook. International Centre for Integrated Mountain Development (ICIMOD), pp 123–163

Chaudhary S, Uddin K, Chettri N et al (2022) Protected areas in the Hindu Kush Himalaya: a regional assessment of the status, distribution, and gaps. Conservat Sci and Prac 4:e12793. https://doi.org/10.1111/csp2.12793

Chauhan J, Purohit VK, Dobhal P et al (2023) Nature interpretation sites: a new hope of ex-situ garden for conservation and cultivation of economically important RET MAPs in higher Himalayan Regions. In: Stress-responsive factors and molecular farming in medicinal plants. Springer, pp 69–87

Chauhan RS, Tiwari D, Bisht AS, Shukla A (2014) *Ex situ* conservation of medicinal and aromatic plants in Bharsar, Uttarakhand, India. Med Plnts Int Jrnl Phyt Rela Ind 6:282. https://doi.org/10.5958/0975-6892.2014.00018.5

Chawla A, Yadav PK, Uniyal SK, et al (2012) Long-term ecological and biodiversity monitoring in the western Himalaya using satellite remote sensing. Curr Sci 1143–1156

Chettri N, Sharma E, Shakya B et al (2010) Biodiversity in the Eastern Himalayas: status, trends and vulnerability to climate change

Chhetri B, Badola HK, Barat S (2021) Modelling climate change impacts on distribution of Himalayan pheasants. Ecol Ind 123:107368. https://doi.org/10.1016/j.ecolind.2021.107368

Chhetri PK, Gaddis KD, Cairns DM (2018) Predicting the suitable habitat of treeline species in the Nepalese Himalayas under climate change. Mt Res Dev 38:153. https://doi.org/10.1659/MRD-JOURNAL-D-17-00071.1

Chitale V, Behera M, Roy P (2015) Global biodiversity hotspots in India: significant yet under studied. Curr Sci 149–150

CITES (1979) Convention on international trade in endangered species of wild fauna and flora. https://lib.icimod.org/record/9656. Accessed 03 April 2024

CITES (2022) Summary of the 19th meeting of the CITES conference of the parties:14–25 November 2022. Earth Negot Bull 21(112):1–21. https://enb.iisd.org/sites/default/files/2021-09/enb09758e.pdf. Accessed 03 April 2024

Dahal KR, Dahal P, Adhikari RK et al (2022) Climate change impacts and adaptation in a hill farming system of the Himalayan Region: climatic trends, farmers' perceptions and practices. Climate 11:11. https://doi.org/10.3390/cli11010011

Dahal N, Lamichhaney S, Kumar S (2021) Climate change impacts on Himalayan biodiversity: evidence-based perception and current approaches to evaluate threats under climate change. J Ind Inst Sci 101:195–210. https://doi.org/10.1007/s41745-021-00237-1

Dahal N, Romine MG, Khatiwara S et al (2023) Gene flow drives genomic diversity in Asian Pikas distributed along the core and range-edge habitats in the Himalayas. Ecol Evol 13:e10129. https://doi.org/10.1002/ece3.10129

de Chazournes LB (2009) Convention on biological diversity and its protocol on biosafety. United Nations Audiovisual Library of International Law

Desai BH, Krishna Prasad KP, Yang YongPing YY et al (2011) Implementation of the convention on biological diversity: a retrospective analysis in the Hindu Kush-Himalayan countries

Dhiman MR, Muthanarasimha GP (2022) Biodiversity conservation of Western Himalayas: a pluralistic approach. IntechOpen

Diduck AP, Pratap D, Sinclair AJ, Deane S (2013) Perceptions of impacts, public participation, and learning in the planning, assessment and mitigation of two hydroelectric projects in Uttarakhand, India. Land Use Policy 33:170–182. https://doi.org/10.1016/j.landusepol.2013.01.001

Dimri AP, Allen S, Huggel C et al (2021) Climate change, cryosphere and impacts in the Indian Himalayan Region. Curr Sci 120:774. https://doi.org/10.18520/cs/v120/i5/774-790

Dimri AP, Dash SK (2012) Wintertime climatic trends in the western Himalayas. Clim Change 111:775–800. https://doi.org/10.1007/s10584-011-0201-y

Diodato N, Bellocchi G, Tartari G (2012) How do Himalayan areas respond to global warming? Intl J Climatol 32:975–982. https://doi.org/10.1002/joc.2340

Dobhal D, Gupta AK, Manish Mehta MM, Khandelwal D (2013) Kedarnath disaster: facts and plausible causes
Dodiya CL, Ahirwar P, Patel SS (2023) Prospective avenues for wildlife conservation in India: challenges and endeavours. Recent Adv Biodivers Res 113–120.
Down to Earth (2023). Climate change severely affecting biodiversity in Hindu Kush Himalayas: ICIMOD. https://www.downtoearth.org.in/news/climate-change/climate-change-severely-affecting-biodiversity-in-hindu-kush-himalayas-icimod-90155. Accessed 12 April 2024
Dullinger S, Dirnböck T, Grabherr G (2004) Modelling climate change-driven treeline shifts: relative effects of temperature increase, dispersal and invasibility. J Ecol 92:241–252. https://doi.org/10.1111/j.0022-0477.2004.00872.x
Duraisami M, Das P, Segarin S, et al (2022) Opportunity for forest protection and landscape restoration in the Indian Himalayan Region. In: Handbook of Himalayan ecosystems and sustainability, vol 1. CRC press, pp 341–368
Dutta M, Dutta PK (2023) Community-based conservation in Eastern Himalayan biodiversity hotspot-a case study. Ind J Tradit Knowl 22:220–229. https://doi.org/10.56042/ijtk.v22i1.33482
Eriksson M, Xu J, Shrestha AB, et al (2009) The changing Himalayas: impact of climate change on water resources and livelihoods in the greater Himalayas. International Centre for Integrated Mountain Development (ICIMOD)
Ewel KC (2001) Natural resource management: the need for interdisciplinary collaboration. Ecosystems 4:716–722
Forrest JL, Wikramanayake E, Shrestha R et al (2012) Conservation and climate change: assessing the vulnerability of snow leopard habitat to treeline shift in the Himalaya. Biol Cons 150:129–135. https://doi.org/10.1016/j.biocon.2012.03.001
Frankham R, Ballou JD, Briscoe DA (2004) A primer of conservation genetics. Cambridge University Press, Cambridge, UK
Frankham R, Briscoe DA, Ballou JD (2002) Introduction to conservation genetics. Cambridge University Press, Cambridge, UK
Gaire NP, Koirala M, Bhuju DR, Borgaonkar HP (2014) Treeline dynamics with climate change at the central Nepal Himalaya. Clim past 10:1277–1290. https://doi.org/10.5194/cp-10-1277-2014
Gairola S, Rawal RS, Todaria NP, Bhatt A (2014) Population structure and regeneration patterns of tree species in climate-sensitive subalpine forests of Indian Western Himalaya. J For Res 25:343–349. https://doi.org/10.1007/s11676-014-0463-0
Garg PK, Shukla A, Yousuf B, Garg S (2022) Temperature and precipitation changes over the glaciated parts of Indian Himalayan Region during 1901–2016. Environ Monit Assess 194:84. https://doi.org/10.1007/s10661-021-09689-5
Gaston AJ, Garson PJ, Hunter ML (1983) The status and conservation of forest wildlife in Himachal Pradesh, Western Himalayas. Biol Cons 27:291–314. https://doi.org/10.1016/0006-3207(83)90088-5
Gatlin PN, Case J, Bell JR et al (2018) Monitoring intense thunderstorms in the Hindu-Kush Himalayan Region, pp NH31E-1020
Ghosh (2018) https://india.mongabay.com/2018/06/two-conservation-communities-from-northeast-india-win-biodiversity-awards/, Mongabay. Accessed 12th April 2024
Ghosh M, Saha A, Pal SC, Ghosal S (2021) Do the collection and uses of NTFPs affect household well-being? evidence from forest villages in eastern Himalaya. In: Forest resources resilience and conflicts. Elsevier, pp 217–225
Global Biodiversity Framework (2021) Summary of the third meeting of the open-ended working group on the post-2020 global biodiversity framework: 23 August–3 September 2021. Earth Negot Bull 9(758). https://enb.iisd.org/sites/default/files/2021-09/enb09758e.pdf. Accessed 03 April 2024

Goodale E, Mammides C, Mtemi W et al (2022) Increasing collaboration between China and India in the environmental sciences to foster global sustainability. Ambio 51:1474–1484. https://doi.org/10.1007/s13280-021-01681-0

Goodwin H (1996) In pursuit of ecotourism. Biodivers Conserv 5:277–291. https://doi.org/10.1007/BF00051774

Gosavi VE, Mukerjee S, Kumar K et al (2020) Water resources of Indian Himalayan Region (IHR): a synthesis in the context of climate change. Environ We Int J Sci Technol 15:81–89

Gupta S, Kumar N, Verma D et al (2021) Impacts of invasive alien plant species on biodiversity in the regions of Western Himalayas, India: an overview. In: Biological diversity: current status and conservation policies. Agro Environ Media—Agriculture and Ennvironmental Science Academy, Haridwar, India, pp 123–135

Gurney GG, Adams VM, Álvarez-Romero JG, Claudet J (2023) Area-based conservation: taking stock and looking ahead. One Earth 6:98–104. https://doi.org/10.1016/j.oneear.2023.01.012

Hilton-Taylor C, Brackett D (2000) 2000 IUCN red list of threatened species

Himachal Tonite (2022) 51 million people practice hill agriculture in Himalayan ecosystem. https://himachaltonite.com/himachal/51-million-people-practice-hill-agriculture-in-himalayan-ecosystem/#:~:text=The%20Himalayan%20ecosystem%20has%2051,Himalayan%20Region%20as%20a%20whole. Accessed 03 April 2024

Hopping KA, Chignell SM, Lambin EF (2018) The demise of caterpillar fungus in the Himalayan region due to climate change and overharvesting. Proc Natl Acad Sci USA 115:11489–11494. https://doi.org/10.1073/pnas.1811591115

Hussain A, Rasul G, Mahapatra B, Tuladhar S (2016) Household food security in the face of climate change in the Hindu-Kush Himalayan region. Food Sec 8:921–937. https://doi.org/10.1007/s12571-016-0607-5

Iannone III BV, McIntyre T, Reisinger AJ et al (2022) Spatial considerations needed when balancing land conservation and development: FOR380/FR451, 3/2022. EDIS 2022

Immerzeel WW, Lutz AF, Andrade M et al (2020) Importance and vulnerability of the world's water towers. Nature 577:364–369. https://doi.org/10.1038/s41586-019-1822-y

IUCN (2015–2016) The IUCN RED list of threatened species. www.iucnredlist.org.

IUCN (2019) Species and climate change. https://www.iucn.org/sites/default/files/2022-04/species_and_climate_change_issues_brief-2019-12.pdf. Accessed 02 April 2024

IUCN-WCPA (2024) Global transboundary conservation network 2024. In: Convention on biological diversity—conference of the parties (CBD-COP), IUCN-WCPA. https://www.tbpa.net/page.php?ndx=70. Accessed 03 April 2024

Jabin G, Dolker S, Joshi BD et al (2023) Lahaul–Zanskar–Sham Valley Corridor in Indian Trans Himalayan Region facilitates dispersal and gene flow in Himalayan Ibex. Biology 12:382. https://doi.org/10.3390/biology12030382

Jiao X, Wu L, Zhang D, et al (2024) Landscape heterogeneity explains the genetic differentiation of a forest bird across the Sino-Himalayan Mountains. Mol Biol Evol 41:msae027. https://doi.org/10.1093/molbev/msae027

Jishtu V, Goraya G (2008) Cold deserts of Himachal Pradesh: Unique habitats and threatened plants. ENVIS Bull: Wildl Prot Areas, Wildl Inst Ind, Dehradun, Ind 11:17–25

Jugran AK, Bhatt ID, Rawal RS et al (2013) Patterns of morphological and genetic diversity of Valeriana jatamansi Jones in different habitats and altitudinal range of West Himalaya, India. Flora—Morphol Distrib Funct Ecol Plants 208:13–21. https://doi.org/10.1016/j.flora.2012.12.003

Kala CP (2000) Status and conservation of rare and endangered medicinal plants in the Indian trans-Himalaya. Biol Cons 93:371–379. https://doi.org/10.1016/S0006-3207(99)00128-7

Kattel GR (2022) Climate warming in the Himalayas threatens biodiversity, ecosystem functioning and ecosystem services in the 21st century: is there a better solution? Biodivers Conserv 31:2017–2044. https://doi.org/10.1007/s10531-022-02417-6

Kc KB, Tzadok E, Pant L (2022) Himalayan ecosystem services and climate change driven agricultural frontiers: a scoping review. Discov Sustain 3:35. https://doi.org/10.1007/s43621-022-00103-9

Khanal L, Chalise MK, Jiang X-L, Kyes RC (2022) Mitochondrial genetic diversity and structure of the langur population in a complex landscape of the Nepal Himalaya. Diversity 14:69. https://doi.org/10.3390/d14020069

Khurana H, Singh DN, Singh A et al (2020) Gut microbiome of endangered *Tor putitora* (Ham.) as a reservoir of antibiotic resistance genes and pathogens associated with fish health. BMC Microbiol 20:249. https://doi.org/10.1186/s12866-020-01911-7

Kohler T, Maselli D (2009) Mountains and climate change. from understanding to action. Geogr Bernensia

Kotru RK, Shakya B, Joshi S et al (2020) Biodiversity conservation and management in the Hindu Kush Himalayan Region: are transboundary landscapes a promising solution? Mt Res Dev 40. https://doi.org/10.1659/MRD-JOURNAL-D-19-00053.1

Krebs CJ (1972) Ecology. The experimental analysis of distribution and abundance

Kripalani RH, Kulkarni A, Sabade SS (2003) Western Himalayan snow cover and Indian monsoon rainfall: a re-examination with INSAT and NCEP/NCAR data. Theoret Appl Climatol 74:1–18. https://doi.org/10.1007/s00704-002-0699-z

Krishnan R, Shrestha AB, Ren G et al (2019) Unravelling Climate Change in the Hindu Kush Himalaya: rapid warming in the mountains and increasing extremes. In: Wester P, Mishra A, Mukherji A, Shrestha AB (eds) The Hindu Kush Himalaya assessment. Springer International Publishing, Cham, pp 57–97

Kumar A, Dewan S, Lochan R, Sharma DK (2020) Spatial genetic structure of black francolin (*Francolinus francolinus asiae*) in the North-Western Himalayan region based on mitochondrial control region. Mitochondrial DNA Part A 31:163–170. https://doi.org/10.1080/24701394.2020.1757664

Kumar M, Savita SH et al (2019) Assessing vulnerability of forest ecosystem in the Indian Western Himalayan region using trends of net primary productivity. Biodivers Conserv 28:2163–2182. https://doi.org/10.1007/s10531-018-1663-2

Kumar P (2012) Assessment of impact of climate change on Rhododendrons in Sikkim Himalayas using Maxent modelling: limitations and challenges. Biodivers Conserv 21:1251–1266. https://doi.org/10.1007/s10531-012-0279-1

Laladhas K, Nilayangode P, Oommen OV (2017) Biodiversity for sustainable development. Springer

Lamichhane S, Joshi R, Poudel B, Subedi P (2020) Role of community in leading conservation: effectiveness, success and challenges of community-based anti-poaching unit in Nepal. Grassroots J Nat Resour 3:94–109. https://doi.org/10.33002/nr2581.6853.03046

Lamsal P, Kumar L, Aryal A, Atreya K (2018) Future climate and habitat distribution of Himalayan Musk Deer (*Moschus chrysogaster*). Eco Inform 44:101–108. https://doi.org/10.1016/j.ecoinf.2018.02.004

Mahesh Y, Rao B, Suman K et al (2011) In Vitro maturation and fertilization in the Nilgai (*Boselaphus tragocamelus*) using oocytes and spermatozoa recovered post-mortem from animals that had died because of foot and mouth disease outbreak. Reprod Domest Anim 46:832–839. https://doi.org/10.1111/j.1439-0531.2011.01751.x

Mainali K, Shrestha BB, Sharma RK et al (2020) Contrasting responses to climate change at Himalayan treelines revealed by population demographics of two dominant species. Ecol Evol 10:1209–1222. https://doi.org/10.1002/ece3.5968

Manish K, Telwala Y, Nautiyal DC, Pandit MK (2016) Modelling the impacts of future climate change on plant communities in the Himalaya: a case study from Eastern Himalaya, India. Model Earth Syst Environ 2:92. https://doi.org/10.1007/s40808-016-0163-1

Mansourian S, Parrotta J (2019) From addressing symptoms to tackling the illness: reversing forest loss and degradation. Environ Sci Policy 101:262–265. https://doi.org/10.1016/j.envsci.2019.08.007

McElwee P (2021) Climate change and biodiversity loss. Curr Hist 120:295–300. https://doi.org/10.1525/curh.2021.120.829.295
McNeely JA, Miller KR, Reid WV, et al (1990) Conserving the world's biological diversity. International Union for conservation of nature and natural resources
Meena RK, Bhandhari MS, Barhwal S, Ginwal HS (2019) Genetic diversity and structure of Dendrocalamus hamiltonii natural metapopulation: a commercially important bamboo species of northeast Himalayas. 3 Biotech 9:60. https://doi.org/10.1007/s13205-019-1591-1
Mehta P, Sekar KC, Bhatt D et al (2020) Conservation and prioritization of threatened plants in Indian Himalayan Region. Biodivers Conserv 29:1723–1745. https://doi.org/10.1007/s10531-020-01959-x
MoEF (1972) Indian Wildlife (Protection) Act
MoEF (1986) The Environment Protection Act, 1986. India Code, 1–15. https://www.indiacode.nic.in/bitstream/123456789/4316/1/ep_act_1986.pdf. Accessed 03 April 2024
Mohapatra KP, Sehgal RN, Sharma RK, Mohapatra T (2009) Genetic analysis and conservation of endangered medicinal tree species *Taxus wallichiana* in the Himalayan region. New For 37:109–121. https://doi.org/10.1007/s11056-008-9112-9
Mondal PP, Zhang Y (2018) Research progress on changes in land use and land cover in the Western Himalayas (India) and effects on ecosystem services. Sustainability 10:4504. https://doi.org/10.3390/su10124504
Mongabay-India (2024) Himalayan treelines might be climbing higher in response to climate change. https://india.mongabay.com/2024/01/himalayan-treelines-might-be-climbing-higher-in-response-to-climate-change/?amp=1. Accessed 02 April 2024
Mora C, Tittensor DP, Adl S et al (2011) How many species are there on earth and in the ocean? PLoS Biol 9:e1001127. https://doi.org/10.1371/journal.pbio.1001127
Muthamizh Selvan K, Lyngdoh S, Gopi GV et al (2014) Population densities, group size and biomass of ungulates in a lowland tropical rainforest forest of the eastern Himalayas. Acta Ecol Sin 34:219–224. https://doi.org/10.1016/j.chnaes.2014.05.003
Nag A, Ahuja PS, Sharma RK (2015) Genetic diversity of high-elevation populations of an endangered medicinal plant. AoB PLANTS 7. https://doi.org/10.1093/aobpla/plu076
Nandargi S, Dhar O (2011) Extreme rainfall events over the Himalayas between 1871 and 2007. Hydrol Sci J 56:930–945
Nayar M, Sastry A (1988) Red data book, vol II. BSI publication, Calcutta, India
Nayar M, Sastry A (1990) Red data book of Indian plants, Botanical survey of India, Culcutta, CAMP
Nayar M, Sastry ARK (1987) Red data book of Indian plants
NBA (1927) The Indian Forest Act, 1927, National Biodiversity Authority 1–26. http://nbaindia.org/uploaded/Biodiversityindia/Legal/3.%20Indian%20forest%20act.pdf Accessed 03 April 2024
Ng D, Carver S, Gotame M et al (2019) Canine distemper in Nepal's Annapurna conservation area—implications of dog husbandry and human behaviour for wildlife disease. PLoS ONE 14:e0220874. https://doi.org/10.1371/journal.pone.0220874
Nie Y, Pritchard HD, Liu Q et al (2021) Glacial change and hydrological implications in the Himalaya and Karakoram. Nat Rev Earth Environ 2:91–106. https://doi.org/10.1038/s43017-020-00124-w
Niyogi R, Shekhar Sarkar M, Shekhar Niyogi V et al (2022) Habitat connectivity for conserving cervids in a multifunctional landscape. J Nat Conserv 68:126212. https://doi.org/10.1016/j.jnc.2022.126212
NMSHE (2010) National mission for sustaining the Himalayan eco-system (NMSHE), National action plan on climate change, Government of India, Ministry of science & technology. https://dst.gov.in/sites/default/files/NMSHE_Mission_document.pdf. Accessed 03 April 2024
Nyaupane GP, Timothy DJ (2022) Introduction: tourism and development in the Himalaya. In: Tourism and development in the Himalaya. Routledge, pp 3–17
Padma T (2014) Himalayan plants seek cooler climes: Race is on to record mountain biodiversity before it is lost. Nature 512:359–360

Pandit MK, Manish K, Koh LP (2014) Dancing on the roof of the world: ecological transformation of the Himalayan landscape. Bioscience 64:980–992. https://doi.org/10.1093/biosci/biu152

Parida BR, Kumar P, Chakraborty JS, et al (2022) Dynamics of land use/land cover over Indian Himalayan region (IHR) using satellite data from 1985 to 2015 coupled with local perceptions. In: Handbook of Himalayan ecosystems and sustainability, vol 1. CRC Press, pp 3–22

Parrotta JA, Wildburger C, Mansourian S (2012) Understanding relationships between biodiversity, carbon, forests and people: the key to achieving REDD+ objectives. A global assessment report prepared by the Global Forest Expert Panel on Biodiversity, Forest Management and REDD+

Pascual U, Adams WM, Díaz S et al (2021) Biodiversity and the challenge of pluralism. Nat Sustain 4:567–572. https://doi.org/10.1038/s41893-021-00694-7

Pathak R, Negi VS, Rawal RS, Bhatt ID (2019) Alien plant invasion in the Indian Himalayan Region: state of knowledge and research priorities. Biodivers Conserv 28:3073–3102. https://doi.org/10.1007/s10531-019-01829-1

Pathak R, Pant V, Negi VS, et al (2023) Introduction to Himalayan region and wild edible diversity. In: Himalayan fruits and berries. Elsevier, pp 1–12

Patley MK, Tiwari A, Kumar K et al (2024) Study of mountain ecosystem accounting in lower Himalaya range in Uttarkhand, India using geospatial technology. Results Eng 21:101811. https://doi.org/10.1016/j.rineng.2024.101811

Patterson T, Bhatta LD, Alfthan B et al (2017) Incentives for Ecosystem Services (IES) in the Himalayas: a 'cookbook' for emerging IES practitioners in the region

PIB (2014) Press Information Bureau, Government of India. Approval for the National Mission for Sustaining Himalayan Ecosystem launched under the National Action plan on climate change. https://pib.gov.in/newsite/PrintRelease.aspx?relid=104351. Accessed 03 April 2024

Poudel RC, Möller M, Liu J et al (2014) Low genetic diversity and high inbreeding of the endangered yews in Central Himalaya: implications for conservation of their highly fragmented populations. Divers Distrib 20:1270–1284. https://doi.org/10.1111/ddi.12237

Prabhu S (2008) 13.0 special habitats and threatened plants of Meghalaya. Spec Habitats Threat Plants Ind 11:83

Prasad N, Semwal M, Roy PS (2015) Remote Sensing and GIS for Biodiversity Conservation. In: Upreti DK, Divakar PK, Shukla V, Bajpai R (eds) Recent advances in lichenology. Springer India, New Delhi, pp 151–179

Qiong L, Zhang W, Wang H, et al (2017) Testing the effect of the Himalayan mountains as a physical barrier to gene flow in *Hippophae tibetana* Schlect. (Elaeagnaceae). PLoS ONE 12:e0172948. https://doi.org/10.1371/journal.pone.0172948

Rana SK, Luo D, Rana HK et al (2021) Geoclimatic factors influence the population genetic connectivity of *Incarvillea arguta* (Bignoniaceae) in the Himalaya-Hengduan Mountains biodiversity hotspot. J Sytem Evol 59:151–168. https://doi.org/10.1111/jse.12521

Rani S, Kumari S, Kumar P, Kumar V (2021) Biological diversity: introduction, values, threats and conservation measures. In: Biological diversity: current status and conservation policies. Agro environ media—Agriculture and Environmental Science Academy, Haridwar, India, pp 1–23

Rashid W, Shi J, Rahim I ur et al (2020) Issues and opportunities associated with trophy hunting and tourism in Khunjerab National Park, Northern Pakistan. Animals 10. https://doi.org/10.3390/ani10040597

Rasul G, Hussain A, Adhikari L, Molden DJ (2022) Conserving agrobiodiversity for sustainable food systems in the Hindu Kush Himalaya. Int J Agric Sustain 20:1117–1135. https://doi.org/10.1080/14735903.2022.2057642

Rattani V (2018) Coping with climate change: an analysis of India's national action plan on climate change. Centre for Science and Environment, New Delhi

Ravindranath N, Sukumar R, Joshi N, Murthy IK (2006) Impact of climate changes on biodiversity in India and adaptation. Ministry of Environments and Forests, Government of India, pp 21–40

Rawal R, Bhatt I, Sekar KC, Nandi S (2013) The Himalayan biodiversity: richness, representativeness, uniqueness and life-support values. GB Pant Institute of Himalayan Environment and Development (GBPIHED), Kosi-Katarmal, Almora, Uttarakhand, India, 84 pp

Rawat N, Thapliyal A, Purohit S et al (2016) Vegetation loss and ecosystem disturbances on Kedargad Mandakini subwatershed in Rudraprayag district of Uttarakhand due to torrential rainfall during June 2013. IJARSG 5:1622–1669. https://doi.org/10.23953/cloud.ijarsg.50

Reddy CS, Sreelekshmi S, Jha CS, Dadhwal VK (2013) National assessment of forest fragmentation in India: Landscape indices as measures of the effects of fragmentation and forest cover change. Ecol Eng 60:453–464. https://doi.org/10.1016/j.ecoleng.2013.09.064

Robinson C, Kawecka B, Füreder L, Peter A (2010) Biodiversity of flora and fauna in alpine waters. Alpine Waters 193–223

Roy P, Murthy M, Roy A, et al (2013) Forest fragmentation in India. Curr Sci 774–780

Sabin TP, Krishnan R, Vellore R et al (2020) Climate change over the Himalayas. In: Krishnan R, Sanjay J, Gnanaseelan C et al. (eds) Assessment of climate change over the Indian region. Springer Singapore, Singapore, pp 207–222

Saini R, Sharma N, Attada R (2023) Delving into recent changes in precipitation patterns in the Western Himalayas under global warming. In: Global warming—a concerning component of climate change. IntechOpen

Salick J, Ghimire SK, Fang Z et al (2014) Himalayan alpine vegetation, climate change and mitigation. J Ethnobiol 34:276. https://doi.org/10.2993/0278-0771-34.3.276

Sarkar MS, Amonge DE, Pradhan N et al (2021) A review of two decades of conservation efforts on tigers, co-predators and prey at the junction of three global biodiversity hotspots in the Transboundary Far-Eastern Himalayan Landscape. Animals 11:2365. https://doi.org/10.3390/ani11082365

Sathyakumar S, Bashir T (2010) Wildlife of the Himalaya: conservation issues and the way forward. Mt Ecosys Man 322–343

Schickhoff U, Bobrowski M, Böhner J et al (2016) Climate change and treeline dynamics in the Himalaya. In: Singh R, Schickhoff U, Mal S (eds) Climate change, glacier response, and vegetation dynamics in the Himalaya. Springer International Publishing, Cham, pp 271–306

Shah RDT, Narayan Shah D, Domisch S (2012) Range shifts of a Relict Himalayan Dragonfly in the Hindu Kush Himalayan region under climate change scenarios. Int J Odonatol 15:209–222. https://doi.org/10.1080/13887890.2012.697399

Sharma E, Chettri N, Tse-Ring K et al (2009) Climate change impacts and vulnerability in the Eastern Himalayas

Sharma E, Tse-ring K, Chettri N et al (2008) Biodiversity in the Himalayas–trends, perception and impacts of climate change

Sharma G, Tambe S, Rawat GS, Arrawatia ML (2016) Yak herding and associated transboundary issues in the Sikkim Himalaya, India. Yak Move transboundary challenges and opportunities for Yak raising in a changing Hindu Kush Himalayan Region, pp 93–112

Sharma R, Kumar R, Pradhan P, Sharma A (2022a) Climate-induced glacier retreats and associated hazards: need for robust glaciers and glacial lake management policy in Sikkim Himalaya, India. Clim Change: Impacts, Respons Sustain Ind Himal 161–182

Sharma S, Chhabra M, Singh SK et al (2022b) Genetic diversity and population structure of critically endangered *Dactylorhiza hatagirea* (D. Don) Soo from North-Western Himalayas and implications for conservation. Sci Rep 12:11699. https://doi.org/10.1038/s41598-022-15742-1

Shekhar C (2021) Quantification and diminution of *Quercus semecarpifolia* forests ecosystem services in himalayan region—an overview. IJAEB 14. https://doi.org/10.30954/0974-1712.01.2021.9

Shivanna KR (2022) Climate change and its impact on biodiversity and human welfare. ProcIndian Natl Sci Acad 88:160–171. https://doi.org/10.1007/s43538-022-00073-6

Shrestha AB, Wake CP, Mayewski PA, Dibb JE (1999) Maximum temperature trends in the Himalaya and Its vicinity: an analysis based on temperature records from Nepal for the period 1971–94. J Clim 12:2775–2786. https://doi.org/10.1175/1520-0442(1999)012%3c2775:MTTITH%3e2.0.CO;2

Shrestha F, Uddin K, Maharjan SB, Bajracharya SR (2016) Application of remote sensing and GIS in environmental monitoring in the Hindu Kush Himalayan region. AIMS Environ Sci 3:646–662. https://doi.org/10.3934/environsci.2016.4.646

Shrestha KB, Hofgaard A, Vandvik V (2015) Recent treeline dynamics are similar between dry and mesic areas of Nepal, C entral Himalaya. JPECOL 8:347–358. https://doi.org/10.1093/jpe/rtu035

Shrestha UB, Gautam S, Bawa KS (2012) Widespread climate change in the Himalayas and associated changes in local ecosystems. PLoS ONE 7:e36741. https://doi.org/10.1371/journal.pone.0036741

Singh C, Chauhan N, Upadhyay SK et al (2021a) The Himalayan natural resources: challenges and conservation for sustainable development. J Pharmacogn Phytochem 10:1643–1648

Singh D, Hajra P (1996) Floristic diversity. Biodiversity Status in the Himalaya New Delhi: British Council 23–38

Singh JS (1998) Sustainable development: an ecological viewpoint. In: Sundriyal RC, Shankar U, Upreti TC (eds) Perspectives for planning and development in Northeast India. Himavikas Occasional Publication, No. 11, G.B. Pant Institute of Himalayan Environment and Development, Kosi-Katarmal, Almora, pp 5–16

Singh J (2006) Sustainable development of the Indian Himalayan region: linking ecological and economic concerns. Curr Sci 784–788

Singh R, Schickhoff U, Mal S (eds) (2016) Climate change, glacier response, and vegetation dynamics in the Himalaya. Springer International Publishing, Cham

Singh R, Sharma RK, Bhutia TU et al (2021b) Conservation policies, eco-tourism, and end of pastoralism in Indian Himalaya? Front Sustain Food Syst 5:613998. https://doi.org/10.3389/fsufs.2021.613998

Singh SP, Bassignana-Khadka I, Karky BS, Eklabya Sharma ES (2011) Climate change in the Hindu Kush-Himalayas: the state of current knowledge

Singh VK, Joshi BD, Singh A et al (2022a) Genetic diversity and population structure of the northern red muntjac (*Muntiacus vaginalis*) in Indian Himalayan region. Mamm Biol 102:537–544. https://doi.org/10.1007/s42991-022-00254-2

Singh VK, Singh SK, Joshi BD et al (2022b) Population genetics of the snow leopards (Panthera uncia) from the Western Himalayas, India. Mamm Biol 102:263–269. https://doi.org/10.1007/s42991-021-00196-1

Singhal M, Chauhan V, Singh R, et al (2022) Glacial lake outburst floods (GLOFs) and the requisite of wireless sensor based early warning system. AIP Publishing

Solomon S (2007) Climate change 2007-the physical science basis: Working group I contribution to the fourth assessment report of the IPCC. Cambridge University Press

Sonam K, Dorjay R, Khanyari M et al (2022) A community-based conservation initiative for wolves in the Ladakh Trans-Himalaya, India. Front Ecol Evol 10. https://doi.org/10.3389/fevo.2022.809817

Stanturf JA (2015) Future landscapes: opportunities and challenges. New Forest 46:615–644. https://doi.org/10.1007/s11056-015-9500-x

Steigerwald E (2021) Impacts of climate change on amphibians. University of California, Berkeley, CA, USA, AmphibiaWeb

Talchabhadel R, Sharma S, Khadka N et al (2022) An outlook on the applicability of satellite precipitation products for monitoring extreme precipitation events in Nepal Himalaya. Weather 77:174–180. https://doi.org/10.1002/wea.4143

Taloor AK, Thapliyal A, Kimothi S et al (2022) Geospatial technology-based monitoring of HAGL in the context of flash flood: a case study of Rishi Ganga Basin, Indaia. Geosyst Geoenviron 1:100049. https://doi.org/10.1016/j.geogeo.2022.100049

Talukdar B, Sinha S (2013) Challenges and opportunities of transboundary rhino conservation in India and Nepal. Pachyderm 54:45–51

Telwala Y, Brook BW, Manish K, Pandit MK (2013) Climate-induced elevational range shifts and increase in plant species richness in a Himalayan biodiversity epicentre. PLoS ONE 8:e57103. https://doi.org/10.1371/journal.pone.0057103

Tewari VP, Verma RK, Von Gadow K (2017) Climate change effects in the Western Himalayan ecosystems of India: evidence and strategies. For Ecosyst 4:13. https://doi.org/10.1186/s40663-017-0100-4

Thapa K, Manandhar S, Bista M et al (2018) Assessment of genetic diversity, population structure, and gene flow of tigers (*Panthera tigris tigris*) across Nepal's Terai Arc Landscape. PLoS ONE 13:e0193495. https://doi.org/10.1371/journal.pone.0193495

Thayyen RJ, Dimri A, Kumar P, Agnihotri G (2013) Study of cloudburst and flash floods around Leh, India, during August 4–6, 2010. Nat Hazards 65:2175–2204

The Third Pole (2019) Biodiversity of Hindu Kush Himalayas will plummet by 2100. https://www.thethirdpole.net/en/nature/hindu-kush-biodiversity/. Accessed 28 March 2024

Tiwari OP, Rana YS, Krishan R et al (2018) Regeneration dynamics, population structure, and forest composition in some ridge forests of the Western Himalaya, India. For Sci Technol 14:66–75. https://doi.org/10.1080/21580103.2018.1447517

Tiwari PC (2000) Land-use changes in Himalaya and their impact on the plains ecosystem: need for sustainable land use. Land Use Policy 17:101–111. https://doi.org/10.1016/S0264-8377(00)00002-8

Tse-ring K, Sharma E, Chettri N, Shrestha A (2010) Climate change impact and vulnerability in the Eastern Himalayas: synthesis report. Climate change vulnerability of mountain ecosystems in the Eastern Himalayas

UNESCO (2014) Great Himalayan National Park conservation area. https://whc.unesco.org/en/list/1406/. Accessed 03 April 2024

U.S. National Park Service (2021) Wildlife and climate change. https://www.nps.gov/articles/000/wildlife-climateimpact.htm. Accessed 02 April 2024

Ved D, Kinhal G, Haridasan K et al (2003a) Report of the conservation assessment and management prioritisation for the medicinal plants of Arunachal Pradesh, Assam, Meghalaya and Sikkim. Foundation for Revitalisation of Local Health Traditions (FLRHT), Bangalore, India, 153 pp

Ved D, Kinhal G, Ravikumar K et al (2003b) Conservation assessment and management prioritization for the medicinal plants of Himachal Pradesh, Jammu & Kashmir and Uttaranchal. Foundation for Revitalisation of Local Health Traditions, Bangalore, India

Waitt G, Lane R, Head L (2003) The boundaries of nature tourism. Ann Tour Res 30:523–545. https://doi.org/10.1016/S0160-7383(02)00104-4

Wang W, Gao X, Zheng S et al (2021) The gut microbiome and metabolome of Himalayan Griffons (*Gyps himalayensis*): insights into the adaptation to carrion-feeding habits in avian scavengers. Avian Res 12:52. https://doi.org/10.1186/s40657-021-00287-0

Wani SA, Ahmad R, Gulzar R et al (2022) Diversity, distribution and drivers of alien flora in the Indian Himalayan region. Glob Ecol Conserv 38:e02246. https://doi.org/10.1016/j.gecco.2022.e02246

Wani SA, Mugal MA, Dar FA et al (2023) Biodiversity data synthesis on trees of Indian Himalayan region: policy and management implications. Glob Ecol Conserv 48:e02698. https://doi.org/10.1016/j.gecco.2023.e02698

Watt A (2020) Land-use intensity and land-use change: impacts on biodiversity. In: Leal Filho W, Azul AM, Brandli L et al (eds) Life on land. Springer International Publishing, Cham, pp 1–13

Werhahn G, Liu Y, Meng Y et al (2020) Himalayan wolf distribution and admixture based on multiple genetic markers. J Biogeogr 47:1272–1285. https://doi.org/10.1111/jbi.13824

Wester P, Chaudhary S, Chettri N, Jackson M, Maharjan A, Nepal S, Steiner JF (2023) Water, ice, society, and ecosystems in the Hindu Kush Himalaya: an outlook. ICIMOD, pp 1–217. https://doi.org/10.53055/ICIMOD.1028

Wood SA, Karp DS, DeClerck F et al (2015) Functional traits in agriculture: agrobiodiversity and ecosystem services. Trends Ecol Evol 30:531–539. https://doi.org/10.1016/j.tree.2015.06.013

WWF (2001) Ecoregion-based conservation in the Eastern Himalaya: identifying important areas for biodiversity conservation. WWF Nepal, Kathmandu

WWF (2024). Habitat loss. https://wwf.panda.org/discover/knowledge_hub/where_we_work/eastern_himalaya/threats/habitat_loss/. Accessed 28 March 2024

Xu J, Badola R, Chettri N et al (2019) Sustaining biodiversity and ecosystem services in the Hindu Kush Himalaya. In: Wester P, Mishra A, Mukherji A, Shrestha AB (eds) The Hindu Kush Himalaya assessment. Springer International Publishing, Cham, pp 127–165

Xu J, Grumbine RE, Shrestha A et al (2009) The melting Himalayas: cascading effects of climate change on water, biodiversity, and livelihoods. Conserv Biol 23:520–530. https://doi.org/10.1111/j.1523-1739.2009.01237.x

Xu J, Wang X, Jing H et al (2020) Identification and genotyping of Enterocytozoon bieneusi in wild Himalayan marmots (*Marmota himalayana*) and Alashan ground squirrels (*Spermophilus alashanicus*) in the Qinghai-Tibetan Plateau area (QTPA) of Gansu Province, China. Parasites Vectors 13:367. https://doi.org/10.1186/s13071-020-04233-9

Yadav A, Singh A (2024) The Himalayas in the anthropocene. In: The Himalayas in the anthropocene: environment and development. Springer, pp 1–31

Yadav RR, Park W, Singh J, Dubey B (2004) Do the western Himalayas defy global warming? Geophys Res Lett 31:2004GL020201. https://doi.org/10.1029/2004GL020201

Yadava AK, Sharma YK, Dubey B et al (2017) Altitudinal treeline dynamics of Himalayan pine in western Himalaya, India. Quatern Int 444:44–52. https://doi.org/10.1016/j.quaint.2016.07.032

Yang L, Wei F, Zhan X et al (2022) Evolutionary conservation genomics reveals recent speciation and local adaptation in threatened takins. Mol Biol Evol 39:msac111. https://doi.org/10.1093/molbev/msac111

Zahoor B, Liu X, Kumar L et al (2021) Projected shifts in the distribution range of Asiatic black bear (*Ursus thibetanus*) in the Hindu Kush Himalaya due to climate change. Eco Inform 63:101312. https://doi.org/10.1016/j.ecoinf.2021.101312

Zeng H, Jia G, Epstein H (2011) Recent changes in phenology over the northern high latitudes detected from multi-satellite data. Environ Res Lett 6:045508. https://doi.org/10.1088/1748-9326/6/4/045508

Zhang J-M, López-Pujol J, Gong X et al (2018) Population genetic dynamics of Himalayan-Hengduan tree peonies, Paeonia subsect. Delavayanae. Mol Phylogenet Evol 125:62–77. https://doi.org/10.1016/j.ympev.2018.03.003

Zhang T, Qiao Q, Novikova PY et al (2019) Genome of *Crucihimalaya himalaica*, a close relative of Arabidopsis, shows ecological adaptation to high altitude. Proc Natl Acad Sci 116:7137–7146

Zomer R, Sharma E (2009) A need for mountain perspectives: impacts of climate change on ecosystem services in the greater Hindu Kush–Himalaya region. IOP Publishing

Zomer RJ, Trabucco A, Metzger MJ et al (2014) Projected climate change impacts on spatial distribution of bioclimatic zones and ecoregions within the Kailash Sacred Landscape of China, India, Nepal. Clim Change 125:445–460. https://doi.org/10.1007/s10584-014-1176-2

Zou D, Tian S, Zhang T et al (2021) Vulture Genomes reveal molecular adaptations underlying obligate scavenging and low levels of genetic diversity. Mol Biol Evol 38:3649–3663. https://doi.org/10.1093/molbev/msab130

Chapter 9
Current Status and Cataloging of Threatened Plants of Uttarakhand

Shaiphali Saxena and Jitendra Prasad

Abstract Uttarakhand, one of the divisions of Indian Himalayan Region (IHR), is an abode of exclusive floral and faunal biodiversity of India. The state is exceptionally bestowed with multifarious natural ecosystems and the very first province to release the report on the conservation of endemic threatened flora to redress the plant blindness in this era of rapidly declining biodiversity. In this order, the present study is a little attempt to compile a list of total threatened flora of Uttarakhand with the help of state, national and international literature. The results concluded a total of 358 plant species under different threat categories within the state. Moreover, nearly half of the total threatened flora emerged out as endemic (40% bryophytes, 100% pteridophytes and 40.3% angiosperms). Collectively, the angiosperms dominated the threatened list of total flora of Uttarakhand followed by gymnosperms, bryophytes and fungi. This may be due to the less attention paid on lower flora to collect the information at different levels. To ignore such type of plant blindness, the United Nations Biodiversity has also insisted the people to use the new term '*funga*' alongside '*flora and fauna*'. The International Union for Conservation of Nature (IUCN) is universally acknowledged to assess the red list of the total threatened floral and faunal species. However, the other threat assessing organizations at state, national and international levels should be given equal importance to maintain the threatened data about all the species of the world as IUCN alone is not the last hope. Further, the collaborative efforts, fundings and engagement of governing bodies and non-governmental organizations can broaden the conservation practices.

Keywords Biodiversity · Current status of threatened species · Endemic plants of Uttarakhand · IUCN red list · Threatened and rare plants

S. Saxena (✉)
Department of Botany, Government Post Graduate College, Kunidhar, Manila, Almora 263667, Uttarakhand, India
e-mail: shefalisaxena1192@gmail.com

J. Prasad
Department of Geography, Government Post Graduate College, Kunidhar, Manila, Almora 263667, Uttarakhand, India

D. Arya et al. (eds.), *Climate Change Impact on Himalayan Biodiversity*, Environmental Science and Engineering, https://doi.org/10.1007/978-3-031-77149-1_9

9.1 Introduction

The world is facing high risk in the number of species considered to be rare or threatened in response to increasing human disturbances and pressure on Earth's resources. Approximately, one-third of flowering plant species are thought to be endangered. The extinction rates are 1000 to 10,000 times higher than the background rates. The biodiversity loss is one of the cardinal problems to be the main operator of Anthropocene. Biodiversity threats are mainly due to the habitat loss, overexploitation, climate change, invasion, fragmentation and pollution. The extinction problem is urgent as many species with limited distribution or endemic nature have not yet been documented (Lasushe et al. 2002; Tali et al. 2019). The Indian Himalayas are hailed as biological and cultural diversity repository due to their rich and unique floral and faunal diversity, ecology and geology. They harbor about half of the flora recorded all over India having 30% endemic plants (Bargali et al. 2022). The Himalayas deserve a special mention among the biodiversity hotspots of the world. They cover an extensive area of bioclimatic regions on the global level, where the vascular plant species are found ranging from 60 to 6400 m above the sea level. The climate change makes Himalayan ecosystem sensitive and vulnerable, leading to altitudinal and latitudinal shifting of plant species with change in environmental conditions (Rawat et al. 2022). Many conservation organizations at national and international level have come forward to assess and manage the threat statuses of biodiversity at global and regional scales. The Global Strategy for Plant Conservation (2011–2020) of the CBD (Convention on Biological Diversity) mainly aims to assess the conservation status of entire flora of the Earth. In that sense, IUCN (International Union for Conservation of Nature) is considered the chief source of maintaining the red list criteria for the threatened organisms under different categories worldwide (CBD 2012; Keith et al. 2014). The IUCN has assigned nine threat ranks to the floral and faunal species on the basis of various data collected from different sources across the world namely extinct (EX), extinct in the wild (EW), critically endangered (CR), endangered (EN), vulnerable (VU), near threatened (NT), least concerned (LC), data deficient (DD) and not evaluated (NE) (Fig. 9.1). The situations for plant conservation before 30 or more decades ago and now have huge difference as in 1980s conservations organizations used to neglect the plants as compared to the animals. Later, IUCN in collaboration with WWF (World Wildlife Fund) launched Plant Conservation Programme to include governments and industries, promote in situ and ex situ strategies, and enhance botanical gardens worldwide to conserve the plants around the world (Heywood 2017).

Being a mountainous state, Uttarakhand lies within the Indian Himalayan global biodiversity (IHR) hotspot with wide area of 53,483 km^2 (Fig. 9.2). The state bears 24,303 km^2 forest area of its total geographical region comprising about 4800 seed plants (wild taxa) coming under 1,400 genera and 125 families. Due to covering 25% of the total Indian floral diversity within only 1.69% of geographical range of the country, the state has been bestowed with the title of 'herbal state of India' (Chaturvedi et al. 2022; Bargali et al. 2022). Additionally, 107 plant species endemic

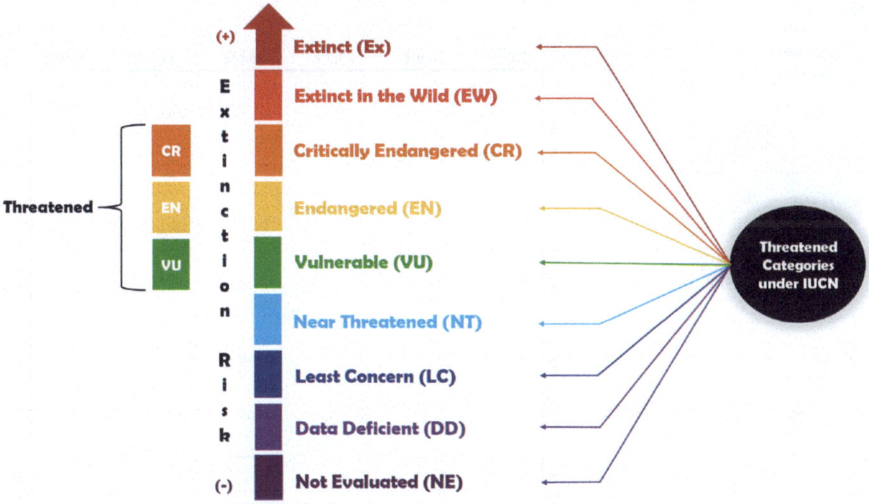

Fig. 9.1 Threat ranks assigned by IUCN with the different levels of risks

(being restrictedly distributed within the state) to Uttarakhand also call for special attention towards their conservation (Singh et al. 2015; Pusalkar and Srivastava 2018). Previous studies state that the flowering plants with 1503 genera and 213 families (93 endemic), 487 fern species (15 endemic), 18 gymnosperm species are reported in different vegetation ranges of Uttarakhand from sub-tropical forests of Upper Gangetic Plain and Shiwalik zone to Trans-Himalayan alpine desert vegetation. The anthropogenic activities like, urbanization, shifting cultivation, mining and migration have resulted in increased habitat fragmentation, decline in forest cover and the loss of biodiversity. Consequently, several floral species in the state have been included under rare, endemic and threatened categories during last few decades (Rawat 2008; Upadhaya et al. 2013). Although, several ecological, taxonomical and conservational studies have been accomplished in the state, but the information related to the threatened flora is still sparse and demands adequate conservational activities. Moreover, the comprehensive checklists of threatened floral species of Uttarakhand are either scanty or mainly emphasized on the higher plants. Therefore, the present study attempts to prepare a catalogue of the threatened flora (including fungi) of Uttarakhand state to help in undertaking the important steps towards their conservation efforts.

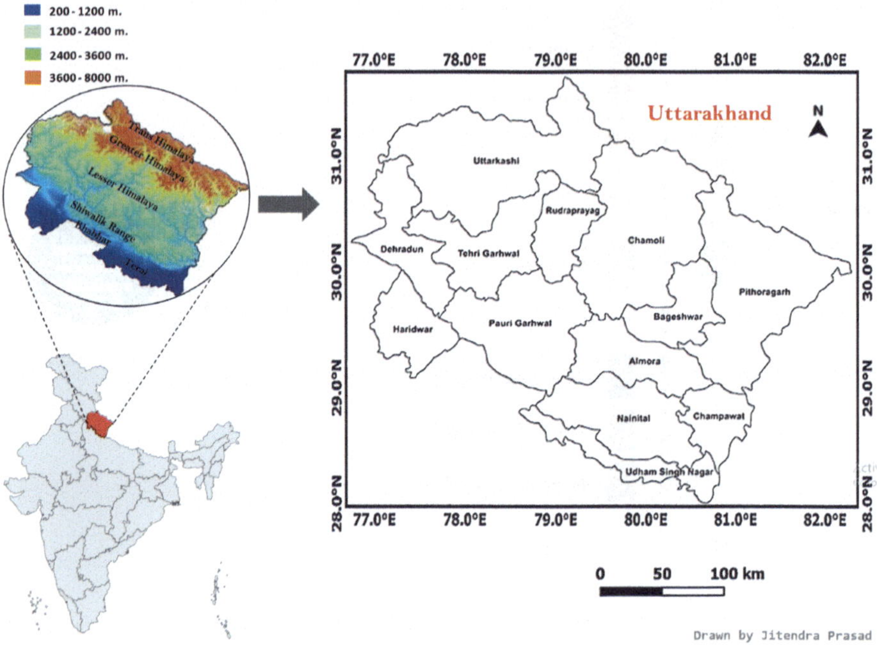

Fig. 9.2 Location map of Uttarakhand

9.2 Materials and Methodology

The current study with main focus upon the threatened taxa of Uttarakhand includes a thorough analysis of the research findings of reputed academic journal websites such as Web of Science (https://wosjournal.com/), Scopus (https://www.scopus.com/home.uri), PubMed (https://pubmed.ncbi.nlm.nih.gov/), Springer (https://link.springer.com/), Google Scholar (https://scholar.google.com/), Research Gate (https://www.researchgate.net/), JSTOR (https://www.jstor.org/), Science Direct (https://www.sciencedirect.com/), Taylor & Francis (https://www.tandfonline.com/), Wiley Online Library (https://onlinelibrary.wiley.com/), etc. from September, 2023 to January, 2024. The extensive search was done by browsing the keywords like biodiversity, threatened and rare plants, endemic plants of Uttarakhand, current status of threatened species, IUCN red list (https://www.iucnredlist.org/), CITES (https://cites.org/eng/app/appendices.php), BSI (https://bsi.gov.in/), BSI Uttarakhand (https://bsi.gov.in/regional-centres/en?rcu=131), Indian Biodiversity Portal (https://indiabiodiversity.org/), Uttarakhand State Biodiversity Board (https://sbb.uk.gov.in/), etc. The authentication of plants names was done by The Plant List (2013) (http://www.theplantlist.org/), Plants of the World Online (POWO 2024 https://powo.science.kew.org/) and International Plant Name Index (IPNI, https://www.ipni.org/). As the research data about the threatened taxa of Uttarakhand is scarce or limited in English, hence manual searches were also opted by using websites and other sources of cited references.

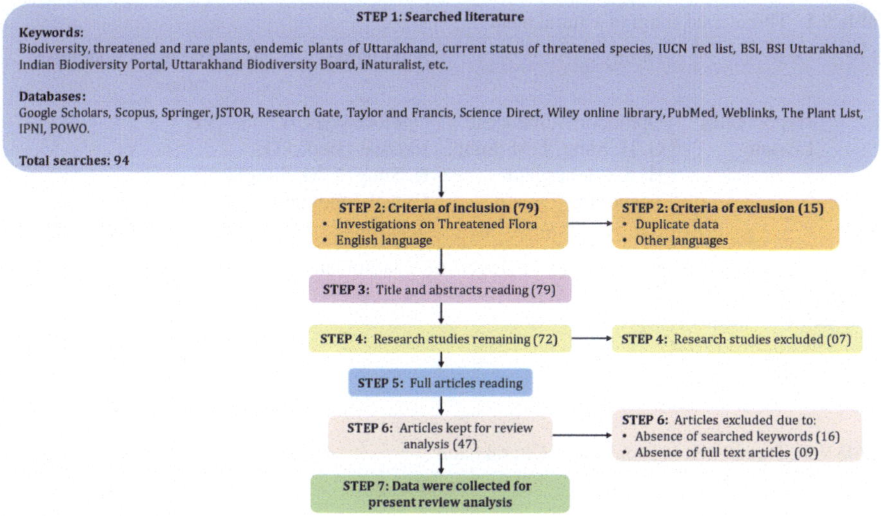

Fig. 9.3 Strategy of literature search and investigation analysis

The duplicated data and papers after search were excluded. Further, the abstracts and full titles were read, screened and selected for the present study (Fig. 9.3).

9.3 Results and Discussion

A comprehensive listing of threatened flora of Uttarakhand categorized under fungi, bryophytes, pteridophytes, gymnosperms and angiosperms was accomplished (Tables 9.1, 9.2, 9.3, 9.4 and 9.5) on the basis of previous literature available in the form of different sources like research articles, reviews, books, floras, web databases, news articles, etc. in the present study. All the plant names are arranged according to Phylogenetic Classification of Fungi (Hibbett et al. 2007; Ashton 2009), Consortium of Bryophyte Herbaria (https://bryophyteportal.org/portal/), Pteridophyte Phylogeny Group Classification, Gymnosperm Phylogeny Group Classification (Luu-Dam et al. 2023), and Angiosperms Phylogeny Group IV Classification (APG IV 2016). The threatened plant list also includes synonyms of the accepted species names which were checked on Plants of the World Online (POWO 2024 https://powo.science.kew.org/).

Accordingly, the present study provides the first consolidated data of threatened flora (including fungi) of Uttarakhand state. After extensive search of the previous literature, it came to conclude that a total of 358 species including 1 fungus, 5 bryophytes, 2 pteridophytes, 15 gymnosperms, and 335 angiosperms have been recorded as threatened in Uttarakhand (Fig. 9.4). They are respectively 0.13% (fungi), 1.52% (bryophytes), 0.25% (pteridophytes), 1.42% (gymnosperms), and 0.52%

Table 9.1 Threatened fungi of Uttarakhand

S.N	Order	Family	Plant name	Threat status	Endemism
1	Hypocreales Lindau	Ophiocordycipitaceae G. H. Sung, J. M. Sung, Hywel-Jones and Spatafora	*Ophiocordyceps sinensis* (Berk.) G. H. Sung, J. M. Sung, Hywel-Jones and Spatafora	VU*	–

Plant families arranged according to Phylogenetic Classification of Fungi (Hibbett et al. 2007; Ashton 2009)
Superscript '*' means IUCN
Source iNaturalist (2024)
VU Vulnerable

Table 9.2 Threatened bryophytes of Uttarakhand

S. N	Order	Family	Plant name	Threat status	Endemism
1	Hypnales W. R. Buck and Vitt	Brachytheciaceae Schimp	*Brachythecium buchananii* (Hook.) A. Jaeger	EN*	–
2	Pottiales Hazsl	Pottiaceae Schimp	*Hyophila involute* (Hook.) A. Jaeger	VU*	–
3	Marchantiales Limpr	Aytoniaceae Cavers	*Plagiochasma appendiculatum* Lehm. and Lindenb	VU*	–
4		Exormothecacaeae Müll. Frib. ex Grolle	*Aitchisoniella himalayensis*	EN*	✓
5			*Stephensoniella brevipedunculata* Kashyap	EN*	✓

Plant families arranged according to Consortium of Bryophyte Herbaria
Superscript '*' means IUCN
Source FRI (2013, 2023)
EN Endangered, *VU* Vulnerable

(angiosperms) as compared to the total threatened world flora compiled by IUCN 2024 (version 2023-1). While to compare with the Indian threatened flora, they are 0.17% (fungi), 6.25% (bryophytes), 0.48% (pteridophytes), 100% (gymnosperms), and 19.71% (angiosperms), respectively (Fig. 9.5). Out of these total threatened flora of Uttarakhand about 40% bryophytes (2 species), 100% pteridophytes (all 2 species), and 40.3% angiosperms (135 species) are endemic in the state. This clearly depicts that less than half of the total threatened flora of Uttarakhand is endemic in nature. All the threat statuses indicated in the tables have been catalogued from different previous studies are also shown (Fig. 9.6).

Overall, angiosperm plant species are on the top in threatened list of the flora of Uttarakhand followed by gymnosperms, bryophytes and fungi in the present study. The order Asparagales is the largest order with 04 families (Orchidaceae with 28

Table 9.3 Threatened pteridophytes of Uttarakhand

S.N	Order	Family	Plant name	Threat status	Endemism
1	Cyatheales A. B. Frank	Cyatheaceae Kaulf	*Alsophila spinulosa* (Wall. ex Hook.) R. M. Tryon [= *Cyathea spinulosa* Wall.]	T[e], CA-2	✓
2	Polypodiales Link	Thelypteridaceae Ching ex Pic. Serm	*Thelypteris* × *kumaunica* (Holttum) Fraser-Jenk. [= *Christella* × *kumaunica* Holttum]	VU[@]	✓

Plant families arranged according to Pteridophyte Phylogeny Group classification (Luu-Dam et al. 2023)
Superscripts USBB (e), ENVIS (@), CITES Appendix-II (CA-2)
Source Checklist of threatened plants of Uttaranchal (2013), Singh and Sahu (2015), FRI (2023)
VU Vulnerable, *T* Threatened

genera and 49 species, Iridaceae with 01 genus and 1 species, Amaryllidaceae with 01 genus and 04 species, and Asparagaceae with 05 genera and 07 species) and 61 threatened plant species occurring in Uttarakhand. Family wise Orchidaceae are the leading plant family (49 species, 14.63%) followed by Fabaceae (28 species, 8.36%), Arecaceae (20 species, 5.97%), Poaceae (20 species, 5.97%), and Apiaceae (17 species, 5.07%) of the total threatened angiosperms recorded. Other orders and families possess smaller number of genera and species than above given order and top 5 families. The *Impatiens* genus possesses highest number (10) of threatened species (*Impatiens devendrae* Pusalkar, *I. duthiei* Hook. f., *I. inayatii* Hook. f., *I. jaeschkei* Hook. f., *I. kaliensis* Grey-Wilson, *I. langeana* Hook. f., *I. podocarpa* Hook. f., *I. polysciadia* Hook. f., *I. reidii* Hook. f., and *I. violoides* Edgew. ex Hook. f.) followed by the genus *Berberis* with 08 species (*Berberis affinis* G. Don, *B. ahrendtii* R. R. Rao and Uniyal, *B. garhwalensis* C. K. Schneid., *B. jaunsarensis* (Ahrendt) Laferr., *B. lambertii* R. Parker, *B. osmastonii* Dunn, *B. pseudumbellata* R. Parker, and *B. rawatii* U. L. Tiwari and B. S. Adhikari), genus *Spiraea* with 07 species (*S. diversifolia* Dunn, *S. duthieana* Zinserl., *S. hypoleuca* Dunn., *S. panchananii* Panigrahi and K M. Purohit, *S. panigrahiana* K. M. Purohit., *S. parkeri* Panigrahi and K. M. Purohit, and *S. raizadae* Panigrahi and K. M. Purohit), and genus *Saussurea* with 07 species (*Saussurea atkinsonii* C. B. Clarke, *S. bracteata* Decne., *S. pterocaulon* Decne., *S. gossypiphora* D. Don, *S. obvallata* (DC.) Edgew., *S. roylei* (DC.) Sch. Bip., and *S. sudhanshui*). While other genera represent lesser number of the threatened plants in Uttarakhand.

As the present assessment clarifies that the lower flora such as fungi, algae, bryophytes and pteridophytes are not as much threatened as the higher flora like phanerogams (gymnosperms and angiosperms). This may be due to the less attention paid on them to collect the information at state, national and world level. Further, high altitudes, land use change (agriculture) and harsh geographical conditions don't allow the researchers and scientists to collect the precise information regarding their population, presence and threat assessment. Threat status and endemism are also

Table 9.4 Threatened gymnosperms of Uttarakhand

S.N	Order	Family	Plant name	Threat status	Endemism
1	Cycadales Pers. ex Bercht. and J. Presl	Cycadaceae Pers	*Cycas beddomei* Dyer	EN*, CA-2	–
2			*Cycas circinalis* L	EN*	–
3		Zamiaceae Horan	*Zamia fischeri* Miq	EN*	–
4			*Zamia furfuracea* L. f	EN*, CA-2	–
5			*Zamia integrifolia*	NT*	–
6			*Zamia pumila* L	VU*	–
7	Ginkgoales Gorozh	Ginkgoaceae Engl	*Ginkgo biloba* L	EN*	–
8	Pinales Gorozh	Pinaceae Spreng. ex F. Rudolph	*Abies spectabilis* (D. Don) Spach	NT*	–
9			*Cedrus atlantica* (Endl.) G. Manetti ex Carriere	EN*	–
10			*Pinus gerardiana* Wall. ex D. Don	NT*	–
11			*Pinus greggii* Engelm. ex Parl	NT*	–
12	Cupressales Link	Cupressaceae Gray	*Cryptomeria japonica* (L.f.) D. Don	NT*	–
13			*Platycladus orientalis* (L.) Franco	NT*	–
14		Taxaceae Gray	*Taxus wallichiana* Zucc	EN*	–
15	Ephedrales Dumort	Ephedraceae Dumort	*Ephedra gerardiana* Wall. ex Florin	VU*	–

Plant families arranged according to GPG classification (Luu-Dam et al. 2023)
Superscripts IUCN (*), CITES Appendix-II (CA-2)
Source FRI (2023)
EN Endangered, *VU* Vulnerable, *NT* Near threatened

somewhere connected with the invasive flora present in particular place or state. The Uttarakhand holds the record of fourth largest number (104) of invasive plant species occurring in India after Karnataka (117), Maharashtra (116), and Bihar (105) (Khuroo et al. 2021). The plant invasion greatly disturbs the native flora to flourish in the wild resulting in their habitat loss. The high-level de-rooting and extraction of threatened flora from the wild having medicinal properties also enhances the habitat loss of the plants which hardly regain their conservation. Some plant and fungal species never get the chance to be discovered to the scientists and lose their existence before they are known to the world. We all talk about the conservation of biodiversity, its value and the benefits, but don't realize to also pay attention towards the proper study, value, benefits and the conservation of lower flora. For example,

Table 9.5 Threatened angiosperms of Uttarakhand

S.N	Order	Family	Plant name	Threat status	Endemism
1	Austrobaileyales Takht. ex Reveal	Schizandraceae Blume	*Schisandra grandiflora* (Wall.) Hook. f. and Thomson [= *Kadsura grandiflora* Wall.]	I^a, NE^*, I^c, $I^@$	–
2			*Schisandra propinqua* (Wall.) Baill. [= *Kadsura propinqua* Wall.]	I^a, NE^*, I^c, $I^@$	–
3	Magnoliales Juss. ex Bercht. and J. Presl	Myristicaceae R. Br	*Myristica malabarica* Lam	VU^*	–
4		Magnoliaceae Juss	*Magnolia doltsopa* (Buch.-Ham. ex DC.) Filger [= *Michelia doltsopa* Buch.-Ham. ex DC.]	DD^*	–
5			*Magnolia kisopa* (Buch.-Ham. ex DC.) Filger [= *Michelia kisopa* Buch.-Ham. Ham. ex DC.]	DD^*	–
6	Laurales Juss. ex Bercht. and J. Presl	Lauraceae Juss	*Alseodaphne himalayana* Kosterm	NE^*, VU^g	–
7			*Cinnamomum glanduliferum* (Wall.) Meisn. [= *Laurus glandulifera* Wall.]	I^a, LC^*	–
8			*Cinnamomum tamala* T. Nees and Eberm	LC^*, VU^h	√
9			*Litsea glutinosa* (Lour.) C. B. Rob	LC^*	–
10	Dioscoreales R. Br. ex Mart	Dioscoreaceae R. Br	*Dioscorea belophylla* (Prain) Voigt ex Haines [= *Dioscorea nummularia* var. *belophylla* Prain]	I^a, NE^*	–
11			*Dioscorea deltoidea* Wall. ex Griseb	V^b, NE^*, EN^h, CA-2, EN^g	–
12	Liliales Perleb	Melanthiaceae Batsch ex Borkh	*Paris polyphylla* Sm. [= *Daiswa polyphylla* (Sm.) Raf.]	NE^*, EN^h, EN^g, HT^e	–
13			*Trillium govanianum* Wall. ex D. Don [= *Trillidium govanianum* (Wall. ex D. Don) Kunth]	NE^*, EN^g, HT^e	√
14		Colchicaceae DC	*Gloriosa superba* L	LC^*, VU^h	–

(continued)

Table 9.5 (continued)

S.N	Order	Family	Plant name	Threat status	Endemism
15		Smilaceae Vent	*Smilax wightii* A.DC	R[b], R[a], NE[*]	–
16		Liliaceae Juss	*Fritillaria cirrhosa* D. Don [= *F. roylei* Hook.]	NE[*], EN[h], EN[g]	–
17			*Lilium polyphyllum* D. Don	CR[*], CR[*], CR[h], CR[g], HT[e]	√
18			*Lilium wallichianum* Schult. and Schult. f	I[a], NE[*], EN[g], I[c], HT[e], I[@]	–
19	Asparagales Link	Orchidaceae Juss	*Aerides odorata* Lour	EN[*]	–
20			*Aphyllorchis gollanii* Duthie	E[b]/ PEx[b], Ex[a]/ E[a], NE[*], CA-2, VU[g], Ex[c], Ex[@]	√
21			*Bulbophyllum reptans* (Lindl.) Lindl. ex Wall. [= *Bulbophyllum raui* Arora; *Tribrachia reptans* Lindl.]	I[a], NE[*], CA-2, I[@]	–
22			*Calanthe alismifolia* Lindl	I[a], NE[*], CA-2, I[c], I[@]	–
23			*Calanthe alpina* Hook. f. ex Lindl	R[b], R[a], NE[*], CA-2, R[c], R[@]	√
24			*Calanthe davidii* Franch. [= *Calanthe pachystalix* Reichb. f. ex Hook. f.]	E[b], NE[*], CA-2	–

(continued)

Table 9.5 (continued)

S.N	Order	Family	Plant name	Threat status	Endemism
25			*Calanthe mannii* Hook. f	R[b], NE*, CA-2, R@	–
26			*Coelogyne cristata* Lindl	R[a], NE*, CA-2	–
27			*Coelogyne flaccida* Lindl	I[a], NE*, CA-2	–
28			*Coelogyne nitida* (Wall. ex D. Don) Lindl. [= *Cymbidium nitidum* Wall. ex D. Don]	R[a], NE*, CA-2	–
29			*Crepidium acuminatum* (D. Don) Szlach. [= *Malaxis acuminata* D. Don]	NE*, CA-2, VU[g]	–
30			*Cymbidium eburneum* Lindl	V[b], NE*, CA-2	–
31			*Cymbidium goeringii* (Rchb. f.) Rchb. f. [= *Cymbidium mackinnonii* Duthie; *Maxillaria goeringii* Rchb. f.]	NE*, CA-2, VU[g]	–
32			*Cymbidium hookerianum* Rchb. f	V[b], NE*, CA-2	–
33			*Cypripedium cordigerum* D. Don	R[b], VU*, VU*, CA-2, EN[g], HT[e], R@	–
34			*Cypripedium elegans* Rchb. f	R[b], EN*, EN*, CA-2, EN[g], HT[e]	–
35			*Cypripedium himalaicum* Rolfe	R[b], EN*, CA-2, EN[g], HT[e], R@	√

(continued)

Table 9.5 (continued)

S.N	Order	Family	Plant name	Threat status	Endemism
36			*Dactylorhiza hatagirea* (D. Don) Soo [= *Orchis hatagirea* D. Don]	NE*, CR[h], CA-2, EN[g], HT[e]	–
37			*Dendrobium macrostachyum* Lindl. [= *Dendrobium gamblei* King and Pantl.]	I[a], LC*, CA-2, I[@]	–
38			*Dendrobium normale* Falc	I[a], NE*, CA-2, I[c], I[@]	√
39			*Dienia muscifera* Lindl. [= *Malaxis muscifera* (Lindl.) Kuntze, *Microstylis muscifera* (Lindl) Ridl.]	VU*, CA-2, EN[h], EN[g], HT[e]	–
40			*Diplomeris hirsuta* (Lindl.) Lindl. [= *Diplochilos hirsutus* Lindl.]	V[b], NE*, CA-2, EN[g], VoE[d], HT[e], HT[f], V[@]	√
41			*Eria occidentalis* Seidenf. [= *Pinalia occidentalis* (Seidenf.) Schuit., Y. P. Ng and H. A. Pedersen]	R[b], R[a], NE*, CA-2, VU[g], R[c], R[@]	√
42			*Eulophia mackinnonii* Duthie	R[b], R[a], NE*, CA-2, R[c], R[@]	–
43			*Eulophia obtusa* (Lindl.) Hook. f. [= *Cyrtopera obtusa* Lindl.]	I[a], NE*, CA-2, VU[g], I[c], I[@]	–

(continued)

Table 9.5 (continued)

S.N	Order	Family	Plant name	Threat status	Endemism
44			*Flickingeria hesperis* Seidenf. [= *Dendrobium hesperis* (Seidenf.) Schuit. and Peter B. Adams]	E[b], E[a], NE[*], CA-2, VU[g], E[c], HT[e], E[@]	√
45			*Galeola falconeri* Hook. f. [= *Cyrtosia falconeri* (Hook. f.) Aver.]	I[a], NE[*], CA-2, I[c], I[@]	–
46			*Galeola lindleyana* (Hook. f. and Thomson) Rchb. f. [= *Cyrtosia lindleyana* Hook. f. and Thomson]	I[a], NE[*], CA-2	–
47			*Gastrochilus calceolaris* (Buch.-Ham. Ex Sm.) D. Don	CR[*]	–
48			*Gastrochilus garhwalensis* Z. H. Tsi	NE[*], CA-2, VU[g]	√
49			*Habenaria edgeworthii* Hook. f. ex Collett [= *Herminium edgeworthii* (Hook. f. ex Collett) X. H. Jin, Schuit., Raskoti and Lu Q. Huang]	NE[*], CA-2, VU[g]	–
50			*Habenaria intermedia* D. Don	NE[*], EN[h], CA-2, VU[g]	–
51			*Herminium kumaunense* Deva and H. B. Naithani	NE[*], CA-2, VU[g]	√
52			*Neottia acuminata* Schltr. [= *Aphyllorchis parviflora* King and Pantl.]	R[b], LC[*], CA-2, R[@]	–
53			*Neottia mackinnonii* Deva and H. B. Naithani	NE[*], CA-2, DD[g], VU[g]	√

(continued)

Table 9.5 (continued)

S.N	Order	Family	Plant name	Threat status	Endemism
54			*Neottia microglottis* (Duthie) Schltr. [= *Archineottia microglottis* (Duthie) S. C. Chen; *Listera microglottis* Duthie]	R[b], R[a], NE[*], CA-2, VU[g], R[c], R[@]	√
55			*Neottia nandadeviensis* (Hajra) Szlach. [= *Listera nandadeviensis* Hajra]	NE[*], CA-2, VU[g]	√
56			*Nervilia gleadowii* A. N. Rao	NE[*], CA-2, VU[g]	√
57			*Nervilia mackinnonii* (Duthie) Schltr. [=Pogonia mackinnonii Duthie]	I[a], NE[*], CA-2, VU[g], I[c], I[@]	–
58			*Nervilia pangteyana* Jalal, Kumar and G. S. Rawat	NE[*], CA-2, VU[g]	√
59			*Nervilia plicata* (Andrews) Schltr. [= *Nervilia biflora* (Wight) Schltr.; *Arethusa plicata* Andrews]	I[a], NE[*], CA-2, I[@]	–
60			*Oreorchis foliosa* (Lindl.) Lindl. var. *indica* (Lindl.) N. Pearce and P. J. Cribb [= *Corallorhiza indica* Lindl., *Oreorchis indica* (Lindl.) Hook. f.]	I[a], NE[*], CA-2, I[c], I[@]	–
61			*Oreorchis micrantha* Lindl	I[a], NE[*], CA-2	–
62			*Pecteilis gigantea* (Sm.) Raf. [= *Orchis gigantea* Sm.]	NE[*], CA-2, VoE[d], HT[e], HT[f]	√
63			*Peristylus elisabethae* (Duthie) R. K. Gupta [= *Peristylus kumaonensis* Renz]	NE[*], CA-2, VU[g]	√

(continued)

Table 9.5 (continued)

S.N	Order	Family	Plant name	Threat status	Endemism
64			*Phaius tankervilleae* (Banks) Blume [= *Limodorum tankervilleae* Banks]	NE*, CA-2, VU[g], VoE[d], HT[e], HT[f]	–
65			*Ponerorchis renzii* Deva and H. B. Naithani	NE*, CA-2, VU[g]	√
66			*Satyrium nepalense* D. Don	NE*, CA-2, VU[g]	–
67			*Tipularia cunninghamii* (King and Prain) S. C. Chen, S. W. Gale and P. J. Cribb [= *Didiciea cunninghamii* King and Prain]	E[b], E[a], NE*, CA-2, E[c], E[@]	√
68		Iridaceae Juss	*Iris milesii* Baker ex Foster	I[a], NE*	–
69		Amaryllidaceae J. St.-Hil	*Allium auriculatum* Kunth	E[a], NE*	–
70			*Allium loratum* Baker	E[a], NE*	–
71			*Allium roylei* Stearn	E[a], NT*	–
72			*Allium stracheyi* Baker	V[b], V[a], NE*, VU[h], VU[g], V[c], V[@]	√
73		Asparagaceae Bercht. and J. Presl	*Asparagus filicinus* Buch.-Ham. ex D. Don	DD*	–
74			*Beaucarnea recurvata* (K. Koch and Filtelm.) Lem	CR*, CA-2	–
75			*Chlorophytum borivilianum* Santapau and R. R. Fern	CR*$	–
76			*Dipcadi reidii* Deb and S. Dasgupta	PEx[b], Ex[a], NE*, CR[g], VU[g], Ex[c], Ex[@]	√

(continued)

Table 9.5 (continued)

S.N	Order	Family	Plant name	Threat status	Endemism
77			*Polygonatum cirrhifolium* (Wall.) Royle [= *Convallaria cirrhifolia* Wall.]	NE*, VU[h], VU[g]	–
78			*Polygonatum graminifolium* Hook	I[a], NE*, I[c], I[@]	–
79			*Polygonatum verticillatum* (L.) All. [= *Convallaria verticillata* L.]	VU[h], VU[g]	–
80	Arecales Bromhead	Arecaceae Bercht. and J. Presl	*Adonidia merrillii* Becc	VU*	–
81			Chamaedorea metallica O. F. Cook ex H. E. Moore	EN*	–
82			*Dypsis decaryi* (Jum.) Beentje and J. Dransf	VU*	–
83			*Dypsis leptocheilos* (Hodel) Beentje and J. Dransf	CR*, CA-2	–
84			*Dypsis lutescens* (H. Wendl.) Beentje and J. Dransf	NT*	–
85			*Elaeis guineensis* Jacq	VU*	–
86			*Hyophorbe lagenicaulis* (L. H. Bailey) H. E. Moore	CR*	–
87			*Hydriastele wendlandiana* (F. Muell.) H. Wendl. and Drude [= *Hydriastele microspadix* (Warb. ex K. Schum. and Lauterb.)]	LC*	–
88			*Hyophorbe verschaffeltii* (W. Bull ex J. Dix) H. Wendl	CR*	–
89			*Latania verschaffeltii* Lem	EN*	–
90			*Licuala ramsayi* (F. Muell.) Domin	LC*	–
91			*Normanbya normanbyi* (A. W. Hill) L. H. Bailey	VU*	–
92			*Phoenix rupicola* T. Anderson	R[b], V[a], NT*	–
93			*Pinanga coronata* Blume	NT*	–
94			*Pritchardia remota* Becc	EN*	–
95			*Ravenea rivularis* Jum. and H. Perrier	VU*, CA-2	–
96			*Thrinax radiata* Lodd. ex Schult. and Schult. f	LC*	–

(continued)

Table 9.5 (continued)

S.N	Order	Family	Plant name	Threat status	Endemism
97			*Trachycarpus takil* Becc	R^{b}, NE^{*}, CR^{g}, VU^{g}, VoE^{d}, HT^{e}, HT^{f}, $R^{@}$	√
98			*Washingtonia filifera* (Lindl.) H. Wendl	LC^{*}	–
99			*Washingtonia robusta* H. Wendl	LC^{*}	–
100	Zingiberales Griseb	Zingiberaceae Martinov	*Cautleya spicata* (Sm.) Baker [= *Cautleya petiolata* Baker]	I^{a}, LC^{*}, I^{c}, $I^{@}$	–
101			*Hedychium spicatum* Sm	NE^{*}, VU^{g}	√
102	Poales Small	Eriocaulaceae Martinov	*Eriocaulon nepalense* J. D. Prescott ex Bong. var. *luzulifolium* (Mart.) Praj. and J. Parn. [= *Eriocaulon pumilio* Hook. f.]	I^{a}, NE^{*}, DD^{g}, I^{c}, $I^{@}$	–
103		Juncaceae Juss	*Juncus duthiei (C. B. Clarke) Noltie [= Microschoenus duthiei C. B. Clarke]*	$I^{@}$	√
104		Cyperaceae Juss	*Carex clavispica* S. R. Zhang [= *Kobresia duthiei* C. B. Clarke]	I^{a}, NE^{*}	–
105			*Carex esenbeckii* Kunth [= *Kobresia esenbeckii* (Kunth) Noltie; *Kobresia trinervis* var. *foliosa* (C. B. Clarke) Kuekenth.]	I^{a}, NE^{*}	–
106			*Carex nandadeviensis* Ghildyal, U. C. Bhattach. and Hajra	NE^{*}, VU^{g}	√
107		Poaceae Barnhart	*Cymbopogon citratus* (DC.) Stapf	$R^{@}$	–
108			*Cymbopogon microstachys* (Hook. f.) Soenarko [= *Cymbopogon flexuosus* var. *microstachys* (Hook. f.) Bor]	R^{a}, NE^{*}, R^{c}, $R^{@}$	–
109			*Cymbopogon osmastonii* R. Parker	V^{a}, NE^{*}, VU^{g}	√

(continued)

Table 9.5 (continued)

S.N	Order	Family	Plant name	Threat status	Endemism
110			*Dendrocalamus somdevae* H. B. Naithani	NE*, EN@	√
111			*Digitaria duthieana* Henrard ex Bor	DD*	–
112			*Elymus duthiei* (Melderis) G. Singh [= *Agropyron duthiei* Melderis]	I[a], NE*, I[c], I@	–
113			*Eulalia madkotiensis* Kandwal, B. K. Gupta and S. K. Srivast	NE*, VU[g]	√
114			*Festuca lucida* Stapf	I[a], NE*, I[c], I@	√
115			*Festuca nandadevica* Hajra	NE*, VU[g]	√
116			*Helictotrichon uniyalii* Kandwal and B. K. Gupta	NE*, VU[g]	√
117			*Microstegium falconeri* (Hook. f.) Clayton [= *Ischnochloa falconeri* Hook. f.]	I[a], NE*, VU[g], I[c], I@	√
118			*Piptatherum hilariae* Pazij [= *Oryzopsis humilis* Bor; *Oryzopsis hilariae* (Pazij) Uniyal]	I[a], NE*, I[c], I@	–
119			*Poa garhwalensis* D. C. Nautiyal and R. D. Gaur	NE*, VU[g]	√
120			*Poa pseudamoena* Bor	I[a], NE*, I[c], I@	–
121			*Poa rhadina* Bor	E[a], NE*, VU[g], E[c], E@	√
122			*Poa royleana* Nees ex Steud	NE*, EN@	√
123			*Pseudodanthonia himalaica* (Hook. f.) Bor and C. E. Hubb. [= *Danthonia himalaica* Hook. f.]	I[a], NE*, VU[g], VU[g]	√
124			*Puccinellia thomsonii* (Stapf ex Hook. f.) R. R. Stewart [= *Glyceria thomsonii* Stapf ex Hook. f.]	I[a], NE*	–

(continued)

Table 9.5 (continued)

S.N	Order	Family	Plant name	Threat status	Endemism
125			*Sehima notatum* (Hack.) A. Camus [= *Ischaemum notatum* Hack.]	NE*, VU[g]	√
126			*Trisetum micans* (Hook. f.) Bor [= *Avena micans* Hook. f.]	I[a], NE*, I[c], I[@]	–
127	Ranunculales Juss. ex Bercht. and J. Presl	Papaveraceae Juss	*Corydalis cashmeriana* Royle	E[a], NE*	–
128			*Corydalis devendrae* Pusalkar	NE*, VU[g]	√
129			*Papaver guilelmi-waldemarii* (Klotzsch) Christenh. and Byng [= *Meconopsis guilemi-waldemarii* Klotzsch; *Meconopsis aculeata* Royle]	E[a], NE*, EN[h]	–
130			*Papaver robustum* (Hook. f, and Thomson) Christenh. and Byng [= *Meconopsis robusta* Hook. f. and Thomson]	NE*, VU[g]	√
131		Berberidaceae Juss	*Berberis affinis* G. Don	R[b], R[a], NE*, VU[g], R[c], R[@]	√
132			*Berberis ahrendtii* R. R. Rao and Uniyal	NE*, EN[g], VU[g]	√
133			*Berberis garhwalensis* C. K. Schneid	NE*, DD[g], VU[g]	√
134			*Berberis jaunsarensis* (Ahrendt) Laferr. [= *Mahonia jaunsarensis* Ahrendt]	I[a], NE*, VU[g], I[c], I[@]	√
135			*Berberis lambertii* R. Parker	V[b]/ E[b], V[a], NE*, CR[g], VU[g], V[@]/ E[@]	√

(continued)

Table 9.5 (continued)

S.N	Order	Family	Plant name	Threat status	Endemism
136			*Berberis osmastonii* Dunn	R[b], R[a], NE[*], VU[g], R[c], R[@]	√
137			*Berberis pseudumbellata* R. Parker	I[a], NE[*], I[c], I[@]	–
138			*Berberis rawatii* U. L. Tiwari and B. S. Adhikari	NE[*], VU[g]	√
139			*Podophyllum hexandrum* Royle [= *Sinopodophyllum hexandrum* (Royle) T. S. Ying]	NE[*], EN[h], CA-2, EN[g]	–
140		Ranunculaceae Juss	*Aconitum heterophyllum* Wall. ex Royle	EN[*], EN[*], CR[h], EN[g], VoE[d], HT[e], HT[f]	√
141			*Aconitum laeve* Royle	NE[*], EN[g]	–
142			*Aconitum lethale* Griff. [= *A. falconeri* Stapf var. *latilobum* Stapf; *A. balfouri* var. *rhombilobatum* Stapf; *A. falconeri* Stapf var. *falconeri*]	V[b], I[a], NE[*], VU[h], EN[g], I[c], VoE[d], HT[f], I[@]	√
143			*Aconitum violaceum* Jacquem. ex Stapf	VU[*], VU[*], VU[h], VU[g], VoE[d], HT[f]	–
144			*Anemone raui* Goel and U.C. Bhattach	NE[*], VU[g]	√
145			*Aquilegia nivalis* (Baker) Falc. ex B. D. Jacks [= *Aquilegia glauca* Lindl. var. *nivalis*; *Aquilegia nivalis* (Baker) Bruehl]	E[a], NE[*]	–

(continued)

Table 9.5 (continued)

S.N	Order	Family	Plant name	Threat status	Endemism
146			*Delphinium koelzii* Munz	I[a], NE[*]	–
147			*Oxygraphis kumaonensis* I. D. Rai and G. S. Rawat	NE[*], VU[g]	√
148			*Ranunculus uttaranchalensis* Pusalkar and D. K. Singh	NE[*], VU[g]	√
149			*Trollius acaulis* Lindl	E[a], NE[*]	–
150	Saxifragales Bercht. and J. Presl	Saxifragaceae Juss	*Bergenia ciliata* (Haw.) Sternb. [= *Megasea ciliata* Haw.]	NE[*], VU[g]	√
151			*Saxifraga jacquemontiana* Decne	E[a], NE[*]	–
152			*Saxifraga meeboldii* Engl. and Irmsch	NE[*], DD[g]	–
153			*Saxifraga minutissima* D. S. Rawat	NE[*], VU[g]	√
154		Crassulaceae J. St.-Hil	*Sedum bhattacharyyae* R. Manik., N. B. Singh and S. K. Srivast. [= *Sedum pedicellatum* N. B. Singh and U. C. Bhattach.]	NE[*], VU[g]	√
155			*Sedum duthiei* Frod	I[a], NE[*], DD[g], VU[g], I[c], I[@]	√
156			*Sedum heterodontum* Hook. f. and Thomson [= *Rhodiola heterodonta* (Hook. f. and Thomson) Boriss.]	NE[*], VU[h]	–
157			*Sedum seelemannii* Raym.-Hamet	NE[*], DD[g], VU[g]	√
158	Fabales Bromhead	Fabaceae Lindl	*Abrus fruticulosus* Wall. ex Wight and Arn	DD[*], DD[*]	–
159			*Astragalus langtangensis* Podlech	DD[*], DD[*]	–
160			*Astragalus nainitalensis* L. B. Chaudhary	NE[*], VU[g]	√
161			*Astragalus stewartii* Baker [= *Astragalus bakeri* Ali]	I[a], NE[*]	–
162			*Astragalus uttaranchalensis* L. B. Chaudhary and J. H. Khan	NE[*], VU[g]	√

(continued)

Table 9.5 (continued)

S.N	Order	Family	Plant name	Threat status	Endemism
163			*Butea pellita* Hook. f. ex Prain [= *Meizotropis pellita* (Hook. f ex Prain) Sanjappa]	NE[*], CR[g], VoE[d], HT[e], HT[f]	√
164			*Caesalpinia bonduc*	EN[*]	–
165			*Dalbergia lanceolaria* L. f	NE[*], CA-2	–
166			*Dalbergia latifolia* Roxb	VU[*], CA-2	–
167			*Dalbergia sericea* G. Don	NE[*], CA-2	–
168			*Dalbergia sissoo* Roxb. ex DC	NE[*], CA-2	–
169			*Dalbergia volubilis* Roxb	NE[*], CA-2	–
170			*Derris kanjilalii* K. C. Sahni and H. B. Naithani	NE[*], VU[g]	√
171			*Desmodium garhwalensis* L. R. Dangwal and R. D. Gaur	NE[*], VU[g]	√
172			*Hedysarum astragaloides* Benth. ex Baker	R[b], R[a], NE[*]	–
173			*Hedysarum cachemirianum* Benth. ex Baker	R[b]/ V[b], NE[*]	–
174			*Hedysarum microcalyx* Baker	V[b], NE[*]	–
175			*Indigofera cedrorum* Dunn	I[a], NE[*], VU[g]	–
176			*Indigofera dosua* Buch.-Ham. ex D. Don var. *simlensis* (Ali) Sanjappa [= *Indigofera simlensis* Ali]	I[a], NE[*], VU[g]	–
177			*Indigofera thothathrii* Sanjappa	NE[*], VU[g]	–
178			*Indopiptadenia oudhensis* (Brandis) Brenan [= *Piptadenia oudhensis* Brandis]	NE[*], EN[g], VoE[d], HT[e], HT[f]	√

(continued)

Table 9.5 (continued)

S.N	Order	Family	Plant name	Threat status	Endemism
179			*Macrotyloma sar-garhwalensis* R. D. Gaur and L. R. Dangwal	NE*, VU[g]	–
180			*Pterocarpus marsupium* Roxb	NT*, CA-2	–
181			*Pterocarpus santalinus* L. f	NT*, CA-2	–
182			*Pueraria garhwalensis* L. R. Dangwal and D. S. Rawat	NE*, VU[g]	√
183			*Saraca asoca* (Roxb.) Willd	VU*$	–
184			*Senna davidsonii* (V. Singh) V. Singh [= *Cassia davidsonii* V. Singh]	NE*, VU[g], DD[g]	√
185			*Thermopsis inflata* Cambess	I[a], NE*	–
186			*Uraria picta* (Jacq.) Desv. ex DC. [= *Hedysarum pictum* Jacq.]	LC*, HT[e]	–
187			*Vigna aconitifolia* (Jacq.) Marechal [= *Phaseolus aconitifolius* Jacq.]	DD*	–
188	Rosales Bercht. and J. Presl	Rosaceae Juss	*Alchemilla palii* Panigrahi and Purohit	NE*, DD[@]	√
189			*Cotoneaster parkinsonii* Panigrahi and Arv. Kumar	NE*, VU[g]	√
190			*Cotoneaster roseus* Edgew. [= *Cotoneaster osmastonii* G. Klotz]	NE*, VU[g]	–
191			*Cotoneaster simonsii* Hort. ex Baker	I[b], I[a], NE*	–
192			*Geum aequilobatum* K. M. Purohit and Panigrahi	NE*, EN[@]	√
193			*Prunus jacquemontii* Hook.f	DD*	–
194			*Rosa hirsuta* Ghora and Panigrahi	NE*, VU[g]	√
195			*Rubus almorensis* Dunn	I[a], NE*, DD[g], VU[g], I[c], I[@]	√
196			*Sibbaldia axilliflora* (Hook. f.) Chatterjee [= *Potentilla axilliflora* Hook. f.]	NE*, DD[g], VU[g]	√

(continued)

Table 9.5 (continued)

S.N	Order	Family	Plant name	Threat status	Endemism
197			*Spiraea diversifolia* Dunn	I^a, NE^*, I^c, $I^@$	√
198			*Spiraea duthieana* Zinserl	NE^*, VU^g	√
199			*Spiraea hypoleuca* Dunn	NE^*, VU^g	√
200			*Spiraea panchananii* Panigrahi and K. M. Purohit	NE^*, VU^g	√
201			*Spiraea panigrahiana* K. M. Purohit	NE^*, VU^g	√
202			*Spiraea parkeri* Panigrahi and K. M. Purohit	NE^*, DD^g, VU^g	√
203			*Spiraea raizadae* Panigrahi and K. M. Purohit	NE^*, VU^g	√
204		Rhamnaceae Juss	*Sageretia devendrae* Pusalkar	NE^*, VU^g	√
205		Ulmaceae Mirb	*Ulmus wallichiana* Planch	E^a, VU^*, NE^*	√
206		Moraceae Gaudich	*Ficus andamanica* Corner	$EN^{*\$}$	–
207			*Ficus cupulata* Haines	$R^@$	–
208			*Ficus palmata* Forssk	LC^*	–
209	Fagales Engl	Betulaceae Gray	*Betula utilis* D. Don	LC^*	–
210			*Carpinus faginea* Lindl	DD^*, DD^*	–
211			*Corylus jacquemontii* Decne	DD^*, DD^*	–
212	Cucurbitales Juss. ex Bercht. and J. Presl	Datiscaceae Dumort	*Datisca cannabina* L	NE^*, EN^h	–
213	Malpighiales Juss. ex Bercht. and J. Presl	Hypericaceae Juss	*Hypericum perforatum* L	LC^*, VU^h	–
214		Violaceae Batsch	*Viola kunawarensis* Royle	I^a, NE^*, I^c, $I^@$	–
215			*Viola repens* Wall. ex Ging	NE^*, VU^g	–
216		Euphorbiaceae Juss	*Euphorbia royleana* Boiss	NE^*, CA-2	√

(continued)

Table 9.5 (continued)

S.N	Order	Family	Plant name	Threat status	Endemism
217	Myrtales Juss. ex Bercht. and J. Presl	Onagraceae Juss	*Epilobium glaciale* P. H. Raven	I[a], NE[*]	–
218	Crossosomatales Takht. ex Reveal	Staphyleaceae Martinov	*Staphylea cochinchinensis* (Lour.) Byng and Christenh. [= *Ticeros cochinchinensis* Lour; *Turpinia cochinchinensis* (Lour.) Merr.; *Turpinia nepalensis* Wall.]	NE[*], VU[g], VoE[d], HT[e], HT[f]	–
219	Sapindales Juss. ex Bercht. and J. Presl	Sapindaceae Juss	*Acer caesium* Wall. ex Brandis	V[b], V[a], NE[*], VU[g], V[c], V[@]	√
220			*Acer oblongum* Wall. ex DC	LC[*], EN[@]	√
221			*Sapindus mukorossi* Gaertn	LC[*]	–
222		Rutaceae Juss	*Zanthoxylum armatum* DC	LC[*], VU[h]	–
223		Meliaceae Juss	*Swietenia mahagoni* (L.) Jacq	NT[*], CA-2	–
224	Malvales Juss. ex Bercht. and J.Presl	Thymelaeaceae Juss	*Aquilaria malaccensis* Lam. [= *Aquilaria agallocha* (Lour.) Roxb. Ex Finl.]	CR[*], CA-2	–
225		Dipterocarpaceae Blume	*Hopea wightiana* Wall	EN[*]	–
226			*Vateria indica* L	VU[*], CR[#]	–
227	Brassicales Bromhead	Brassicaceae Burnett	*Eutrema purii* (D. S. Rawat, L. R. Dangwal and R. D. Gaur) Al-Shehbaz, G. Q. Hao and J. Quan Liu [= *Dilophia purii* D. S. Rawat, L. R. Dangwal and R. D. Gaur]	NE[*], VU[g]	√
228	Santalales R. Br. ex Bercht. and J. Presl	Santalaceae R. Br	*Santalum album* L	VU[*$]	–
229	Caryophyllales Juss. ex Bercht. and J. Presl	Polygonaceae Juss	*Koenigia binsarii* (Silas and R. D. Gaur) R. D. Gaur [= *Polygonum binsarii* Silas and R. D. Gaur]	NE[*], VU[g]	√
230			*Rheum australe* D. Don [= *Rheum emodii* Wall. ex Meisn.]	NE[*], EN[h], EN[g]	–

(continued)

Table 9.5 (continued)

S.N	Order	Family	Plant name	Threat status	Endemism
231			*Rheum moorcroftianum* Royle	NE*, VU^g	–
232			*Rheum webbianum* Royle	NE*, VU^h, VU^g	–
233			*Rumex gangotrianus* Aswal and S. K. Srivast	NE*, VU^g	√
234		Caryophyllaceae Juss	*Arenaria neelgherrensis* Wight and Arn	I^a, NE*	–
235			*Cerastium thomsonii* Hook. f	I^a, NE*, I^c, I$^@$	–
236			*Eremogone curvifolia* (Majumdar) Pusalkar and D. K. Singh [= *Arenaria curvifolia* Majumdar]	E^b, E^a, NE*, CR^g, VU^g, HT^e, E$^@$	√
237			*Eremogone ferruginea* (Duthie ex F. N. Williams) Pusalkar and D. K. Singh [= *A. ferruginea* Duthie ex F. N. Williams]	E^b, E^a, NE*, DD^g, VU^g, I^c, I$^@$	√
238			*Odontostemma thangoense* (W. W. Sm.) Rabeler and W. L. Wagner [= *Arenaria thangoensis* W. W. Sm.]	V^b, V^a, NE*	–
239			*Silene kumaonensis* F. N. Williams	R^b, R^a, NE*, R$^@$	√
240			*Silene stracheyi* Edgew	NE*, DD^g	–
241			*Stellaria depressa* Em. Schmid	I^a, NE*	–
242		Cactaceae Juss	*Kroenleinia grusonii* (Hildm.) Lode [=*Echinocactus grusoniiI* Hildm.]	EN*	–

(continued)

Table 9.5 (continued)

S.N	Order	Family	Plant name	Threat status	Endemism
243	Cornales Link	Cornaceae Bercht. and J. Presl	*Alangium lamarckii* Thwaites	R[@]	–
244	Ericales Bercht. and J. Presl	Balsaminaceae A. Rich	*Impatiens devendrae* Pusalkar	NE[*], VU[g]	√
245			*Impatiens duthiei* Hook. f	NE[*], VU[g]	√
246			*Impatiens inayatii* Hook. f	NE[*], DD[g], VU[g]	√
247			*Impatiens jaeschkei* Hook. f	NE[*], EN[@]	√
248			*Impatiens kaliensis* Grey-Wilson	NE[*], VU[g]	√
249			*Impatiens langeana* Hook. f	NE[*], DD[g], VU[g]	√
250			*Impatiens podocarpa* Hook. f	NE[*], EN[@]	√
251			*Impatiens polysciadia* Hook. f	NE[*], DD[g], VU[g]	√
252			*Impatiens reidii* Hook. f	NE[*], DD[g], VU[g]	√
253			*Impatiens violoides* Edgew. ex Hook. f	NE[*], DD[g], VU[g]	√
254		Primulaceae Batsch ex Borkh	*Embelia tsjeriam-cottam* (Roem. and Schult.) A. DC. [= *Ardisia tsjeriam-cottam* Roem. and Schult.]	NE[*], VU[h]	–
255			*Primula garhwalica* (Balodi and S. Singh) K. K. Khanna and An. Kumar [= *Androsace garhwalicum* Balodi and S. Singh]	NE[*], VU[g]	√
256			*Primula drummondiana* Craib	I[a], NE[*]	–
257			*Primula minutissima* Jacquem. ex Duby	E[a], NE[*]	–

(continued)

Table 9.5 (continued)

S.N	Order	Family	Plant name	Threat status	Endemism
258		Symplocaceae Desf	*Symplocos paniculata* Miq	NE*, VU[h]	–
259		Ericaceae Juss	*Rhododendron rawatii* I. D. Rai and B. S. Adhikari	NE*, VU[g]	√
260	Gentianales Juss. ex Bercht. and J. Presl	Rubiaceae Juss	*Coffea arabica* L	EN*	–
261			*Clarkella nana* (Edgew.) Hook. f. [= *Ophiorrhiza nana* Edgew.]	R[b], R[a], NE*, R[c], R[@]	–
262			*Leptodermis riparia* R. Parker	NE*, VU[g]	√
263			*Rubia edgeworthii* Hook. f	V[b], V[a], NE*, V[c], V[@]	–
264		Gentianaceae Juss	*Gentiana cachemirica* Decne	E[a], NE*	–
265			*Gentiana crassuloides* Bureau and Franch	R[a], NE*, R[c], R[@]	–
266			*Gentiana kurroo* Royle	CR*, CR[h], EN[g], VoE[d], HT[e], HT[f]	√
267			*Gentiana saginoides* Burkill	R[a], NE*, CR[g], VU[g], R[c], R[@]	√
268			*Gentiana tetrasepala* Biswas	NE*, VU[g], VU[g]	√
269			*Kuepferia infelix* (C. B. Clarke) Adr. Favre [= *Gentiana infelix* C. B. Clarke]	R[a], NE*, VU[g], R[c], R[@]	–

(continued)

Table 9.5 (continued)

S.N	Order	Family	Plant name	Threat status	Endemism
270			*Swertia alpina* U. C. Bhattach. and S. Agrawal	NE*, VU[g]	√
271			*Swertia chirayita* (Roxb.) H. Karst. [= *Gentiana chirayita* Roxb.]	NE*, EN[h], EN[g]	–
272		Apocynaceae Juss	*Ceropegia angustifolia* Wight	V[b], V[a], NE*	–
273			*Ceropegia bulbosa* Roxb	V[a], NE*, EN[h], VU[g]	–
274			*Rauvolfia serpentina* (L.) Benth. ex Kurz	NE*, VU[h], CA-2	–
275	Boraginales Juss. ex Bercht. and J. Presl	Boraginaceae Juss	*Arnebia benthamii* (Wall. ex G. Don) I. M. Johnst. [= *Echium benthamii* Wall. and G. Don]	E[a], NE*, VU[g], CR[h], HT[e]	–
276			*Arnebia euchroma* (Royle ex Benth.) I. M. Johnst. [= *Lithospermum euchromon* Royle ex Benth.]	NE*, EN[h], VU[g], HT[e]	–
277			*Cynoglossum jaunsarensis* (Kazmi) Pusalkar [= *Ivanjohnstonia jaunsariensis* Kazmi]	NE*, DD[g], VU[g]	√
278	Solanales Juss. ex Bercht. and J. Presl	Convolvulaceae Juss	*Ipomoea laxiflora* H. J. Chowdhery and M. R. Debta	NE*, VU[g]	–
279		Solanaceae Juss	*Hyoscyamus niger* L	NE*, VU[h]	–
280	Lamiales Bromhead	Oleaceae Hoffmanns. and Link	*Fraxinus micrantha* Lingelsh	DD*	–
281			*Schrebera swietenioides* Roxb	NE*, EN[g], VoE[d], HT[e], HT[f]	√
282		Gesneriaceae Rich. and Juss	*Didymocarpus aromaticus* Don	NE*, VU[g]	–
283			*Didymocarpus pedicellatus* R. Br	NE*, VUv	–

(continued)

Table 9.5 (continued)

S.N	Order	Family	Plant name	Threat status	Endemism
284		Plantaginaceae Juss	*Kashmiria himalaica* (Hook. f.) D. Y. Hong [= *Falconeria himalaica* Hook. f.; *Wulfenia himalaica* (Hook. f.) Pennell]	NE[*], VU[g], VU[g]	√
285			*Neopicrorhiza scrophulariiflora* (Pennell) D. Y. Hong [= *Picrorhiza scrophulariiflora* Pennell]	NE[*], EN[g]	–
286			*Picrorhiza kurroa* Royle ex Benth	V[b], NE[*], CR[h], CA-2, EN[g]	–
287			*Picrorhiza tungnathii* Pusalkar	NE[*], VU[g]	√
288		Scrophulariaceae Juss	*Scrophularia obtusa* Edgew. ex Hook. f	NE[*], VU[g]	√
289		Acanthaceae Juss	*Phlogacanthus lambertii* Raizada	NE[*], DD[g], VU[g]	√
290		Bignoniaceae Juss	*Incarvillea emodi* (Royle ex Lindl) Chatterjee [= *Amphicome emodi* Royle ex Benth.]	NE[*], VU[g], HT[e]	–
291			*Jacranda mimosifolia* D. Don	VU[*]	–
292		Lentibulariaceae Rich	*Pinguicula alpina* L	NE[*], VU[g], VoE[d], HT[e], HT[f]	–
293		Lamiaceae Martinov	*Elsholtzia densa* Benth	I[a], NE[*]	–
294			*Nepeta campestris* Benth	NE[*], I[a]	–
295			*Phlomoides superba* (Royle ex Benth.) Kamelin and Makhm. [= *Eremostachys superba* Royle ex Benth.]	NE[*], VU[h], EN[g], VoE[d], HT[e], HT[f]	√

(continued)

Table 9.5 (continued)

S.N	Order	Family	Plant name	Threat status	Endemism
296			*Roylea cinerea* (D. Don) Baill. [= *Ballota cinerea* D. Don]	NE[*], VU[h]	–
297		Orobanchaceae Vent	*Gleadovia ruborum* Gamble and Prain	NE[*], EN[g]	–
298	Asterales Link	Campanulaceae Juss	*Campanula wattiana* B. K. Nayar and Babu	R[b], R[a], NE[*], R[c], EN[@], R[@]	–
299			*Cyananthus integer* Wall. ex Benth	R[b], R[a], NE[*], R[c], EN[g], R[@]	√
300		Asteraceae Bercht. and J. Presl	*Artemisia austrohimalayaensis* Y. R. Ling and H. S. Puri [= *Artemisia austrohimalayana* Y. R. Ling and H. S. Puri]	NE[*], VU[g]	√
301			*Aucklandia costus* Falc. [= *Saussurea costus* (Falc.) Lipsch.; *Saussurea lappa* (DC.) Sch. Bip.]	E[b], E[a], CR[*], CA-1	√
302			*Catamixis baccharoides* Thomson	V[b], V[a], NE[*], CR[g], VU[g], V[@]	√
303			*Cremanthodium arnicoides* (DC. ex Royle) R. D. Good [= *Ligularia arnicoides* DC. ex Royle]	E[a], NE[*]	–
304			*Dolomiaea macrocephala* DC. ex Royle [= *Jurinea dolomiaea* Boiss.]	NE[*], EN[h], EN[g]	–
305			*Himalaiella foliosa* (Edgew.) Raab- Straube [= *Aplotaxis foliosa* Edgew.; *Saussurea foliosa* (Edgew.) Hook. f.]	I[a], NE[*], I[@]	–

(continued)

Table 9.5 (continued)

S.N	Order	Family	Plant name	Threat status	Endemism
306			*Melanoseris filicina* (Stebbins) N. Kilian [= *Lactuca filicina* Duthie ex Stebbins; = *Cicerbita filicina* (Duthie ex Stebbins) Mumgain and R.R. Rao]	E[b], E[a], NE[*], CR[g], VU[g], E[@]	√
307			*Saussurea atkinsonii* C. B. Clarke	I[a], NE[*]	–
308			*Saussurea bracteata* Decne	R[b], R[a], NE[*], R[c], R[@]	–
309			*Saussurea pterocaulon* Decne. [= *Saussurea clarkei* Hook. f.]	R[b], R[a], NE[*]	–
310			*Saussurea gossypiphora* D. Don	NE[*], VU[h]	–
311			*Saussurea obvallata* (DC.) Edgew. [= *Aplotaxis obvallata* DC.]	NE[*], EN[h], VU[g]	–
312			*Saussurea roylei* (DC.) Sch. Bip. [= *Aplotaxis roylei* DC.]	I[a], NE[*]	–
313			*Saussurea sudhanshui* Hajra	NE[*], VU[g]	√
314	Dipsacales Juss. ex Bercht. and J. Presl	Caprifoliaceae Juss	*Nardostachys jatamansi* (D. Don) DC. [= *Patrinia jatamansi* D. Don; *N. grandiflora* DC.]	CR[*], CR[h], EN[g], HT[e], HT[f]	√
315			*Valeriana jatamansi* Jones	NE[*], VU[h], VU[g]	–
316			*Valeriana mussooriensis* Ved Prakash, Aswal and Mehrotra	NE[*], VU[g]	√
317			*Valeriana roylei* Klotzsch	NE[*], VU[g]	√
318	Apiales Nakai	Pittosporaceae R. Br	*Pittosporum eriocarpum* Royle	I[b], I[a], EN[*], VU[g], VU[g], I[c], R[@]	√

(continued)

Table 9.5 (continued)

S.N	Order	Family	Plant name	Threat status	Endemism
319		Araliaceae Juss	*Panax pseudoginseng* Wall	V[b], NE[*], CR[g]	–
320		Apiaceae Lindl	*Angelica archangelica* L	LC[*], EN[g]	–
321			*Angelica glauca* Edgew	EN[*], EN[h], EN[g]	√
322			*Angelica indica* Pimenov and Kljuykov	NE[*], VU[g]	√
323			*Bupleurum maddenii* C. B. Clarke	NE[*], VU[g]	√
324			*Ferula jaeskeana* C. B. Clarke	NE[*], VU[h]	–
325			*Heracleum candicans* Wall. ex DC	NE[*], VU[h]	–
326			*Heracleum jacquemontii* C. B. Clarke [= *Heracleum jacquemontii* C. B. Clarke ex Hook. f.]	I[b], I[a], NE[*], I[c], I[@]	–
327			*Hymenidium dentatum* (DC.) Pimenov and Kljuykov [= *Pleurospermum erosa* (DC.) P. K. Mukh.; *Hymenolaena dentata* var. *erosa* DC.]	NE[*], VU[g]	√
328			*Kailashia robusta* Pimenov and Kljuykov	NE[*], VU[g]	√
329			*Kedarnatha meifolia* Pimenov and Kljuykov	NE[*], VU[g]	√
330			*Kedarnatha sanctuarii* P. K. Mukh. and Constance	NE[*], EN[@]	√
331			*Oreocome aegopodioides* Pimenov and Kljuykov	NE[*], VU[g]	√
332			*Peucedanum dehradunense* Babu	NE[*], VU[g]	√
333			*Pimpinella stracheyi* C. B. Clarke	NE[*], VU[g]	√

(continued)

Table 9.5 (continued)

S.N	Order	Family	Plant name	Threat status	Endemism
334			*Pleurospermum angelicoides* (Wall. ex DC.) Benth. ex C. . Clarke [= *Hymenolaena angelicoides* DC.]	NE*, VU[h]	–
335			*Trachyspermum falconeri* (C. B. Clarke) H. Wolff. [= *Carum falconeri* C. B. Clarke]	NE*, VU[g]	√

Plant families arranged according to APG IV (2016)

Superscripts IUCN (*), ENVIS (@), listed in CITES Appendix-I, II (CA-1, CA-2), Indian Biodiversity Portal (#), Rao et al. (2003) (a), Nayar and Sastry (1987–1990) (b), Uniyal et al. (2007) (c), National Biodiversity Authority for Uttarakhand (d), Uttarakhand State Biodiversity Board (e), eBook Uttarakhand State Biodiversity Board (f), Pusalkar and Srivastava (2018) (g), Ved et al. (2003) (h)

Source Checklist of threatened plants of Uttaranchal (2013), Pandey and Pandey (2013), Acharya et al. (2018), Bhatt et al. (2020), Gowthami et al. (2021), Chaturvedi et al. (2022), FRI (2013, 2023), BSI (2021), CITES (2023), IUCN (2024), Nayar and Sastry (1987), Singh (2022), ENVIS BSI (2020)

EX Extinct, *EW* Extinct in the wild, *CR* Critically endangered, *EN* Endangered, *VU* Vulnerable, *NT* Near threatened, *LC* Least concerned, *DD* Data deficient, *NE* Not evaluated, *I* Indeterminate, *PEx* Presumed extinct, *HT* Highly threatened, *VoE* Verge of Extinction

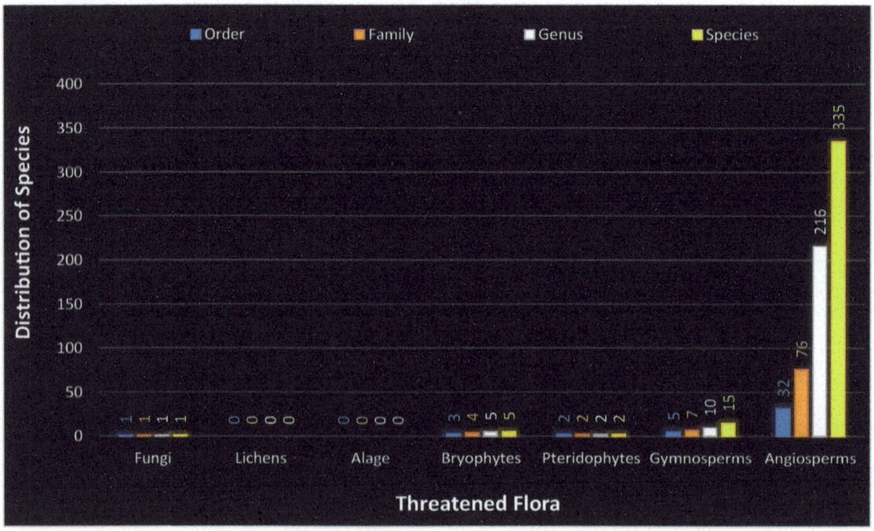

Fig. 9.4 Numerical distribution of threatened plant species according to taxonomy

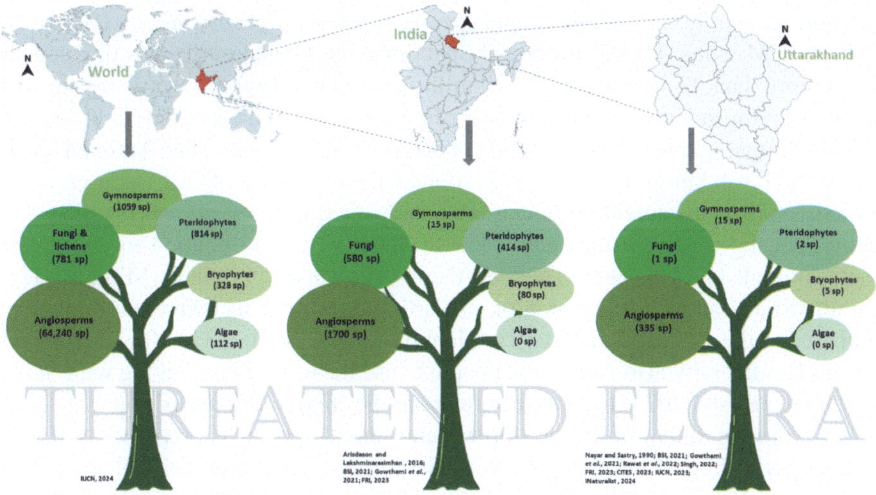

Fig. 9.5 Status of threatened flora of Uttarakhand as compared to India and World. *Source* Nayar and Sastry (1990), Arisdason and Lakshminarasimhan (2016), BSI (2021), Gowthami et al. (2021), Chaturvedi et al. (2022), Singh (2022), FRI (2023), CITES (2023), IUCN (2024), iNaturalist (2024)

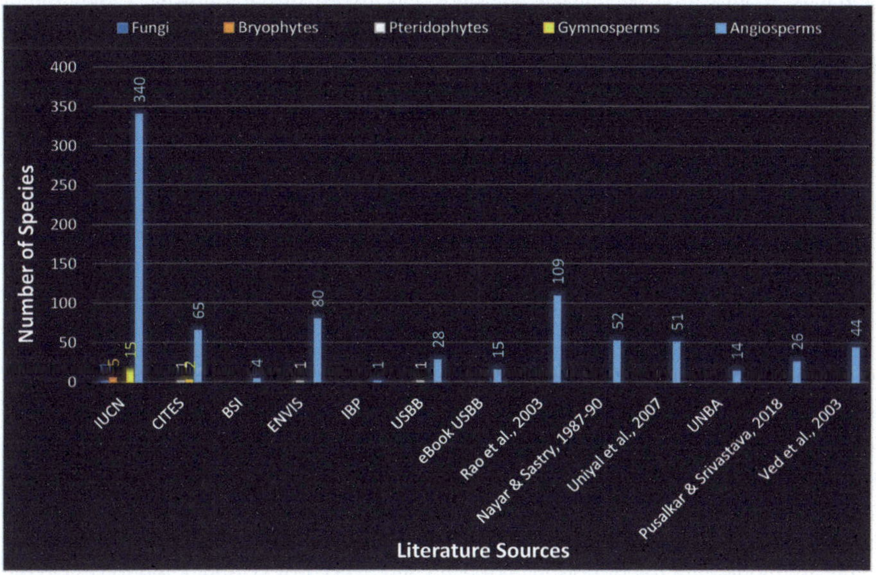

Fig. 9.6 Number of plant species denoted by different threat statuses according to previous literature sources.
Abbreviations: IUCN (International Union for Conservation of Nature), CITES (Convention on International Trade in Endangered Species of Wild Fauna and Flora), BSI (Botanical Survey of India), ENVIS (Environmental Information System), IBP (Indian Biodiversity Portal), USBB (Uttarakhand State Biodiversity Board), UNBA (National Biodiversity Authority for Uttarakhand)

in the present study only single fungal species (*Ophiocordyceps sinensis* (Berk.) G. H. Sung, J. M. Sung, Hywel-Jones and Spatafora) has been described as threatened. About more than 92–95% out of 2.5–9 million estimated fungi of the world are still unknown to the science. Moreover, according a report from Kew, only a little over 1,55,000 fungal species are named and got proper classification so far (Antonelli et al. 2023). The Earth Summit in Rio (1992) adopted the objectives of CBD (conservation of biodiversity), sustainable use, equitable and fair sharing of all aspects of biodiversity, but the world misunderstood the term 'diversity' in relation to the higher plants, animals and microbes. They left the lower flora as an orphan (Minter 2013). A limited interest in the conservation of particular flora in its own environment as compared to the other is completely a 'plant blindness' (FRI 2023). Now the United Nations Biodiversity also incited the people to use the term '*funga*' along with '*flora and fauna*' (Busby 2023) to make aware about the '*fungal blindness*' too.

The risk assessment of all the floristic and faunal species is prepared by IUCN (International Union for Conservation of Nature) red list at international level. The list is updated in every three years to assess the threatened species under different categories (taxonomy wise and country wise) and makes efforts to add more new species. However, to compile the information related to threatened species only from IUCN red list is insufficient as it is not necessary that the species evaluated by IUCN would also be recognized under its given threat statuses. Other threat evaluating agencies and organizations at state, national and international levels are also equally important to maintain the possibly correct threatened data about the species (Chaturvedi et al. 2022).

9.4 Conclusion

In the present study all possible attempts are maintained to compile the information regarding threatened flora of the state Uttarakhand from various authoritative sources. The present comprehensive information will help in filling the gaps to collect the information of threatened flora of Uttarakhand from different previous existing literatures till date. This will also provide a taxonomic insights and practical applications of threatened flora of the state so far. The correct assessment of obscured threat status related information about the flora is the very need for updating and making acceptable the IUCN red list at global scale. The proper awareness to conserve the threatened flora should be raised at different levels and to engage the governing bodies to understand the importance of each and every flora (higher and lower) to the society. The IUCN red list tool is not the last hope of gathering the risk assessment related information about the organisms (flora and fauna). Other collaborative efforts, proper fundings and engagement of governing bodies, non-governmental organizations, universities, botanical gardens, research institutes and stakeholders are also necessary for broadening the conservation practices.

Acknowledgements The present study received no specific grant from any funding agency in the public, commercial, or not-for-profit sectors. The authors are highly indebted to the amateur & professional researchers and the scientists, whose research findings, books, floras, reviews and articles helped in compiling the present study knowledge on the threatened plant species of Uttarakhand.

Authors Contribution SS hypothesized, written and drafted the manuscript. JP drawn the location map of Uttarakhand and helped in drafting the manuscript.

References

Ashton H (2009) Ainsworth and Bisby's dictionary of the fungi. Ref Rev 23(5):42–42. https://doi.org/10.1108/09504120910969104

Arisdason W, Lakshminarasimha P (2016) Status of plant diversity in India: an overview. Central National Herbarium, Botanical Survey of India, Howrah. https://bsienvis.nic.in/Content/StatusofPlantDiversityinIndia_17566.aspx?format=Print#:~:text=Family%20Poaceae%20is%20the%20largest,Lamiaceae%20(72%2F454)

Acharya B, Anupam S, Bhasker J, Prajapati UB (2018) Etymology of endemic plants of Uttarakhand, India. Int J For Usufructs Manag 19:66–69. https://patanjali.res.in/assets/pdf/13.pdf

Antonelli A, Fry C, Smith RJ, Eden J, Govaerts RHA, Kersey (2023) State of the World's Plants and Fungi 2023. Royal Botanic Gardens, Kew. https://doi.org/10.34885/wnwn-6s63

APG IV (2016) An update of the Angiosperm Phylogeny Group classification for the orders and families of flowering plants: APG IV. Bot J Linn Soc 181:1–20. https://doi.org/10.1111/boj.12385

BSI (2021) Biodiversity. EnviStats India 2022, Vol II: Environment Accounts. Ministry of Statistics and Programme Implementation. Government of India. https://www.mospi.gov.in/sites/default/files/reports_and_publication/statistical_publication/EnviStats/Chap7-Biodiversity_envst22.pdf

Busby M (2023) Flora, fauna and … funga: campaigners call for new term for conservation talks. https://www.theguardian.com/environment/2023/aug/30/flora-fauna-and-funga-un-backs-new-term-for-conservation-discussions

Bhatt D, Chandra Sekar K, Kumar K (2020) Tree diversity of Indian Himalayan Region. G.B. Pant National Institute of Himalayan Environment, Almora, Uttarakhand, India

Bargali H, Kumar A, Singh P (2022) Plant studies in Uttarakhand, Western Himalaya—a comprehensive review. Trees, For People 8:100203. https://doi.org/10.1016/j.tfp.2022.100203

CBD (2012) The global strategy for plant conservation: 2011–2020. Botanic Gardens Conservation International. https://www.bgci.org/files/Plants2020/GSPCbrochure/gspc_english.pdf

CITES (2023) Convention on international trade in endangered species of wild fauna and flora (Appendices I, II and III). https://cites.org/eng/app/appendices.php

Checklist of threatened plants of Uttaranchal (2013). https://indiabiodiversity.org/datatable/show/264?offset=0

Chaturvedi P, Chandra S, Rawat DS (2022) Threatened flora of Uttarakhand: an update. J Threatened Taxa 14(12):22309–22328. https://doi.org/10.11609/jott.6330.14.12.22309-22328

Consortium of bryophyte herbaria. https://bryophyteportal.org/portal/. Accessed 03 Jan 2023

ENVIS BSI (2020). http://www.bsienvis.nic.in/Database/RedlistedPlants_3940.aspx

eBook USBB. Threatened species of Uttarakhand. https://sbb.uk.gov.in/files/books/Threatened_Species_Book-CTP.pdf

FRI (2013) Details of species conserved. Uttarakhand Forest Research Institute. https://forest.uk.gov.in/uploads/downloads_details/1684753396.pdf

FRI (2023) Details of species conserved. Uttarakhand Forest Research Institute. https://forest.uk.gov.in/uploads/downloads_details/1630480829.pdf

Gowthami R, Sharma N, Pandey R, Agrawal A (2021) Status and consolidated list of threatened medicinal plants of India. Genet Resour Crop Evol 68: 2235–2263.
Heywood VH (2017) Plant conservation in the Anthropocene–challenges and future prospects. Plant Divers 39(6):314–330. https://doi.org/10.1016/j.pld.2017.10.004
Hibbett DS, Binder M, Bischoff JF, Blackwell M, Cannon, PF, Eriksson et al (2007) A higher-level phylogenetic classification of the Fungi. Mycol Res 111(5):509–547. https://doi.org/10.1016/2Fj.mycres.2007.03.004
IPNI (2023) International plant name index. https://www.ipni.org/. Accessed 28 Dec 2023
IUCN (2024) The IUCN red list of threatened species (version 2023-1). https://www.iucnredlist.org/. Accessed 15 Jan 2024
iNaturalist (2024). https://www.inaturalist.org/places/uttarakhand#page=2&taxon=47170
IBP (Indian Biodiversity Portal). https://indiabiodiversity.org/
Keith DA, Mahony M, Hines H, Elith J, Regan TJ, Baumgartner JB et al (2014) Detecting extinction risk from climate change by IUCN Red List criteria. Conserv Biol 28(3):810–819. https://doi.org/10.1111/cobi.12234
Khuroo AA, Ahmad R, Hamid M, Rather ZA, Malik AH, Rashid I (2021) An annotated inventory of invasive alien flora of India. Invasive Alien Species: Obs Issues Around World 2:16–37. https://doi.org/10.1002/9781119607045.ch14
Lasushe KU, Mir AH, Singh PP, Chaudhary KL, Choudhury H, Deori C, Roy DK, Singh B, Upadhaya K (2022) A comprehensive checklist of threatened plants of Meghalaya, Northeast India. J Asia-Pacific Biodivers 15(3):435–441. https://doi.org/10.1016/j.japb.2022.03.008
Luu-Dam NA, Lu NT, Pham TH, Do TV (2023) Classification of vascular plants in vietnam according to modern classification systems. Plants 12(4):967. https://doi.org/10.3390/plants12040967
Minter D (2013) A future for fungi: the orphans of rio. International Society for Fungal Conservation, UK. http://www.fungal-conservation.org/blogs/orphans-of-rio.pdf
Nayar MP, Sastry ARK (eds) (1987, 88, 90) Red data book of Indian plants, vols 1–3. Botanical Survey of India, Calcutta, pp 268–271
POWO (2024) Plants of the world online. https://powo.science.kew.org/. Accessed 15 Dec 2023
Pandey S, Pandey S (2013) Status and distribution pattern of native and endemic species in Uttarakhand Western Himalayan Region, India. Int J Sci Eng Res 4(11)
Pusalkar PK Srivastava SK (2018) Flora of Uttarakhand: gymnosperms and angiosperms (Ranunculaceae-Moringaceae). Botanical Survey of India, Kolkata, 1099 pp
Rawat GS (2008) Special habitats and threatened plants of India. ENVIS Bull Wildl Protect Areas 11(1):239. https://scholar.google.com/scholar?q=Special%20habitats%20and%20threatened%20plants%20of%20India
Rao CK, Geetha BL, Suresh G (2003) Red list of threatened vascular plant species compiled from the 1997 IUCN red list of threatened plants. ENVIS, Botanical Survey of India, Howrah. file:///C:/Users/HP/Downloads/Redlistbook_2003%20(3).pdf
Rawat N, Purohit S, Painuly V, Negi GS, Bisht MPS (2022) Habitat distribution modeling of endangered medicinal plant *Picrorhiza kurroa* (Royle ex Benth) under climate change scenarios in Uttarakhand Himalaya, India. Ecol Inf 68:101550. https://doi.org/10.1016/j.ecoinf.2021.101550
Singh S, Sahu TR (2015) Tree ferns of pachmarhi biosphere reserve, Madhya Pradesh, India: taxonomy, ethnobotany and conservation. Int J Adv Res 3(8):566–577. https://www.journalijar.com/uploads/716_IJAR-6764.pdf
Singh P, Karthigeyan K, Lakshminarasimhan P, Dash SS (2015) *Endemic vascular* plants of India. Botanical Survey of India, Kolkata, 339 pp. https://www.researchgate.net/publication/280132681_Endemic_Vascular_Plants_of_India
Singh SS (2022) Three Himalayan medicinal plants enter IUCN red list.https://www.thehindu.com/sci-tech/energy-and-environment/three-himalayan-medicinal-plants-enter-iucn-red-list/article66243601.ece
The Plant List (2013). http://www.theplantlist.org/. Accessed 25 Nov 2023

Tali BA, Khuroo AA, Nawchoo IA, Ganie AH (2019) Prioritizing conservation of medicinal flora in the Himalayan biodiversity hotspot: an integrated ecological and socioeconomic approach. Environ Conserv 46(2):147–154. https://doi.org/10.1017/S0376892918000425

UNBA (2009) National biodiversity authority for Uttarakhand. https://faolex.fao.org/docs/pdf/ind189888.pdf

USBB (2009) Uttarakhand state biodiversity board. https://sbb.uk.gov.in/files/Documents/AR-2010-11_2011-12.pdf

Uniyal BP, Sharma JR, Choudhery U, Singh DK (2007) Flowering plants of Utarakhand (A Checklist). Bishen Singh Mahendra Pal Singh, Dehradun, 404 pp

Upadhaya K, Thapa N, Lakadong NJ, Barik SK, Sarma K (2013) Priority areas for conservation in northeast India: a case study in Meghalaya based on plant species diversity and endemism. Int J Ecol Environ Sci 39(2):125–136. https://doi.org/10.13140/2.1.4360.0004

Ved DK, Kinhal GA, Ravikumar K, Prabhakaran V, Ghate U, Sankar RV, Indresha JH (2003) Conservation assessment & management prioritisation for the medicinal plants of Jammu & Kashmir, Himachal Pradesh & Uttarakhand. FRLHT, Bangalore, India

Chapter 10
Ecology and Conservation Issues of Avifauna in the Ganges Floodplain of Bangladesh

Ashikur Rahman Shome, Mohammad Firoj Jaman, Md. Mahabub Alam, Aleya Begum, Aminul Islam, Nabila Zaman Joty, Joy Roy, Mishon Chandra Deb, and Md. Fazle Rabbe

Abstract The Ganges floodplain of Bangladesh is one of the prominent bio-ecological zones that is enriched with wildlife resources. The natural wetlands and modified habitat in this area sustain a diverse range of avifauna. To uncover the present status, seasonality, habitat usages, existing threats, and conservation issues of avifauna, an eight year-long (2016–2023) scientific study was conducted in the Ganges floodplain of Bangladesh by direct field observation along with reviewing the published literatures of the research team of the area. A total of 260 avian species were observed under 20 orders and 62 families. Non-passerine bird species (167 species, 64.23%) specially wetland specialist birds were dominant in this area. From the study area, 98 (37.69%) migratory bird species were observed. In the avian communities among three study sites, wetland specialist bird species and individuals under the order Charadriformes, Anseriformes, Accipitriformes and Pelecaniformes was higher in the wetlands and associated habitat, whereas in terrestrial habitat Passeriformes was higher. A significant difference was found among the species richness in five types of major habitats. The highest number of bird species richness

A. R. Shome (✉) · M. F. Jaman · Md. M. Alam · A. Begum · N. Z. Joty · Md. Fazle Rabbe (✉)
Department of Zoology, University of Dhaka, Dhaka 1000, Bangladesh
e-mail: shomear61@gmail.com

Md. Fazle Rabbe
e-mail: fazle_zool@du.ac.bd

A. R. Shome
Center for Environmental and Geographic Information Services (CEGIS), Sher-E-Bangla Nagar, Dhaka 1207, Bangladesh

Ministry of Environment, Forest and Climate Change & Ministry of Water Resources, Dhaka 1000, Bangladesh

A. Islam
Bangladesh Environmental and Biodiversity Conservation Alliance, Dhaka, Bangladesh

J. Roy · M. C. Deb
Department of Zoology, MC College, Tilagor, Sylhet 3100, Bangladesh

D. Arya et al. (eds.), *Climate Change Impact on Himalayan Biodiversity*,
Environmental Science and Engineering, https://doi.org/10.1007/978-3-031-77149-1_10

and abundance was observed during the winter season with the highest number of season-specific unique species. The wetland habitats are more important for holding the highest number of threatened bird species and important bird species. The 19 nationally threatened species along with 11 newly recorded bird species reflect the conservation necessity of this area. The observation status of avifauna from the study area shows that 116 species (44.61%) were very common, 15 (5.76%) common, 88 (33.84%) uncommon and 41 (15.76%) were few. At present, illegal bird hunting, trade of wild birds, water pollution, agricultural nets, tourism, and existing fishing practices are major threats to birds. This study will help take future conservation initiatives for the conservation of birds in the study area.

Key Words Wetlands · Birds · Habitat · Conservation · Migration · Threats · Ganges · Floodplain

10.1 Introduction

Areas of low-lying terrain that are vulnerable to flooding by lateral overflow water from nearby rivers or lakes are known as floodplains (Tockner et al. 2010). Throughout the world, floodplains are considered as a dynamic ecosystem covering 15% area of wetlands which play a significant role in the ecology, economy as well as in the human culture (Latterell and Naiman 2007; Fischlin et al. 2007). Floodplains in Bangladesh play a vital role in conserving wetlands biodiversity, specially avifauna by providing livelihood support (Rahman et al. 2015; Kibria and Yousuf Haroon 2017).

Among the bio-resources, avifauna- the warm-blooded, feathered vertebrate wildlife group-holds more than 10,000 species globally (del Hoyo et al. 2014). They are important bio-indicators for the environment to determine the pollution level, impacts of climate change, status of the ecological processes, pollination of plants, and gene flow by seed dispersal (Sekercioglu 2006). They have an exceptional role in controlling the population of dangerous insects and other pests (Şekercioğlu et al. 2004). Additionally, insectivorous birds and raptors also control disease vectors such as mosquitos (Gatesire et al. 2014).

Bangladesh is situated in the intersection of two tremendously enriched bio-geographical realms named as Indo-Himalayan and Indo-China sub-region (Nishat et al. 2002). It holds a great diversity of flora and fauna residing in three major vegetation types that inhabit exceptionally diversified ecosystems (IUCN Bangladesh 2015a; Khan 2018). Besides, this country is primarily a deltaic country in the Ganges, Brahmaputra and Meghna (GBM) river system, with a large number of wetlands constituting a waterway of about 22,155 km (IUCN Bangladesh 2015b). All of these habitat structures aid this country in holding more than 700 species of avifauna (Shome et al. 2022).

Ganges is one of the major and most important rivers of Bangladesh stretching about 326 km long and passing through 13 districts of Bangladesh. It is an exceptional wetland ecosystem covering the southwestern districts (Kushtia, Meherpur, Rajbari, Chuadanga, Jhenaidah, Jessore, Narail, Magura, Khulna, Satkhira, Pirojpur, Bagherhat, Faridpur, Gopalganj, Madaripur, Shariatpur, Barishal, Patuakhali, and Barguna) and slightly northwestern portion (Chapainawabganj, Rajshahi) of Bangladesh (Nishat et al. 2002). The Ganges floodplain with char land, riverine grassland, *beel*, *baor*, ditch, canal, tributary and distributary river is known to harbor a good number of wetlands bird (Rahman et al. 2015; Reza et al. 2014).

The natural and modified habitats of the Ganges floodplain are home to different group of wildlife resources specially avifauna. Some scattered research works are present on some particular area of Ganges floodplain (Jaman et al. 2023; Biswas et al. 2022; Shome et al. 2022, 2020) but overall scenario is still unknown. At the same time, there is no designated protected area in this region. Some research indicates about the threats (e.g., illegal bird hunting, agricultural net, trade of birds) to avian species from the area (Shome et al. 2022). This chapter represents the overall scenario of avifauna in this area along with their major habitat usage, seasonality, threats and conservation aspects. Besides, this study will be helpful in taking future conservation initiatives in the study area.

10.2 Materials and Method

Study area: Bangladesh is divided into 22 bio-ecological zones according to different physical condition, climatic status, environmental condition, ecological characteristics, floral and faunal composition, and biological diversity (Nishat et al. 2002). Among them, the Ganges floodplain is one of the most prominent bio-ecological zones where this study was conducted (Fig. 10.1).

The Ganges floodplain is the meandering floodplain of the Ganges river. The morphology of the floodplain is primarily rounded, with historic rivers, basins, and terrain. In every flooding season, the Ganges channel continuously moves, erodes and deposits vast stretches of char land inside its dynamic floodplain (Banglapedia 2024). The floodplain is distinguished by its' diverse vegetation and to a certain degree, it supports a wealth of biodiversity because of the quantity of tributaries, channels, rivers, *beels* (lake), *baor* (Oxbow lake) and still water bodies. In *beels* and other bodies of water, a considerable proportion of aquatic vegetation floats freely (Nishat et al. 2002). Different types of habitats were surveyed for better understanding about the composition of the birds in the study area (Table 10.1, Fig. 10.2).

Data collection: This chapter detailed out all studies (both published and unpublished) done by the authors from January 2016 to December 2023 in the study areas. The authors conducted this eight year-long study primarily by direct field observation in the study area along with reviewing the published literature related to it. For understanding the seasonal variation of the bird, survey was conducted in three seasons

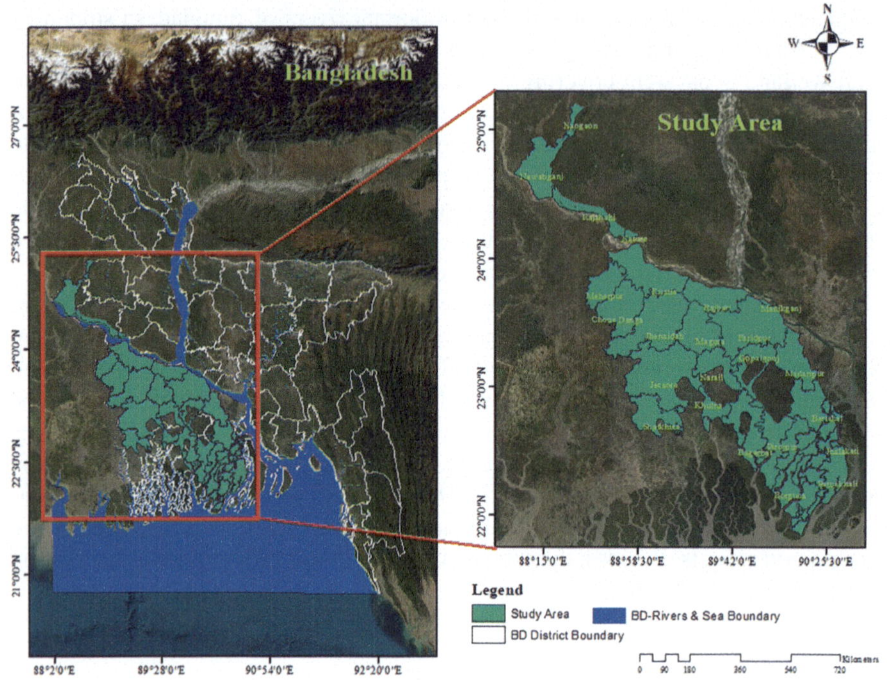

Fig. 10.1 Map of the study area showing the Ganges floodplain of Bangladesh

Table 10.1 Habitat type and the details of habitats surveyed in the study area

Habitat	Type	Description
Major river associated wetlands (MR)	Aquatic	Sandbars, floodplain, marshland, open-water body and the habitats associated with the major river
Inland wetlands (IW)	Aquatic	Beel, *baor* (Oxbow lake), canal, ditch, small river outside the major river
Human planted forest (HS)	Terrestrial	Human planted trees, green-spaces, homestead forest, garden
Agricultural land (AG)	Terrestrial	Agricultural land, crop field
Grassland (GL)	Terrestrial	Grassy area

(Summer: March to June; Rainy: July to October; Winter: November to February). Transect line survey and opportunistic survey method were followed for collecting data. Each observation was done early in the morning (from 6:00am to 10:30am) and afternoon (from 3:00 pm to 6:30 pm). For collecting data of nocturnal birds, night survey was conducted using torch light and head lamp. Photographs of birds was taken using Nikon D500 DSLR Camera with 200–500 mm VR lens.

Data analysis: Spreadsheets (MS Excel), and PAST software were used for statistical analysis. To estimate the observation status of bird, Khan (2015) was followed as very common (VC) 80–100%, common (C) 50–79%, fairly common (FC) 20–49% and few (F) 10-19% which was calculated based on total sighting per survey attempt.

10.3 Results

10.3.1 Species Composition and Status

This study recorded the existence of 260 species of avifauna under 20 orders and 62 families from the Ganges floodplain of Bangladesh (Table 10.2). Among the recorded avian species, the highest species number was under the order Passeriformes (93 species, 35.76%) which is followed by Charadriformes (40 species, 15.38%), Anseriformes (21 species, 8.07%) and Accipitriformes (21 species, 8.07%) (Fig. 10.3a). Considering family comparison, the family Anatidae (21 species, 8.07%) had the highest number of bird species followed by Accipitridae (20 species, 7.69%),

Fig. 10.2 Different types of habitats observed in the study area

Fig. 10.2 (continued)

Scolopacidae (14 species, 5.38%) and Ardeidae (12 species, 4.61%) (Fig. 10.3b). Among the recorded bird species, 157 species were resident, 98 were migratory (94 winter migratory, four summer migratory) and the remaining five were vagrant. Observation status of avifauna from the study area showed that, 116 species (44.61%) was very common, 15 (5.76%) common, 88 (33.84%) uncommon and 41 (15.76%) was few.

According to the literature review and field observation, wetlands holds the highest number of bird species in the study area (Table 10.3). At the same time sandbars of river, urban and rural areas play significant role in holding different species of bird in the study area (Table 10.3). Research findings also showed that bird diversity in and around the wetlands of Faridpur and Rajshahi region is comparatively higher than any other area of the Ganges floodplain (Table 10.3).

Seasonal variation: Among the recorded species, 133 (51%) species observed throughout the year. During the winter season, the highest number of unique species (118 species, 44.5%) observed (Fig. 10.4) which were found particularly in the winter season from the study area. No unique species found during the rainy season (Fig. 10.4).

Table 10.2 List of avifauna recorded from the Ganges floodplain from January 2016 to December 2023. (Note: OS- Observation Status; VC- Very Common; C-Common, UC- Uncommon, Few- F; IBD- IUCN Bangladesh status 2015, LC- Least Concern, CR- Critically Endangered, EN- Endangered, VU- Vulnerable, NT- Near Threatened, NA- Not Assessed, DD- Data Deficient, RE- Regionally Extinct)

Scientific name	English name	IBD-2015	OS	Scientific name	English name	IBD-2015	OS
Coturnix coromandelica	Rain Quail	DD	F	*Aquila nipalensis*	Steppe Eagle	LC	F
Francolinus pondicerianus	Grey Francolin	RE	F	*Aquila heliaca*	Eastern Imperial Eagle	VU	F
Dendrocygna bicolor	Fulvous Whistling Duck	LC	UC	*Hieraaetus pennatus*	Booted Eagle	LC	UC
Dendrocygna javanica	Lesser Whistling Duck	LC	VC	*Circus aeruginosus*	Western Marsh-harrier	LC	F
Anser indicus	Bar-headed Goose	LC	UC	*Circus spilonotus*	Eastern Marsh-harrier	LC	F
Anser anser	Greylag Goose	LC	F	*Circus cyaneus*	Hen (Northern) Harrier	DD	UC
Anser erythropus	Lesser White-fronted Goose	VU	F	*Circus melanoleucos*	Pied Harrier	LC	UC
Tadorna tadorna	Common Shelduck	LC	UC	*Accipiter badius*	Shikra	LC	C
Tadorna ferruginea	Ruddy Shelduck	LC	UC	*Ichthyophaga ichthyaetus*	Grey-headed Fish- eagle	NT	UC
Sarkidiornis melanotos	African Comb Duck	NT	UC	*Milvus migrans*	Black Kite	LC	VC
Nettapus coromandelianus	Cotton Pygmy-goose	LC	VC	*Haliastur indus*	Brahminy Kite	LC	VC
Netta rufina	Red-crested Pochard	LC	UC	*Butastur teesa*	White-eyed Buzzard	LC	F
Aythya ferina	Common Pochard	LC	UC	*Buteo rufinus*	Long-legged Buzzard	LC	UC
Aythya nyroca	Ferruginous Duck	NT	UC	*Upupa epops*	Common Hoopoe	LC	VC
Aythya fuligula	Tufted Duck	LC	UC	*Merops orientalis*	Asian Green Bee-eater	LC	VC
Spatula querquedula	Garganey	LC	VC	*Merops leschenaulti*	Chestnut-headed Bee-eater	LC	UC
Spatula clypeata	Northern Shoveler	LC	UC	*Merops philippinus*	Blue-tailed Bee-eater	LC	VC
Mareca falcata	Falcated Duck	NT	UC	*Coracias benghalensis*	Indian Roller	LC	VC
Mareca strepera	Gadwall	LC	VC	*Alcedo meninting*	Blue-eared Kingfisher	LC	F

(continued)

Table 10.2 (continued)

Scientific name	English name	IBD-2015	OS	Scientific name	English name	IBD-2015	OS
Mareca penelope	Eurasian Wigeon	LC	UC	*Alcedo atthis*	Common Kingfisher	LC	VC
Anas poecilorhyncha	Indian Spot-billed Duck	LC	C	*Ceryle rudis*	Pied Kingfisher	LC	VC
Anas acuta	Northern Pintail	LC	UC	*Pelargopsis capensis*	Stork-billed Kingfisher	LC	VC
Anas crecca	Common Teal	LC	UC	*Halcyon smyrnensis*	White-breasted Kingfisher	LC	VC
Tachybaptus ruficollis	Little Grebe	LC	VC	*Todiramphus chloris*	Collared Kingfisher	LC	C
Podiceps grisegena	Red-necked Grebe	LC	F	*Psilopogon (Megalaima) haemacephala*	Coppersmith Barbet	LC	VC
Podiceps cristatus	Great Crested Grebe	LC	C	*Psilopogon lineatus*	Lineated Barbet	LC	VC
Columba livia	Rock Dove	LC	VC	*Psilopogon asiaticus*	Blue-throated Barbet	LC	VC
Streptopelia decaocto	Eurasian Collared Dove	LC	VC	*Jynx torquilla*	Eurasian Wryneck	LC	UC
Streptopelia tranquebarica	Red Turtle Dove	LC	VC	*Chrysocolaptes guttacristatus (lucidus)*	Greater Flameback	LC	UC
Spilopelia chinensis	Eastern Spotted Dove	LC	VC	*Dinopium benghalense*	Black-rumped Flameback	LC	VC
Chalcophaps indica	Grey-capped Emerald Dove	LC	UC	*Picus xanthopygaeus*	Streak-throated Woodpecker	LC	VC
Treron phoenicopterus	Yellow Footed Green Pigeon	LC	VC	*Dendrocopos macei*	Fulvous-breasted Woodpecker	LC	VC
Caprimulgus macrurus	Large-tailed Nightjar	LC	VC	*Falco tinnunculus*	Common Kestrel	LC	UC
Caprimulgus mahrattensis	Sykes's nightjar	NA	F	*Falco peregrinus*	Peregrine Falcon	LC	UC
Cypsiurus balasiensis	Asian Palm Swift	LC	VC	*Psittacula cyanocephala*	Plum-headed Parakeet	LC	UC

(continued)

Table 10.2 (continued)

Scientific name	English name	IBD-2015	OS	Scientific name	English name	IBD-2015	OS
Apus nipalensis (A. affinis)	House Swift	LC	VC	*Psittacula eupatria*	Alexandrine Parakeet	LC	UC
Centropus sinensis	Greater Coucal	LC	VC	*Psittacula krameri*	Rose-ringed Parakeet	LC	F
Centropus bengalensis	Lesser Coucal	LC	C	*Pitta brachyura*	Indian Pitta	LC	UC
Phaenicophaeus tristis	Green-billed Malkoha	LC	VC	*Artamus fuscus*	Ashy Woodswallow	LC	VC
Clamator jacobinus	Jacobin Cuckoo	LC	C	*Aegithina tiphia*	Common Iora	LC	VC
Eudynamys scolopaceus	Western Koel (Asian koel)	LC	VC	*Tephrodornis gularis*	Large Woodshrike	LC	C
Cacomantis merulinus	Plaintive Cuckoo	LC	VC	*Tephrodornis pondicerianus*	Common Woodshrike	LC	C
Hierococcyx varius	Common Hawk- Cuckoo	LC	VC	*Coracina macei*	Large Cuckooshrike	LC	UC
Cuculus micropterus	Indian Cuckoo	LC	VC	*Coracina melaschistos*	Black-winged Cuckooshrike	LC	UC
Zapornia pusilla	Baillon's Crake	LC	UC	*Coracina melanoptera*	Black-headed Cuckooshrike	LC	UC
Amaurornis phoenicurus	White-breasted Waterhen	LC	VC	*Pericrocotus cinnamomeus*	Small Minivet	LC	VC
Gallicrex cinerea	Watercock	LC	UC	*Lanius cristatus*	Brown Shrike	LC	VC
Porphyrio porphyrio	Purple Swamphen	LC	UC	*Lanius schach*	Long-tailed Shrike	LC	VC
Gallinula chloropus	Common Moorhen	LC	UC	*Lanius tephronotus*	Grey-backed Shrike	LC	UC
Fulica atra	Eurasian Coot	LC	UC	*Oriolus oriolus*	Eurasian Golden Oriole	LC	UC
Leptoptilos javanicus	Lesser Adjutant	VU	F	*Oriolus chinensis*	Black-naped Oriole	LC	UC
Mycteria leucocephala	Painted Stork	CR	F	*Oriolus xanthornus*	Black-hooded Oriole	LC	VC

(continued)

Table 10.2 (continued)

Scientific name	English name	IBD-2015	OS	Scientific name	English name	IBD-2015	OS
Anastomus oscitans	Asian Openbill	LC	VC	*Dicrurus macrocercus*	Black Drongo	LC	VC
Ciconia nigra	Black Stork	VU	F	*Dicrurus leucophaeus*	Ashy Drongo	LC	UC
Ciconia episcopus	Asian Woolyneck	CR	F	*Dicrurus aeneus*	Bronzed Drongo	LC	C
Ephippiorhychus asiaticus	Black-necked Stork	EN	F	*Dicrurus remifer*	Lesser Racket-tailed Drongo	LC	F
Threskiornis melanocephalus	Black-headed Ibis	VU	UC	*Rhipidura albicollis*	White-throated Fantail	LC	C
Plegadis falcinellus	Glossy Ibis	LC	F	*Hypothymis azurea*	Black-naped Monarch	LC	C
Pseudibis papillosa	Red-napped Ibis	NA	F	*Terpsiphone paradisi*	Asian Paradise-flycatcher	LC	UC
Ixobrychus sinensis	Yellow Bittern	LC	VC	*Dendrocitta vagabunda*	Rufous Treepie	LC	VC
Ixobrychus cinnamomeus	Cinnamon Bittern	LC	C	*Corvus splendens*	House Crow	LC	VC
Ixobrychus flavicollis	Black Bittern	NT	UC	*Corvus levaillantii*	Jungle Crow	LC	VC
Nycticorax nycticorax	Black-crowned Night Heron	LC	VC	*Parus major*	Great Tit	LC	VC
Butorides striata	Green-backed (Striated) Heron	LC	VC	*Riparia riparia*	Sand Martin	LC	UC
Ardeola grayii	Indian Pond Heron	LC	VC	*Riparia paludicola*	Plain (Brown-throated) Martin	LC	UC
Bubulcus ibis	Cattle Egret	LC	VC	*Hirundo rustica*	Barn Swallow	LC	VC
Ardea cinerea	Grey Heron	LC	UC	*Hirundo daurica*	Red-rumped Swallow	LC	UC
Ardea purpurea	Purple Heron	LC	VC	*Mirafra assamica*	Rufous-winged Lark	LC	VC

(continued)

Table 10.2 (continued)

Scientific name	English name	IBD-2015	OS	Scientific name	English name	IBD-2015	OS
Ardea alba (Casmerodius albus)	Great White Egret (Great Egret)	LC	VC	*Calandrella raytal*	Indian Short-toed (Sand) Lark	LC	F
Ardea (Egretta) intermedia	Intermediate Egret	LC	VC	*Eremopterix grisea*	Ashy-crowned Sparrow-lark	LC	F
Egretta garzetta	Little Egret	LC	VC	*Alauda gulgula*	Oriental Skylark	LC	F
Phalacrocorax carbo	Great Cormorant	LC	C	*Cisticola juncidis*	Zitting Cisticola	LC	VC
Phalacrocorax fuscicollis	Indian Cormorant	LC	UC	*Prinia hodgsonii*	Grey-breasted Prinia	LC	F
Microcarbo (Phalacrocorax) niger	Little Cormorant	LC	VC	*Prinia gracilis*	Graceful Prinia	LC	VC
Anhinga melanogaster	Oriental Darter	NT	F	*Prinia inornata*	Plain Prinia	LC	VC
Burhinus indicus (oedicnemus)	Indian Thick-knee	LC	F	*Pycnonotus cafer*	Red-vented Bulbul	LC	VC
Recurvirostra avosetta	Pied Avocet	LC	UC	*Orthotomus sutorius*	Common Tailorbird	LC	VC
Himantopus himantopus	Black-winged Stilt	LC	VC	*Megalurus palustris*	Striated Grassbird	LC	UC
Pluvialis squatarola	Grey Plover	LC	UC	*Acrocephalus aedon*	Thick-billed Warbler	LC	F
Pluvialis fulva	Pacific Golden Plover	LC	VC	*Acrocephalus agricola*	Paddyfield Warbler	LC	VC
Charadrius dubius	Little Ringed Plover	LC	VC	*Acrocephalus dumetorum*	Blyth's Reed-warbler	LC	F
Charadrius alexandrinus	Kentish Plover	LC	VC	*Acrocephalus orinus*	Large-billed Reed-warbler	DD	F
Charadrius mongolus	Lesser Sandplover	LC	UC	*Acrocephalus stentoreus*	Clamorous Reed-warbler	LC	VC
Vanellus duvaucelii	River Lapwing	NT	VC	*Phylloscopus affinis*	Tickell's Leaf-warbler	LC	F

(continued)

Table 10.2 (continued)

Scientific name	English name	IBD-2015	OS	Scientific name	English name	IBD-2015	OS
Vanellus malarbaricus	Yellow-wattled Lapwing	NT	UC	*Phylloscopus fuscatus*	Dusky Warbler	LC	VC
Vanellus cinereus	Grey-headed Lapwing	LC	UC	*Phylloscopus inornatus*	Inornate Warbler	LC	F
Vanellus indicus	Red-wattled Lapwing	LC	VC	*Phylloscopus trochiloides*	Greenish Warbler	LC	VC
Rostratula benghalensis	Greater Painted Snipe	LC	C	*Malacocincla abbotti*	Abbott's Babbler	LC	VC
Hydrophasianus chirurgus	Pheasant-tailed Jacana	LC	UC	*Turdoides earlei*	Striated Babbler	LC	VC
Metopidius indicus	Bronze-winged Jacana	LC	VC	*Turdoides striata*	Jungle Babbler	LC	VC
Numenius phaeopus	Whimbrel	LC	UC	*Zosterops palpebrosus*	Oriental White-eye	LC	VC
Limosa limosa	Black-tailed Godwit	NT	UC	*Sturnus contra*	Asian Pied Starling	LC	VC
Calidris (Philomachus) pugnax	Ruff	LC	F	*Sturnus malabaricus*	Chestnut-tailed Starling	LC	VC
Calidris temminckii	Temminck's Stint	LC	VC	*Sturnus pagodarum*	Brahminy Starling	LC	F
Calidris alpina	Dunlin	LC	UC	*Acridotheres fuscus*	Jungle Myna	LC	VC
Calidris minuta	Little Stint	LC	VC	*Acridotheres ginginianus*	Bank Myna	LC	VC
Gallinago stenura	Pin-tailed Snipe	LC	VC	*Acridotheres tristis*	Common Myna	LC	VC
Gallinago gallinago	Common Snipe	LC	VC	*Ficedula albicilla*	Taiga Flycatcher	LC	VC
Lymnocryptes minimus	Jack Snipe	DD	UC	*Eumyias thalassina*	Verditer Flycatcher	LC	UC
Actitis hypoleucos	Common Sandpiper	LC	VC	*Luscinia calliope*	Siberian Rubythroat	LC	UC
Tringa ochropus	Green Sandpiper	LC	VC	*Luscinia svecica*	Bluethroat	LC	UC
Tringa nebularia	Common Greenshank	LC	UC	*Copsychus saularis*	Oriental Magpie-robin	LC	VC
Tringa totanus	Common Redshank	LC	UC	*Phoenicurus ochruros*	Black Redstart	LC	UC

(continued)

Table 10.2 (continued)

Scientific name	English name	IBD-2015	OS	Scientific name	English name	IBD-2015	OS
Tringa glareola	Wood Sandpiper	LC	UC	*Saxicola caprata*	Pied Bushchat	LC	VC
Turnix suscitator	Barred Buttonquail	LC	UC	*Zoothera citrina*	Orange-headed Trush	LC	VC
Glareola maldivarum	Oriental Pratincole	LC	UC	*Dicaeum erythrorhynchos*	Pale-billed Flowerpecker	LC	UC
Glareola lactea	Little Pratincole	LC	VC	*Anthreptes singalensis*	Ruby-cheeked Sunbird	LC	UC
Rynchops albicollis	Indian Skimmer	CR	F	*Nectarinia zeylonica*	Purple-rumped Sunbird	LC	VC
Larus brunnicephalus	Brown-headed Gull	LC	UC	*Nectarinia asiatica*	Purple Sunbird	LC	VC
Larus ridibundus	Black-headed Gull	LC	UC	*Arachnothera longirostra*	Little Spiderhunter	LC	UC
Larus ichthyaetus	Pallas's Gull	LC	UC	*Passer domesticus*	House Sparrow	LC	VC
Larus fuscus (includes L. heuglini)	Lesser Black-backed Gull	LC	UC	*Ploceus benghalensis*	Black-breasted Weaver	LC	F
Sterna albifrons	Little Tern	LC	UC	*Ploceus manyar*	Streaked Weaver	LC	F
Chlidonias hybrida	Whiskered Tern	LC	UC	*Ploceus philippinus*	Baya Weaver	LC	VC
Sterna aurantia	River Tern	NT	UC	*Amandava amandava*	Red Avadavat	LC	UC
Tyto alba	Common Barn Owl (Barn Owl)	LC	VC	*Lonchura malabarica*	White-throated Munia	LC	VC
Ninox scutulata	Brown Boobook (Hawk Owl)	LC	VC	*Lonchura atricapilla*	Chestnut Munia	LC	VC
Athene brama	Spotted Owlet	LC	VC	*Lonchura punctulata*	Scaly-breasted Munia	LC	VC
Otus lettia (bakkamoena)	Collared Scops Owl	LC	UC	*Lonchura striata*	White-rumped Munia	LC	VC
Otus sunia	Oriental Scops Owl	LC	F	*Dendronanthus indicus*	Forest Wagtail	LC	UC

(continued)

Table 10.2 (continued)

Scientific name	English name	IBD-2015	OS	Scientific name	English name	IBD-2015	OS
Asio flammeus	Short Eared Owl	LC	F	*Motacilla alba*	White Wagtail	LC	VC
Ketupa zeylonensis	Brown Fish Owl	LC	C	*Motacilla cinerea*	Grey Wagtail	LC	UC
Pandion haliaetus	Osprey	LC	VC	*Motacilla citreola*	Citrine Wagtail	LC	UC
Elanus caeruleus	Black-winged Kite	LC	VC	*Motacilla flava*	Yellow Wagtail	LC	VC
Pernis ptilorhyncus	Oriental Honey Buzzard	LC	VC	*Anthus hodgsoni*	Olive-backed Pipit	LC	VC
Spilornis cheela	Crested Serpent Eagle	LC	VC	*Anthus richardi*	Richard's Pipit	LC	UC
Gyps bengalensis	White-rumped Vulture	CR	F	*Anthus roseatus*	Rosy Pipit	LC	UC
Nisaetus (Spizaetus) cirrhatus	Changeable Hawk- eagle	LC	VC	*Anthus rufulus*	Paddyfield Pipit	LC	VC
Clanga (Aquila) hastata	Indian Spotted Eagle	EN	UC	*Emberiza pusilla*	Little Bunting	LC	UC
Clanga (Aquila) clanga	Greater Spotted Eagle	VU	UC	*Emberiza melanocephala*	Black-headed Bunting	NA	F

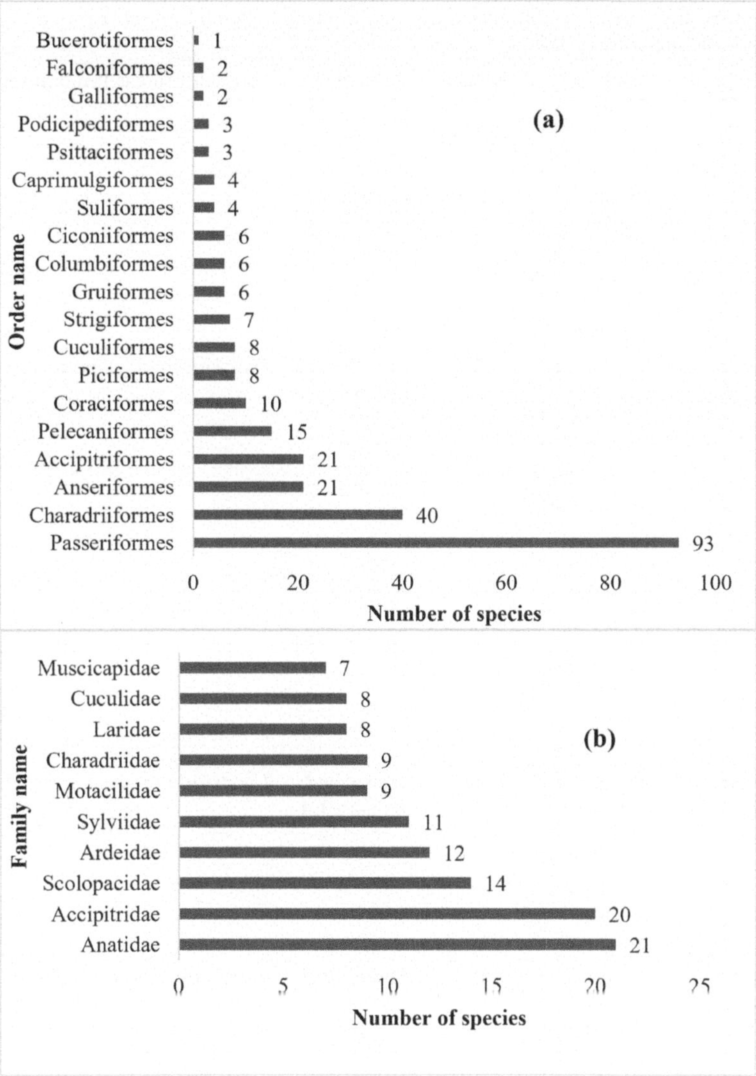

Fig. 10.3 **a** Order-wise species diversity and **b** ten most abundant family with species number in the Ganges floodplain, Bangladesh

Habitat usage: In the study area, aquatic habitat holds the highest number of bird species compared to other habitats. Major rivers associated with wetlands habitat hold the highest number of avian species followed by inland wetlands whereas, the lowest number of species was observed in grassland (61 species) (Fig. 10.5). A significant difference ($\chi^2 = 73.163, df = 4, p < 0.0001$) was found among the species richness in five types of major habitats.

Table 10.3 Bird diversity in different areas of the Ganges floodplain according to the field observation and literature review

Location	Habitat Type	Number of Species	Reference	Nationally threatened bird species
Sreepur Upazila, Magura	Rural planted forest, agricultural land, Small inside wetlands	84	Mandal et al. (2021)	*Clanga hastata*
Magura Sadar Upazila, Magura	Urban and rural planted forest, agricultural land, small inside wetlands	140	Shome et al. (2020)	*Threskiornis melanocephalus*
Keshabpur, Jessore	Rural planted forest, agricultural land, small inside wetlands	105	Jaman et al. (2015)	–
Kashipur Union, Barishal	Rural planted forest, agricultural land, small inside wetlands	141	Shome and Jaman 2021	*Threskiornis melanocephalus*
Faridpur Sadar Upazila, Faridpur	Urban planted forest, agricultural land, inside wetlands, wetlands associated major river	168	Shome et al. 2022a	*Clanga hastata, Threskiornis melanocephalus, Clanga clanga, Emberiza melanocephala*

(continued)

Table 10.3 (continued)

Location	Habitat Type	Number of Species	Reference	Nationally threatened bird species
Madhukhali Upazila, Faridpur	Urban and rural planted forest, agricultural land, small inside wetlands	109	Biswas et al. (2022)	–
Marjat *baor*, Jhenaidah	Inside wetlands (Oxbow lake), rural planted forest, agricultural land	131	Jaman et al. (2023)	*Clanga hastata*, *Clanga clanga*, *Threskiornis melanocephalus*
Kaliganj Upazila, Jhenaidah	Urban and rural planted forest, agricultural land, inside wetlands	153	Scientific observation of the research team	*Clanga hastata*, *Clanga clanga*
Sandbars in Ganges River (Rajbari, Faridpur, Madaripur)	Wetlands associated major river, agricultural land, grassland	193	Scientific observation of the research team	*Ciconia episcopus*, *Gyps bengalensis*, *Mycteria leucocephala*, *Clanga hastata*, *Ciconia nigra*, *Clanga clanga*, *Leptoptilos javanicus*, *Threskiornis melanocephalus*

(continued)

Table 10.3 (continued)

Location	Habitat Type	Number of Species	Reference	Nationally threatened bird species
Wetlands in Faridpur, Magura, Jhenaidah, Gopalganj	Wetlands associated major river, Inland wetlands agricultural land, grassland	202	Scientific observation of the research team	*Ciconia episcopus, Gyps bengalensis, Mycteria leucocephala, Clanga hastata, Ciconia nigra, Clanga clanga, Leptoptilos javanicus, Rynchops albicollis, Threskiornis melanocephalus*
Trans-boundary area of Rajshahi and Chapainawabganj	Wetlands associated major river, Inland wetlands agricultural land, grassland	232	Scientific observation of the research team	*Ciconia episcopus, Gyps bengalensis, Mycteria leucocephala, Rynchops albicollis, Clanga hastata, Ephippiorhychus asiaticus, Anser erythropus, Aquila heliacal, Ciconia nigra, Clanga clanga, Leptoptilos javanicus, Threskiornis melanocephalus, Anhinga melanogaster, Aythya nyroca, Ichthyophaga ichthyaetus, Ixobrychus flavicollis, Limosa limosa, Mareca falcate, Sarkidiornis melanotos, Sterna aurantia, Vanellus duvaucelii, Vanellus malarbaricus, Acrocephalus orinus, Circus cyaneus, Coturnix coromandelica, Lymnocryptes minimus, Caprimulgus mahrattensis, Pseudibis papillosa, Francolinus pondicerianus*

Among the five types of habitats, 24 species were observed in all types of habitats. Thirty-nine species was found in human planted forest habitat (HS) that is the maximum number of unique species in a single habitat while no unique species observed in agricultural area (AG) but the wetlands associated habitat holds the highest number of unique species (Fig. 10.6).

Conservation status: Among the recorded bird species, four were Critically Endangered (*Ciconia episcopus, Gyps bengalensis, Mycteria leucocephala, Rynchops albicollis*), two Endangered (*Clanga hastata, Ephippiorhychus asiaticus*), six Vulnerable (*Anser erythropus, Aquila heliacal, Ciconia nigra, Clanga clanga, Leptoptilos javanicus, Threskiornis melanocephalus*), 10 Near Threatened (*Anhinga melanogaster, Aythya nyroca, Ichthyophaga ichthyaetus, Ixobrychus flavicollis, Limosa limosa, Mareca falcata, Sarkidiornis melanotos, Sterna aurantia, Vanellus duvaucelii, Vanellus malarbaricus*), four Data Deficient (*Acrocephalus orinus, Circus cyaneus, Coturnix coromandelica, Lymnocryptes minimus*), three Not Assessed (*Caprimulgus*

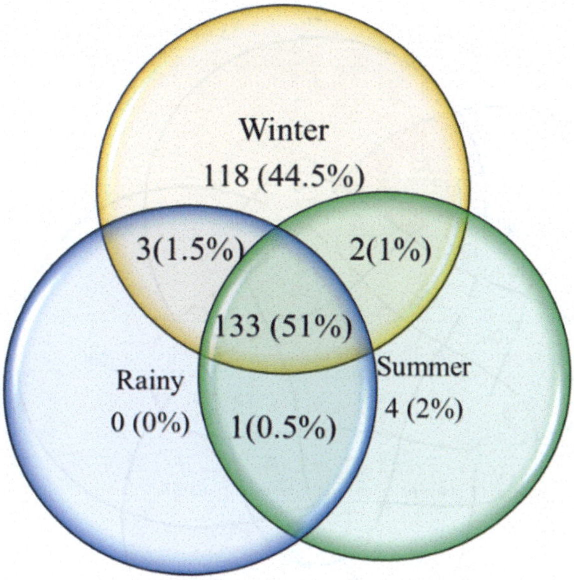

Fig. 10.4 Venn diagram showing the number of shared and unique species in three seasons from the study area

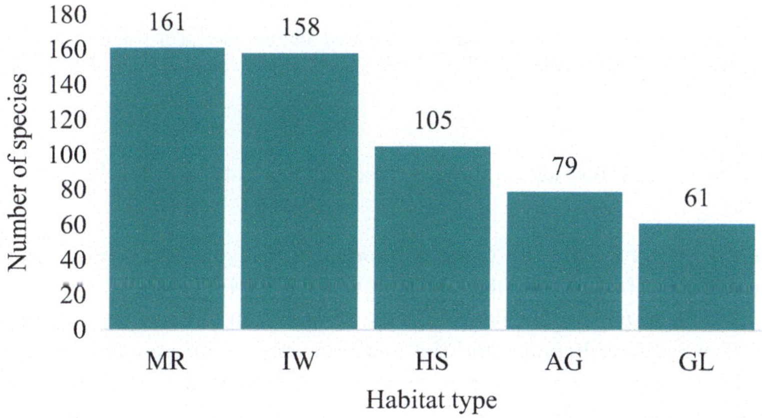

Fig. 10.5 Habitat-wise species richness from the study area (Major rivers associated wetlands-MR, Inland wetlands-IW, Human planted forest- HS, Agricultural land- AG, Grassland- GL)

mahrattensis, *Emberiza melanocephala*, *Pseudibis papillosa*), one Regionally Extinct (*Francolinus pondicerianus*) and rest are Least Concern according to IUCN Bangladesh (2015a, b).

The habitat associated with major river is significant for holding the highest number of threatened bird species (18 species). The inland wetlands habitat is also

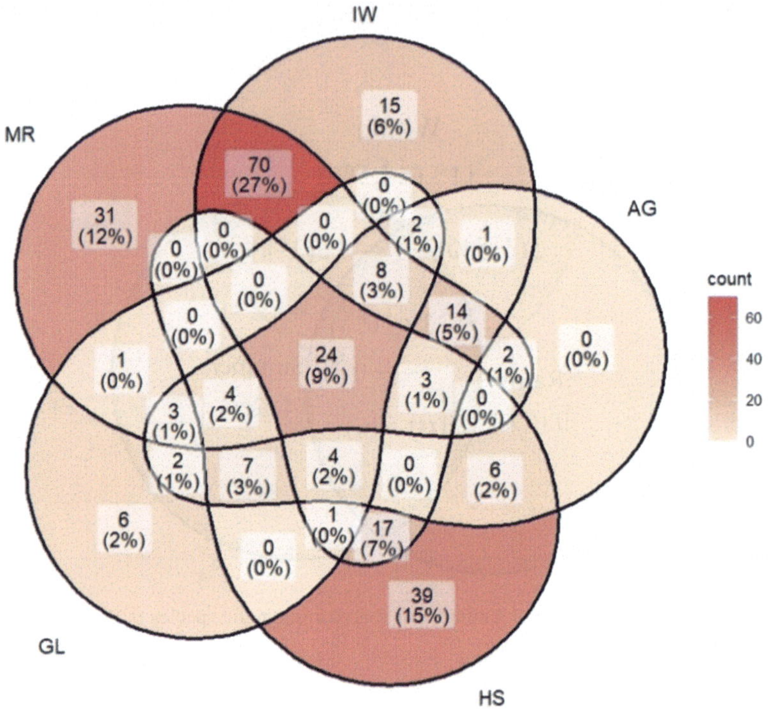

Fig. 10.6 Venn diagram showing the number of shared and unique species among five habitats. (Major rivers associated wetlands-MR, Inland wetlands-IW, Human planted forest- HS, Agricultural land- AG, Grassland- GL)

potential for holding threatened bird species and this habitat holds 11 bird species (Fig. 10.7).

New country record from the Ganges floodplain: Eleven new country record of birds have been enlisted during the last eight years according to literature review and national daily newspaper information (Table 10.4). Seven of them has been recorded from river or wetland habitats and the rest four were from human-planted forest habitat. All of them were observed from the transboundary areas which are beyond the nationally protected landscape. Moreover, in 2018, *Francolinus pondicerianus* was rediscovered from Bangladesh (transboundary area of Rajshahi) which was previously considered as nationally extinct bird species (IUCN Bangladesh 2015a).

Threats to avifauna in the Ganges floodplain: Illegal bird hunting constitutes a significant threat to avian community in the study area (Table 10.5). Furthermore, the deployment of agricultural net within crop fields and fruit gardens emerges as a prominent hazard to avifauna in this locality. Within inland wetlands, water pollution alongside escalating fishing and tourism activities pose substantial threats to birds'

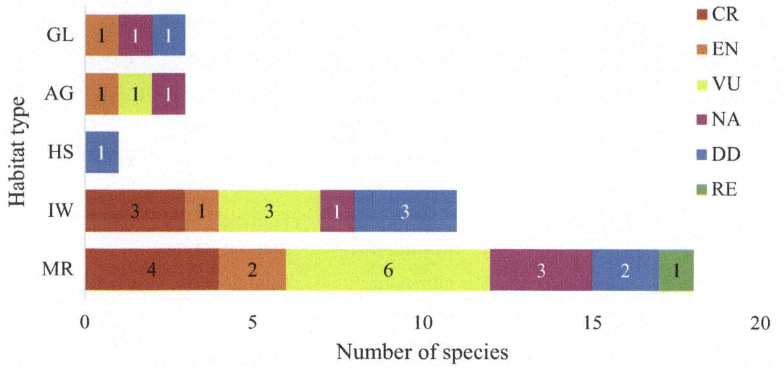

Fig. 10.7 Threatened species recorded in different habitats of the Ganges floodplain. (Major rivers associated wetlands-MR, Inland wetlands-IW, Human planted forest- HS, Agricultural land- AG, Grassland- GL; LC- Least Concern, CR- Critically Endangered, EN- Endangered, VU- Vulnerable, NT- Near Threatened, NA- Not Assessed, DD- Data Deficient, RE- Regionally Extinct)

survival in aquatic habitats. The gradual decline of native plant species within human planted forests is observed, exerting an influential role in altering avian communities.

10.4 Discussion

This study attempted to uncover the avifaunal status and community structure in the Ganges floodplain of Bangladesh. The Ganges floodplain of Bangladesh is one of the most productive bio-ecological sites in Bangladesh (Nishat et al. 2002). The species composition in this bio-ecological zone represents more than one-third (38%) of the bird species in Bangladesh. There has been some scattered research on avifauna in different districts within this area (e.g., Emrose et al. 2024; Shome et al. 2022; Mandal et al. 2022, Shome et al. 2020, Jaman et al. 2015). This study focused on the entire floodplain zone and represented the overall avian diversity of the area.

The whole floodplain zone does not have any natural forest or nationally designated protected area, but it is enriched with natural wetlands (e.g., major rivers, small river, beel, oxbow lake, floodplain marsh, canal, ditch) (Jaman et al. 2023; Shome et al. 2022). The nature of the wetlands and their associated habitats support a plethora of wetland birds (wild ducks, waterfowl, waders, raptors), contributing to high avian diversity in the study area (Fig. 10.3a, b). The greater Faridpur region located near the Ganga-Jamuna river confluence, witnessed an increase in wetland diversity following the deposition of sediments from the Ganges and Jamuna rivers. This phenomenon facilitated the formation of various wetlands, such as floodplains, marshlands, *beels*, oxbow lakes, and sandbars (Islam 2016), which in turn, attracted a multitude of wetland avifauna (Shome et al. 2022; Kumar and Gupta 2013). Oxbow lakes, locally referred to as *baors*, constitute a significant wetland habitat type within

Table 10.4 Recent country record of avifauna from the Ganges floodplain, Bangladesh

Order	Family	Scientific name	Habitat type	Season	Year
Strigiformes	Strigidae	*Otus bakkamoena*	Human planted forest	Rainy	2018
Passeriformes	Dicruridae	*Dicrurus caerulescens*	Human planted forest	Winter	2018
Strigiformes	Strigidae	*Glaucidium radiatum*	Human planted forest	Winter	2021
Charadriiformes	Glareolidae	*Glareola pratincola*	Inland wetlands	Winter	2020
Charadriiformes	Charadriidae	*Vanellus leucurus*	Inland wetlands	Winter	2019
Charadriiformes	Caprimulgidae	*Caprimulgus mahrattensis*	Major river associated wetlands	Winter	2017
Piciformes	Picidae	*Picoides nanus*	Human planted forest	Summer	2017
Procellariiformes	Oceanitidae	*Oceanites oceanicus*	Major river associated wetlands	Summer	2020
Charadriiformes	Laridae	*Onychoprion fuscatus*	Major river associated wetlands	Summer	2020
Charadriiformes	Laridae	*Onychoprion anaethetus*	Major river associated wetlands	Summer	2020
Charadriiformes	Stercorariidae	*Stercorarius longicaudus*	Major river associated wetlands	Summer	2020
Passeriformes	Chloropseidae	*Chloropsis jerdoni*	Human planted forest	Rainy	2016

the Ganges floodplain, with a total of 87 designated *baors* identified in this region (Rahman 2005). These *baors* serve as crucial habitats for diverse species of wetland birds (Jaman et al. 2023). At the same time, the human planted forest, agricultural land and grasslands act as significant habitats for Passerine bird species in the study area. The Ganges floodplain, with no designated protected forest but with homestead forests, urban green spaces, parks, residential areas, rural forests and fruit gardens serve as habitat for different species of doves, nightjar, swift, malkoha, cuckoo, coucal, woodpecker, owl, raptor, bee-eater, barbet, parakeet, wood swallow, shrike, minivet, oriole, drongo, fantail, monarch, flycatcher, crow, warbler, white eye, starling, redstart, thrush, sunbird, spider hunter, flowerpecker, munia and many more in the study area (Shome and Jaman 2021).

Table 10.5 Existing threats to different bird groups in the study area collected from previous studies and observed by the research team

Location	Threat types	Affected avian group	References
Sreepur Upazila, Magura	Deforestation, Illegal hunting	Passerine birds, dove, heron, egret, owls, parakeets	Mandal et al. (2021)
Magura Sadar Upazila, Magura	Deforestation, Agricultural net, Urbanization	Dove, woodpeckers, nightjar, owl, passerine birds	Shome et al. (2020)
Keshabpur, Jessore	Deforestation, Urbanization	Dove, passerine birds	Jaman et al. (2015)
Kashipur Union, Barishal	Deforestation, Illegal hunting,	Heron, egret, bittern, waterfowl, dove, cuckoo	Shome and Jaman (2021)
Faridpur Sadar Upazila, Faridpur	Illegal hunting and trade, Deforestation, Urbanization	Waterfowl, wild duck, dove, heron, egret, bittern, parakeet, munia	Shome et al. (2022)
Madhukhali Upazila, Faridpur	Illegal hunting, Deforestation, Urbanization	Waterfowl, dove, egret, heron, bittern	Biswas et al. (2022)
Marjat *baor*, Jhenaidah	Illegal hunting and trade, Water pollution, Fishing practice, Tourism, Agriculture net	Waterfowl, wild duck, dove, heron, egret, bittern, dove, woodpecker, nightjar, owl, passerine birds	Jaman et al. (2023)
Kaliganj Upazila, Jhenaidah	Deforestation, Illegal hunting Agricultural net	Wild duck, heron, egret, bittern, dove, woodpeckers, nightjar, owl, passerine birds	Scientific observation by research team
Sandbars in Ganges River (Rajbari, Faridpur, Madaripur)	Illegal hunting, Fishing, Tourism	Waterfowl, wild duck, dove, heron, egret, bittern, wader, sparrow, waver	Scientific observation by research team
Wetlands in Faridpur, Magura, Jhenaidah, Gopalganj	Illegal hunting, Fishing, Tourism	Waterfowl, wild duck, dove, heron, egret, bittern, wader	Scientific observation by research team
Transboundary area of Rajshahi and Chapainawabganj	Illegal hunting Agricultural net	Wild duck, heron, egret, bittern, dove, woodpeckers, nightjar, owl, passerine birds	Scientific observation by research team

The Ganges floodplain zone comprises transboundary areas at the northern part which play significant role in bird conservation. The area close to Chapainawabganj, Rajshahi, Kushtia and Meherpur districts of the Ganges floodplain is located along the international boundary. Transboundary areas are less disturbed habitat, thus providing more opportunities and security for different groups of bird species (Emrose et al. 2024). Wetlands and associated habitats in the transboundary areas become more enriched with diverse types of avian species, specially threatened and rare bird species.

Every year during the winter season, more than 220 bird species visit in Bangladesh and among them, a large number are wetland-associated bird species (Khan 2018). The Ganges floodplain is full of wetlands habitat alongside planted forests, agricultural lands, and grassland habitats. During the winter season, 94 migratory species visit the study area (Fig. 10.4), which increases the species richness in the study area (Jaman et al. 2023). Wetlands are the major natural habitat of the Ganges floodplain and this area is blessed with the Ganges river and its distributaries, small rivers, sandbars, floodplains, marshes, *beels*, *baors*, canals, ditches, temporary water-bodies in the rainy season, which support as a habitat wetland birds. Thus, bird diversity in wetlands habitats is higher in the study area (Fig. 10.5), and wetland specialist bird species are also higher in number (Fig. 10.6). Additionally, the habitat structure of wetlands contributes to the higher number of threatened bird species in this habitat of the study area (Fig. 10.7). The newly recorded bird species, as well as threatened birds, indicates that this study area is a potential hotspot for avifauna where rapid conservation initiatives are essential for proper conservation (Table 10.4).

Major river-associated wetlands are important habitats for different groups of wild ducks, grebes, waterfowls, waders, kingfishers, raptors, doves, bee-eaters, swallows, weavers, martins, larks, munias and wagtails. The benthos, planktons, crustaceans, insects, fishes, amphibians, reptiles, small birds, rodents, small mammals and aquatic plants of sandbars, mudflats, floodplain marshes are the major feeding resources of these bird species, thus their number is higher in this area. Furthermore, remote sandbars are more secure for different groups of large sized wetland specialist birds like herons, storks, ibises, adjutants, and wild ducks. Consequently, their presence is comparatively higher in these areas (Reza et al. 2014).

Different species of waterfowl (*Tachybaptus ruficollis*, *Zapornia pusilla*, *Amaurornis phoenicurus*, *Gallicrex cinerea*, *Porphyrio porphyria*, *Gallinula chloropus*, *Fulica atra*), herons, egrets, bitterns, ibises, storks, adjutants, cormorants, jacanas, waders, raptors, kingfishers, coucals, bee-eaters, warblers, munias and wagtails are important bird species in inland wetlands. The presence of diverse aquatic plants, algae, benthos, mollusks, crustaceans, insects, small fishes, frogs increases the sources of food for these bird species (Jaman et al. 2023). Thus, this group of birds is more abundant in the inland wetlands.

In the Ganges floodplain, agricultural lands are one of the most important modified habitats for birds. This area supports dove, swift, coucal, cuckoo, egret, bittern, heron, lapwing, kite, buzzard, eagle, owl, hoopoe, shrike, kestrel, and different groups of passerine birds (Fig. 10.8). Among different raptorial birds, the abundance of *Elanus caeruleus* was higher in the agricultural fields of the study area. Soil-dwelling nematodes, mollusks, crustaceans, insects, frogs, reptiles are the major feeding sources for these birds (Shome et al. 2020).

Scattered grasslands are present in the Ganges floodplain around different rivers which support different species of quail, francolin, dove, swift, coucal, cuckoo, kite, lapwing, eagle, bee-eater, kestrel, shrike, oriole, drongo, crow, lark, grassbird, martins, cisticola, prinia, bulbul, starling, redstart, bush-chat, weaver, warbler, munia, bunting etc.

According to the published literature and present observation, wetland bird species richness and abundance were higher in number. The abundance of wild ducks (e.g., *Dendrocygna javanica*, *Mareca strepera*, *Anas crecca*), herons (e.g., *Ardeola grayii*), egrets (e.g., *Bubulcus ibis*, *Egretta garzetta*), cormorants (e.g., *Microcarbo niger*), waders (e.g., *Himantopus himantopus*) was comparatively higher in different study sites. Due to the availability of wetland birds in higher numbers, these bird groups face existential crisis for illegal hunting and trade than any other groups of birds (Table 10.5). Wetlands in remote areas are in more sensitive state due to illegal bird

Elanus caeruleus

Ardea cinerea

Bubulcus ibis

Himantopus himantopus

Mycteria leucocephala

Ciconia episcopus

Fig. 10.8 Birds from the study area

Ciconia nigra

Clanga hastata

Buteo rufinus

Francolinus pondicerianus

Anser indicus

Wader

Fig. 10.8 (continued)

hunting and trade. In the greater Jessore and Chapainawabganj area, fruit cultivation has become a popular practice, where nets are used to protect fruits from birds. This practice poses a major threat for forest dwelling bird species, specially dove, nightjar, swift, owl, woodpeckers and other passerine birds. Existing fishing practices are already one of the major threats to different groups of wetland bird species. The fishing practices are increasing gradually, causing disturbance to aquatic birds and destroying bird habitats. Besides, the present use of the 'China duari' nets causes critical problems for wetland birds. This net is not eco-friendly. In winter, some fisherman use net for bird hunting for consumption, specially the wetlands bird.

Currently, the movement of water vessels has increased in the hotspots of water birds. This can cause a rapid decline in water birds if not controlled properly. In the char land areas, agricultural practices are expanding rapidly due to the increasing population of the char lands, with local people expanding their agricultural activities. For this reason, birds are losing their habitat. Upland tourism has gradually increased in the wetlands of the Ganges floodplain which sometimes causes disturbance and contributes to the illegal hunting and trade of avifauna.

10.5 Conclusion

The Ganges floodplains are enriched with avian biodiversity, specially wetland specialist birds, as revealed by this eight year-long study, the first of its kind in Bangladesh. The study area is enriched with natural and modified habitats, which are ideal for different groups of migratory and resident bird species. Winter is the most significant period for migratory birds. However, some threats exist. Bird hunting is the major problem for the avifauna in the study area. As a popular recreation spot on holidays, large numbers of people visit different areas of the study area for refreshment, picnics and shooting. Unfortunately, most of them are un aware of the biodiversity of this area and the disturbances they create. Currently, bird hunting by the tourist has become a major concern for bird conservation. Strict monitoring systems, implementation of existing laws, conservation of all habitat types, tourism management, cooperation among different stakeholders are important to save the avifauna of Ganges floodplain. Awareness programs, workshops, and celebrations such as World Migratory Bird Day are crucial for raising awareness among fisherman. Proper tourism management is equally important. Local people have a positive attitude toward birds and bird conservation. Therefore, community-based conservation must be emphasized in the management plan. Detailed research is necessary, focusing on behavioral ecology, migration patterns of birds, effects of environmental abiotic factors and water quality, relationships with vegetation and hunting patterns. Proper planning and management for bird species is an urgent need along with strict maintenance of these measures.

References

Banglapedia (2024). https://en.banglapedia.org/index.php/Floodplain

Biswas D, Jaman MF, Alam MM, Rabbe MF, Shome AR (2022) Community structure, ecology, and conservation issues of avifauna of two human-dominated landscapes in Faridpur district, Bangladesh during COVID-19 pandemic. J Asiat Soc Bangladesh Sci 48:53–66

del Hoyo J, Elliot A, Christie DA (2014) Handbook of the Birds of the World. Vol. 11. Old world flycatchers to old world warblers. Lynx Edicions: Barcelona

Emrose A, Shome AR, Alam MM, Rabbe MF, Hawladar N, Jaman MF (2024) Northwestern trans-boundary area of Bangladesh: a key to sustaining a quarter of the country's avian species. Ornithol. Res, 1–10

Fischlin A, Midgley GF, Price JT, Leemans R, Gopal BC, Turley C et al (2007) Ecosystems, their properties, goods, and services. In: Parry ML, Canziani OF, Palutikof JP, Van der Linden PJ, Hanson CE (eds) Climate change 2007: impacts, adaptation and vulnerability. Contribution of working group II to the fourth assessment report of the intergovernmental panel on climate change. Cambridge University Press, Cambridge, pp 211–272

Gatesire T, Nsabimana D, Nyiramana A, Seburanga JL, Mirville MO (2014) Bird diversity and distribution in relation to urban landscape types in Northern Rwanda. Sci World J 2014:1–12

Islam SN (2016) Deltaic floodplains development and wetland ecosystems management in the Ganges–Brahmaputra–Meghna Rivers Delta in Bangladesh. Sustain Water Resour Manag 2:237–256

IUCN Bangladesh (2015a) Red List of Bangladesh Volume 3: Birds. IUCN, International union for conservation of nature, Bangladesh Country Office, Dhaka, Bangladesh

IUCN Bangladesh (2015b) Red List of Bangladesh Volume 5: Freshwater Fishes. IUCN, International union for conservation of nature, Bangladesh Country Office, Dhaka, Bangladesh

Jaman MF, Alam MM, Rabbe MF, Shome AR (2023) Temporal variation, habitat heterogeneity and anthropogenic stressors influencing wildlife communities in an oxbow shaped wetland of lower Ganges floodplain. Bangladesh. J Biodiv Con Biores Manage 9(2):83–98

Jaman MF, Uddin M, Alam MM, Rahman MM, Khatun MT, Alam SMI (2015) Species diversity and population status of wildlife in Keshabpur. J Biodiv Con Biores Manage 1(2):9–21

Khan MAR (2015) Wildlife of Bangladesh: checklist and guide. Chyabithi, Dhaka, Bangladesh

Khan MM (2018) Photograhic guide to the wildlife of Bangladesh. Arranayk foundation, Dhaka, Bangladesh

Kibria G, Yousuf Haroon AK (2017) Climate change impacts on wetlands of Bangladesh, its biodiversity and ecology, and actions and programs to reduce risks. In: Prusty B, Chandra R, Azeez P (eds) Wetland Science. Springer, New Delhi, pp 189–204

Kumar P, Gupta SK (2013) Status of wetland birds of Chhilchhila Wildlife Sanctuary, Haryana. India. J Threat Taxa 5(5):3969–3976

Latterell JJ, Naiman RJ (2007) Sources and dynamics of large logs in a temperate floodplain river. Ecol Appl 17(4):1127–1141

Mandal AK, Jaman MF, Alam MM, Rabbe MF, Shome AR (2021) Vertebrate wildlife diversity of Sreepur upazila, Magura, Bangladesh. J Biodivers Conserv Bioresour Manag 7:51–62

Nishat A, Huq SMI, Barua SP, Reza AHMA, Khan AR, Moniruzzaman AS (2002) Bio-ecological Zones of Bangladesh. IUCN Bangladesh Country Office, Dhaka, Bangladesh

Rahman AKA (2005) Freshwater fishes of Bangladesh. Zoological Society of Bangladesh, Department of Zoology, University of Dhaka, Bangladesh

Rahman HM, Hickey GM, Sarker SK (2015) Examining the role of social capital in community collective action for sustainable wetland fisheries in Bangladesh. Wetlands 35(3):487–499

Reza AS, Al-Amin M, Rahman SMA, Islam MS, Hasan MA, Parween S (2014) A checklist of waterbirds of the Padma River charland (Godagari to Charghat), Rajshahi, Bangladesh. Univ J Zool Rajshahi Univ 33:19–23

Sekercioglu CH (2006) Ecological significance of bird populations-handbook of the birds of the world. Birdlife international

Şekercioğlu ÇH, Daily GC, Ehrlich PR (2004) Ecosystem consequences of bird declines. PANS 101(52):18042–18047

Shome AR, Alam MM, Rabbe MF, Rahman MM, Jaman MF (2020) Diversity, status and habitat usage of avifauna at Magura Sadar upazila. Bangladesh. Bangladesh J Zool 48(2):441–456

Shome AR, Jaman MF (2021) Diversity and seasonal occurrence of vertebrate wildlife at a rural site of Bangladesh: Threats and conservation issue. J Biodivers Conserv Bioresour Manag 7(2):61–72

Shome AR, Rabbe MF, Alam MM, Emon SF, Islam MM, Setu RS, Jaman MF (2022) Avifauna in an urban landscape of a Lower Ganges District of Bangladesh: community structure, seasonality, habitat preference and conservation issues. Dhaka Univ J Biol Sci 31(2):261–274

Tockner K, Bunn SE, Gordon C, Naiman RJ, Quinn GP and Stanford JA (2010) Flood plains: critically threatened ecosystems. In: Polunin NVC, ed. Aquatic Ecosystems: Trends and Global Prospects. Cambridge University Press

Chapter 11
Transition in Traditional Cropping Pattern Led to Engagement in Alternative Economic Activity

Lakpa Tamang and Mayank Joshi

Abstract Himalaya represents a unique blend of nature's resource repository. Many resources in this belt have not been unveiled, valued or being efficiently used but on the other hand in-used resources are also depleting. In the hills of Indian Himalaya region, agriculture is the prime economic activity; which is being carried out by majority of inhabitants. The effects of Climate Change have been felt globally such as the instances of frequent drought, changing climatic pattern, floods etc. The biodiversity of Himalayan region is also very sensitive to Climatic Change which results into decreasing many endemic species and depleting natural resources. Traditionally in Western and Eastern Himalaya, various mountain specific traditional crops are significantly grown, but due to climate change agricultural output has been significantly declined. To accommodate with this rising situation, people started to shift their means of livelihood and few changes in existing agricultural practice has been observed. Similarly, in Sikkim Himalaya also few changes have been observed in the agriculture sector due to direct impact of Climate Change. Transition in livelihood practices among farmers in Sikkim is witnessed with the decline of agricultural output. The inherent fragility and the emerging climate induced challenges put the people of this region under great livelihood stress.

Keywords Transition · Livelihood · Himalaya · Agricultural crops · Sikkim

11.1 Introduction

Hindu Kush Himalaya (HKH) extends over Afghanistan, Bangladesh, Bhutan, China, India, Myanmar, Nepal and Pakistan with an area expansion of 4.2 million km^2. This belt is geo-ecologically vibrant for world assets and considered as the fourth largest global biodiversity hotspot. It supports around 240 million people through

L. Tamang · M. Joshi (✉)
G. B. Pant National Institute of Himalayan Environment, Sikkim Regional Centre, Pangthang, Gangtok, Sikkim, India
e-mail: mayank.geo@gmail.com

D. Arya et al. (eds.), *Climate Change Impact on Himalayan Biodiversity*,
Environmental Science and Engineering, https://doi.org/10.1007/978-3-031-77149-1_11

various ecosystem services (Wester et al. 2019). The effect of Climate Change is also evident from HKH due to its inherent fragile nature such as glacier retreat, rise in permafrost temperatures and depleting snow field. The complex hydrological cycle of the HKH (sustain by permanent snow, ice, frozen soil of the river basin) are also sensitive to Climate Change (You et al. 2017). Indian Himalayan Region (IHR) is a narrow southern concave extension of HKH and stretches between the degree of $26^0 20'-35^0 40'$ N and $74^0 50'-95^0 40'$ E (Dikshit et al. 2020). IHR is being considered as one of the most diversified areas in terms of geology, lithology, rainfall distribution, land use/land cover, soil properties and stream networks. This region suffers a lot from various natural hazards and the Climate Change may trigger abnormal floods, landslides, drought and loss of biodiversity. Thus potential threat to food security. Around 51 million people inhabited in the fragile and diverse ecosystem of IHR, which constitute 11% of India's population (Tewari et al. 2017). Nearly 80% of the population earn their livelihood from the farm related economic activities in the region (Nautiyal et al. 2007). Traditional livestock-based farming system is mainly practice by the people considering the available ecosystem services. The inhabitants of the region are engaged in subsistence agricultural system on small and fragmented land holding following topographic stability. In a recent past, the transition in the Himalayan Region from traditional subsistence agriculture to cash-crop based agriculture has been observed (Nautiyal et al. 2007). Due to this transition, many traditional crops have vanished and some other crops have moved towards extinction (Nautiyal et al. 2007). There is a heavy burden among the farmers to accommodate with the changing climatic condition and to sustain their livelihood through various alternative measures. The entire IHR is divided into Western, Central and Eastern Himalaya. On which Sikkim marks the western limit of the Eastern Himalayan Region and the topographic elevation ranges from 250 to 8598 m (Subba 2008). The great variation in elevation and climatic condition over the short distance leads to development of distinct agro-ecosystem to support the livelihood of hill community. In Sikkim Himalaya, only about 11% of area is under cultivation on which more than 60% of total population is dependent for their livelihood (Mishra et al. 2021). Emerging interconnected issues of environmental, social and economic has led to changes in agrarian livelihood practises. Rural communities can adapt to changing environment by reviving traditional ecological knowledge, agro-biodiversity, various land uses, and diversification of livelihood (Rahman et al. 2012).

11.2 Study Area

The present study has been carried out in villages under Gangtok district, Sikkim. Sikkim is a small hilly state on North-Eastern flank of India extending between 27.08^0N and 28.13^0N Latitude and 88.08^0–88.92^0E Longitude (Fig. 11.1). Geographically, Sikkim State covers an area of 7096 sq km and the entire state extend over approximately 114 km from north to south and 64 km from east to west within altitudinal variation of 250–8598 m (Subba 2008). The northern part of Sikkim falls in

Higher Himalayan Zone and the southern slope of the state falls in Lesser Himalayan Zone. The area experience four main seasons spring (March to April), summer (May to September), autumn (October to November) and winter (December to February). The elevation differences mark the area experiencing sub-tropical humid at lower hills and valley to alpine cold at higher altitude within state geographical expansion. The average temperature during summer range between 5^0 to 25^0C and during winter month's temperature drops upto -5^0 to 7^0C (Joshi, 2022). For the study five villages namely Samdong, Lingdok, Pangthang Forest Block, Kambal and Raley-Khesay were selected. Agriculture is extensively practiced in these villages and located at a negligible distance to supply agricultural products to urban hub, Gangtok (Tables 11.1 and 11.2).

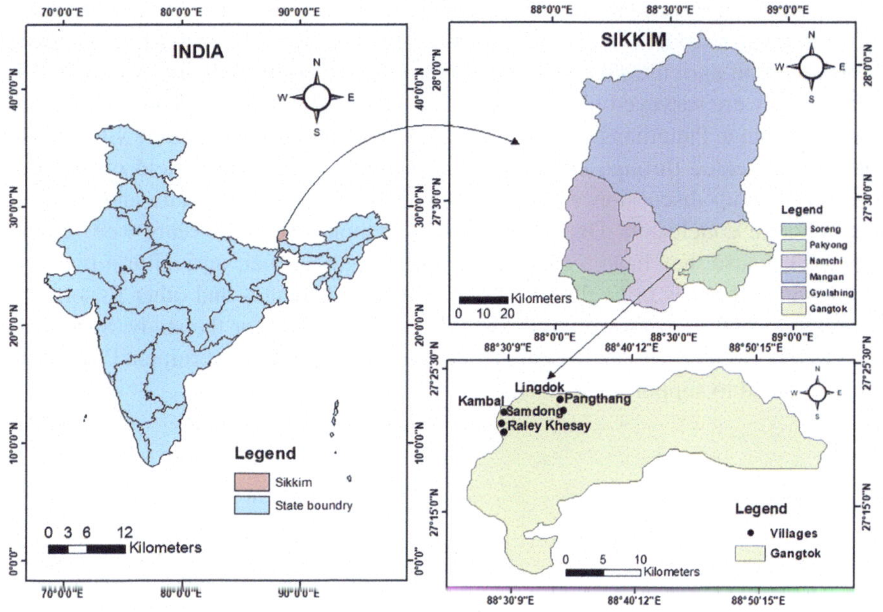

Fig. 11.1 Location map of the study area

Table 11.1 List of study area with demographic details (Census 2011)

Village name	District	No. of households	Total female	Total male	Total population
Samdong	Gangtok	481	1035	1090	2125
Lingdok	Gangtok	182	427	469	896
Pangthang	Gangtok	87	191	208	399
Kambal	Gangtok	202	507	542	1049
RaleyKhese	Gangtok	391	859	853	1712

Table 11.2 Total surveyed households

Si.no	Village name	Sample
1	Samdong	76
2	Lingdok	14
3	Pangthang	28
4	Kambal	21
5	Rarely Khesay	53
Total		192

11.3 Methodology

In the present study primary field data has been collected. The detailed questionnaire survey was carried out in 5 villages under Gangtok District, Sikkim to gather inclusive information on agricultural crops and performing economic activity. A total of 192 households were surveyed for the study. The surveyed village are Samdong village, Lingdok village, Pangthang village, Kambal village and Rarely-Khesay village. The data was collected through simple random sampling technique. Field observation and focus group discussion was also conducted to validate the data and to arrive at a concrete conclusion. During the survey, people above 18 years were mostly targeted because they have a clear understanding about their agricultural practices and adopted changes in livelihood due to Climatic Changes and other factors. All the respondents' views were treated equally and analysed for the study. Secondary sources like Google scholar, Research gate and Sikkim State Government Portal have been accessed to support this study.

11.4 Results

In this study particular focus has been laid to understand the emerging changes in cropping pattern among the farmers of Sikkim. The following results highlight the changes in crop production over the years and their alternate livelihood approach. Total household surveyed in the area is 192 households and the data on crops they are growing is been collected. Figure 11.2 shows the common varieties of crops grown in the study area. Percentage wise maize is cultivated by 32.63% of households, paddy is cultivated by 22.54% households, green vegetables is cultivated by 14.32% households, buckwheat is cultivated by 11.03% households, ginger is cultivated by 6.81% households, cardamom is grown by 5.16% households, mustard is grown by 3.52% households, broom is grown by 1.88% households, potato is grown by 1.88% households and millet is cultivated by 0.23% households only. These are the main grown crops in the area to meet their needs at a subsistence level. Along with agriculture, people usually keep livestock to strengthen their economic status in rural areas. Likewise Fig. 11.3 indicates the types of livestock raised in the study area to

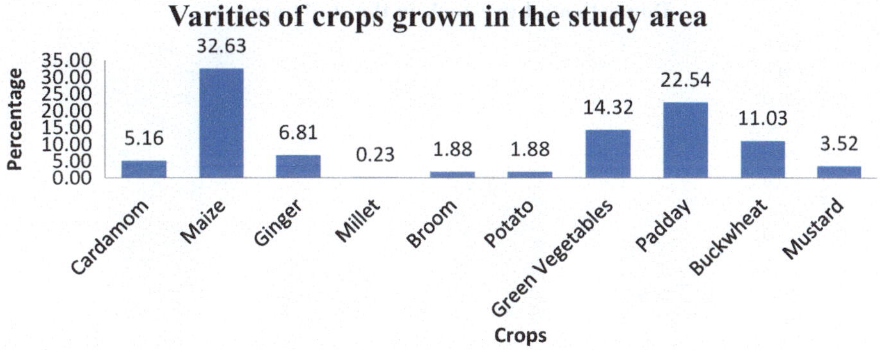

Fig. 11.2 Showing common variety of crops grown in the study area

generate additional income. Around 45.89% households are raising poultry, 27.73% households are keeping goats, 15.48% households are keeping cattle and 10.9% households are keeping pig to support the family income and to meet food diet. This are common practices to keep livestock to sustain their livelihood and as well to support their agriculture functioning. People are also engaged in various economic activities to sustain their livelihood. In rural areas people are mostly engage in primary activities. However, over the time economic sphere are also changing in rural areas. In Fig. 11.4 we can see about 61.46% households derive their income from agriculture directly as a primary source and out of which 34.58% households relying on some kind of secondary source income, 28.13% households earn their income from service sector alone, 1.56% households earn from self employed activities as a primary source of income and 28.04% households are earning from self employed activities as a secondary source, 5.73% households are dependent on labour activities as a primary source and 16.82% households have taken labour activities as a secondary source and remaining 2.6% are earning from any other activities like their engagement in secondary sector or any kind of activity.

11.5 Discussion

Farmers in Sikkim usually carry agricultural activities on terraced farm lands along the hill slopes. These are prepared in three ways as levelled, sloping and reversed terraces (Mishra et al. 2020). Study finds that the reversed terraces are effective in protecting soil loss, nutrient loss, and moisture retention and in increasing crop yield (Mishra and Rai 2011). In higher altitude between 3000 to 4300 m crops like barley, wheat, potato, cabbage, apple, maize, peas, dry land paddy, beans, pulses and some medicinal plants are commonly cultivated by the people. In lower flat valleys 300–1800 m traditional varieties of rice is most extensively grown (Sharma and Dhakal 2011). Crops like soya bean, black gram are grown along with rice on

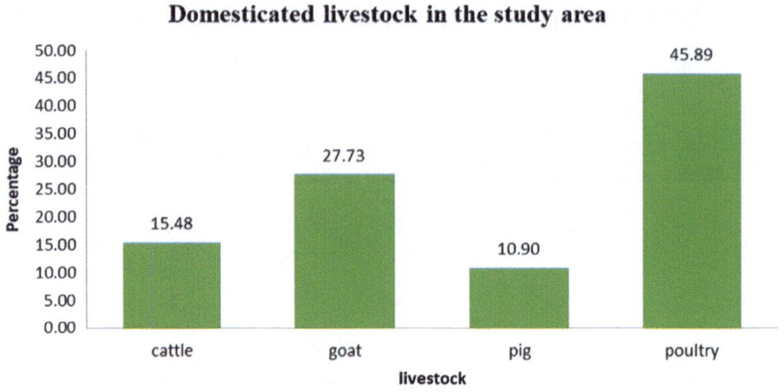

Fig. 11.3 Showing livestock reared in study area

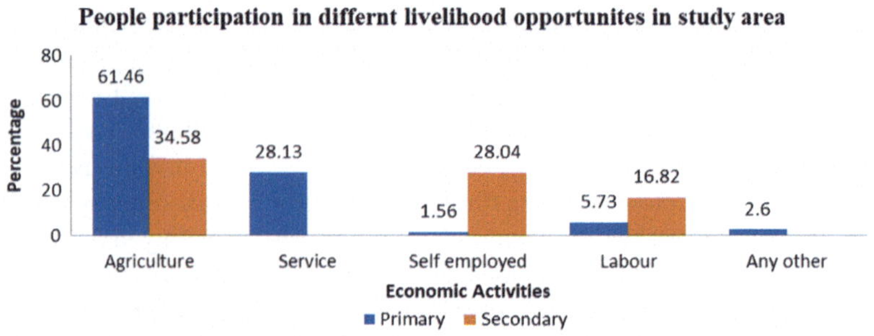

Fig. 11.4 Showing people participation in different livelihood opportunities

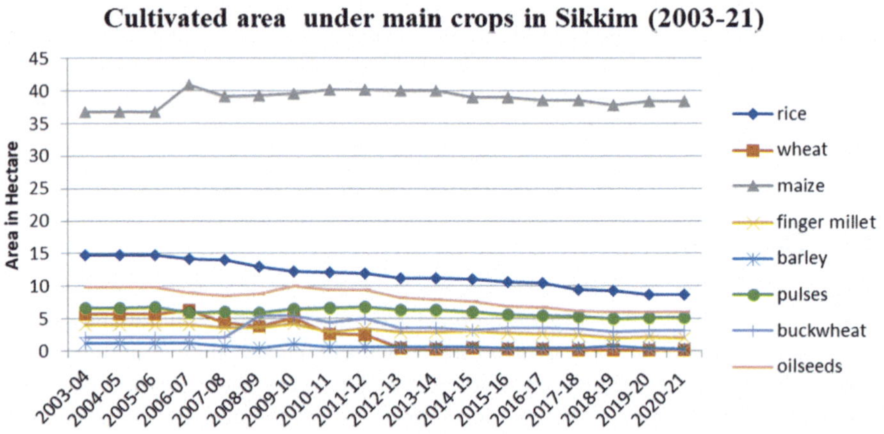

Fig. 11.5 Cultivated area under main crops in Sikkim from 2003–21. *Source* http://sikenvis.nic.in/

the prepared bunds (Chettri and Sathyanarayana 2021). Mixed farming system is commonly practised by majority of the households in which maize, ginger, large cardamom and vegetables are grown in combination by the farmers of the Sikkim (Mishra et al. 2021). In 2005–2006 to 2010–2011 area under irrigated holdings has decreased by 0.15% and unirrigated holdings decrease by 0.06% (Gurung and Hannan 2021). Farmers usually build small gully structure to drain water into paddy field from nearby perennial stream or spring. Majorly they are dependent on irrigation to cultivate rabi crops during non-monsoonal season. The main crops usually grown in Sikkim are listed in Fig. 11.2. The Fig. 11.5 depicts the proportion of area covered under different crops grown between 2003 to 2021. These 8 varieties of crops are a traditional crop which is very much prevalent in the area, but over the due course of time the area under these crops are in declining trend. When we take 2003–2004 as the base year for the following crops then in 2003–2004 area under rice is 14.74 h, wheat is 5.74 h, maize is 36.7 h, finger millet is 4.15 h, barley is1.23 h, pulses is 6.71 h, buckwheat is 2.01 h and oilseeds is 9.95 h. Almost after two decades in 2020–2021 the area under rice has come down to 8.7 h, wheat is 0.1 h, maize is 38.39 h, finger millet is 2.05 h, barley is 0.35 h, pulses is 5.14 h, buckwheat is 3.11 h and oilseeds is 6.13 h. Figure 11.5 clearly indicates that the cultivated area under this traditional crop has significantly declined over two decades except maize and buck wheat which has either remain constant or increase by certain value. The productions of traditionally grown crops in Sikkim are going down with decreasing area under cultivation. So the area and production of these crops are positively related and going down over the decades. The productions in 2003–2004 of rice is 21.19ton, wheat is 8.09ton, maize is 57.05ton, finger millet is 3.57ton, barley is 1.51ton, pulses is 6.36ton, buckwheat is 1.55ton and oilseed is 7.44ton. when we see in 2020–2021 the production of rice is 16.19ton, wheat is 0.12ton, maize is 67.94ton, finger millet is 2.13ton, barley is 0.4ton, pulses is 4.95ton, buckwheat is 3.07ton and oilseeds is 5.67ton. In Fig. 11.6 we can clearly see the decreasing trend line of all crop production except maize and buckwheat which is constant. On comparing overall Sikkim and study area data indicate similar trend in the production of traditional crops. In the study area maize is most commonly grown followed by paddy, green vegetables and buckwheat. It gives the indication that due to climatic effect people are shifting towards the cultivation of less water intensive and short seasonal crops. Rice is the staple food of Sikkim but its production and cultivated area is going down year by year. In general there is expanding gap between the available food resources and the increasing demand coincidence with population growth. Now a day's people are more inclined towards the production of market oriented crops like green vegetables and cash crops. To support the activity and to strength their economic status in between the planting and harvesting crops farmers domesticate livestock. In the study area 45.89% people are raising poultry, 27.73% people are raising goat, 15.48% people are raising cattle and 10.90% people are raising pig to sustain livelihood. Cattle are the core of the farming system and the income generation. In reference to the study area we can make out that people engagement in cattle rearing is less in comparison to other livestock. From this point of view we can say that less cattle means shortage in manure supply to field and relatively less area under utilization. The decrease in the area and production under

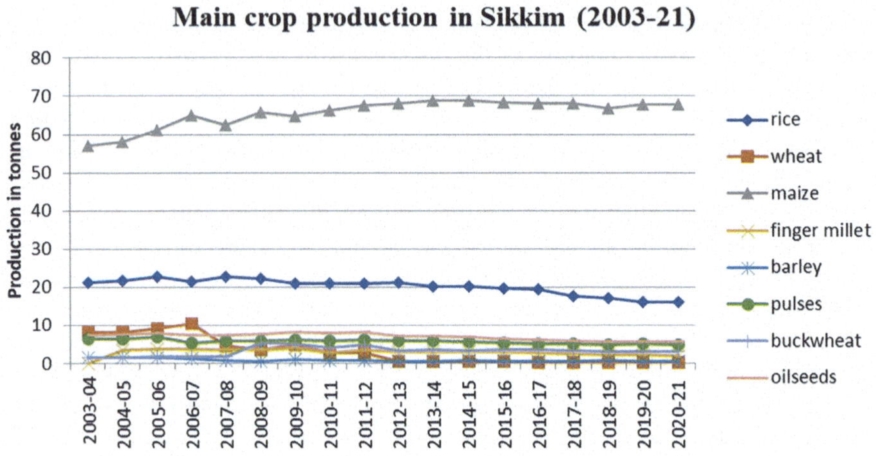

Fig. 11.6 Main crop production of Sikkim during 2003–21

traditional crops has been evident in Figs. 11.2, 11.5, 11.6. Due to which certain changes can be observe in the economic activities carried out by the people to secure their daily means of sustenance. Figure 11.4 indicates the economic engagement of people in the study area. It is observed that people are engage in primary activities along with some households deriving their income from secondary sources as well. Almost 79.44% households are also engage in some kind of secondary activities. Out of which 34.58% households have taken agriculture, 28.04% households are engage in self-employed activity and 16.82% are working as labour to derive additional income as secondary sources. The major concern of this study is the 5.73% of people who have already started to perform labour as their primary activity, earlier who used to be self independent farmers. Due to agriculture and economic constraints these vulnerable group of people are more likely to increase in coming days. It indicates people are getting more inclined towards other economic activities then their participation in agricultural activity. So, in near future we are going to witness a massive shift in the internal structure of the economy.

11.6 Conclusion

Indian Himalayan Region is considered one of the most diversified area in terms of geology, lithology, rainfall distribution, land use/land cover, soil properties and stream networks. It holds huge repository of forest resources, bio-diversity and source of perennial rivers. In the recent past this region has suffer from numerous natural hazards and the ongoing climatic changes may trigger abnormal floods, landslides, drought, loss of biodiversity and potential threat to food security. Around 80% of the inhabitant population are deriving livelihood from the agriculture related activates.

Similarly, in Sikkim in about 11% of cultivated area around 60% of people are directly sustaining their livelihood from the utilization. Changes in the cultivation of area and in production of traditional crops have been observed in the state. In this study similar trend of decreasing in the cultivation of traditional crop has been observed. To adapt with changing condition shift in the cultivation of traditional crop has been observed and adaptation of cash crops cultivation. As per the study people are engage in primary activities and more than half of them are also indulge in certain secondary activities for the livelihood. Those engaged in agriculture had started to take up various available non-farm activities for the livelihood. They started to engage themselves in economic activities like self-employment and labour to earn hand cash. As per this study in near future Sikkim is going to witness economic sectoral changes with emerging new vulnerable group. Therefore, there is a great scope to do study on sectoral change and how it is going to impact in the existing economic sphere.

Acknowledgements We would like to thank the Director of G.B. Pant "National Institute of Himalayan Environment" (NIHE) for providing the facilities and encouragement. This study is partially funded by In-House Project-1 of the Institute and NMHS- Spring project.

References

Chettri D, Sathyanarayana N (2021) Traditional rice landraces and indigenous knowledge associated with the rice cultivation in Sikkim Himalaya: An analysis based on local farming practices. Indian J Hill Farming 34(1):1–15

Dikshit A, Sarkar R, Pradhan B, Segoni S, Alamri AM (2020) Rainfall induced landslide studies in Indian Himalayan region: a critical review. Appl Sci 10(7):24–66

Gurung A, Hannan A (2021) Dynamics of land use and trends of agriculture in Sikkim. J North East India Stud 11(1)

Joshi M (2022) Co-seismic landslides in the Sikkim Himalaya during the 2011 Sikkim earthquake: lesson learned from the past and inference for the future. Geol J 57(12):5039–5060

Mishra PK, Rai SC (2011) Cost-Benefit Analysis of Terrace Cultivation in Sikkim Himalaya. Governing Council of the Indian Geographical Society, India, p 29

Mishra PK, Rai A, Abdelrahman K, Rai SC, Tiwari A (2021) Analysing challenges and strategies in land productivity in Sikkim Himalaya India. Sustainability 13(19):11112

Mishra PK, Rai A, Rai SC (2020). Indigenous knowledge of terrace management for soil and water conservation in the Sikkim Himalaya, India

Nautiyal S, Kaechele H, Rao KS, Maikhuri RK, Saxena KG (2007) Energy and economic analysis of traditional versus introduced crops cultivation in the mountains of the Indian Himalayas: A case study. Energy 32(12):2321–2335

Rahman H, Karuppaiyan R, Senapati PC, Ngachan SV, Kumar A (2012) An analysis of past three decade weather phenomenon in the mid-hills of Sikkim and strategies for mitigating possible impact of climate change on agriculture. Climate change in Sikkim: Patterns, impacts and initiatives, 1–18

Sharma G, Dhakal T (2011) Opportunities and challenges of the globally important traditional agriculture heritage systems of the Sikkim Himalaya. Biodiversity of Sikkim: exploring and conserving a global hotspot. IPR Department, Government of Sikkim, Gangtok, 411–440

Subba JR (2008) History, culture and customs of Sikkim. Gyan Publishing House

Tewari VP, Verma RK, Von Gadow K (2017) Climate change effects in the Western Himalayan ecosystems of India: evidence and strategies. Forest Ecosystems 4(1):1–9
Wester P, Mishra A, Mukherji A, Shrestha AB (2019) The Hindu Kush Himalaya assessment—Mountains, climate change, sustainability and people Springer Nature Switzerland AG, Cham
You QL, Ren GY, Zhang YQ, Ren YY, Sun XB, Zhan YJ, Shrestha AB, Krishnan R (2017) An overview of studies of observed climate change in the Hindu Kush Himalayan (HKH) region. Adv Clim Chang Res 8(3):141–147

Chapter 12
Status of Vascular Plant Endemism in Western Himalaya, India

Himani Tiwari, Asheesh Pandey, K. Chandra Sekar, and Dhani Arya

Abstract Diversity of endemic plants is a prime criterion for determining the global biodiversity hotspot. Endemic species are distributed in a restricted range which often vulnerable to climate and anthropogenic disturbances, and thus require immediate effective conservation action. Considering the importance, we investigated the diversity and distribution pattern of vascular endemic plants across Western Himalaya (Ladakh, Jammu & Kashmir, Himachal Pradesh and Uttarakhand) of Indian Himalayan Region (IHR). A total of 227 endemic taxa (207 species, 5 subspecies and 15 varieties) belonging to 115 genera and 43 families were recorded. Among the different life-forms, most of the species were herbaceous (184), followed by shrub (34), tree (4), climber (1), and bamboo (1). Asteraceae (49) was the dominant family followed by Fabaceae (17) and Poaceae (11). Endemic richness was recorded highest (113) in Uttarakhand, followed by Jammu & Kashmir (84), Himachal Pradesh (61) and Ladakh (15) respectively. Further studies are required for population assessment and conservation action of these species along with development of propagation protocols and reintroduction in natural habitat.

Keywords Endemic vascular plants · Diversity · Distribution · Conservation · Western Himalaya

H. Tiwari · A. Pandey
Centre for Biodiversity Conservation and Management, G.B. Pant National Institute of Himalayan Environment, Kosi-Katarmal, Almora, Uttarakhand 263643, India

K. C. Sekar (✉)
Garhwal Regional Centre, G.B. Pant National Institute of Himalayan Environment, Upper Bhaktiyana, Srinagar Garhwal, Uttarakhand, 246174, India
e-mail: kcsekar1312@rediffmail.com

D. Arya
Department of Botany, Soban Singh Jeena University, Almora, Uttarakhand 263601, India

D. Arya et al. (eds.), *Climate Change Impact on Himalayan Biodiversity*,
Environmental Science and Engineering, https://doi.org/10.1007/978-3-031-77149-1_12

12.1 Introduction

Endemism is the principal criterion for hotspot status determination because endemic species are distributed in a restricted range, and are often the most vulnerable to climate changes (Myers et al. 2000). These species, confined to highly vulnerable ecosystems, will almost certainly be the first to be hit by extinction processes, and hence need rapid and effective conservation action (Heywood & Watson, 1995). Towards the efforts of conserving the endemic species, a total of 36 global biodiversity hotspots established, covering 2.4% of the earth's total land area for securing around 50% population (Hrdina and Romportl, 2017). As compared to common and native species, endemics are more impacted by climate change. Further, 84% of mountain species were projected to face high extinction risks under changing climate (Manes et al. 2021). India is one among the mega biodiversity countries of the world and consists four biodiversity hotspots viz., the Himalaya, Western Ghats, Indo-Burma, and Sundaland. These regions also having high degree of endemism as well as higher number of threatened plants (Hajra and Mudgal 1997). A total of 11.4% of global floral biodiversity is represented by India, which include 47,513 plant species of which 11,058 (23.8%) species are endemic to the Indian region (Singh and Dash, 2014; Singh and Thadani 2015).

Himalayan biodiversity hotspot, due to varied topographical features, resulting in enormous diversity and habitat, which harbors rich repository of biodiversity elements (Pandey et al. 2018). The species diversity of the Himalayan biodiversity hotspot includes 10,000 plants (31.6% endemic), 300 Mammals (4.0%), 105 Amphibians (40.0%) and 977 birds (1.5%), 176 reptiles (27.3%), 269 fresh water fishes (12.3%) (Rawal et al. 2013). The Indian Himalayan Region (IHR) constitutes a large portion of the Himalayan biodiversity hotspot and significantly contributes to the richness and representativeness of its biodiversity elements. The IHR covers 11 States / Union Territories (i.e., Jammu & Kashmir, Ladakh, Himachal Pradesh, Uttarakhand, Sikkim, Arunachal Pradesh, Nagaland, Manipur, Mizoram, Tripura, and Meghalaya) and two states partially (Hilly districts of Assam and West Bengal) (Mehta et al. 2023). According to Singh and Pusalkar (2020), a total species of 8700 angiosperm, 51 gymnosperm, 766 pteridophyte, and 1955 bryophyte are represented in the IHR.

Understanding diversity and distribution of plant species is crucial for conservation of species (Tobler et al. 2007) and using existing database for analyzing the distribution of species diversity can be beneficial for identifying conservation demanding areas over large vegetated areas (Mehar et al. 2009). Recently, we recorded a total of 1076 endemic vascular plants, including 1061 angiosperms, 03 gymnosperms and 12 pteridophytes belonging to 432 genera and 100 families in Indian Himalayan Region (Tiwari et al. 2024). As far as endemic vascular plant diversity in Western Himalaya is concerned, there is no complete database available. In view of importance and identifying the hot-spot of vascular plant diversity, the present study is an attempt to document the diversity and analyze the distribution patterns of endemic vascular plants of Western Himalaya.

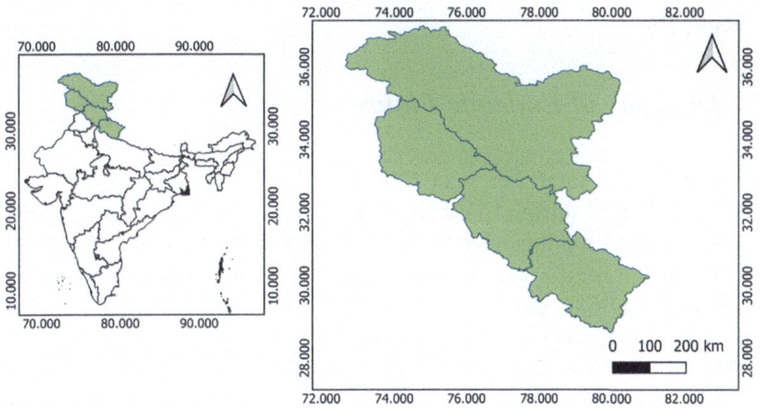

Fig. 12.1 Map of study area

12.2 Materials and Methods

12.2.1 Study Area

The Western Himalaya comprising two states [Himachal Pradesh (HP) and Uttarakhand (UK)] and two union territories [Jammu and Kashmir (JK), Ladakh (LA)] lies between 28.72° N to 37.14° N latitudes and 72.68° E and 81.03° E longitudes was the study area to document the endemic plant diversity (Fig. 12.1).

12.2.2 Inventorization of Endemic Taxa

The inventory of endemic taxa was prepared by reviewing online literature and datasets such as (www.gbif.org, indiabiodiversity.org, www.efloras.org, www.theplantlist.org, www.ipni.org, plants.jstor.org, efloraofindia, www.tropicos.org), regional floras, e-floras, etc. Authentic scientific names, family wise distribution, geographic distribution of endemic taxa, etc. were accessed for the taxa. The taxonomic description and distribution of species were validated following powo.science.kew.org, www.gbif.org. To identify and map the endemic rich areas of Western Himalaya, location specific details of the taxa were collected by consulting herbarium records housed at Botanical Survey of India (BSD), Forest Research Institute (DD), virtual herbarium (bsi.gov.in, www.kew.org) and various published literature (research papers, eflora of India, articles)) (Singh et al. 2015; Bhatt et al. 2021; Singh and Dash 2014) and specific flora of the region.

12.3 Results

12.3.1 Diversity of Endemic Plants

In the present study, total 227 taxa (207 sp., 5 subsp., 15 var.) including 224 angiosperms (205 sp., 4 subsp., 15 var.) and 3 pteridophytes (2 sp., 1 subsp.) belonging to 115 genera and 43 families were documented as endemic to Western Himalaya. Among the different life-forms, endemic plants were dominated by herbaceous layer (184 taxa) followed by shrub (34 taxa), tree (4 taxa), undershrub (3 taxa), climber (1 taxa), and bamboo (1 taxa) respectively (Fig. 12.2 and Table 12.1).

Out of total 224 angiosperms, 27 (12%) taxa were monocotyledons and 197 (87.9%) taxa were dicotyledons. Considering the familial distribution of endemic taxa, Asteraceae was the dominant family with 49 (43 sp., 3 subsp., 3 var.) followed by Fabaceae 17 (16 sp., 1 var.), and Poaceae (11 species). Total 16 families were monotypic with representing only 1 endemic species. *Taraxacum* was the dominant genera while considering endemics of Western Himalaya, representing 24 endemic taxa (22 sp., 2 subsp.), followed by *Impatiens* (10 species) and *Astragalus* (10 species) (Fig. 12.3). Analysis of use value of species revealed, 8 taxa (6 sp., 1 subsp., 1 var.) were being used by local people for various purposes such as medicinal (6), fibre (1), fodder (2), dye, furniture (1). Mehta et al. (2023) reported total 85 threatened species (high risk categories) from IHR. However, the present study recorded only 2 species (*Butea pellita*- CR, *Pittosporum eriocarpum*- EN) fall under IUCN threat

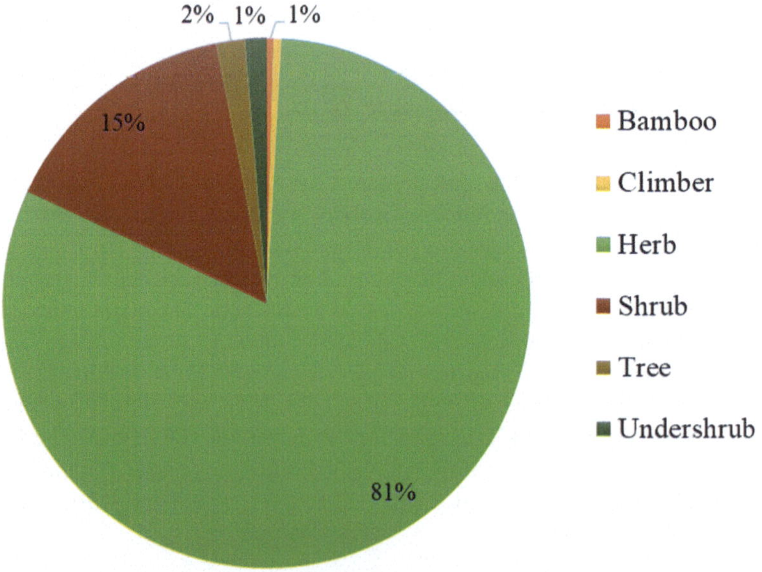

Fig. 12.2 Different life forms of endemic taxa

Table 12.1 Inventory and distribution of endemic taxa of Western Himalaya, India

Name of the taxa	Family	Habit	Elevation range	Distribution in states of IHR			
				J&K	LK	HP	UK
Barleria somdevae	Acanthaceae	Ush	–				+
Seseli petrosciadium	Apiaceae	H	3000–3700	+			+
Trachyspermum falconeri	Apiaceae	H	2000–2800				+
Angelica indica	Apiaceae	H	3000–3300				+
Ferula thomsonii	Apiaceae	H	1600–1900	+			
Heracleum jacquemontii	Apiaceae	H	–	+		+	+
Kedarnatha garhwalica	Apiaceae	H	3500				+
Kailashia robusta	Apiaceae	H	3700–3850				+
Kedarnatha meifolia	Apiaceae	H	3300				+
Kedarnatha sanctuarii	Apiaceae	H	3600				+
Oreocome aegopodioides	Apiaceae	H	3000–3300				+
Ceropegia kumaonensis	Apocynaceae	H	–				+
Ceropegia attenuatula	Apocynaceae	H	–			+	
Ceropegia parvissima	Apocynaceae	H	–			+	
Aralia devendrae	Araliaceae	Sh	–	+			
Trachycarpus takil	Arecaceae	T	1800–2600				+
Aristolochia dilatata	Aristolochiaceae	C	2800				+
Taraxacum harbhajan-singhii subsp. pahalgamense	Asteraceae	H	2000–2600	+			
Saussurea atkinsonii	Asteraceae	H	3000–4500	+		+	+
Taraxacum xanthophyllum	Asteraceae	H	–	+			
Tricholepis nakaoi	Asteraceae	H	–	+			
Anaphalis himachalensis	Asteraceae	H	2400			+	
Artemisia austrohimalayaensis	Asteraceae	H	4000			+	+
Artemisia filiformilobulata	Asteraceae	H	2600–3600	+		+	+

(continued)

Table 12.1 (continued)

Name of the taxa	Family	Habit	Elevation range	Distribution in states of IHR			
				J&K	LK	HP	UK
Artemisia indica var. kumaonensis	Asteraceae	H	1500				+
Aster molliusculus var. minor	Asteraceae	H	–			+	
Bidens tetraspinosa	Asteraceae	H	1500	+			
Chondrilla setulosa	Asteraceae	H	2500–2800	+			
Crepis dachhigamensis	Asteraceae	H	1500–2000	+			
Doronicum kamaonense	Asteraceae	H	2600–2800	+			
Dubyaea oligocephala subsp. latifolia	Asteraceae	H	1200–1500			+	
Inula acuminata	Asteraceae	H	1700–2900	+		+	
Cicerbita benthamii	Asteraceae	H	3000–5500	+			
Melanoseris decipiens var. multifida	Asteraceae	H	2700	+			
Melanoseris filicina	Asteraceae	H	1800–2500				–
Melanoseris kashmiriana	Asteraceae	H	2599–4000	+			
Melanoseris lahulensis	Asteraceae	H	3000–4000			+	
Ligularia jacquemontiana	Asteraceae	Sh	3000–4000	+		+	
Olgaea thomsonii	Asteraceae	H	2800–4000		+		
Saussurea sudhanshui	Asteraceae	H	4500–4800				+
Senecio Kundaicus	Asteraceae	H	3300		+		
Tanacetum robustum	Asteraceae	H	3150–4000	+		+	+
Taraxacum amblylepidocarpum	Asteraceae	H	3600	+			
Taraxacum aurorum	Asteraceae	H	3500–4000	+		+	
Taraxacum banhyhalense	Asteraceae	H	1900–2500	+			
Taraxacum coronatum	Asteraceae	H	3000	+			

(continued)

Table 12.1 (continued)

Name of the taxa	Family	Habit	Elevation range	Distribution in states of IHR			
				J&K	LK	HP	UK
Taraxacum erioscapum	Asteraceae	H	4000		+		
Taraxacum flavum	Asteraceae	H	2000–3000	+			
Taraxacum fulvescens	Asteraceae	H	2200–2500	+			
Taraxacum fulvobrunneum	Asteraceae	H	2000–2600	+			
Taraxacum harbhajan-singhii subsp. harbhajan-singhii	Asteraceae	H	2600–2700	+			
Taraxacum helianthum	Asteraceae	H	2600–2700	+			
Taraxacum heybroekii	Asteraceae	H	2500–3600	+		+	
Taraxacum latibasis	Asteraceae	H	–	+			
Taraxacum lobbichleri	Asteraceae	H	–	+			
Taraxacum nigrum	Asteraceae	H	–	+			
Taraxacum parvuliforme	Asteraceae	H	3500		+		
Taraxacum phoenicolepis	Asteraceae	H	–	+			
Taraxacum pseudobicorne	Asteraceae	H	1350	+		+	
Taraxacum pseudoeriopodum	Asteraceae	H	2650	+			
Taraxacum pseudostenolepium	Asteraceae	H	4500–5000	+			+
Taraxacum stewartii	Asteraceae	H	3600	+			
Taraxacum tenebristylum	Asteraceae	H	2700–3000	+			
Taraxacum tricuspidatum	Asteraceae	H	3900		+	+	
Tragopogon kashmirianus	Asteraceae	H	1500–2000	+			
Oreoseris lacei	Asteraceae	H	2438	+		+	
Impatiens devendrae	Balsaminaceae	H	2400–3200				+

(continued)

Table 12.1 (continued)

Name of the taxa	Family	Habit	Elevation range	Distribution in states of IHR			
				J&K	LK	HP	UK
Impatiens jaeschkei	Balsaminaceae	H	2700–3100				+
Impatiens langeana	Balsaminaceae	H	2500–2700				+
Impatiens pahalgamensis	Balsaminaceae	H	2400–3000	+			
Impatiens podocarpa	Balsaminaceae	H	2100–2500				+
Impatiens polysciadia	Balsaminaceae	H					+
Impatiens reidii	Balsaminaceae	H	1800–2400			+	+
Impatiens stoliczkae	Balsaminaceae	H	–			+	
Impatiens vexillaria	Balsaminaceae	H	–			+	+
Impatiens violoides	Balsaminaceae	H	2700–3000			+	+
Berberis affinis	Berberidaceae	Sh	2300–3000				+
Berberis ahrendtii	Berberidaceae	Sh	2200–3000				+
Berberis huegeliana	Berberidaceae	Sh	1700–2000		+		
Berberis kashmirana	Berberidaceae	Sh	3000	+			
Berberis lamberti	Berberidaceae	Sh	2650–2900				+
Berberis osmastonii	Berberidaceae	Sh	2200–2800				+
Berberis petiolaris var. grahwalana	Berberidaceae	Sh	2400–3300				+
Berberis rawatii	Berberidaceae	Sh	2340				+
Berberis jaunsarensis	Berberidaceae	Sh	2000–2600				+
Arnebia bhattacharyyae	Boraginaceae	H	3200			+	
Arnebia ladakhensis	Boraginaceae	H	4148		+		
Arnebia nandadeviensis	Boraginaceae	H	3600–4000				+
Cynoglossum stewartii	Boraginaceae	H	–	+			
Hackelia meeboldii Brand	Boraginaceae	H	–	+			
Ivanjohnstonia jaunsariensis	Boraginaceae	H	–				+
Maharanga egregia	Boraginaceae	H	–				+
Trigonotis rotundifolia	Boraginaceae	H	3600–5600				+
Draba alshehbazii	Brassicaceae	H	5300–6000		+		

(continued)

Table 12.1 (continued)

Name of the taxa	Family	Habit	Elevation range	Distribution in states of IHR			
				J&K	LK	HP	UK
Draba aubrietioides	Brassicaceae	H	1800	+			
Draba stenobotrys	Brassicaceae	H	1800	+			
Draba tenerrima	Brassicaceae	H	4300		+		
Campanula wattiana	Campanulaceae	H	2200–3900			+	+
Cyananthus integer	Campanulaceae	H	3000–4500				+
Eremogone curvifolia	Caryophyllaceae	H	3340–3650				+
Eremogone ferruginea	Caryophyllaceae	H	2000–3000				+
Dianthus cachemiricus	Caryophyllaceae	H	2500–3000	+			
Sagina purii	Caryophyllaceae	H	–				+
Silene cancellata	Caryophyllaceae	H	3300–3700	+		+	
Silene gangotriana	Caryophyllaceae	H	3500–4000			+	+
Silene kunawarensis	Caryophyllaceae	H	3200–4000		+	+	
Silene stewartii	Caryophyllaceae	H	3000–4000		+	+	
Stellaria pinvalliaca	Caryophyllaceae	H	3800–3900			+	
Calystegia hederacea. var. longipedicellata	Convoculaceae	H	100–3500	+			
Sedum sedoides	Crassulaceae	H	–	+		+	+
Sedum seelemanni	Crassulaceae	H					+
Carex hokarsarensis	Cyperaceae	H	1600	+			
Carex hypsobates	Cyperaceae	H	4572–5182	+			
Carex munroi	Cyperaceae	H	4000			+	
Carex muirii	Cyperaceae	H	–	+			
Carex nandadeviensis	Cyperaceae	H	3000				+
Kobresia rcsrivastavae	Cyperaceae	H	–				+
Scirpus dialgamensis	Cyperaceae	H	1600–1700	+			
Dryopteris khullarii	Dryopteridaceae	Sh	2900				+
Henckelia dimidiate subsp. mehrae	Dryopteridaceae	Sh	1800				+
Rhododendron rawatii	Ericaceae	Sh	3120				+

(continued)

Table 12.1 (continued)

Name of the taxa	Family	Habit	Elevation range	Distribution in states of IHR			
				J&K	LK	HP	UK
Euphorbia sharmae	Euphorbiaceae	Ush	4200–5000				+
Euphorbia thyrsoidea	Euphorbiaceae	Ush	2300–2750	+			
Butea pellita	Fabaceae	Sh	1500				+
Senna davidsonii	Fabaceae	Sh	–				+
Astragalus drasianus	Fabaceae	H	3200		+	+	
Astragalus gooraiensis	Fabaceae	H	2348	+			
Astragalus himalayanus	Fabaceae	H	3300–4400			+	
Astragalus kashmirensis	Fabaceae	H	1800–3800	+			
Astragalus isabellae	Fabaceae	H	2100–4000	+			
Astragalus maxwellii	Fabaceae	H	2100–4000	+			
Astragalus nainitalensis	Fabaceae	H					+
Astragalus sherriffii	Fabaceae	H	4100		+		
Astragalus tenuicaulis	Fabaceae	H	2100–2700				+
Astragalus uttaranchalensis	Fabaceae	H	3200–3500				+
Indigofera dosua var. simlensis	Fabaceae	Sh	800–2500			+	+
Indigofera gangetica	Fabaceae	Sh	–			+	+
Indigofera himachalensis	Fabaceae	Sh	2664–4500			+	+
Oxytropis sanjappae	Fabaceae	H	3500–4200			+	
Oxytropis shivae	Fabaceae	H	–		+		
Corydalis amarnathiana	Fumariaceae	T	3300–3500	+			
Corydalis devendrae	Fumariaceae	H	3600–4000				+
Corydalis duthiei	Fumariaceae	H	3000–4300			+	+
Corydalis govaniana var. kedarensis	Fumariaceae	H	3500–4300				+
Corydalis vaginans	Fumariaceae	H	2800–4500				+
Corydalis violacea	Fumariaceae	H	3900–5200			+	+
Gentiana decemfida	Gentianaceae	H	2100			+	+

(continued)

Table 12.1 (continued)

Name of the taxa	Family	Habit	Elevation range	Distribution in states of IHR			
				J&K	LK	HP	UK
Gentiana capitata var. hemkundiana	Gentianaceae	H	4250–4350				+
Swertia kashmirensis	Gentianaceae	H	2650	+			
Swertia pahalgamensis	Gentianaceae	H	–	+			
Gentiana capitata var. harwanensis	Gentianaceae	H	3400–4000	+			+
Gentiana saginoides	Gentianaceae	H	3686				+
Gentiana spitiensis	Gentianaceae	H	–			+	
Gentiana tetrasepala	Gentianaceae	H	4500–4800				+
Swertia alpina	Gentianaceae	H	3000–4000				+
Geranium clarkei	Geraniaceae	H	2100–4200	+			
Didymocarpus oblongus	Gesneriaceae	H	1800–3000				+
Dipcadi reidii	Hycinthaceae	H	2000–2300				+
Chelonopsis cashmerica	Lamiaceae	H	2000	+			
Chrysojasminum parkeri	Oleaceae	Sh	1600–2300			+	
Epilobium rhynchospermum	Onagraceae	H	3800–5200	+			
Bulbophyllum raui	Orchidaceae	H	1700–2000				+
Pinalia occidentalis	Orchidaceae	H	1500–2000				+
Herminium kumaunense	Orchidaceae	H					+
Neottia mackinnonii	Orchidaceae	H					+
Neottia nandadeviensis	Orchidaceae	H	3000–3500				+
Nervilia gleadowii	Orchidaceae	H	1000				+
Nervilia pangteyana	Orchidaceae	H	900				+
Peristylus kumaonensis Renz	Orchidaceae	H	2178				+
Meconopsis robusta	Papaveraceae	H	2800–4300				+
Papaver kachroianum	Papaveraceae	H	1600	+			
Papaver pamporicum	Papaveraceae	H	1600	+			

(continued)

Table 12.1 (continued)

Name of the taxa	Family	Habit	Elevation range	Distribution in states of IHR			
				J&K	LK	HP	UK
Pittosporum eriocarpum	Pittosporaceae	T	900–2100			+	+
Trisetopsis pirpanjalensis	Poaceae	H	3000			+	
Trisetopsis himalayensis	Poaceae	H	2500				+
Tzveleviochloa yadavii	Poaceae	H	3450			+	
Dendrocalamus somdevae	Poaceae	B	600				+
Elymus kuramensis	Poaceae	H	3400		+		
Helictotrichon uniyalii	Poaceae	H	2300–2700				+
Muhlenbergia rakhchamensis	Poaceae	H	2950			+	
Poa rhadina	Poaceae	H	3500–4000				+
Pseudodanthonia himalaica	Poaceae	H	2500–2800			+	+
Sehima notata	Poaceae	H	1800–2100				+
Koeleria micans	Poaceae	H	2400–2750				+
Primula garhwalica	Primulaceae	H	4200–4800				+
Aconitum moschatum	Ranunculaceae	H	1600	+			
Anemone raui	Ranunculaceae	H	2900–3200				+
Delphinium hajrae	Ranunculaceae	H	1600		+		
Delphinium koelzii	Ranunculaceae	H	–			+	
Isopyrum ludlowii	Ranunculaceae	H	3000–3100	+			
Paraquilegia gangotriana	Ranunculaceae	H	–			+	+
Ranunculus bikramii	Ranunculaceae	H	4600			+	
Ranunculus uttaranchalensis	Ranunculaceae	H	–				+
Berchemia jainiana	Rhamnaceae	Sh	1800				+
Rhamnus collettii	Rhamnaceae	Sh	2000–2500			+	
Sageretia devendrae	Rhamnaceae	Sh	–	+			+
Sageretia kashmirensis	Rhamnaceae	Sh	600–2000	+			

(continued)

Table 12.1 (continued)

Name of the taxa	Family	Habit	Elevation range	Distribution in states of IHR			
				J&K	LK	HP	UK
Sageretia kishtwarensis	Rhamnaceae	Sh	2200	+			
Sageretia santapaui	Rhamnaceae	Sh	1400–2500				+
Ziziphus oxyphylla var. pedicellaris	Rhamnaceae	Sh	1200–1500	+		+	+
Cotoneaster obovatus	Rosaceae	Sh	2400	+			
Cotoneaster parkinsonii	Rosaceae	Sh	–				+
Potentilla karakoramica	Rosaceae	H	2300–2600	+			
Argentina sericophylla	Rosaceae	Sh	2000–3000	+			
Spiraea duthieana	Rosaceae	H	2700–3050				+
Spiraea emarginata	Rosaceae	Sh	1700	+			
Spiraea hypoleuca	Rosaceae	H	2300–3000				+
Spiraea panchananii	Rosaceae	Sh	2400				+
Spiraea vacciniifolia	Rosaceae	Sh	1300–2200			+	+
Galium confertum	Rubiaceae	H	2300–3000			+	
Galium duthiei	Rubiaceae	H	2500–4500				+
Galium falconeri	Rubiaceae	H	1800–3000				+
Galium kulluense	Rubiaceae	H	1500–2000			+	
Salix calostachya	Salicaceae	T	–	+			
Saxifraga chadwellii	Saxifragaceae	H	4100		+		
Saxifraga minutissima	Saxifragaceae	H	4600–4800				+
Saxifraga poluniniana var. mucronata	Saxifragaceae	H	3100				+
Pedicularis pectinata var. rosea	Scrophulariaceae	H	3300–4200			+	+
Alectra parasitica subsp. chitrakutensis	Scrophulariaceae	H	–				+
Kashmiria himalaica	Scrophulariaceae	H	3400				+
Pedicularis pushpangadanii	Scrophulariaceae	H	4000			+	
Valeriana pyrolifolia	Valerianaceae	H	2500–3000	+			+

(continued)

Table 12.1 (continued)

Name of the taxa	Family	Habit	Elevation range	Distribution in states of IHR			
				J&K	LK	HP	UK
Valeriana jaeschkei var. jaeschkei	Valerianaceae	H	2600–4600	+		+	
Valeriana jaeschkei var. kaschmiriensis	Valerianaceae	H	2600–4600	+		+	
Viola jangiensis	Violaceae	H	–			+	
Viola rupestris var. himalayensis	Violaceae	H	1000–4000	+			
Deparia macdonellii	Woodsiaceae	Sh	1800–2700	+		+	+

categories. However, 24 taxa (23 sp., 1 var.) were included in Red list of Indian Plants in various categories such as Endangered (4), Rare (7), Vulnerable (2), Indeterminate (11) (Tiwari et al. 2024). Besides, a total of 8 species of Orchidaceae were included in the Convention on International Trade in Endangered Species of Wild Fauna and Flora (CITES).

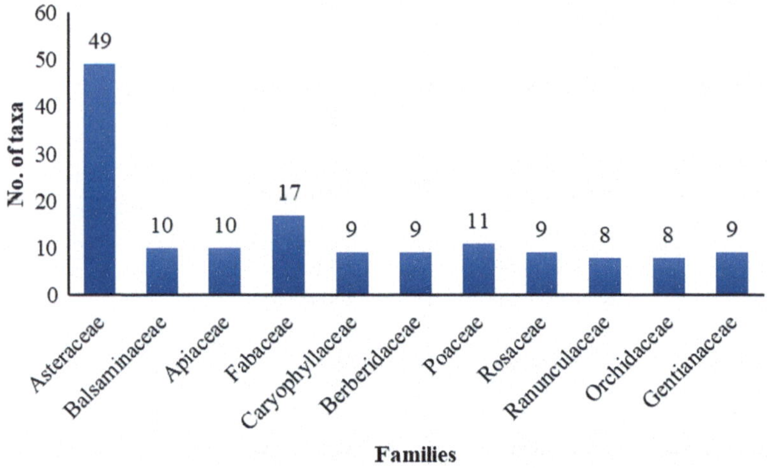

Fig. 12.3 Familial distribution of endemic taxa

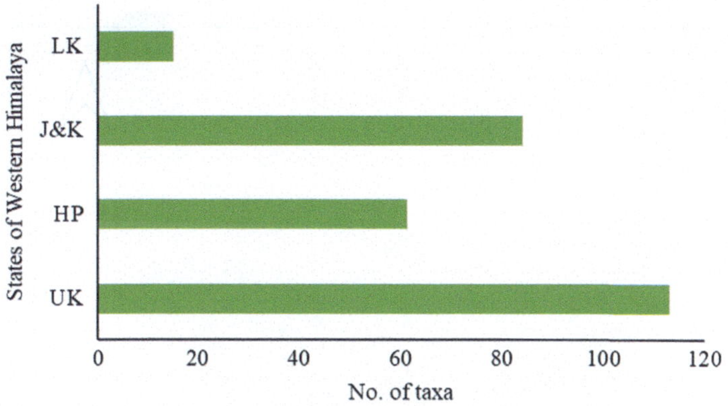

Fig. 12.4 State wise distribution of endemic taxa

12.3.2 Distribution of Endemic Plants Across Western Himalayan States

The State wise distribution of endemic plants show unequal pattern with respect to the geographical area of the states. Among Western Himalayan States, maximum number of endemic plants (103 sp., 1 subsp., 9 var.) were documented in Uttarakhand followed by Jammu and Kashmir (75 sp., 2 subsp., 7 var.), Himachal Pradesh (54 sp., 1 subsp., 6 var.), and Ladakh (15 species) respectively (Fig. 12.4).

12.3.3 Endemic Rich Areas in Western Himalaya

While compiling the endemic taxa, the locations of the species were also documented and Endemic rich areas were identified on the basis of species locations in Western Himalayan States (Fig. 12.5). State wise, in Uttarakhand (Khaliyatop, Munsyari, Kalamuni, Namik, Sahastradhara, Mussoorie, Dhanaulti, Chakrata, Bhyudar, Ghangaria, Bhojbasa-Chirbasa, Gangotri, Valley of flowers NP, and Alaknanda valley), in Jammu & Kashmir UT, localities namely Gulmarg, Sonmarg, Tangmarg, Khilanmarg, Baramulla, Banihal, Pahalgam and Verinag, and in Himachal Pradesh, the Lahaul & Spiti district were identified as endemic rich areas.

12.4 Discussion

Endemism and IUCN red list of threatened species serves as a dispensable tool for the conservation of biodiversity. Many high value plant species are harvested from Himalaya for trading at local, regional and international levels in which the

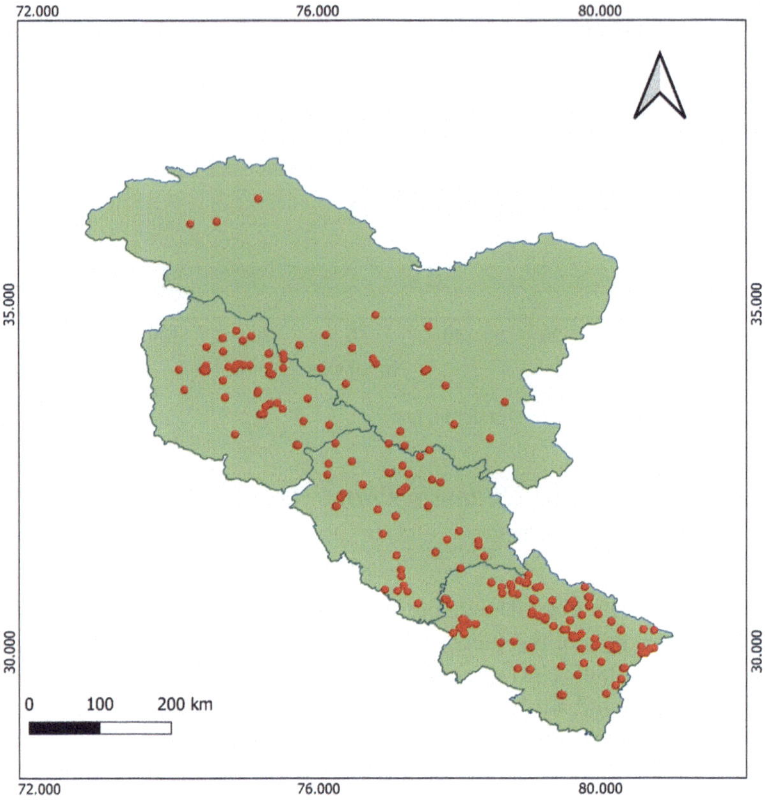

Fig. 12.5 Distribution map of endemic taxa across western Himalaya, India

IHR contributes more than 50% (Pandey et al. 2019). Several medicinal plants are collected by local people for cash income, which results in over-exploitation and thus population decline of these species (Dhar et al. 1998, Gaur 1999, Mehta et al. 2021). The dominance of endemic elements at high altitudes, especially in the northwest and the West Himalayas suggests the high conservation value of these areas (Dhar, 2002).

The present study reveals that most of the endemic taxa have unrevealed population status and rarity status which effects the conservation implications, particularly for narrow-endemic taxa. Analysis of life forms in the western Himalayan states (LA, J&K, HP, UK) show a rich diversity of herbs. This may be due to higher number of endemic species in higher elevations. Since there are more endemism-rich regions above the timberline in the western Himalayan states, which are suitable for herbaceous species, thus contributing in greater variety of endemic herb taxa (Singh et al., 2015).

In the Western Himalayan states, distribution of endemic taxa shows positive correlation with forest cover but negative correlation with the total geographical area of the states which can be explained by the fact that forest provides habitat to larger number of species and influences species as well as endemic taxa richness of an area. While states having larger geographical areas may have larger desert cover (i.e. Ladakh) and other harsh climatic conditions that influence the species richness negatively.

Areas with high endemism may include many species within a small geographic distribution. Anthropogenic activities such as urbanization, deforestation and unsustainable harvesting of medicinal plants can be a threat for these plants (Mehta et al. 2020, 2023). Many endemic-rich areas were observed outside the boundaries of protected areas (i.e. Sonamarg, Alaknanda valley, Sahastradhara, Kalamuni, Munsyari) which require attention and inclusion under protected area network for their effective conservation and management. Since many narrow endemic plant taxa are also present in these outside areas their importance for developing future conservation plans/strategies is significantly increases.

12.5 Conclusion

Endemism of an area can be used as important criteria while identifying priority sites for conservation implications. The present study reveals distribution pattern of endemic taxa in Western Himalaya, India. Total 227 taxa have been documented as endemic to Western Himalaya. The distribution map of the species reveals that many areas have high species richness of endemic taxa in a relatively small area, which can be further included under protected area boundary. Assessing present distribution of endemic plants would be crucial contribution for their conservation planning and management. The study recommends inclusion of endemic rich areas in the Protected Area Netwok (PAN) in order to ensure their sustainability in the Western Himalaya. Further, the population of threatened endemic species should be monitored and promoted for proper ex-situ conservation. Raising awareness should be promoted to the local communities for the importance as well as the conservation need of endemic plants.

References

Bhatt D, Sekar KC, Rawal RS (2021) Tree diversity, congruence and endemism: Himalaya 'The land of diversity.' Biodivers Conserv 30(10):2633–2654

Dhar U (2002) Conservation implications of plant endemism in high-altitude Himalaya. Curr Sci 82(2):141–148

Dhar U, Rawal RS, Samant SS (1998) Endemic plant diversity in the Indian Himalaya IV: Poorly represented primitive families. Biogeographica (Paris) 74(1):27–39

Gaur RD (1999) Flora of the District Garhwal, North West Himalaya. Transmedia

Hajra PK, Mudgal V (1997) Plant diversity hotspots in India, Botanical Survey of India, Dehradun

Heywood VH, Watson RT (1995) Global biodiversity assessment, vol 1140. Cambridge University Press, Cambridge

Hrdina A, Romportl D (2017) Evaluating global biodiversity hotspots–very rich and even more endangered. J Landsc Ecol 10(1):108–115

Mahar G, Dhar U, Rawal RS, Bhatt ID (2009) Implications of location specific data and their usefulness in conservation planning: an example from Indian Himalayan Region (IHR). Biodivers Conserv 18:1273–1286

Manes S, Costello MJ, Beckett H, Debnath A, Devenish-Nelson E, Grey KA, Vale MM (2021) Endemism increases species' climate change risk in areas of global biodiversity importance. Biol Cons 257:109070

Mehta P, Bisht K, Chandra Sekar K (2021) Diversity of threatened medicinal plants of Indian Himalayan Region. Plant Biosyst 155(6):1121–1132

Mehta P, Bisht K, Chandra Sekar K, Tewari A (2023) Mapping biodiversity conservation priorities for threatened plants of Indian Himalayan Region. Biodivers Conserv 32:2263–2299. https://doi.org/10.1007/s10531-023-02604-z

Mehta P, Chandra Sekar K, Bhatt D, Tewari A, Bisht K, Upadhyay S, Negi VS, Soragi B (2020) Conservation and prioritization of threatened plants in Indian Himalayan Region. Biodivers Conserv 29(6):1723–1745

Myers N, Mittermeier RA, Mittermeier CG, Da Fonseca GA, Kent J (2000) Biodiversity hotspots for conservation priorities. Nature 403(6772):853–858

Pandey A, Chandra Sekar K, Joshi B, Rawal RS (2019) Threat assessment of high-value medicinal plants of cold desert areas in Johar valley Kailash Sacred Landscape. India Plant Biosyst 153(1):39–47

Rawal RS, Bhatt ID, Chndra Sekar K, Nandi SK (2013) The himalayan biodiversity: richness, representativeness, uniqueness and life-support values. In: GB Pant Institute of Himalayan Environment and Development (GBPIHED), Kosi-Katarmal, Almora, Uttarakhand, India, pp 84

Singh P, Dash SS (2014) Plant discoveries 2013-New Genera, species and New Records. Botanical Survey of India, Kolkata

Singh DK, Pusalkar PK (2020) Floristic diversity of the Indian Himalaya. Biodiversity of the Himalaya: Jammu and Kashmir State, pp 93–126

Singh P, Karthigeyan K, Lakshminarasimhan P, Dash SS (2015) Endemic vascular plants of India. Botanical Survey of India, Kolkata

Singh SP, Thadani R (2015) Complexities and controversies in Himalayan research: a call for collaboration and rigor for better data. Mt Res Dev 35(4):401–409

Tiwari H, Chandra Sekar K, Pandey A, Tiwari A, Mehta P, Kanwal KS, Arya D (2024) Diversity, distribution and need of urgent conservation of endemic plants in Himalaya. Biodivers Conserv 33:2285–2303

Tobler M, Honorio E, Janovec J, Reynel C (2007) Implications of collection patterns of botanical specimens on their usefulness for conservation planning: an example of two neotropical plant families (Moraceae and Myristicaceae) in Peru. Biodivers Conserv 16:659–677

Chapter 13
Climate Change and Its Impacts on Rural Areas of Central Himalaya: Some Experiences from Almora District, Uttarakhand, India

Puran Chandra Joshi

Abstract Climate change is one of the biggest challenges in present era. Entire earth is facing this problem. Thus, the Himalayas are not untouched with this global phenomenon. The Himalayas are divided in three ranges according to their origin. Likewise, the other parts of the globe Himalayas are also facing human induced climate change. The middle part of this long mountain system is known as central Himalaya. For the past few years, this part of the Himalayas has been affected from erratic rainfall and rise in temperature. Forest fire disaster is very common in throw-out the year. As a result of this water scarcity is very common in past few years. Agricultural system is also affected, even in some areas people are not growing the wheat crop because of lack of rainfall in winter season. It is a well-known fact that climate change has severely affected the poor countries and their people. In rural areas of the central Himalayas, people who are not capable to earn much for their livelihood, they have become completely dependent on the food grains given by the government. People are forced to migrate in big cities where they lead a low standard of living due lack of resource. Present paper discusses the impacts of climatic change in culture, traditions and life style in rural Himalaya. Paper also traces out the situation of gender discourse, livelihood and out -migration in the research area due to poor situation of agriculture. Present paper will help in understand how developing societies are facing problems due to climate change.

Keywords Central Himalaya · Climate change · Migration · Indigenous knowledge · Livelihood

P. C. Joshi (✉)
Department of Geography, Soban Singh Jeena University Campus, Almora, Uttarakhand, India
e-mail: puran.joshi3@gmail.com

D. Arya et al. (eds.), *Climate Change Impact on Himalayan Biodiversity*,
Environmental Science and Engineering, https://doi.org/10.1007/978-3-031-77149-1_13

13.1 Introduction

Himalaya is one of the most sensitive as well as fragile parts of the earth. It is the newest as well as the highest mountain system too. In order of their origin Himalayas are divided into three parallel ranges. In a sequence from north to south, they are known as great Himalaya, lesser Himalaya and Shivalik (outer most) respectively. The middle part of this lofty mountain also known as Central Himalaya. This part is the most populous region of the Himalayas. In this region, the impression of human and their work is visible from the prehistoric period (Joshi 2001). Thus, Central Himalaya is the focal point of culture and civilization since ancient time. Indian Himalayan region (IHR) covers an area of total 530,795 sq. km., about 16.16% of total geographic area of India (Singh 2006). Kumaun Himalaya, where the study area is situated, is the significant part of entire Himalayan region in terms of population, ancient culture and diversity, consisting 21,035 sq. km. area (Jalal 2001).

If we talk about climate change, it is a variable aspect in the history of our earth (Krishnan et al. 2020). It is a natural as well as long term process but after the industrial revolution the world population grew very rapidly. Scientific inventions played a major role to increase the temperature. Heat budget of the earth has become imbalanced. Therefore, climate change is the matter to study now.

Himalayan glaciers are retreating. The retreatment rate of Gangotri glacier, origin of river Ganga, is 19 m per year (Rawat 2012a).

13.2 Study Area

The study area falls in the northern part of India, in the state of Uttarakhand, Having an area of 3139 sq. km. area. The area of investigation lies between 29° 25′ 39.5" N to 29° 59′ 33" N latitudes and 79° 03′ 03"E to 80° 04′ 45"E longitudes. Entire area is situated in the lesser Himalaya. Altitudinal range varies between 1200 to 2200 m from mean sea level. Geologically, the region is set between Main Central Thrust (MCT) & Main Boundary Thrust (MBT) (Heim and Gansser 1939) and geologically divided into the northern and southern sedimentary regions, made up of poorly fossiliferous, thoroughly deformed rocks of the middle Proterozoic to Cambrian age5. Almora Nappe passes through the area of investigation4. The study area comes under temperate climate. Highest and lowest average temperatures recorded 31.5°c and 0.1°c respectively in 2013 (District Disaster Management Authority n.d). Ramganga west, Gagas and Kosi are major rivers flow in the region. These rivers are rainfed. If we take look at the history of the earth and see the sequence of human development, the impression of human feet and its work is visible from pre historic time (Joshi 2001) (Fig. 13.1).

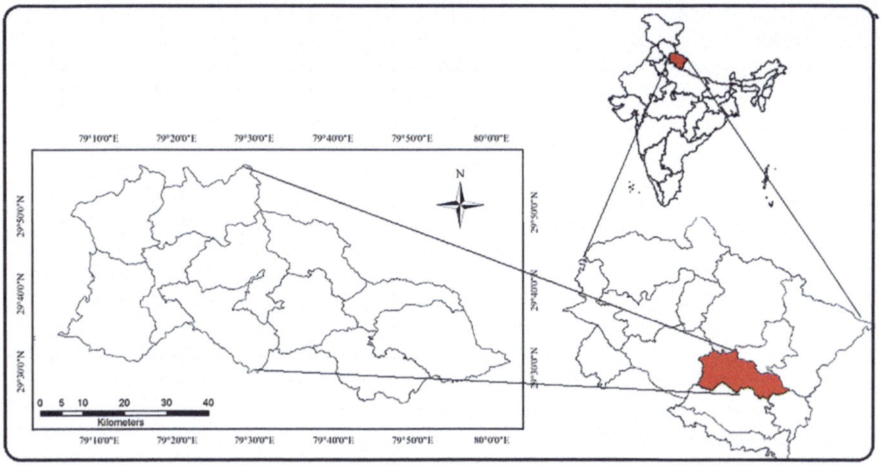

Fig. 13.1 Location map of the study area

13.3 Methodology

Present paper is based on primary and secondary data. Primary data is collected after extensive field work. Secondary data is collected from various government officials. Some information in based on observations of local people. A series of interview has been taken from senior citizen living in villages.

13.4 Objectives

1. To find out people perception on climate change.
2. To find out the patterns of precipitation in the study area.
3. To investigate the impacts of climate change on rural people.
4. To understand the dynamics of climate change in the terms of gender inequality livelihood and migration.

13.5 Precipitation Patterns in the Study Area

Precipitation is water released from clouds in the form of rain, freezing rain, sleet, snow, or hail. It is the primary connection in the water cycle that provides for the delivery of atmospheric water to the Earth. Most precipitation falls as rain (https://www.usgs.gov/special-topics/water-science-school/science/precipitation-and-water-cycle) (USGS). It is observed that precipitation patterns have been changed in recent past in the study area. Present region receives maximum rainfall

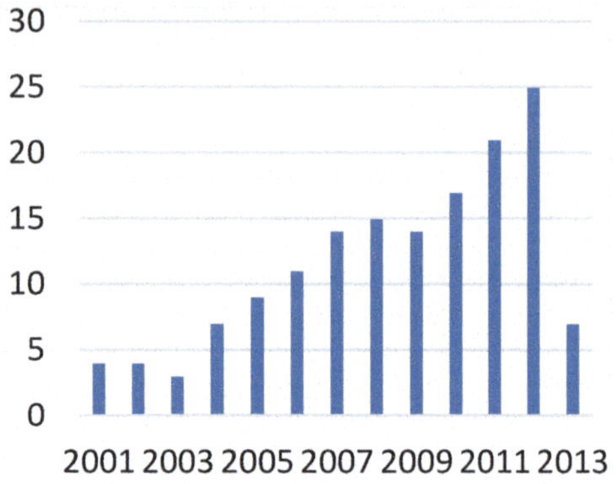

Fig. 13.2 Days of high intensity rainfall

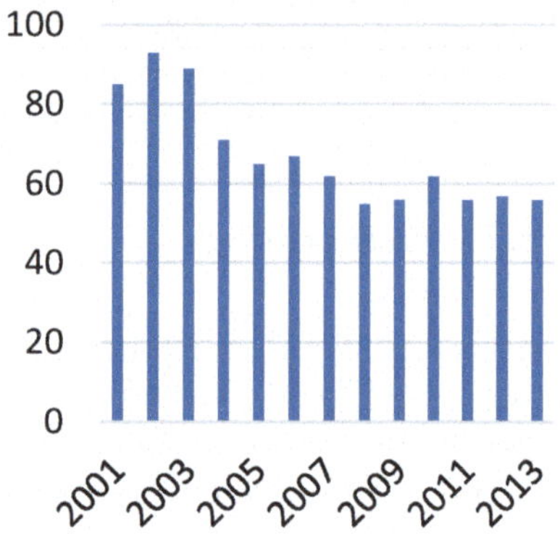

Fig. 13.3 Annual number of rainy days

in the month of July (Rawat 2012b). Total numbers of rainy days have decreased and intensity of rainfall has increased (Figs. 13.2, 13.3 and 13.4 and Tables 13.1 and 13.2).

In the study area the south-west monsoon is dominant for rainfall (Tiwari and Joshi 2015). According to the table number 1, average rainfall in the study area is approximately same but the total number of rainfall days are decreasing while the number of high intensity rainfall days are increasing.

13.6 People Perception About Climate Change

A survey was also conducted to understand the effects of climate change among people. This survey was conducted in 25 villages of Almora district, where a random sampling of one-third of the total population was selected, consisting of people aged 60 years and above. As we know, climate change is the study of long-term changes in the climate, and therefore, people over the age of 60 can perceive this change through their observations at a certain level.

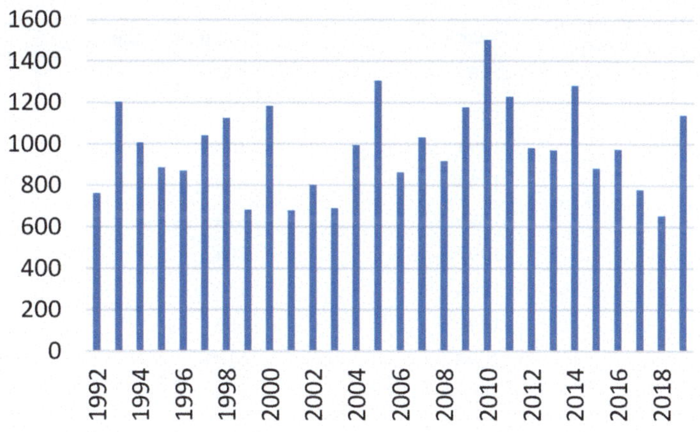

Fig. 13.4 Rainfall in mm

Table 13.1 Past years rainfall data of Mukteshwar near to Almora

Year	Total rainfall (MM)	Annual number of rainy days	Days of high intensity rainfall
2001	2400	85	4
2002	2125	93	4
2003	2010	89	3
2004	2200	71	7
2005	1915	65	9
2006	1870	67	11
2007	1700	62	14
2008	2970	55	15
2009	1635	56	14
2010	3570	62	17
2011	1610	56	21
2012	1605	57	25
2013	1155	56	7
Average	2059	67	12

Source Meteorological Station, Muktshwer, Nainital, Uttarakhand

First part of this survey is based on the demographic profile of respondents and second part is divided into four types of questions related to climate change perception. The first question was about the understanding of climate change, The second question was about its causes and third one was about impacts on different categories of society, finally, the question of what measures can be taken to cope with it.

Table 13.2 Rainfall data of Almora region 1992–2019

Year	Rainfall in mm
1992	764
1993	1205
1994	1008
1995	889
1996	874
1997	1042
1998	1127
1999	683
2000	1185
2001	679
2002	803
2003	690
2004	996
2005	1307
2006	865
2007	1033
2008	920
2009	1178
2010	1505
2011	1230
2012	982
2013	972
2014	1282
2015	883
2016	974
2017	779
2018	654
2019	1140

Source NRDMS Centre Almora

13.7 Demographic Profile of Respondents

All the respondents were chosen from rural area of central Himalaya in Almora district (Table 13.3).

Table shows these people are living in rural areas and they are basically from a low-income group. When researcher interacted with them in the field many facts were coming out. They told to the researcher that only poor people are living in village otherwise a person gets little opportunity, he migrates to the cities (Fig. 13.5 and Table 13.4).

Table 13.3 Demographic profile

Sr. N	Name of the demographic indicator	Value		
1	Male Female	Male		Female
		75		75
2	Age Group	Above 60		
3	Education	Up to higher secondary		
4	Occupation	Agriculture	Job in another sector	Manrega
		120	22	08
5	Income	3000 INR/month		

Source Primary survey

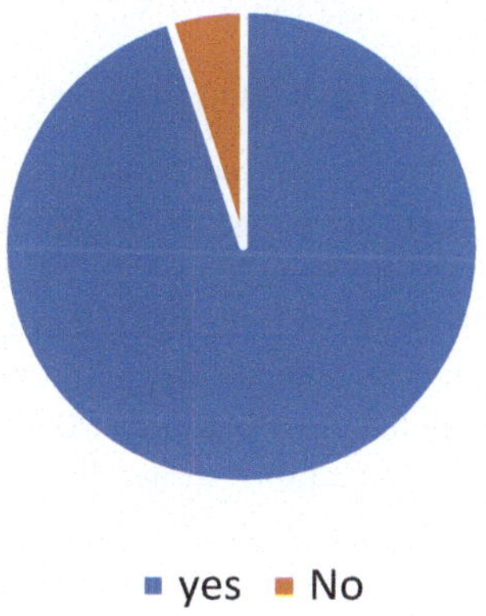

Fig. 13.5 Perception about climate change

Table 13.4 Responses

Sr.N	Responses	Percent
1	Yes	95
2	No	5

Source Primary survey

1. Are you aware with the term 'Climate change'
2. What are the causes of climate change?

Increasing Population, Environmental Pollution, Urbanisation, Deforestation,

3. What are the impacts of climate change?

Irregular rainfall, Inadequate snowfall in high hills, Cloud burst, Flood, Low agricultural productivity, Water scarcity.

4. What are the measures can be taken to cope with climate change?

Save forest, Plastic should be banned, Save water.

Even though people may not be familiar with the term "climate change," they do understand it. Most of the respondents believe that the climate has changed in the past 30–40 years, and this has affected their agriculture and way of life. They are aware with the causes and impacts of this global phenomenon.

13.8 Impacts of Climate Change in the Study Area

I. Extreme Weather Events

The frequency of extreme weather events in the study area is also increasing rapidly. Examples of destructive events such as in 2010, 2013, and 2021 are evident before us (Table 13.5).

II. Increasement in Forest Fire Disaster

Forest fire has become a major disaster in the study area in recent past. Some years ago, forest fire incidents were less as compare to present. In 2002, there were 1,401 such incidents. In 2016, the figure went up to 12,958, and then fell to 2,158 in 2019. As of June 2022, there have been 2,131 forest fires, affecting more than 3,348 hectares and causing a loss of Rs87.31 lakh. Uttarakhand has a forest cover of 53.48 lakh hectares (Awasthi 2022). (The Week Magazine) (Table 13.6).

Above table shows forest fire incidents in the study area. In 2011 frequency is lowest while in 2012 it is highest. Forest fires are a significant threat to wildlife, birds, and plants. Small plants and trees do not even get a chance to grow. The birds' eggs are destroyed in large quantities (Fig. 13.6 and Plate 1).

The reason behind the increasing forest fire incidents is decreasing rainfall days in the study area. Moisture in soil stops fire in the forest, and when soil is dry due to

Table 13.5 Damage due to extreme weather events like flash-flood and landslide in 2021

Sr. N	Category of damaged property	Frequency of damaged property	Value (In Million INR)
1	Agricultural field	20	01
2	House (Partially)	10	02
3	House (Completely)	05	10
4	School	01	0.1
5	Hotel	02	10
6	Cow Shed	02	0.01
7	Animals	03	0.1
8	Ornaments	06	10
9	Furniture	01	10
10	Hen	01	0.005

Source Primary survey

Table 13.6 Forest Fire incidents in the study area

Year	Forest fire incidents
2005	5
2006	7
2007	6
2008	39
2009	264
2010	100
2011	3
2012	307
2013	11
2014	72
2015	37
2016	195
2017	175

Source Forest department, Uttarakhand State

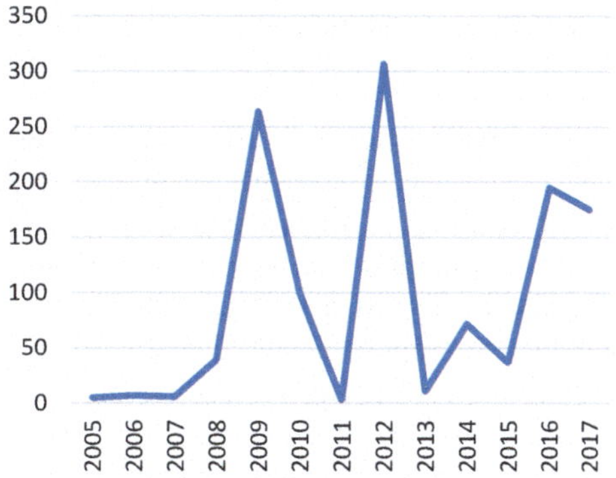

Fig. 13.6 Forest fire incidents

irregular pattern of precipitation, it becomes more vulnerable for forest fire disaster. Respondents told to the researcher that some years ago May and June are the main months of forest fire but now forest fire occurs even in winters. According to a report of forest department in 2022, 140 forest fire incidents were reported in February and march (Report Forest (n.d.)).

III. Impacts on Hill Agriculture

We can see two forms of agriculture can be seen in the study area, first is based on irrigation in the river valleys, and the other rain-fed agriculture which has been the basis of life here for centuries. 98 percent of the people believed that the decrease in rainfall as well rainfall days have had a very bad effect on agriculture, and reasoning the trend towards migration has increased, The production of wheat, which used

Plate 1 Forest fire in study area

to grow well in rain-fed areas, has now decreased significantly. People admit that 30–35 years ago, along with wheat, lentils and mustard were also grown abundantly.

During the preliminary survey, it was found that even in the non-irrigated areas, known as 'uparāṇv' in the Kumaon Himalayas, where crops were grown without irrigation in the past, a good amount of grain production was possible. However, today people have stopped growing wheat in those areas. In many of the villages selected for the survey, respondents admitted that wheat cultivation was ending in unsaturated areas due to a lack of winter rainfall. This fact has also forced people to migrate, although there are certainly other reasons for migration. As well as a major reason is that poor people in the villages are dependent on government-provided rations. These people are also at risk of malnutrition. Agriculture not only provides grain but also many other supportive things, such as mustard is grown with wheat, which is eaten as a vegetable, and lentils are consumed in various forms in mountainous regions. People have stopped producing all of these things. People have adopted consuming cheap, packaged food, which is not good for their health and safety.

Table 13.7 shows, there has been a huge decline in production of crops in non-irrigated areas. Thirty years ago, people grew a significant amount of wheat in winter in non-irrigated land. One more thing which is very important, these are completely organic products. People use animal dung as manure in their agriculture fields. Respondents told this fact to the researcher during field work.

People believe that due to changing seasons in recent years, the timing of the rainy season has been affected, causing a significant impact on agriculture in the central Himalayan region. Most of the farming in this area depends on rainfall. Although

Table 13.7 Production of winter crops in non-irrigated areas in the study area

Sr N	Name of the crop	Average production before 30 years (in Kilogram) in per nali (Rawat 2012a)	Average production in present (in Kilogram) in per nali (Rawat 2012a)
1	Wheat	30	2
2	Lentil	10	1
3	Mustard	04	1

Source Primary survey

monsoon season agriculture is still somewhat manageable, the cultivation of wheat during winters has been severely affected. Many villages have stopped planting wheat altogether.

IV. Livelihood and Migration

In recent years, people have been forced to leave their homes due to climate change. These individuals move to a larger city, which leads to an increase in the population and urban problems in big cities in developing societies. During the covid-19 period, when migratory people returning to their native land, we observed this thing thoroughly.

Climate change has become one of the major reasons for migration in the study area. Agricultural production is reduced (Table 13.7) and traditional sources of water are drying up fast (Plate 2).

Plate 2 Traditional method of farming called 'Terrace Farming'

[1]'Nali' is a Local Unit for Land Measurement & 1 Nali is Equivalent to 2160 Sq. Feet.

These people have their own indigenous knowledge. They utilised this knowledge in agriculture and conducting their daily life. They talk about the agricultural production which they used to grow in the fields, but in recent past entire system has changed due to the change in weather cycle. Livelihood is the biggest issue in front of them. They are forced to migrate from their native regions. If we analyse a report prepared by Rural Development and Migration Commission, Uttarakhand, in Almora, total 8% revenue villages have become uninhabited due to migration. The total number of villages/toks/majra that have seen more than 50% decline in the population in last 10 years is 80. Main reasons behind this out migration are employment, education and health facilities. Maximum percentage migrated due to the reason of employment which is 47.78% in the study area. Before 30–40 years ago agriculture was the basic occupation but due to low productivity, people are leaving it.

V. Water Scarcity

Climate change has led to an increase in water-related issues for drinking and other household uses in the study area. During the initial survey, it was found that many villages had abundant natural sources of water. Now these sources are on the verge of drying up, with many completely dried up. The chosen village for the survey, Chhatgulla, was found to have 9 natural springs ('naulas') in Kumaoni, which refer to natural sources of water that emerge from the ground and are channelled for cleaning and safety purposes).

Most of these springs are now at risk of disappearing. Many of them are drying up. Small rivers which feed the big rivers are also at risk. They are drying up rapidly (Plate 13.9).

VI. Gender Inequality

Even today in rural areas, the responsibility of filling water still lies with women. Due to the drying up of water sources, women have to travel long distances to fetch water for domestic needs, which takes up a significant portion of their time. As a

Plate 13.9 'Naula' Traditional Sources of water in study area

Table 13.8 Women participation in fetching water

Sr. N	Number of women	Distance to fetch water (km)
1	15	2
2	05	3
3	03	4

Source Primary survey

result, their education and health have been adversely affected. The drying up of water sources has led to gender inequality and exacerbating climate change. Most of the villages have their pipelines for tap water but they get water after four or five days (Table 13.8).

Table shows that 10% women in the study area travel 2 km daily for fetching water. 2% women in the study area have travel 4 km to fetch water for their use.

13.9 Conclusion

To sum up, it can be said that most people in the study area are familiar with the concept of climate change. They may not directly understand the term but they are aware of the changing weather patterns, declining agricultural productivity, and the impacts on their daily lives. People in the area of investigation are concerned about erratic rainfall and increasing temperatures. Today, the entire Himalayan region is experiencing this crisis. They know the measures also. This study clearly indicates that there is a decreasing trend in people's inclination towards agriculture, and they are seeking employment opportunities elsewhere. As a result, their focus is shifting away from the traditional economy. The uncontrolled regions are experiencing a significant decrease in agricultural production, leading to food security crises. In the study area, the consequences of climate change have also manifested in natural disasters. One example is the untimely forest fires. On the other hand, there has been an increase in events like floods and landslides due to irregular rainfall. Traditional sources of water are drying up. Water scarcity and gender inequality have also been observed in the area of investigation.

Acknowledgements The author is thankful to the local people of Almora district and officials of different government agencies. The author is also very much thankful to the department of geography, SSJ University, Almora, specially, to the head of the department Dr. Jyoti Joshi.

References

Awasthi P (2022) Why forest fire are ravaging Uttarakhand

District disaster management authority, Almora, Work Plan, 2020–2021, pp 41–45

Heim A, Gansser A (1939) Central Himalaya-geological observations of Swiss expedition 1936. Mem Sqc Helv Sci Nat, 73:1–245
Jalal DS (2001) Geographical perspective of Kumaun, Kumaun land and people. In: Valdiya KS (ed) Gyanoday, Nainital pp 13
Joshi MP (2001) Kumaun: Archaeological and Historical Perspective, Kumaun Land and people. In: Valdiya KS (ed), *Gyanoday*, Nainital pp 73
Krishnan R, Gnanaseelan C, Sanjay J, Swapna P, Dhara C, Sabin TP, Jadhav J, Sandeep N, Dey Choudhury A, Singh, Mujumdar M, Anant P, Tewari A, Mehajan R (2020) Introduction to climate change over the indian region. In: Krishnan R, Sanjay J, Gnanaseelan C, Mujumdar M, Kulkarni A, Chakraborty S, Springer open, pp 1
Rawat JS (2012) Climate Change, Uttarakhand State of the Environment Report-2012 pp 251
Rawat JS (2012) Climate change, Uttarakhand state of the environment report-2012 pp 247
Singh JS (2006) Sustainable development of the Indian Himalayan region: Linking ecological and economic concerns. Curr Sci 90(6):785
https://www.usgs.gov/special-topics/water-science-school/science/precipitation-and-water-cycle
Tiwari PC, Joshi B (2015) Climate change and rural out-migration in Himalaya, Change Adaptation Socioecol. Syst 2(8–25):5

Chapter 14
Impacts of Climate Warming on Biodiversity, Ecosystem Functionality, and Services in the Himalayan Landscape

Vinod Chandra Joshi, R. C. Sundriyal, and Dhani Arya

Abstract The Himalayas are not merely a geological wonder but a biodiversity hotspot of global significance, housing unique flora and fauna. The unprecedented pace of climate warming in the Himalayan region poses significant threats to biodiversity, ecosystem functionality, and the critical services these ecosystems provide, ultimately affecting the livelihoods of the people inhabiting this region. It was scrutinizing the fragmentation of habitats, emphasizing the need for holistic conservation efforts. Furthermore, the study underscores the importance of Himalayan ecosystems in providing essential provisioning services, such as freshwater supply, agriculture, and forest resources. It demonstrates how climate warming is altering hydrological regimes, affecting crop yields, and influencing the distribution of forest species. In conclusion, this study highlights the interconnectedness of ecological systems in the Himalayas and the urgent need for collaborative efforts to address the consequences of climate warming. It serves as a clarion call for governments, conservation organizations, and communities to safeguard this ecologically significant region and ensure the sustainability of its services for current and future generations. The challenges are great, but with concerted action, there is hope for preserving the Himalayas' invaluable biodiversity, functionality, and services in the face of climate change.

Keywords Climate changes · Biodiversity · Ecosystem services · Indian Himalayan region

V. C. Joshi (✉)
Wildlife Institute of India, Chandrabani, Dehradun, Uttarakhand, India
e-mail: vinodjoshi537@gmail.com

R. C. Sundriyal
Department of Forestry and Natural Resources, HNB Garhwal University Srinagar, Srinagar, Uttarakhand, India

D. Arya
Department of Botany, Soban Singh Jeena University Campus, Almora, India

D. Arya et al. (eds.), *Climate Change Impact on Himalayan Biodiversity*, Environmental Science and Engineering, https://doi.org/10.1007/978-3-031-77149-1_14

14.1 Introduction

The Himalayan Mountain range, often referred to as the "Roof of the World," is not only a geological wonder but also a biodiversity hotspot of global significance (Khuroo and Dar 2020). Its unique topography and climatic diversity have led to the evolution of a rich and diverse array of species, making it one of the most biologically vibrant regions on Earth. However, the Himalayas are not immune to the effects of climate change. This chapter delves into the multifaceted impact of climate change on the biodiversity, ecosystem function and structure of the Himalayan region, drawing upon a wide range of references and research to elucidate the current state of affairs. Climate change stands as a formidable global challenge, encompassing a far-reaching threat that is destined to present substantial environmental, economic, social, and political hurdles for many years, if not decades, ahead (Parry 2007). It has firmly established itself as a worldwide environmental concern that commands the attention of the global community due to its impact on the shared global atmosphere. Scientifically, it remains one of the least predictable phenomena, and its repercussions are poised to disproportionately affect the most vulnerable and impoverished individuals, those who have contributed the least to the primary drivers of climate change (Kattel 2022).

Mountains, in this context, serve as early indicators of climate change (Krishnan et al. 2019). As glaciers retreat and snowlines ascend, the flow of rivers is poised to undergo alterations, potentially giving rise to a host of social challenges. These changes may have far-reaching consequences, encompassing threats to hydropower generation, the endangerment of biodiversity, impacts on livelihoods dependent on forestry and agriculture, and the overall well-being of the affected populations (Negi et al. 2017). Before we can examine the impact of climate change on biodiversity, it is essential to understand the changing climate patterns in the Himalayas. In the Himalayan region it was observed a change in temperature, precipitation, and glacial melt rates, substantiating these insights with relevant scientific studies (IPCC 2019; Chandel and Ghosh 2021). The rising temperatures in the Himalayas have triggered significant shifts in vegetation zones and changing distribution of flora in the region. Some of the highlighting studies that have observed alterations in alpine and subalpine ecosystems (Baniya et al. 2009). Climate change poses a substantial threat to many species in the Himalayas, particularly those with specialized habitat requirements or narrow altitudinal ranges (Manish et al. 2016). As climate change drives species to seek more suitable habitats, it leads to habitat fragmentation. Climate warming influenced the composition, structure and function (Negi et al. 2021).

It has been proven that the Himalayan mountain is one of the ecosystems that is most susceptible to anthropogenic stress and climate change, and the local extinction of species may have more detrimental ecological effects. However, the structure and function of mountain ecosystem are going to threatened due to various anthropogenic and natural changes (i.e., land-use and land cover changes to over harvesting of natural resources and increasing impacts of climatic changes. Therefore, the collective impact of these two factors (i.e., anthropogenic and climate change) significantly alters the capability of Himalayan regions to offer perilous services. Himalaya region, besides these changes in land use, habitat fragmentation and environmental stresses often lead to reduction of biodiversity in natural ecosystem (Schulze and Mooney 2012; Negi et al. 2018). The ecosystem services depend on species composition, diversity, richness, distribution pattern, and function, which is believed to be changing over time (Manral et al. 2018). In the present study we asses the impact of climate warming of biodiversity, structure, function and services of Himalayan landscape.

14.2 Climatic Warming and Biodiversity Losses in Himalayan Region

Climate warming is poised to emerge as one of the most significant catalysts for biodiversity declines in the in the Himalayan region during the twenty-first century. Limited research has been conducted on the impacts of climate change on biodiversity losses in this region. Natural fluctuations in climate, such as reduced precipitation and escalating temperatures, have manifested in the retreat of Himalayan glaciers in response to recent warming trends, imposing severe threats to terrestrial biodiversity. For instance, rapid climatic warming has led to advances in vegetation phenology and an increase in greenery. Simultaneously, it has caused moisture stress, affecting factors like plant growth, seedling survival, and vegetation browning (Sigdel et al. 2018). Montane forests, located at higher altitudes in the Himalayas, display high sensitivity to climate change, thereby placing the species inhabiting these regions at risk due to rising temperatures. The warming trend is expected to be more pronounced at higher altitudes, with potential future impacts on the structure and functions of ecosystems (Behera et al. 2019). Continued warming has brought about transformations in plant assemblages. In the Sikkim Himalaya, for instance, up to 87% of 124 endemic species have undergone geographical range shifts driven by warming, resulting in species extinction (Telwala et al. 2013).

Changes in species richness and beta diversity have exhibited robust responses to climatic warming (Bhattacharjee et al. 2017), particularly affecting species in the treeline ecotone. Species like *Betula utilis* (Himalayan birch) have shown a tendency to shift towards the Eastern Himalayas while declining in numbers towards the west (Hamid et al. 2018). The combined and interactive effects of photosynthetically active radiation (PAR) and increased CO_2 levels in the treeline ecotone have led to species redundancy, causing a shift in dominance away from deciduous *Betula utilis* (birch). According to Behera et al. in 2019, the transition from deciduous to evergreen *Rhododendrons* has been noted. In the lesser Himalayas, there have been reports of a negative net ecosystem exchange (NEE) in Sal forests due to an increased susceptibility to climate-induced wildfires, as highlighted by Behera et al. in the same year. The far-reaching consequences of alterations in water resources, agricultural practices, and the rising prosperity of Himalayan communities under the influence of climatic warming have further exacerbated the loss of biodiversity, as indicated by Negi et al. (2022). While anthropogenic threats continue to pose a high risk to the flora and fauna of the Himalayas, it is projected that climatic warming will exert the most severe and direct impacts on biodiversity loss and a subsequent reduction in ecosystem services within the HKH region in the future (Behera et al. 2019). Empirical evidence suggests that the loss of biodiversity can detrimentally affect the functioning of ecosystems and the services they provide, as demonstrated by (Negi et al. 2021).

Climate warming has led to noticeable changes in the phenology of Himalayan trees (Negi et al 2021). The temperatures are altering the timing of bud burst, leaf emergence, flowering, and fruiting in these forests. Relevant studies are cited to support these observation (Shrestha et al. 2012). Changes in tree phenology have cascading effects on Himalayan ecosystems. Altered flowering and fruiting patterns can impact pollinators and seed dispersers, which, in turn, affect plant and animal species (Dahal et al. 2022). The changing tree phenology in Himalayan forests, driven by climate change, is a clear indicator of the region's vulnerability to global environmental shifts. To preserve the ecological balance and protect the rich biodiversity of the Himalayas, it is imperative to prioritize sustainable land management, conservation initiatives, and collaborative research efforts. The insights gained from these studies can inform adaptive.

14.3 Ecosystem Function and Ecosystem Services: Case of Himalayan Landscape

Ecosystem function and ecosystem services in the Himalayas are intimately connected and play a crucial role in sustaining both the environment and the livelihoods of the region's inhabitants. The ecosystem services broadly categorized into four group; Provisioning; such as food, fodder, fuelwood, fresh water, plant genetic resources, fiber, biochemicals and freshwater to the human wellbeing; regulating; carbon sequestration, climate regulation, pollination, seed dispersal and paste and control; cultural, such as spiritual and researchable benefit, educational and inspirational and recreational and aesthetic values; supporting; such as primary production, nutrient cycling (Soil nutrient, decomposition), provision of habitat, biogeochemical cycling, and production of atmospherics oxygen (Table 14.1).

Human being is part of ecosystem and they operate within it their collective e effect however can influence external factor like climate. The functionality of the Himalayan ecosystem is predominantly influenced by the inherent variations in factors such as temperature, soil composition, nutrient content, substrate quality, elevation, pH levels, light availability, water clarity, mineral presence, and conductivity. For instance, within the high-altitude Himalayas of South Asia, in regions like the rainfed wheat-paddy cropping system, the sustainability of agriculture greatly benefits from the interactions between soil and microorganisms. Ecosystems in the Himalayas, including forests and wetlands, play a vital role in recycling nutrients.

Table 14.1 Profile and explanation of various ecosystem services emanated from the Himalayan forests

Provisioning services: Direct use products derived from forest ecosystems	
• Genetic resources (including crop improvement genes, and health care)	Forests act on reservoir of biodiversity comprising herbs, shrubs, trees, and other genetic material of tangible or future value. All genetic resources support diverse livelihood and income need of communities
• Raw materials	Timber, pulp, pole, firewood, fodder, organic matter, fertilizer, material for fencing, forest floor litter for animal bedding etc. for commercial and/or households needs
• Water resource	Forests support provision of water supply, rain interception and fog (mist) and captured by soil. This influence surface runoff, interflow, and ground flow, extent of woody debris, litter, and humus impact water interception and storage
• Medicinal resources	Forest ecosystems provide a large variety of medicinal plants to cure various health aliments along with supplying raw materials for producing pharmaceuticals
• Other non-timber forest products (NTFPs)	Edible fruits, flowers, vegetables, nuts, berries, foliage, pith, honey, edible fungi, spices, beverages, bark, wax, gums, resins, oils, etc
Regulating services: Regulatory functions that forest ecosystems provide	
• Climate regulation	Forests capture carbon (CO_2) and store it (sequestration) thus help to reduce climate change. Also, protect and regulate effects of extreme temperature, precipitation, wind, ultra-violet light
• Control of flood and erosion	Forests control flood, stream-flow and soil erosion though moderation of rainfall, thus play key role in managing hazards
• Maintenance of soil fertility	Forests maintain fertile soils through accumulation of leaves and litter that decompose at the soil surface. In hills, forest litter is frequently collected for animal bedding that act as a source of farm yard manure
• Control of disease and pest regulation	Varied species composition and age class distribution of tree individuals greatly reduce the risk of pest and pathogens
• Pollination	Trees, woodlands and forests provide habitat for pollinator species
• Purifying air, water, soil, and noise	Forest helps to absorb pollution from air, water and soil, as well control noise pollution

(continued)

Table 14.1 (continued)

Provisioning services: Direct use products derived from forest ecosystems	
Supporting services: Ecosystem processes that support and underpin provisioning, regulating and cultural services	
• Habitat for species	Forest support habitats to species comprising plants, animals, microbes, and also support their needs of food, air, water, and shelter to survive
• Primary production	By the process of photosynthesis, plants fix CO_2 and produces organic matter and O_2, and also result in plant growth
• Biodiversity	Biodiversity support flora and fauna thus contribute immensely in forest dynamics
• Soil formation	Accumulation of organic matter and its decomposition along with the base material support soil formation
• Nutrient cycling	Forest helps in nutrient cycling through leaf fall, decomposition, and uptake. Thus, nutrients translocate from soil to plant and again back to soil through
• Water cycling	Forest helps in hydrological cycle
Cultural services: Non-material well-being benefits which people gain from forest ecosystems	
• Aesthetic and cultural value	Comprise scenic value such as purified air meditation, relaxing, refreshing, stimulating, and spiritual environment
• Mental and physical health	Forest helps controlling anxiety by enhancing enjoyment, recreation, and leisure actions
• Tourism and landscape	Forests provide important tourist destinations particularly in protected areas to enjoy the wildlife and landscapes brings local communities
• Education and learning	Forests and woodlands form an important sector of formal leaning for diverse array of people, from naturalists, foresters, educationists, sociologists, and academicians
• Spiritual experience	Local communities exhibit strong spiritual relationship and historical associations with forests
• Economy	Forests helps livelihood and income by selling forests produces as well as recreation and enterprises

Trees, shrubs, and other vegetation take up nutrients from the soil and release them back into the ecosystem through processes like litter decomposition and nutrient cycling, supporting plant growth and productivity (Joshi et al. 2022a, b). The intricate exchange of nutrients between plants and soil encompasses a range of vital parameters, such as soil organic carbon, microbial carbon, microbial nitrogen, microbial phosphorus, soil respiration, leaf nitrogen (N), leaf phosphorus (P), and leaf N/P ratios. These elements have been observed to play a role in facilitating the processes and functionality of forest ecosystems, ultimately resulting in increased biomass productivity. This connection between leaf characteristics and soil attributes has a profound impact on both above-ground and below-ground ecosystem processes (Joshi et al. 2021). Himalayan forests serve as crucial carbon sinks, effectively capturing and retaining substantial amounts of carbon dioxide from the atmosphere, thereby contributing significantly to climate change mitigation (Joshi et al. 2021). Numerous factors influence the carbon stock within Himalayan forests, including the type of forest, elevation, and forest management practices.

Climate change poses both challenges and opportunities for carbon stock and sequestration in Himalayan forests. Altered precipitation patterns, temperature changes, and shifting vegetation zones can impact the carbon balance (Poudel et al. 2019). In addition, when evaluating the biodiversity and biomass within the forests of the Dachigam National Park (DNP) in the Indian Kashmir Himalaya, it becomes apparent that pine forests, particularly those comprising *Pinus wallichiana*, are of paramount importance. These pine forests are recognized as a forest type with the highest ecological value, as substantiated by empirical evidence (Haq et al. 2020). However, the Himalayan ecosystem's functionality faces continuous threats from climate change and other human disturbances, leading to alterations in ecosystem services. As temperatures rise, species adapted to or favouring higher temperatures gradually shift towards higher altitudinal gradients, while those accustomed to lower temperatures struggle to maintain their ecological functions (Regina 2022). A case in point is the recorded baseline vegetation dynamics in the Kashmir Himalayas from 1960 to 1990, which revealed significant shifts in ecosystem functioning and subsequently, ecosystem services. Grasslands and tropical deciduous forests have been severely impacted by recent climate warming, resulting in the colonization of savannahs, shrublands, temperate evergreen broadleaf forests, boreal evergreen forests, and mixed forests primarily in the cold desert, rock, and ice land cover areas within the region, as reported by Rashid et al. (2015).

It is widely acknowledged that changes in temperature and precipitation patterns, including seasonal variations and intensity, have substantial effects on long-term water availability, cryosphere conditions, and evapotranspiration, thereby influencing the functioning of the Himalayan ecosystem and ecosystem services (Kattel 2022). Altered timing of the monsoon (July–September) and snowpack accumulation or melt seasons (March-June), as highlighted by Ali et al. 2023), has disrupted hydrological regimes, impacting the overall hydrology of Himalayan watersheds. The region's hydropower projects in Bhutan, China, India, Nepal, and Pakistan, for instance, are critical ecosystem services generated by Himalayan river systems. However, the decreasing freshwater ecosystem functionality due to reduced discharge poses a severe threat to the region's hydropower potential, as discussed by Khatiwada et al. (2016). Dependence on runoff patterns and the increasing unpredictability of seasonal flows, characterized by drier winters and shorter, more intense monsoons, heightens the potential risks to drinking water supplies, hydropower reservoirs, and crop irrigation in the area, as emphasized by Gautam et al. (2014).

14.4 Vulnerable Himalayan Ecosystem Services in the Face of Climate Warming

Governments worldwide have the responsibility of crafting strategies to attain nationally prioritized objectives that encompass social, economic, and environmental aspects of sustainability. Building upon the achievements made during the era of the UN Millennium Development Goals, the Sustainable Development Goals (SDGs) constitute a universally endorsed framework comprising 17 goals, 169 interconnected targets, and over 200 related indicators that delineate the roadmap for addressing the challenges of sustainable development by 2030 (Wood et al. 2018). It is imperative that the global community, with a particular emphasis on the governments of Himalayan nations and the multilateral organizations operating in the region, take proactive measures and assume accountability for evaluating and safeguarding various ecosystem services threatened by climate change. Such actions are crucial to secure the well-being of future generations.

14.5 Provisioning Services

The Indian Himalayan region, encompassing a vast and diverse landscape, is crucial for its provisioning services that sustain millions of lives across South Asia. These services include freshwater supply, agriculture, and forest resources, all of which are susceptible to the impacts of climate change. This comprehensive review examines the evidence of climate change-induced alterations in provisioning services within the Indian Himalayan region, drawing upon a range of references and studies (Pachauri et al. 2014). Millennium Ecosystem Assessment (MEA) 2005, emphasized the critical role of water as a fundamental provisioning ecosystem service (Carpenter et al. 2009). In the Himalayas, water resources are currently facing substantial exposure to the impacts of climate change, necessitating immediate action for effective management. Several countries in the Himalayan region have embraced the sixth Sustainable Development Goal (SDG-6), which aims to ensure universal access to water and sanitation by 2030. Beyond water, the importance of forests and other vital provisioning services is increasingly crucial for resource-scarce communities in the Himalayas. Forests serve as the primary source of household energy, satisfying approximately 97% of the region's energy needs (Thorn et al. 2020). In Nepal alone, the ecosystem services derived from community forests contribute to over 20% of rural households' income (Acharya et al. 2020). Collectively, forests and water resources in the Himalayan region play a pivotal role in providing sustenance and income. Medicinal plants constitute an integral component of indigenous medical systems and livelihoods, having long been harvested, consumed, and conserved by indigenous populations throughout the Himalayan region (Kuniyal et al. 2021). Medicinal plants contribute to healthcare for up to 80% of the population (Kuniyal et al. 2021). These plants are employed in treating a wide range of ailments, including those related to the digestive, respiratory, urinary, blood-circulatory, and reproductive systems, as well as skin-related conditions (Tiwari et al. 2023). The Indian Himalayas are the source of major rivers that provide water to millions of people downstream. Changes in temperature and precipitation patterns are affecting the timing and availability of freshwater (Yao et al. 2021). The impacts of climate change on provisioning ecosystem services in the Indian Himalayan region are complex and multifaceted. Changes in temperature, precipitation, and glacial melt rates are significantly affecting freshwater resources, forests, agricultural productivity, livelihoods, and communities. It is imperative that policymakers, scientists, and local stakeholders collaborate to develop adaptive strategies that protect and enhance these vital services in the face of ongoing climate change.

14.5.1 Regulating Services

Himalayan forests are important sink of CO_2 and play a significant role in mitigation of climate change impacts. The C pool in Indian Himalaya forests (tree biomass + soil) has been estimated at 5.4 billion t and these forests can sequester C @ 65 mt/yr in aboveground biomass alone worth Rs. 375 million/yr, and total value of C stock in forests being appx. Rs. 9400 million (Singh 2007). The Central Himalayan forests sequester carbon @ 5–9 t C ha^{-1} yr^{-1} for good forests to 1.5–3 t C ha^{-1} yr^{-1} for poor quality forests (Singh and Singh 1992). Shrestha et al. (2017) Pollination, the intricate process responsible for transferring pollen from male flowers to female counterparts, ultimately leading to fertilization and reproduction, stands as a vital regulatory service within the Himalayan ecosystems (Mukherjee et al. 2018). The significance of pollinators is deeply intertwined with human well-being, as they play a pivotal role in upholding ecosystem health and functionality, facilitating the reproduction of wild vegetation, crop production, and thus, sustaining our food sources (Negi et al. 2017). Both abiotic factors like wind and water, as well as a rich array of invertebrates (including bees, flies, butterflies, moths, wasps, beetles, and thrips), and vertebrates (such as birds, bats, other mammals, and lizards), actively participate in pollination, ensuring the continued functioning of the ecosystem in this region. However, the impacts of climate change and extensive land-use intensification practices, such as monocultures with heavy chemical usage, have detrimentally affected the connectivity, abundance, and diversity of these essential pollinators (Kovács-Hostyánszki et al. 2017). Consequently, the Himalayan region currently grapples with severe pollen limitations in its flowering plants (Jiang and Xie 2020), stemming from the decline in the populations of the Eastern honeybee (*Apis cerana*), which poses a serious threat to the regulation services provided (Potts et al. 2016). However, the high-mountain climate change and pollution from wastewater in the downstream river basins have diminished the self-purification abilities of these lakes and rivers (Rashid and Romshoo 2013). The interconnected network of lakes and rivers across the region holds immense potential for providing essential food and climate-regulating services (Kummu et al. 2014). Wetlands and river channels not only aid in retaining substantial water resources in lowland floodplains but also serve as ecological focal points within the glaciated landscapes (Jiang and Xie 2020). In the central Kashmir Himalaya of India, integrated cropping practices that include legumes, cereals, and apple cultivation have demonstrated their potential to mitigate the impacts of climate change by sequestering substantial soil carbon, with rates reaching as high as 46 t ha − 1 (Zahoor et al. 2021). Notably, several wetlands in the HKH region have been integrated into national greenhouse gas (GHG) emissions assessment programs due to their remarkable carbon sequestration potential. These wetlands, as integral components of the broader ecosystem, contribute to the intricate balance between ecological health, climate regulation, and human well-being in the Himalayas."

14.5.2 Supporting Services

The Indian Himalayan region, with its vast expanse of diverse ecosystems, serves as a crucial source of supporting services, contributing to the ecological balance, regional livelihoods, and the well-being of communities. However, the escalating impacts of climate change are challenging the resilience of these ecosystems and their capacity to provide essential services. The Himalayan terrestrial and aquatic ecosystems play a vital role in providing essential services, such as nutrient recycling, soil respiration, and stream metabolism, all the while preserving the region's rich biodiversity (Negi 2022) also found that forests contained higher soil organic carbon than other ecosystem types in Nepal. This ecosystem service is especially important in the context of climate change mitigation as it helps decrease atmospheric greenhouse gases. It is estimated that benthic invertebrates in this area contribute significantly, accounting for over 50% of these supporting services through their role in processing riparian leaf litter inputs into headwater streams. However, it's crucial to recognize that climate change and riparian deforestation are leading to adverse effects in the region. These changes are causing stream channels to narrow, increasing water velocity, reducing bed roughness, and altering nitrogen and phosphorus processing, as well as dissolved organic matter processing and overall net stream metabolism. Consequently, these alterations are diminishing the functional capacity of benthic stream ecosystems and the services they provide (Sweeney et al. 2004).

14.6 Cultural Services

The Indian Himalayan region, renowned for its breath-taking landscapes, diverse cultures, and rich traditions, is currently undergoing substantial transformations due to the influence of climate change (Wang et al. 2019). Mountainous areas cater to a growing demand for picturesque landscapes, offering a multitude of cultural ecosystem services to both their local residents and tourists (NITI Aayog 2018). Numerous national parks and conservation areas in the Himalayas not only safeguard biodiversity but also play a crucial role in providing cultural services, including recreation and environmental education for visitors, while also supporting sustainable livelihoods for local communities. National parks effectively balance the preservation of natural resources with the needs of the local community for their fundamental livelihoods (Rawal et al. 2013). Cultural ecosystem services within the Himalayas encompass a wide range of experiences, such as knowledge-based services like bird watching, religious activities like pilgrimage, and aesthetic or recreational pursuits like boating, rafting, swimming, hiking, fishing, off-road driving, skiing, and paragliding (Ament et al. 2017). Above all, the Himalayan region holds profound sacred and spiritual significance for Hinduism, Buddhism, Jainism, and Sikhism. For instance, Mount Kailash (6714 m) and Lake Mansarovar are revered by devotees from various faiths, including Hindu, Buddhist, Bon, and Jain, who make pilgrimages to these sites from all corners of the world. The Kailash Sacred Landscape (KSL) program by the International Canter for Integrated Mountain Development (ICIMOD) estimates the annual average use value of its cultural services to visitors and households in the KSL area to be USD 2.9 million (Das et al. 2017). The nature-based tourism industry in the Himalaya supports the local population, and, as a result, mountain communities actively engage in landscape design and protection, resource circulation and consumption, creating a social space for themselves by making their regions attractive to tourists (Apollo et al. 2022). Traditional Ecological Knowledge is deeply rooted in Himalayan cultures, offering a repository of wisdom on resource management, weather forecasting, and sustainable practices. However, climate change is challenging the relevance of this knowledge as shifting weather patterns and extreme events disrupt traditional practices (Ogra and Badola 2008). Additionally, traditional agricultural practices are intricately linked to local cultures (Joshi et al. 2022a, b). Climate change-induced shifts in temperature and precipitation patterns are affecting crop yields and the availability of specific crop varieties, thereby challenging both food security and agricultural traditions (Rai 2023). The impact of climate change on cultural services in the Indian Himalayan region presents a multifaceted challenge that deeply affects the core of local traditions and identities.

14.7 Conclusion and Way Forward

In conclusion, our comprehensive study on "The Consequences of Climate Warming on Himalayan Biodiversity, Ecosystem Functionality, and Ecosystem Services" has shed light on the intricate web of interactions between climate change and the Himalayan ecosystems. We have documented the profound impact of climate warming on the Himalayan biodiversity, with shifts in vegetation zones, endangerment of iconic species, disruption of avian migration patterns, and the fragmentation of habitats. These changes not only threaten the region's unique flora and fauna but also disrupt the delicate balance of its ecosystems. Also, our study underscores the importance of Himalayan ecosystems in providing essential services, including nutrient cycling, soil respiration, and stream metabolism. In the face of climate change, it is imperative to prioritize strategies that not only protect iconic species but also preserve the functional integrity of ecosystems. This includes initiatives aimed at habitat restoration, sustainable land management, and the reduction of greenhouse gas emissions. Furthermore, international collaboration and the sharing of knowledge and best practices are essential to address the complex challenges presented by climate warming in the Himalayas. Only through joint efforts can we hope to mitigate the impacts of climate change, protect the region's invaluable biodiversity, and secure the continued provision of ecosystem services upon which both local and global well-being depend.

References

Aayog NITI (2018) Report of working Group II sustainable Tourism in the Indian Himalayan Region https://www.niti.gov.in/writereaddata/files/documentpublication. *Doc2. pdf (last access: 25 January 2022).*

Acharya RP, Maraseni T, Cockfield G (2020) Assessing the financial contribution and carbon emission pattern of provisioning ecosystem services in Siwalik forests in Nepal: Valuation from the perspectives of disaggregated users. Land Use Policy 95:104647

Ali S, Kim BH, Akhtar T, Kam J (2023) Past and future changes toward earlier timing of streamflow over Pakistan from bias-corrected regional climate projections (1962–2099). J Hydrol 617:128959

Ament JM, Moore CA, Herbst M, Cumming GS (2017) Cultural ecosystem services in protected areas: understanding bundles, trade-offs, and synergies. Conserv Lett 10(4):440–450

Apollo M, Andreychouk V, Rawat K, Mostowska J, Jones TE, Rettinger R, Maciuk K (2022) Himalayan nature-based tourism. Challenges for tourism and protected areas. Int J Conserv Sci 13(1):249–266

Baniya CB, Solhøy T, Vetaas OR (2009) Temporal changes in species diversity and composition in abandoned fields in a trans-Himalayan landscape, Nepal. Herbaceous plant ecology: recent advances in plant ecology, 19–35

Behera MD, Behera SK, Sharma S (2019) Recent advances in biodiversity and climate change studies in India. Biodivers Conserv 28(8–9):1943–1951

Bhattacharjee A, Anadón J, Lohman D, Doleck T, Lakhankar T, Shrestha B, Thapa P, Devkota D, Tiwari S, Jha A, Siwakoti M, Devkota N, Jha P, Krakauer N (2017) The impact of climate change on biodiversity in Nepal: current knowledge, lacunae, and opportunities. Climate 5(4)

Chandel VS, Ghosh S (2021) Components of Himalayan river flows in a changing cimate. Water Resour Res 57(2):2020WR027589
Dahal KR, Dahal P, Adhikari RK, Naukkarinen V, Panday D, Bista N, Marambe B (2022) Climate change impacts and adaptation in a hill farming system of the Himalayan region: climatic trends. Farmers' perceptions and practices. Climate 11(1):11
Das S, Rai RK, Bhatta LD, Somanathan E, Kotru R, Khadayat MS, Negi GCS (2017) Valuation of ecosystem services in the Kailash Sacred Landscape. International centre for integrated mountain development (ICIMOD)
Gautam BR, Li F, Ru G (2014) Climate change risk for hydropower schemes in Himalayan region. Environ Sci Technol 48(14):7702–7703
Hamid M, Khuroo AA, Charles B, Ahmad R, Singh CP, Aravind NA (2018) Impact of climate change on the distribution range and niche dynamics of Himalayan birch, a typical treeline species in Himalayas. Biodivers Conserv 28(8–9):2345–2370
Haq SM, Calixto ES, Kumar M (2020) Assessing biodiversity and productivity over a small-scale gradient in the protected forests of Indian Western Himalayas. J Sustain Forestry, 1–20
IPCC (2019) Special report on the ocean and cryosphere in a changing climate
Jiang H, Xu X, Guan M, Wang L, Huang Y, Jiang Y (2020) Determining the contributions of climate change and human activities to vegetation dynamics in agro-pastural transitional zone of northern China from 2000 to 2015. Sci Tot Environ 718:134871
Joshi VC, Bisht D, Sundriyal RC, Pant H (2022) Species richness, diversity, structure, and distribution patterns across dominating forest communities of low and mid-hills in the Central Himalaya. Geo Eco Landscapes, 1–11
Joshi VC, Negi VS, Bisht D, Sundriyal RC, Arya D (2021) Tree biomass and carbon stock assessment of subtropical and temperate forests in the Central Himalaya, India. Trees, Forests and People 6:100147
Joshi VC, Pant H, Sundriyal RC (2022b) Traditional Nutrient management practices in the Marginalized agriculture sector of Kumaun Hills. ENVIS Bull Him Eco 27:56–58
Kattel GR (2022) Climate warming in the Himalayas threatens biodiversity, ecosystem functioning and ecosystem services in the 21st century: is there a better solution? Biodi Conser 31(8–9):2017–2044
Khuroo AA, Dar GH (2020) Biodiversity of the Himalaya: Jammu and Kashmir State. Springer Singapore
Krishnan R, Shrestha AB, Ren G, Rajbhandari R, Saeed S, Sanjay J, Ren Y (2019) Unravelling climate change in the Hindu Kush Himalaya: rapid warming in the mountains and increasing extremes. The Hindu Kush Himalaya assessment: Mountains, climate change, sustainability and people, 57–97
Kovács-Hostyánszki A, Espíndola A, Vanbergen AJ, Settele J, Kremen C, Dicks LV (2017) Ecological intensification to mitigate impacts of conventional intensive land use on pollinators and pollination. Ecol Lett 20(5):673–689
Kuniyal JC, Kanwar N, Bhoj AS, Rautela KS, Joshi P, Kumar K, Singh HB et al (2021) Climate change impacts on glacier-fed and non-glacier-fed ecosystems of the Indian Himalayan Region: people's perception and adaptive strategies. Curr Sci 120(5):888–899
Manish K, Telwala Y, Nautiyal DC, Pandit MK (2016) Modelling the impacts of future climate change on plant communities in the Himalaya: a case study from Eastern Himalaya, India. Mode Ear Sys Envir 2:1–12
Manral V, Padalia K, Karki H (2018) Plant diversity and regeneration potential of three different forests of Central Himalaya. Curr World Environ: Int Res J Environ Sci 13(1):113–123
Mukherjee S, Aadhar S, Stone D, Mishra V (2018) Increase in extreme precipitation events under anthropogenic warming in India. Weather Clim Extremes 20:45–53
Negi GCS (2022) Trees, forests and people: the central Himalayan case of forest ecosystem services. Trees, Forests and People 8:100222
Negi GCS, Singh P, Singh SP (2021) Atmospheric warming-associated phenological earliness does not increase the length of growing season in Himalayan trees. For Sci 67(6):694–700

Negi VS, Maikhuri RK, Pharswan D, Thakur S, Dhyani PP (2017) Climate change impact in the Western Himalaya: people's perception and adaptive strategies. J Mount Sci 14(2):403–416
Negi HS, Kanda N, Shekhar MS, Ganju A (2018) Recent wintertime climatic variability over the North West Himalayan cryosphere. Curr Sci 760–770
Negi VS, Tiwari DC, Singh L, Thakur S, Bhatt ID (2022) Review and synthesis of climate change studies in the Himalayan region. Environ Dev Sustain 24(9):10471–10502
Ogra M, Badola R (2008) Compensating human–wildlife conflict in protected area communities: ground-level perspectives from Uttarakhand, India. Hum Ecol 36:717–729
Pachauri RK, Allen MR, Barros VR, Broome J, Cramer W, Christ R, van Ypserle JP (2014) Climate change 2014: synthesis report. In: Contribution of Working Groups I, II and III to the fifth assessment report of the Intergovernmental Panel on Climate Change. IPCC, p 151
Parry ML (Ed) (2007) Climate change 2007-impacts, adaptation and vulnerability: working group II contribution to the fourth assessment report of the IPCC (Vol 4). Cambridge University Press
Poudyal BH, Maraseni T, Cockfield G (2019) Impacts of forest management on tree species richness and composition: Assessment of forest management regimes in Tarai landscape Nepal. Appl Geogr 111:102078
Rai SC (2023) Food and livelihood securities in changing climate of the Himalaya (Vol 9). Springer Nature
Rashid I, Romshoo SA (2013) Impact of anthropogenic activities on water quality of Lidder River in Kashmir Himalayas. Environ Monit Assess 185:4705–4719
Rashid I, Romshoo SA, Chaturvedi RK, Ravindranath NH, Sukumar R, Jayaraman M, Lakshmi TV, Sharma J (2015) Projected climate change impacts on vegetation distribution over Kashmir Himalayas. Clim Change 132(4):601–613
Rawal RS, Bhatt ID, Sekar KC, Nandi SK (2013) The himalayan biodiversity: richness, representativeness, uniqueness and life-support values. GB Pant Institute of Himalayan Environment and Development (GBPIHED), Kosi-Katarmal, Almora, Uttarakhand, India, 84 pp
Regina K (2022) The direct influence of climate change on marginal populations: a review. Aqu Sci 84(24) https://doi.org/10.1007/s00027-022-00856-5
Schulze ED, Mooney HA (eds) (2012) Biodiversity and ecosystem function. Springer Science & Business Media
Shrestha HL, Bhandari TS, Karky BS, Kotru R (2017) Linking soil properties to climate change mitigation and food security in Nepal. Environments
Shrestha UB, Gautam S, Bawa KS (2012) Widespread climate change in the Himalayas and associated changes in local ecosystems. PLoS ONE 7(5):e36741
Sigdel SR, Wang Y, Camarero JJ, Zhu H, Liang E, Penuelas J (2018) Moisture-mediated responsiveness of treeline shifts to global warming in the Himalayas. Glob Chang Biol 24(11):5549–5559
Singh JS, Singh SP (1992) Forests of Himalaya: Structure, functioning and impact of man. Gyanodaya Prakashan
Singh SP (2007) Himalayanforest ecosystem services. Central Himalayan Environment Association, Nainital, Uttarakhand, India, 53
Sweeney BW, Bott TL, Jackson JK, Kaplan LA, Newbold JD, Standley LJ, Hession WC, Horwitz RJ (2004) Riparian deforestation, stream narrowing, and loss of stream ecosystem services. Proc Natl Acad Sci USA 101(39):14132–14137
Telwala Y, Brook BW, Manish K, Pandit MK (2013) Climate-induced elevational range shifts and increase in plant species richness in a Himalayan biodiversity epicentre. PLoS ONE 8(2):e57103
Tiwari A, Khanduri VP, Rawat D, Singh B, Riyal MK, Mishra G, Kumar M (2023) Vegetation composition, soil properties, and carbon stock of montane forests along a disturbance in the Garhwal Himalaya, India. Front Forests Glob Change 6:1193792
Wang Y, Wu N, Kunze C, Long R, Perlik M (2019) Drivers of change to mountain sustainability in the Hindu Kush Himalaya. The Hindu Kush Himalaya assessment: Mountains, climate change, sustainability and people, 17–56

Zahoor B, Liu X, Ahmad B, Kumar L, Songer M (2021) Impact of climate change on Asiatic black bear (Ursus thibetanus) and its autumn diet in the northern highlands of Pakistan. Glob Change Biol 27(18):4294–4306

Chapter 15
Impact of Climate Change Adaptation on the Millet Crops: As a Climate Resilient and Sustainable Crops of Uttarakhand Himalaya

Prema Kaira, Nisha Tamta, N. C. Pant, Jyoti Joshi, and Deepak

Abstract Millets are sustainable and climate resilient crop in Uttarakhand. Millets are important staple cereal crop for the poor which are known as "cereals of the poor". These are very rich of nutrient, health conscious (low GI index), climate resilient crop and sustainable crop which have high efficiency in absorbing and utilizing carbon dioxide. Today, climate change has become a significant issue at the global level which has impact on major grain crops such as rice and wheat. Millets are adapted to high temperatures and little rainfall, are more resilient to climate change and millet crop cultivation has the potential to become a sustainable cereal crop. This paper focuses on impact of climate change adaptation on millet crops of Uttarakhand. The paper also describes how millets are climate resilient crops of Uttarakhand as a sustainable cereal crop. In this study compound annual growth rate (CAGR) has been calculated the area, production and productivity of the finger, barnyard and amaranth from 2000 to 2022 and these millets productivity ranking has been calculated by Kendall method. The main objectives of the study are identification of high yield productivity areas, examine of impacts of climate change and analyses of production and productivity of millet crops in different districts of Uttarakhand. Finger, barnyard and amaranth are major crops of Uttarakhand Himalaya in which finger has highest area, production and productivity from 2000 to 2022. The total CAGR of different millets is finger −1.24%, barnyard −1.05% and amaranth 8.15%. Uttarakashi, Rudraprayag, Chamoli, Deharadun (Hilly Area) and Champawat districts have high productivity ranking of Uttarakhand from 2000–2022. In this study the status of millet crops productivity from the time of formation of Uttarakhand till present (2022).

Keywords Millet crops · Climate change · CAGR · Productivity rank · Sustainable crops

---

P. Kaira (✉) · N. Tamta · N. C. Pant · J. Joshi · Deepak
Department of Geography, Soban Singh Jeena University Campus, Almora 263601, Uttarakhand, India
e-mail: premakaira655@gmail.com

D. Arya et al. (eds.), *Climate Change Impact on Himalayan Biodiversity*,
Environmental Science and Engineering, https://doi.org/10.1007/978-3-031-77149-1_15

15.1 Introduction

Millets are the member of *Poaceae* family that is small seeded cereal grown extensively in dry and tropical parts of Eurasia and Africa (Kheya et. al. 2023). Millets are one of the oldest foods known to human kind. Their charred grains have been found in the Harappen sites and they are also in mentioned in the Vedas. Millets were domesticated and cultivated as early as 10,000 years ago in Northern China (Yang et al 2005). In ancient time Millet have been used both for our consumption and as animal feed. In the western, central and eastern Himalaya, agriculture is practiced as terraced cultivation while, in the eastern extension of Himalaya, shifting cultivation is the main form of farming system (Sati 2015).

Millets are considered to be the most significant crop of cereal grains worldwide, behind wheat, rice, maize and barley (Ramashia et. al. 2021). Millets are important crop in relation to health/nutritional security and climate change and are well adopted in the farming system of most of farmers with small land holdings (McDonough et al. 2000). Relative to other cereals, millets have higher pest and disease resistance (Verma and Patel 2013).

The Himalayan regions of Uttarakhand have rich bio-resources consisting many medicinal plants, different species of cereals such as rice, maize, different major or small millets, wheat and pulses etc. Among these cereal crops millets is useful and important crop for human and animals. Millets as future gold crops will require well-planned and long-term public sector investment in multidisciplinary research activities by major growing countries (Meena et al. 2021). Millets are described as "Miracle Crops" because of their numerous advantages, including their use as food and food products with added value, forage, contribution to agro-diversity, low nutrient requirements, greater C sequestration (C4 plants), ability to prevent erosion in arid regions and confirmation of adequate supply of food and nutrition for smallholders who live in harsh environmental condition (Kumar and Singh 2015). Health problems are increasing due to people's interest in western food. Therefore, millets are the perfect food for people who care about their health because millets are gluten-free (Kumar and Singh 2015).

Uttarakhand is one of the leading states in the production of small millets producing 89 thousand tonnes (Indian Agriculture Statistical Report 2011). Various types of millet crops are produced in Uttarkhand Himalaya in which finger (Eleusine coracana), barnyard (Echinochloa frumentacea) and amaranth millets are mainly cultivated. These are traditional food of local people which is grown in both Kumaun and Garhwal divisions. About 80% of people in the Garhwal hills of the central Himalayas practice subsistence agriculture (Maikhuri et al. 2001).

Millets are traditional as well as become commercial crops of Uttarakhand. India is the leading producer of millets in the world followed by Niger and China (Weber Steve 1998; Takei 2013) Uttarakhand is one of the millet producing state after Rajasthan, Karnataka, Maharashtra, Uttar Pradesh, Haryana, Gujarat, Madhya Pradesh, Tamil Nadu, and Andhra Pradesh. The highest producing districts of millets in Uttarakhand are Almora (finger), Tehri (barnyard) and Chamoli (amaranth). The maximum millet production is consisting in hilly areas and these farmers are practiced predominantly subsistence farming because the hilly area did not favorable for cereal crops.

Climate change is refers to the long term shifts in temperature and weather condition. Its cause of increasing is natural and anthropogenic. Lithosphere, atmosphere, hydrosphere, biosphere, cryosphere are major components of climate change which affected millet crop. There are a few regions in the world where climate change might be as rapid as found in case of Himalayan region, and even very low where the penalty of climate change are likely to be as severe for biodiversity, ecosystems services and human well-being (Chaudhary and Bawa 2011).

Magrin et al (2005) in a study on Argentina find a positive impact of climate change on crop yields. Overall, the biophysical effects of climate change on agricultural production will be positive in some agricultural systems and regions, and negative in others, and these effects will also vary through time (Parry et al. 2004). This negative effect of temperature rise is especially pronounced in semitropical and tropical conditions because many crops are already at their tolerance limits of temperature in those regions (Jayaraman 2011).

The maximum millets area of Uttarakhand is unirrigated. Here maximum agricultural land is depending on climate which main components are temperature and rainfall patterns. These have more importance in determining of crop production and productivity. Changes in regional climate are will thus impact agricultural systems (Gornall et al. 2010). The climate and impact of climate is in different from one region to another. In areas where temperatures are already close to the physiological maxima of crops i.e. tropical regions, heat stress and water loss due to evapotranspiration may lead to decline of yields while areas in mid and high latitudes the suitability and productivity of crops are expected to increase (Olesen 2007).

Millets are known for their climate-resilient features including adaptation to a wide range of ecological conditions, less irrigational requirements, better growth and productivity in low nutrient input conditions, less reliance on synthetic fertilizers, and minimum vulnerability to environmental stresses (Kole et al. 2015). Also, millets are nutritionally superior to other major cereals as they are rich in dietary fibers, resistant starches, vitamins, essential amino acids, storage proteins and other bioactive compounds (Amadou et al. 2013). These attributes have made millets a crop of choice for cultivation in arid and semi-arid regions of the world.

Millets' climate-resilient characteristics continue to support sustainable agriculture without requiring chemical fertilizers (Thilakarathna and Raizada 2015). Traditionally, millets have been produced as rain-fed crops with little to no fertilizer input over a period of thousands of years (Devkota et al. 2016). It is cultivated mostly as a rainfed crop for its valued food grains, dry fodder and adaptability to a wide range of geographical areas (Ulaganathan and Nirmalakumari 2014; Negi et al. 2017a; Negi et al. 2017b).

A crop variety that has been specially bred or chosen to withstand the difficulties presented by climate change is known as a climate-resilient crop. Drought, heat exhaustion, harsh weather, and increased pest and disease pressures are some of these difficulties. The creation of climate-resilient crops is crucial because it can safeguard food security and enhance farmers' livelihoods, especially in areas where climate change has the potential to have a major influence on agricultural output.

Promoting sustainable agriculture, zero hunger and ensuring food security are main objectives of the SDGs'. The Food and Agricultural Organization of the United Nations (FAO) has emphasized millets' potential to support food and nutrition security, particularly in areas with high rates of malnutrition and poverty (FAO 2022).

15.2 Material and Methods

Present study is entirely based on secondary data compiled from Statistical Section of Statistical Agriculture Directorate Dehradun, Uttarakhand from period 2000–2022. The study is basically focused on finger, barnyard and amaranth millet crops in 13 districts with hilly and plain area of Uttarakhand. Therefore, 13 districts of area, production and productivity of these millets were studied. The study is focus on Compound Annual Growth Rate (CAGR) and millets ranking on based on Kendall method.

(i) **Compound Annual Growth Rate (CAGR)**: The CAGR was calculated using the following lon-lin model as given in-

$$Y_t = AB^t$$

where, Y_t = area/production/ yield of millets in time period

t = time.

A = constant.

B = (1 + r).

r = compound growth rate.

After log transformation and estimation of the above function as:

$$\ln Y_t = \ln A + t \ln B$$

Compound growth rate was estimated as: r = (Antilog B–1) × 100.

(ii) **Kendall's Ranking Coefficient method:**

The method used to arrange in sequence any given number of units growing the same range of crops (Mane and Gaikwad 2019). In the study rank are given according to yield per unit of area for millet in different districts of Uttarakhand. The method is useful for average rank of crops. In the study per hectare yields of finger, barnyard and amaranth of 13 districts with hilly and plain area in Uttarakhand are arranged according to ranks.

$$\textbf{Formularankingcoefficientindex} = \frac{\boldsymbol{R}1 + \boldsymbol{R}2 + \boldsymbol{R}3 \ldots\ldots\ldots\ldots \boldsymbol{R}15}{N}$$

where, R- Ranking of yield of individual Crop and N-Number of Crop

Objectives: The main objectives of study are:

1. To identify high yield productivity areas of millet crops.
2. To analyses the impacts of climate change on the millet crops.
3. To analyses the production, productivity and trends of millet crops.

15.3 Study Area

Uttarakhand is a Himalayan state of India which shares international and national boundaries with China and Tibet in the north, Nepal in the east part, Himachal Pradesh in the west, and Uttar pradesh in the south. Uttarakhand has an area of 53,483 km^2 and lies between 28°43′N to 31°27′ N latitudes and 77°34′ E to 81°02′ E longitudes and the recorded forest area in the state is 34,651 km^2, which constitute 64.79% of its total geographical area covered by the state (SFR, 2011). The total geographical area of Uttarakhand is 53,483 km^2 and it has 86% hilly (46,035 km^2) and 13.93% (7448 km^2) plain area. The state is divided into two administrative divisions, Garhwal and Kumaun, with a total of 13 districts which 10 districts are entirely hilly and 3 districts are plain. Physiographically, the state can be divided into five zones- the Grater Himalayas, the Lesser Himalaya, the Shiwaliks, the Bhabar and the Tarai region. The temperature of high altitude has −16.09 °C and 46.43 °C in low altitude areas (Fig. 15.3). The average rainfall in the state is 1550 mm (SFR 2009). The state consists of major occurring forest types such as Tropical Moist Deciduous, Tropical Dry Deciduous, Sub Tropical Pine, Himalayan Moist temperature, Himalayan Dry temperature, Sub Alpine and Alpine forests (SFR 2005). The study area is shown in Fig. 15.1. The total population of the Uttarakhand is 10,086,292 with a population density of 488 per Km^2.

15.4 Results

The main objective of the study is focused on impacts of climate change adaptation on finger, barnyard and amaranth millet of different districts in Uttarakhand. Temperature and rainfall climate as well as terrain, soil are the main component of millets which affects the agricultural production and productivity of specific region.

The results of the study analysis three key findings- First finger, barnyard and amaranth crops are heat tolerant millet crops of Uttarakhand. Second, Impact of climate change due to which the temperature has increased. Third, Millets are sustainable and climate resilient crop of Uttarakhand which has affected the productivity of millets. The highest millets productivity ranks of Uttarkhand

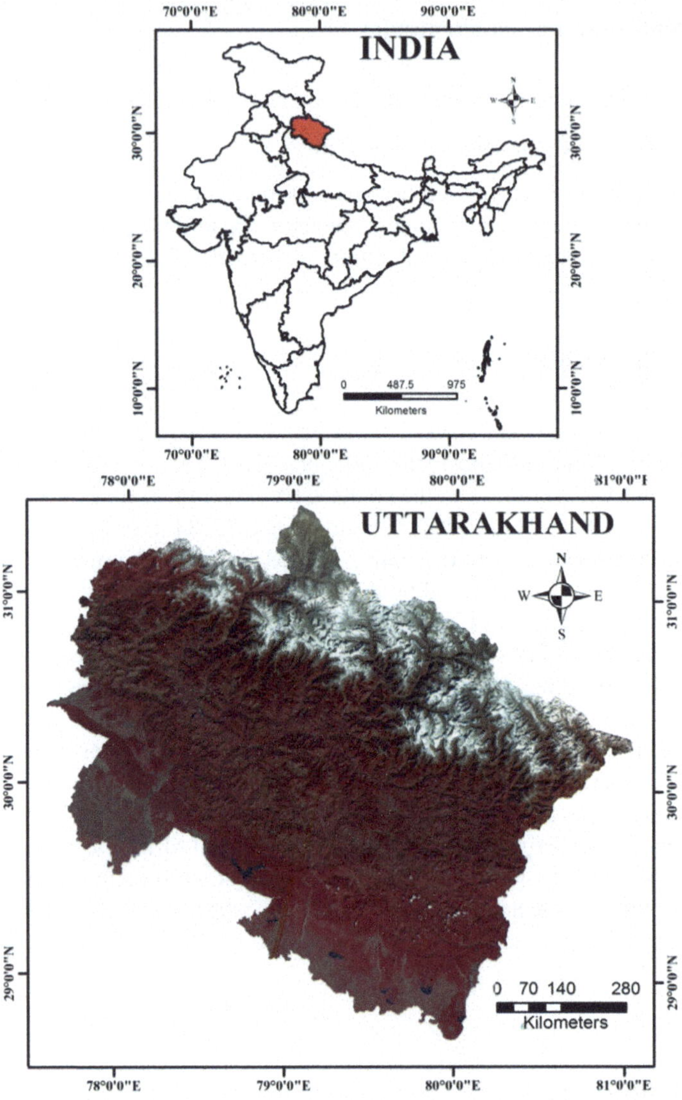

Fig. 15.1 Location map of the study area

are Uttarakashi, Rudraprayag, Chamoli, Deharadun (Hilly Area) and Champawat districts in 2000–2022 (Table 15.5 (b)).

Figure 15.2 depicts average temperature of different districts of Uttarkhand. The millets require minimum 8–15 °C to 26–29 °C mean temperature. The maximum average temperature is 23–26 °C U.S. Nagar and minimum average temperature is −5 to −12 °C in different year from 2000–2022 in Uttarakhand. Despite being maximum coldest district of Uttarkashi has the highest productivity of millet because

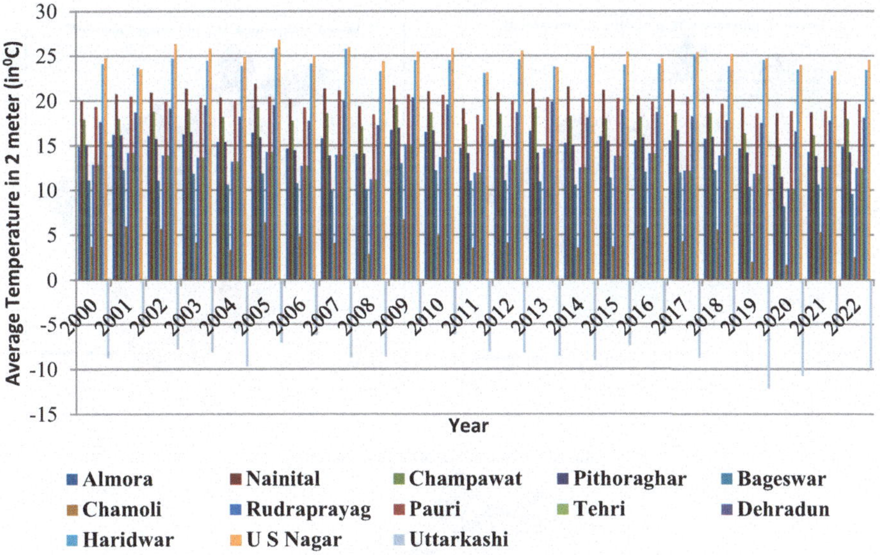

Fig. 15.2 Average temperature of different districts of Uttarkhand from 2000–2022 (*Source* NASA/POWER CERES/MERRA2)

millets are summer season crop. Uttarkashi is situated in Grater Himalayan region due to which the temperature is very low in winter season. The area and production has minimum in U.S. Nagar due to which their production has also minimum.

Climate change has caused increase heat, drought and glacial outbreak. Figure 15.3 depicts the annual temperature map of different decades based on physiographic regions of Uttarakhand. There has been a continuous increase in the temperature of Uttarakhand from 2000–2022. The annual temperature of Uttarakhand in 2000 was highest 37.85 °C in Tarai region. After this, the temperature increased continuously in year 2010 and 2020. According to which, the highest temperature in 2010 is 41.37 °C and 2020 is 46.43 °C in Tarai region of Uttarakhand, respectively. The maximum high temperature Tarai region is included plain districts of Uttarkhand. Thus, in the last 20 years, there has been continuous increase of about 5 °C at an interval of 10 years. The main reason for which is urbanization and industrialization in plain region of Uttarakhand.

15.5 Discussion

(I) **Millets as a Sustainable and Climate Resilient Crop:**

Millets are grown in low fertile soil, rainfed condition, high temperature, short duration, low efforts, hilly area with rocky and gravel soil, both low and high altitude,

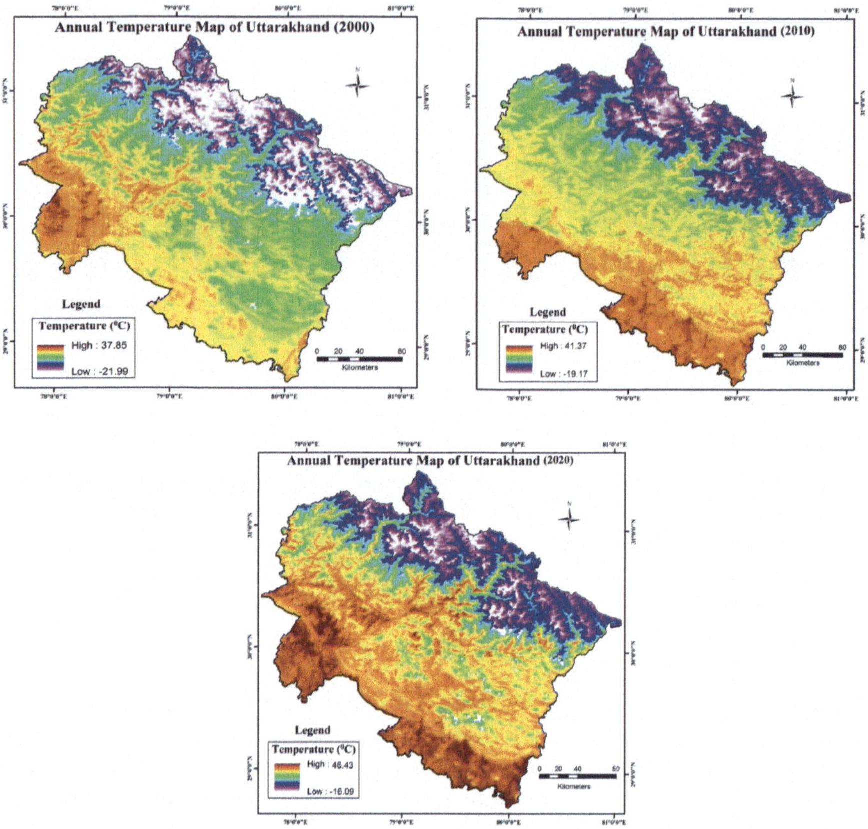

Fig. 15.3 Average temperature of different physiographic regions of Uttarakhand Himalaya from 2000–2022

non-irrigated land etc. Uttarkhand is a mountainous state, hence all these conditions are found here. Millets can become a sustainable and climate resilient crop for Uttarakhand. In recent years, there has been a renewed interest in millets as a climate-resilient and nutrition-rich crop, and as a means of enhancing food security and livelihoods in regions affected by climate change (Bhaskar et al. 2021). Therefore, due to the following characteristics, millets are a sustainable and climate resilient crop.

(i) **Nutritional Values:**

Millets are rich in nutrients. Millets are nutritionally superior to major cereals for carbohydrate and energy, and serve as a healthy balanced diet with good source of protein, high dietary fiber, vitamins, minerals in balanced proportion, antioxidants and micronutrients which helps to maintain our immune system which may act as a crucial shield against any disease (Malleshi 2001).

(ii) **Tolerant of Drought and High Temperature:**

Today, climate change remains a main problem for farmers. High temperature, erratic rainfall pattern, flood and drought are having an impact on cultivation up to a great extent. The major cereals crops depend on favourable climatic conditions and grow in long season. Millets are unique from major cereals crops due to their short growing season (Michaelraj and Shanmugam 2013; Sarkar and Dutta 2014) and it also tolerate high temperature and dry conditions. The drought tolerance of finger millet may be attributed to an efficient antioxidant potential and increased signal perception (Wambi et al. 2020).

(iii) **Higher Efficiency in Absorbing and Utilizing Carbon Dioxide:**

Climate change is occurring due to increase of carbon dioxide. Millets have higher efficiency in absorbing and utilizing carbon dioxide. These are the ideal crops for climate crisis because millets are C4 carbon sequestrating crops contributes to the reduction of CO_2 in the atmosphere; besides being water efficient so can withstand low soil moisture conditions in rainfed areas (Shalini Brijbhooshan and Didal 2019). Therefore, millets contribute to adaptation of climate change as it helps reduce the atmospheric carbon pressure CO_2.

However, the prevalence of stress conditions and their consequences are circumvented by several traits such as short stature, small leaf area, thickened cell walls, and the capability to form dense root system (Li and Brutnell 2011). In the C4 system, carbon dioxide (CO2) is concentrated around ribulose-1, 5- bisphosphate carboxylase/oxygenase (RuBisCO), which in turn suppresses ribulose 1, 5 bisphosphate (RuBP) oxygenation and photorespiration (Aubry et al. 2011). Thus, C4 mechanism enhances the concentration of CO_2 in bundle sheath, which suppresses photorespiration (around 80%) depending on the temperature and increases the in plant catalytic activity of RuBisCO (Sage et al. 2011).

(iv) **Health Conscious Food:**

Millets are rich in iron, zinc and other essential nutrients which are more important for pregnant women and children. Millets are low in Glycemic Index (GI) but nowadays people are interested in Western and Chinese food, which is mostly made flour. It is dangerous for diabetic patients. They are producing high percentage of glucose. Dietary fiber present in millets has water absorbing and bulking property which increases transit time of food in the gut thus help in reducing risk of inflammatory bowel disease and acts as detoxifying agent in the body (Bera 2021).

(v) **Soil and Water Conservation:**

Soil is one of main component of agriculture. Intercropping systems control soil erosion by preventing rain drops from hitting the bare soil where they tend to seal surface pores, prevent water from entering the soil and increase surface runoff (Seran and Brintha 2010). The roots of millet plants are longer and stronger compared to rice, wheat and pulses which soil erosion and land degradation can be mitigated. For example, in Burkina Faso, a mixed crop of sorghum and cowpea reduced runoff by 20–30% compared to sorghum alone and by 5–10% compared to cowpea alone, resulting in a reduction in soil erosion of 80 and 45–55%, respectively (Zougmore et al. 1998). The studies by Kujur and his associates on cropping system show that intercropping with specific plant geometry and selection of compatible crop with suitable varieties is profitable practice and to make proper use of available soil moisture and nutrients more efficiently and thus improve the productivity of rainfed crops.

(II) **Major Millets of Uttarakhand Himalaya:**

15.6 Finger

Finger belongs to the *Chenopodiaceae* family and it is also known as *Madua, Koda* and *Ragi* in different area of Uttarakhand. It is cultivated basically 1200–2800 m elevation and growing period of June to October in Kharif season. It is adaptable and resilient for climate change because it is heat loving plant. It requires minimum temperature is 8–15 °C for germination and 26–29 °C mean temperature during the growth of plant development and high productivity. It is drought-tolerant and can survive on low soil fertility (Purohit and Palai 2020). It has ability to grow in poor soil with low water resources areas. In Uttarakhand maximum districts depend on dryland/rainfed farming system. Millets are cultivated in all districts of Uttarakhand except Haridwar. The highest millets cultivate in districts of Uttarakhand are Almora, Pauri, Tehri, Uttarkashi and Chamoli.

15.7 Barnyard

Barnyard millet belongs to the *Chenopodiaceae* family. *Shownk, Sawa*, and *Jhingora* are the local name of barnyard in different regions of Uttarakhand. It is cultivated in between 1200–2800 m. It is grown in Kharif season (June to October). It has a rapid growth rate and is known to be drought-tolerant, making it well-suited to grow in areas with low rainfall (Maithani et al. 2022). Its basic climatic conditions and physical factors are same as finger millet. The plant's remarkable climatic resilience qualities allow it to thrive in tough and vulnerable locations while requiring minimum agricultural inputs (Maithani et al. 2022). Barnyard millet is cultivated only hilly district of Uttarakhand. U.S. Nagar and plain region of Nainital, Almora, Pauri and Tehri are containing highest area, production and productivity which Haridwar is the only district where in not produced in Uttarakhand.

15.8 Amaranth

Amaranth is a hardy crop that can grow both high and low altitudes of Uttarakhand. It is belongs to the Amaranthaceae family and cultivated in up to 2500 m elevation. Amaranth has the same climatic condition and physical factors as finger and barnyard. In different places of Uttarakhand it known by other names such as *Saliara, Seul,* and *Chaulai.* It has a number of properties that make it well-suited to survive in drought, hot and cold and dry environments (Al-Whaibi 2011). It is famous for local dishes *'Chaulai ladoo'* which is rich of protein iron and calcium. These offered to lord Shiva in Kedarnath. It is cultivated low area as compare finger and barnyard but its productivity has highest. The plain region of Nainital, U.S. Nagar, Haridwar and plain region of Dehradun are not cultivated amaranth in Uttarakhand. Uttarakashi, Chamoli, Tehri and hilly region of Dehradun are highly cultivated districts of amaranth.

Figure 15.4 represents district wise area, production and productivity of finger, barnyard and Amaranth millets in Uttarakhand from 2000 to 2022. The total area of millets in different district of Uttarakhand is 9,933,879 ha, production 4,997,151 mt. and productivity is 41.99 mt/ha. In which the finger millet total production is 33,16,607 mt under 24,93,772 ha area with the highest productivity millet crop of Uttarakhand at 19.85 mt/ha, barnyard is the second highest production millet crop 15,94,146 mt under 12,98,345 ha area with 15.55 mt/ha productivity and amaranth total production 86,398 mt under 1,41,762 ha area with 6.59 mt/ha productivity. The area under major millets is maximum in Almora (1,030,875 ha) and maximum production is also in Almora district (1,179,154 mt). The maximum productivity of millets is Uttarkashi (3.76 mt/ha). The highest production of finger recorded in Almora 8,26,509 mt and the lowest in U.S. Nagar 75 mt, barnyard in Tehri with 4,84,650 mt and Dehradun plain region is the lowest production district of Uttarakhand. Amaranth maximum production in Chamoli with 28,304 mt and the lowest production district Champawat with 104 mt.

Figure 15.5 represents area production and productivity of Finger in different district of Uttarakhand. The finger is first rank of millets in Uttarakhand. The area and production of finger is maximum in Almora (7,19,969 ha and 8,26,509 mt). Uttarkashi and hilly region of Dehradun district has the highest productivity (1.63 mt/ha) of the finger in the state. Finger is a Kharif season crop of Uttarakhand which is versatile millet grown in rainfed areas of hilly regions.

Figure 15.6 represented the area production and productivity of Barnyard in different district of Uttarakhand. The area and production of barnyard is maximum in Tehri (3,51,786 ha and 484,650 mt). Tehri produce 30.40% of barnyard in Uttarakhand and then Pauri 24.53%.

Figure 15.7 presented the area, production and productivity of amaranth in different district of Uttarakhand. Chamoli district has maximum area and production (45,929 ha and 28,304 mt) under Amaranth in the state. Chamoli produce 32.76% of amaranth in Uttarakhand. Uttarkashi and plain area of Dehradun are the major producers contributing 32.17% and 11.25% after Chamoli districts.

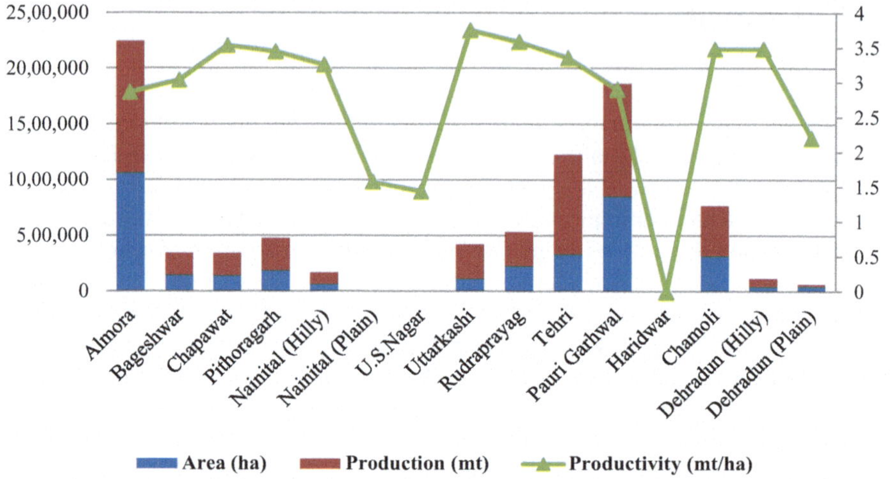

Fig. 15.4 District Wise total area, production and productivity of Millets (Finger, Barnyard and Amaranth) from 2000 to 2022 in Uttarakhand (*Source* Agriculture Directorate Dehradun, Uttarakhand)

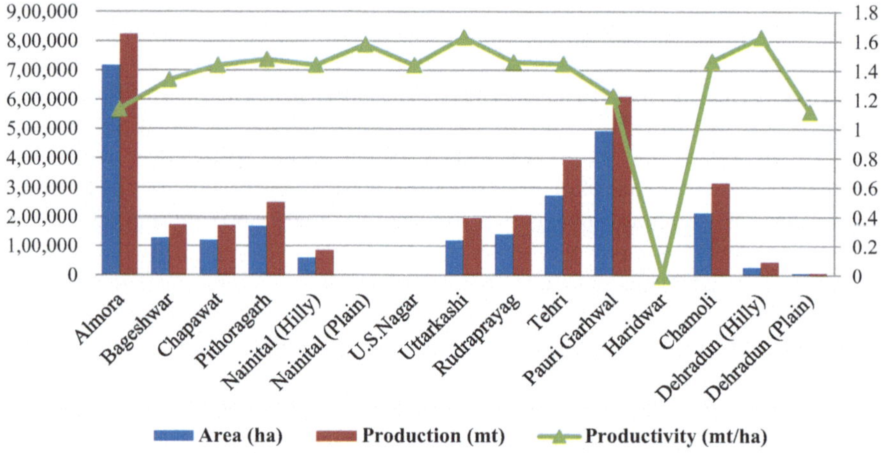

Fig. 15.5 District Wise total area, production and productivity of Finger Millet from 2000 to 2022 in Uttarakhand (*Source* Agriculture Directorate Dehradun, Uttarakhand)

Table 15.1 shown compound annual growth rate (CAGR) of area, production and productivity of millets in Uttarakhand. The area and production of millets have declined with −2.08% and −1.02% per annum from 2000 to 2022 but the productivity has increased with the growth rate of 2.42% per annum (Fig. 15.8). The status of decadal growth of millets in the state area under finger, barnyard and amaranth has in decreased with −1.29% and −3.32% during two decades but in 2021–2022 there

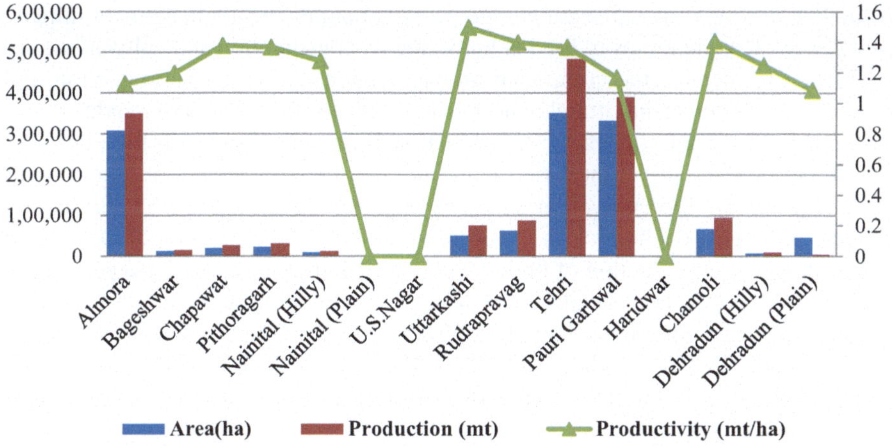

Fig. 15.6 District Wise total area, production and productivity of Barnyard Millet from 2000 to 2022 in Uttarakhand (*Source* Agriculture Directorate Dehradun, Uttarakhand)

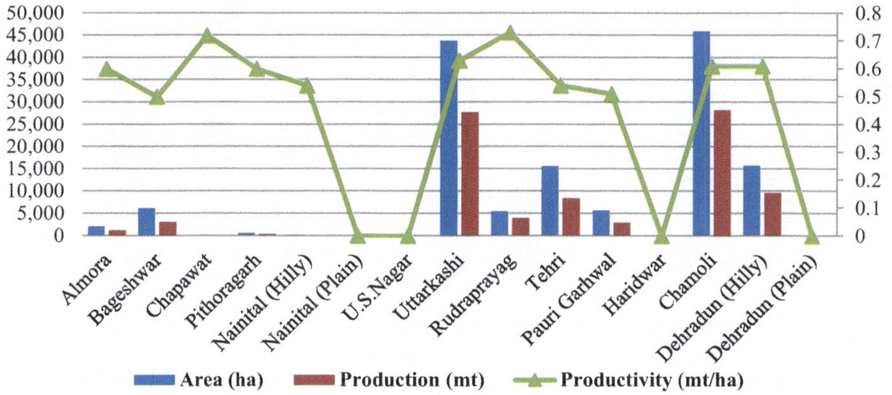

Fig. 15.7 District Wise total area, production and productivity of Amaranth Millet from 2000 to 2022 in Uttarakhand (*Source* Agriculture Directorate Dehradun, Uttarakhand)

has been an increase of 1.82%. The production status of different millets in the state is positive as compare in area. The production CAGR of decadal growth of millets has increased with 0.21% in 2000–2010 and 5.16% in 2021–2022 but 2011–2020 decadal growth of millet is decreased with the rate of −2.66%. The main reasons for the decline in the area and production of millets has been huge level of migration, construction activities in agricultural land and PDS scheme (rice, wheat, sugar etc.) through the mission of department of food and civil supplies. As a result, farmers are disinterested in cultivation and the total cultivable land is changing into barren. The cultivation of millets declined from 1,80,882 ha to 1,31,577 ha during 22 years. Apart from this, lack of knowledge among the people about the health benefits of millets, preference of staple grains in food is hindering its consumption and cultivation.

Table 15.2 depicts the compound annual growth rate of area, production and productivity of finger in Uttarakhand during 2000 to 2022. The area under finger declined with the rate of −2.04%, production is also declined with the rate of −1.24% but productivity is increased with the rate of 0.81% annum during 2000 to 2022 (Fig. 15.8). The cultivation area declined with 127,733 ha to 85,880 ha and production is also declined with 162,286 mt to 126,916 mt during 2000–2022 in Uttarakhand. The production of finder declined with the rate of −1.24% per annum during 2000–2022. The decadal growth in the production of finger revealed that production grew with growth rate of 0.61% in 2000–2010 and 0.81% during 2021–2022. It declined −3.2% during 2011–2020. The major reasons for the decline in area and production were migration, increase barren land, low of interest in agriculture etc.

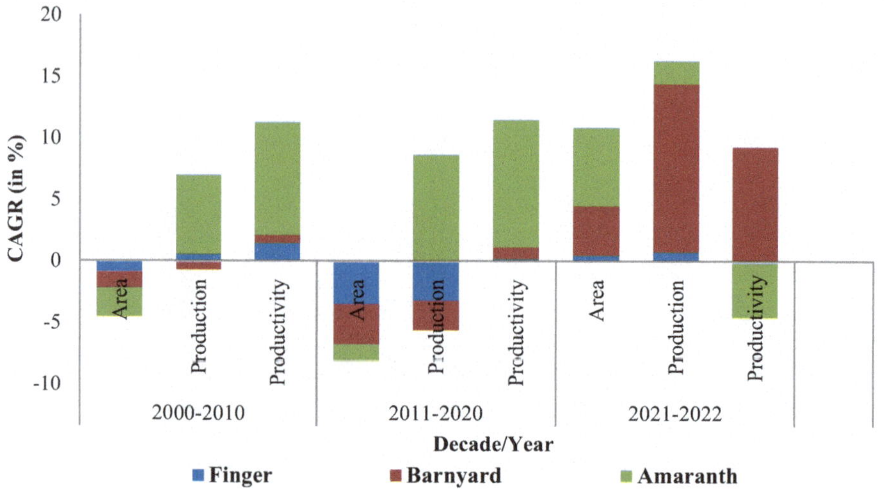

Fig. 15.8 Compound annual growth rate (in %) of area, production and productivity of Major Millets in different decades/Year of Uttarakhand

Table 15.1 Compound annual growth Rate (in %) of area, production and productivity of Millets in Uttarakhand

S. n	Year	Area	Production	Productivity
1	2000–2010	−1.07	0.21	1.88
2	2011–2020	−3.32	−2.66	2.53
3	2021–2022	1.82	5.16	2.21
4	2000–2022	−2.08	−1.02	2.42

Source Agriculture Directorate Dehradun, Uttarakhand

Table 15.2 Compound annual growth rate (in %) of area production and productivity of Finger in Uttarakhand

S. n	Year	Area	Production	Productivity
1	2000–2010	−0.89	0.61	1.53
2	2011–2020	−3.47	−3.2	0.25
3	2021–2022	0.53	0.81	0
4	2000–2022	−2.04	−1.24	0.81

Source Agriculture Directorate Dehradun, Uttarakhand

Table 15.3 shown CAGR area, production and productivity of barnyard of different decades. The CAGR in 2000–2022 of area and production of barnyard is decreased with −2.26% and −1.05%, respectively but the productivity is increased with 1.26% (Fig. 15.8). The area under barnyard has declined by −1.29% per annum during 2000–2010 and production have also declined by −0.68% but the productivity have increased by 0.63%. Next decade in 2011–2020 the area and production of barnyard also declined with −3.27% and −2.38%, respectively but the productivity is increased with 0.92%. The area, production is increased with 4%, 13.64% and 9.3, respectively in 2021–2022. The barnyard millet has seen growth in present year cause of awareness of people, motivation and policy by government.

Table 15.4 depicts the compound annual growth rate of area, production and productivity of amaranth in different decades of Uttarakhand. After, the formation of Uttarakhand over the period of 22 years the area under amaranth has declined by −1.25% per annum and the production and productivity have increased 8.15% and 9.62%, respectively (Fig. 15.8). The area under amaranth declined in two decades (2000–2010 and 2011–2020) except present year (2021–2022) and production and productivity is increased (Table 15.4). The area and production of amaranth in 2021–2022 has increased by 6.34% and 1.86%, respectively but productivity has decreased by −4.51%.

Table 15.5 (a, b) shows the Ranking of Yield Productivity of Millet Crops from 2000 to 2022 based on Kendall's ranking coefficient method in Uttarakhand. Mainly, the high millet crop productivity is found in five districts, namely Uttarakashi (0.4), Rudraprayag (0.66), Chamoli (0.8), Dehradun (Hilly Area) (0.86), and Champawat

Table 15.3 Compound annual growth rate (in %) of area, production and productivity of Barnyard in Uttarakhand

S. n	Year	Area	Production	Productivity
1	2000–2010	−1.29	−0.68	0.63
2	2011–2020	−3.27	−2.38	0.92
3	2021–2022	4	13.64	9.3
4	2000–2022	−2.26	−1.05	1.26

Source Agriculture Directorate Dehradun, Uttarakhand

Table 15.4 Compound annual growth rate (in %) of area, production and productivity of Amaranth in Uttarakhand

S. n	Year	Area	Production	Productivity
1	2000–2010	−2.33	6.4	9.15
2	2011–2020	−1.32	8.68	10.34
3	2021–2022	6.34	1.86	−4.51
4	2000–2022	−1.25	8.15	9.62

Source Agriculture Directorate Dehradun, Uttarakhand

Table 15.5 (a) Millet Productivity Rank Calculate from 2000 to 2022 Based on M. C. Kendell's Method in Uttarakhand

S. n	District		Finger	Barnyard	Amaranth	Total Rank	Ranking Coefficient	Productivity Rank
1	Almora		13	11	6	30	2	9
2	Bageshwar		11	9	11	31	2.06	10
3	Chapawat		10	4	2	16	1.06	5
4	Pithoragarh		4	6	7	17	1.13	6
5	Nainital	Hilly	8	7	8	23	1.53	8
		Plain	3	15	15	33	2.2	12
6	U.S. Nagar		9	15	15	39	2.6	13
7	Uttarkashi		2	1	3	6	0.4	1
8	Rudraprayag		6	3	1	10	0.66	2
9	Tehri		7	5	9	21	1.4	7
10	Pauri Garhwal		12	10	10	32	2.13	11
11	Haridwar		15	15	15	45	3	15
12	Chamoli		5	2	5	12	0.8	3
13	Dehradun	Hilly	1	8	4	13	0.86	4
		Plain	14	12	15	41	2.73	14

(b) Top Five Ranking of Yield Productivity of Millet Crops Based on Kendall's Ranking Coefficient Method in Uttarakhand

S. n	Productivity Grade	Name of Districts
1	High	Uttarakashi, Rudraprayag, Chamoli, Deharadun (Hilly Area), Champawat
2	Medium	Pithoragarh, Tehri, Nainital (Hilly area), Almora, Bageshwar
3	Low	Pauri Garhwal, Nainital (Plain area), U.S. Nagar, Dehradun (Plain area), Haridwar

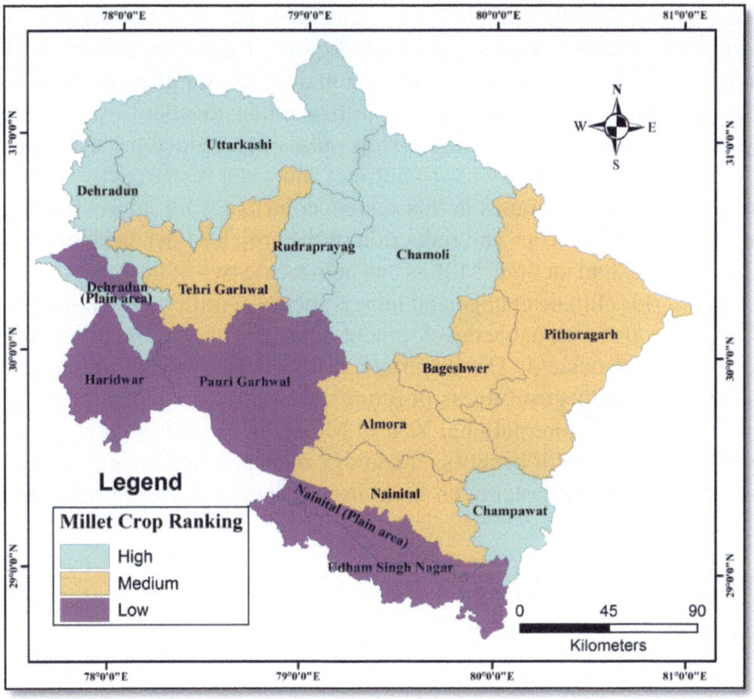

Fig. 15.9 High, medium and low productivity of Millet ranking on besed on Kendall's Method

(1.06) (Fig. 15.9). Concerning the study area, all high productivity regions have favorable climatic conditions for finger, barnyard, and amaranth millet crops. The medium millet productivity has been observed in the five districts mainly, Pithoragarh (1.13), Tehri (1.4), Nainital (Hilly area) (1.53), Almora (2) and Bageshwar (2.06) favorable conditions for all millet crops (Fig. 15.9). The very low millet crop productivity is found in the study area mainly, Pauri Garhwal (2.13), Nainital (Plain area) (2.2), U.S Nagar (2.6), Dehradun (Plain area) (2.3) and Haridwar (3.0) (Fig. 15.9).

15.9 Conclusion

The study indicated that the compound annual growth rate (CAGR) of millets productivity has increased of Uttarakhand in 2000–2022. The finger, barnyard and amaranth can become potential cereal crops of Uttarakhand. These crops CAGR have positive productivity. Almora is high production district of Uttarakhand in millets. Uttarakashi, Rudraprayag, Chamoli, Deharadun (Hilly Area) and Chapawat are Top five districts which have high productivity rank in millets. These millets are sustainable and climate resilient crop because the impact of climate change are low compare as other cereal crops. Millets can become viable option in dry land, minimum water ability, unfertile soil and high temperature regions. Uttarakhand hills are largely dominated by millets diversity contributing a major part in millet production in the country hence; there is a great scope to utilize millet diversity by intercropping with legumes. Intercropping of millets with pulses can reduce the use of external inputs due to the complementary use of nutrient and water resources by the intercrop components. The use of legumes in this system contributes for improving the nutrients status for the millet crops and reduction of the crop loss risk besides increasing overall productivity and income. Millets can help tackle two of the most Challenging issues of the world- climate change and hunger, because millets are climate resilient crop. Therefore, our Sustainable development goals and climate-resilient traditional crops are closely associated. The government have taken special initiative for the benefits of millets and its usefulness in future. Therefore, the Indian government is declared 2023 to be the International Year of Millets in an effort to promote millets' nutritional value and health benefits, encourage their production and consumption, and draw attention to their potential role in sustainable agriculture, food security and climate change adaptation.

References

Al-Whaibi MH (2011) Plant heat-shockproteins: a mini review. J KingSaud Univ Sci 23(2):139–150

Aubry S, Brown NJ, Hibberd JM (2011) The role of proteins in C3 plants prior to their recruitment into the C4 pathway. J Exp Bot 62:3049–3059

Amadou I, Gounga ME, Le GW (2013) Millets: nutritional composition, some health benefits and processing—a review. Emir J Food Agric 25:501–508

Bhaskar VV, Venkateswarlu B, Devi MD, Reddy GB (2021) Millets: potential for climate-resilient agriculture and food security. J Agric Meteorol 23(1):1–9

Bera A (2021) Importance of millets cultivation in the context of climate change, pp 1–5

Chaudhary P, Bawa KS (2011) Local perceptions of climate change validated by scientific evidence in the Himalayas. Biol Let 7:767–770

Devkota R, Khadka K, Gartaula H, Shrestha A, Karki S, Patel K, Chaudhary P (2016) Gender and labour efficiency in finger millet pro-duction in Nepal. In: Transforming gender and food security in the Globalsouth, Routledge

FAO (Food and Agriculture Organization), Millets, Retrieved from, http://www.fao.org/millet/en/, 2019. Accessed 20 December 2022

Gornall J, Betts R, Burke E, Clark R, Camp J, Willett K, Wiltshire A (2010) Implications of climate change for agricultural productivity in the early twenty-first century. Philos Trans Royal Soc 365:2973–2989

Jayaraman T (2011) Climate change and agriculture: a review article with special reference to India. Rev Agrarian Studies 1(2):16–78

Kole C, Muthamilarasan M, Henry R, Edwards D, Sharma R, Abberton M (2015) Application of genomics-assisted breeding for generation of climate resilient crops: progress and prospects. Front Plant Sci, 563

Kheya AS, Talukder KS, Datta P, Yeasmin S, Rashid H, Hasan KA, Anwar P, Islam A, Islam M (2023) Millets: the future crops for the tropics-Status, Challenges and future prospects, Heliyon 2

Kumar A, Singh P (2015) Millets: nutritional composition, some health benefits and processing—a review. J Food Sci Technol 52(5):2481–2495
Li P, Brutnell TP (2011) Setaria viridis and Setaria italica, model genetic systems for the panicoid grasses. J Exp Bot 62:3031–3037
McDonough CM, Rooney LW, Serna-Saldivar SO (2000) The Millets". Food science and technology: handbook of cereal science and technology. CRC Press. 99 (2nd ed), pp 177–210
Michaelraj PSJ, Shanmugam A (2013) A study on millets-based cultivation and consumption in India. Int J Marketing, Financ Serv Manag Res, 49–58
Mane SP, Gaikwad SB (2019) Agriculture productivity calculate based on M.G. Kendall's Method in Malshiras Tahsil, pp 145–151
Meena RP, Joshi D, Bisht JK, Kant L (2021) Global scenario of millets cultivation. In: Kumar A, Tripathi MK, Joshi D, Kumar V, (eds) Millets and millet technology. Springer, Singapore
Malleshi NG (2001) Nutritional qualities of millets and pseudo cereals, pp 17–26. In: Course material; of short-term course on processing and value addition to millets and pseudo-cereals, 19th-23rd Nov.2001. CFTRI, Mysore, pp 17–26
Maikhuri RK, Rao KS, Semwal RL (2001) Changing scenario of Himalayan agroecosystems: loss of agrobiodiversity- an indicator of environmental change in Central Himalayas. Environmentalist 21:23–39
Magrin GO, Travasso MI, Rodriguez GR (2005) Changes in climate and crop production during the 20th century in Argentina. Clim Change 72:229–249
Maithani D, Sharma A, Gangola S, Bhatt P, Bhandari G, Dasila H (2022) Barnyard millet (Echinochloa spp.): a climate resilient multipurpose crop. Vegetos 1–15
Negi S, Bhatt A, Kumar V (2017a) Character association and path analysis for yield and its related traits in finger milllet Eleusine coracana (L.) Gaertn genotypes. J Appl Natural Sci 9 (3):1624—1629
Negi S, Kumar V, Bhatt A (2017b) Morphological characterization and genetic analysis of finger millet (Eleusine coracana (L.) Gaertn) germplasm. Int J Bio-res Stress Manage 8 (3):469–472
Olesen JE (2007) Uncertainties in projected impacts of climate change on European agriculture and terrestrial ecosystems based on scenarios from regional climate models. Clim Change 81:123–143
Parry ML, Rosenzweig C, Iglesia LM, Fischer G (2004) Effects of climate change on global food production under SRES emissions and socio-economic scenarios. Glob Environ Chang 14:53–67
Purohit S, Palai JB (2020) Finger millet a boon to food security under changing climatic condition. I J Chem Studies 8(3):1385–1388
Ramashia S, Mashau M, Onipe O (2021) Millets Cereal Grains: Nutritional Composition and Utilisation in Sub-Saharan Africa
SFR (2005) State of Forest Report. Forest Survey of India, Dehradun
SFR (2009) State of Forest Report. Forest Survey of India, Dehradun
SFR (2011) State of Forest Report. Forest Survey of India, Dehradun
Sarkar S, Dutta SC (2014) Influence of fertilizer loaded nanoclay superabsorbent polymer composite (NCPC) on dynamics of P and N availability and their uptake by Pearl millet (Pennisetum glaucum) in an inceptisols. Int J Bio-Res Manage 5(2):221–227
Seran TH, Brintha I (2010) Review on maize based intercropping. J Agron 9(3):135–145
Sage RF, Christin PA, Edwards EA (2011) The lineages of C4 photosynthesis on planet Earth. J Exp Bot 62:3155–3169
Sati VP (2015) Natural resources potentials and socio-economic status in the Indian Himalayan region 20(2):48–58
Shalini Brijbhooshan, Didal VK (2019) Sustaining minor millet production in hilly areas of Uttarakhand through Intercropping of minor millets and pulses- a review 8(11):397–406
Takei E (2013) Millet culture and indigenous cuisine in Taiwan. The 2013 international conference on Chinese food culture, Kunming, Yunnan, China. 10
Thilakarathna MS, Raizada MN (2015) A review of nutrient man- agement studies involving finger millet in the semi-arid tropics of Asia and Africa, 262–290

Ulaganathan V, Nirmalakumari A (2014) Genetic variability and correlation studies for quantitative traits in finger millet (*Eleusine coracana* (L.) Gaertn germplasm. The Ecoscan 5(6):21–25

Verma V, Patel S (2013) Value added products from nutria-cereals, Finger millet (*Eleusine coracana*). Emir J Food Agric 25(3):169–176

Weber Steve SA (1998) Out of Africa: the initial impact of Millets in South Asia. Curr Anthropol 39(2):267–274

Wambi W, Otienno G, Tumwesigye W, Mulumba J (2020) Genetic and genomic resources for finger millet improvement: Opportunities for advancing climate-smart agricult ture. J Crop Improvem 35(2):204–233

Yang L et al (2005) Handbook of Chinese mythology. New York: Oxford University Press 70(131–135):198

Zougmore RF, Kamboun K, Ouattara S, Guillobez S (1998) The cropping system of sorghum-cowpea in the prevention of runoff and erosion in the Sahel of Burkina. International Development Research Centre, Ottawa, pp 217–224

Chapter 16
Role of Bryophyes in Various Ecosystems Services and Their Function as a Green Brain

Nisha and Ravindra Kumar

Abstract Bryophytes comprise nearly 23,000 species and are part of the most diverse group of terrestrial plants. They are the most important components of biodiversity and play a significant role in ecosystem dynamics. They are able to grow in a variety of habitats, including the upper part of the ground containing humus under the forest, on soil, on rocks, in crevices, on tree barks, dead logs, and on the branches of trees. They are adapted to survive on surfaces that have the least amount of humus. The bryophytes play an important role in the ecosystem, as they fill the gaps and maintain the moisture. They can render various ecosystem services, such as oil formation, which provides habitat to small organisms; nutrient cycling; soil erosion management; pest and disease control; and bio-monitoring of heavy metals, which is an indicator of pollution.

Keywords Brophytes · Ecosystem services; soil formation · Nutrient cycling · Heavy metals · Bioindicator

16.1 Introduction

Human civilization shares an incredible relationship with the ecosystem since time immemorial, they deriving services from the ecosystem (Combertia et al. 2015). These ecosystem services are "the functions and products of ecosystems beneficial for humans, or yield welfare to society" (Millennium Ecosystem Assessment 2005), this term 'ecosystem services' was coined in 1981 by Ehrlich and Ehrlich. However, the concept shot to prominence only in 1997 (Gomez-Baggethun et al. 2010) when ecosystem services were highlighted in the literature by Costanza and Daly (1992),

Nisha (✉)
Department of Botany, S.S.J. Govt. P.G. College Syalde, Almora 263661, Uttarakhand, India
e-mail: nishaarya24@gmail.com

R. Kumar
Department of Botany, Soban Singh Jeena University Campus, Almora 263601, Uttarakhand, India

D. Arya et al. (eds.), *Climate Change Impact on Himalayan Biodiversity*,
Environmental Science and Engineering, https://doi.org/10.1007/978-3-031-77149-1_16

Costanza et al (1997) and Daily (1997a, 1977b) (Nonya et al. 2022). As climate change and urbanization have a marked effect on biodiversity which going to change the ecosystems, it is important to understand the relationships among species and how these relationships will change is important. One powerful tool for estimate these effects of climate change on biodiversity has been use of empirical niche models to identify exposure to climate change (Schwartz 2012). However, because climate change and its effects on biodiversity is a complex issue, examining vulnerability, exposure, sensitivity, and adaptive capacity of a species or population is necessary to accurately predict losses and plan conservation methods (Dawson et al. 2011). The capacity of the ecosystem to provide goods and services to human society is directly interlinked with biodiversity (Mori et al. 2017). Biodiversity maintains a broad range of ecosystem services ES, such as climate regulation, food production, pollination, pest control etc. (Cardinale et al. 2012). In this regard the bryophytes are archegoniate, non-vascular, homosporous simple cryptogams which have no true stem, leaves and roots (Majumdar and Dey 2020). There are approximately 28,000 species of bryophytes around the globe thus constitute the second largest group in plant kingdom (Shaw et al. 2011). They are present in almost all habitats except sea and oceans and provide significant ecosystem services ES in each of them (Yayintas and Irkin 2018). Bryophytes are divided into three different phyla, namely, Marchantiophyta (liverworts), Anthocerophyta (hornworts) and Bryophyta (mosses) (Majumdar and Dey 2020).

Bryophytes include mosses, liverworts, and hornworts, are one understudied group of plants and expected to be negatively impacted by climate and land use changes. These are non-vascular plants poikilohydric in nature (they have the inability to regulate water content independently of their environment). However, research regarding effects of climate change on plants and land use changes is heavily skewed towards vascular plants (e.g., flowering plants and conifers). Due to differences between bryophytes and vascular plants, such as size differences and the life strategy of bryophytes, studies on vascular plants cannot be used to predict response of bryophytes to climate change (He et al. 2016; Turetsky 2003; Zanatta et al. 2020) Although bryophytes may be small and inconspicuous compared to many vascular species, that does not imply that they are less important than vascular plants in their communities. Many bryophyte species, such as *Sphagnum* (Jassey et al. 2013), are termed as ecosystem engineers (organisms that can modify their environment to create a particular habitat) which help to maintain community biodiversity (Gavini et al. 2019). In addition, bryophyte provides important ecosystem services such as altering nutrient availability and cycling of carbon and water (Turetsky 2003). They benefit humans directly as well because of they are the main component of garden substrate in countries like Japan and have been used by Indigenous peoples as medicine to treat various health ailments (Glime 2017a, b, c, d; Martin 2015). Although more research is needed, bryophytes may be useful as green roof vegetation (Glime 2017a, b, c, d), as bio-indicators (Govindapyari 2010; Koroleva et al. 2020; Oishi and Hiura 2017; Shi et al. 2017), in horticulture (Glime 2017a, b, c, d; Martin 2015), and also in modern medicine (Decker and Reski 2020).In this regard, it is worthy to note bryophytes as one of important components of biodiversity that provide enormous ecosystem services.

16.2 Ecosystem Services Maintaining by Bryophyte

Bryophytes are considered ecosystem engineers because they able to strongly influence ecosystem processes. In natural systems, these services include but are not limited to global nutrient and carbon cycling (Cornelissen et al. 2007; Crowley and Bedford 2011; Porada et al. 2013; Turetsky 2003), regulation of soil temperature (Sun et al. 2017), water retention (Gignac 2011), and maintenance of community level biodiversity (Cornelissen et al. 2007; Crowley and Bedford 2011; Gavini et al. 2019; Glime 2017a, b, c, d). It is important to understand the extent to which bryophytes influence these processes and how losses to diversity and biomass of bryophytes may cascade effects for communities.

16.3 Role of Bryophytes in Different Ecosystem Services

16.3.1 Soil Formation and Community Development

Bryophytes play a crucial role in soil formation and maintenance of its fertility. They promote soil formation by accelerating physical and chemical weathering; trapping inorganic and organic material from the atmosphere, and adding un-decomposed organic matter (Vanderpoorteen and Goffinet 2009). Mosses, some time along with lichens are the first organisms to colonise rocks—rhizoids of bryophytes and rhizines of lichens break the rock surface, release different organic acids and chelating agents and thus helped in dissolution of minerals (Porada et al. 2016). In addition, accumulation of death and decay of older parts of mosses on substratum also help soil formation (Majumdar and Dey 2020). They also inhabit a wide range of substratum, such as volcanic soil, karst areas, bog soil, etc. and promote succession in them. *Nardia succulenta* (Rich. ex Lehm. and Lindenb.) Spreng. a pioneer liverwort can inhabit bare volcanic soil, where along with volcanic ash, forms layered deposits up to 15 cm thick. These deposits then form a continuous carpet by adhering to vertical cliffs and bridges across volcanic boulders and facilitate initiation of growth of vascular plants (Jongmans et al. 2001). Moss mats can absorb sufficient moisture and increase fertility, which help in development of rock loving herbaceous plants, later succeeded by shrubs and trees (Majumdar and Dey 2020). Bryophytes can also stabilize karst rocky desertification areas and promote vegetation. Along with microorganisms such as fungi and bacteria, they stabilize the physico-chemical properties of surface; create a balanced acidic and alkaline condition, trap dust and nutrients on their substratum, consequently forming soil and promoting establishment and succession of other plant species (Cao et al. 2020).

16.3.2 Habitat

Bryophytes provide habitat to wide range of organisms, such as insects, earthworms, microorganisms, molluscs, nematodes, protozoans, rotifers, spiders, birds as well as other plants (Bahuguna et al. 2013; Glime 2017d, 2017e). For small animals and invertebrates, bryophytes provide food, shelter and nesting material. For birds, bryophytes provide a niche for foraging and gathering nest material (Chmielewski and Eppley 2019). In turn, birds help bryophytes in spore dispersal. Bryophytes also serve as a refuge for species of order Odonata (e.g., dragonflies and damselflies). Some species like naiads use bryophytes as an emergence site where they shed their exoskeleton, leaving it behind as an exuvia (Glime 2017c). For higher plants, they offer suitable substratum and seed beds for seedling germination (Tavili et al. 2017).

16.3.3 Nutrient Cycling

Nutrient cycle is a sequence of events consisting uptake of elements by living organisms from nature, their transfer among organisms and finally, releasing these elements back to nature (Yang et al. 2021). Bryophytes have unique twin features of rapid acquisition and slow release of elements back to environment; thus, they play a significant role in nutrient cycle (Oluwole 2019). They can mineralize nutrients from the substratum (P) or actively accumulate from atmosphere and canopy leachates (Ca, K, Mg and N) on their thallus, slowly release these nutrients to soil and make them available for other plants (Bates 2009; Alam and Sharma 2015). In addition, bryophytes also control nutrients flow in environment through biomass production; especially in high latitude and high elevation ecosystems where their dominance is high (Lett et al. 2021). For example, bryophytes contribute more than 50% of the primary production and standing biomass in many tundra ecosystem types (Wielgolaski 1971; Huemmrich et al. 2010). Also, in black spruce forest, North America, net primary productivity (NPP) of understory (*Sphagnum* dominated) was 51% of the total NPP (understory, overstory, bryophytes and underground) (O'Connell et al. 2003). Therefore, bryophytes act as a primary form of carbon storage in such ecosystems (Singh 2006) and help in nutrient cycle in the ecosystem.

16.3.4 Water Cycling

Due to their poikilohydric nature, bryophytes act as a temporary water storage space and alter water cycling within their community (Gignac 2011). For example, in semiarid climates, bryophytes significantly increase soil moisture content at shallow depths (5 cm) (Xiao et al. 2016). Bryophyte also regulates soil moisture through evaporation; Chen et al. (2019) reported that soil water evaporation increased underneath mosses with increasing temperatures. These results suggest that the mosses regulate soil temperature through evaporation similarly to how vascular plant regulates temperatures by the process of transpiration. However, they have a greatest impact on water cycling is through rainfall interception. Reported that including lichens and bryophytes in a model of global rainfall interception increase total evaporation by 61% compared to a model that only included vascular plants.

16.3.5 Carbon Cycling

Carbon cycling occurs as carbon from the atmosphere moves to reservoirs e.g. ocean, organisms, soil, and rocks. Like vascular plants, bryophytes influence this cycle through their growth and metabolic processes such as photosynthesis and respiration. Net primary production (the rate at which biomass is stored; NPP) represents the difference between the carbon gained through photosynthesis and carbon lost through respiration; this difference can be used to determine the amount of carbon stored by plants (Bowman et al. 2017). A global estimate of NPP is an important parameter to determine because it quantifies the carbon cycle and storage of plants and informs predictions of the effects of climate change (Ito 2011).

16.4 Carbon Sequestration and Climate Change Indicator

One of the most important roles played by bryophytes is sequestration of carbon from the atmosphere. Carbon dioxide (CO_2) is one of greenhouse gas that is responsible for the earth warming. Increasing CO_2 in atmosphere contributes to global warming which is a major threat to survival of all life forms. Bryophytes play an important role and act as an efficient CO_2 sequestering agent (Salimi et al. 2021). Hence, they help in mitigation of global warming issues. They can also regulate carbon budget by absorbing CO_2 from different sources at a rapid pace and trapping them in biomass due to slow decomposition rate (Smith et al. 2017; Newman et al. 2018; Sun et al. 2017; Vitt and House 2021). In fact, removal of bryophytes resulted in reduced floor carbon and efflux of mineral soil CO2, soil organic carbon, microbial biomass carbon and dissolved organic carbon (Sun et al. 2017). In addition, their presence in a wide range of habitats also allows them to influence carbon flux in different ecosystems. Peat lands are one of the ecosystems where bryophytes dominate and play a significant role in carbon Sequestration.

16.4.1 Nitrogen Fixation

Bryophytes are also very important for global nitrogen fixation and can fix nitrogen through multiple pathways. Two significant forms of nitrogen fixation by bryophytes occur through atmospheric deposition and biological fixation (symbiosis with cyanobacteria) (Lindo and Whiteley 2011, Markham and Fernandez Otarola et al. 2021, Turetsky 2003). In artic systems, mosses influence microorganisms that fix nitrogen by regulating soil temperature and moisture. Similarly, Sun et al. (2017) reported that the removal of bryophytes from the understory of shrub land and coniferous habitats decreased biomass of microbial community in the terrestrial ecosystem; nitrogen (N) is one of the macronutrients that limit primary productivity. Bryophytes help in nitrogen fixation via atmospheric wet or dry deposition, absorption from the substratum, and biological nitrogen fixation. Bryophytes form symbiotic association with N_2-fixing cyanobacteria, such as *Nostoc* spp. and diazotrophs.

16.4.2 Restoration of Habitat

Restoration is improvement of degraded land or disturbed ecosystems to rebuild ecological integrity and enhance people's lives. Bryophytes are one of the suitable restoration agents in the plant kingdom for heavily disturbed areas such as post-fire substratum, karst rocky areas, etc. Bryophytes help in restoration of forests by colonising the substratum. Species like *Funaria hygrometrica* Hedw., *Ceratodon purpureus* (Hedw.) Brid., *Marchantia polymorpha* L., *Bryum argenteum* Hedw., *Breutelia diffracta* Mitt. Can easily colonise the post-fire substratum (Ryoma and Lindberg 2005; Hylander et al. 2021); facilitating seedling germination of other plant groups (Rehm et al. 2019). E.g, in restoration corridors of *Acacia koa* A.Gray and *Metrosideros polymorpha* J.R.Forst. ex Hook.f., bryophyte mats enhanced seedling germination rate by ten times as compared to woody litter and nurse logs (Rehm et al. 2019);

16.4.3 Bryophytes as Heavy Metal Bioindicators

The expansion and lack of environmental regulation has lead to a large input of various toxic pollutants into the atmosphere. There is a need of continuous monitoring of air quality. This is accomplished either by the construction or employment of bio indicators/biomonitors. Bioindicators refers to all organisms that depict the quality of the environment on the basis of changes in morphology; physiology etc. Bio-indicators are reliable tools used to detect the changes in natural environment (Parmar et al. 2016). It refers to any living organism that can reveal status of pollutants based on specific symptoms, morphological and anatomical changes and population of species in ecosystem (Govindapyari et al. 2010; Barkan and Lyanguzova 2018). Bryophytes grow on a variety of habitats and obtain nutrients from the substances dissolved in ambient moisture. The substances are directly absorbed from the substrata by diffusion through the cells. Pollutants reach in the bryophyte tissues from the dry deposition in the form of gases and particles. Bryophytes are considered as good bioindicator to assess atmospheric deposition of metals. Amongst the three different phyla (Marchantiophyta, Anthocerotophyta and Bryophyta), species belonging to Bryophyta are most commonly used as bioindicators of heavy metal pollution and radioisotopes in air (Oishi and Hiura 2017; Kłos et al. 2018). They are good bioindicator of metals because they are ectohydric and the minerals including heavy metals.

List of Some Bryophytes Accumulate Heavy Metals

S.n	Species name	Type of accumulated heavy metal	References
1	*Abietinella abietina* (Hedw.) M. Fleisch	As, Cd, Co, Cr, Cu, Fe, Hg, Mo, Ni, Pb, V, Zn	Zechmeister (1998)
2	*Asterella angusta* Beauv	Cu, Zn, Pb, Ni, Cd Cr	Ghatge et al. (2011)
3	*Atrichum undulatum* (Hedw.) P. Beauv	Cd, Pb, Zn	Oda (1995)
4	*Barbula constricta* Mitt	Fe, Cu, Pb	Vats et al. (2010)
5	*Bryum capillare* Hedw	Zn, Pb, Cu	Vukojević et al.
6	*Cyathodium tuberosum* Kash	Cu, Zn, Pb, Ni, Cd Cr	Ghatge et al. (2011)
7	*Funaria hygrometrica* Hedw	Pb	Itouga et al. (2017)
8	*Haplocladium angustifolium* (Hampe & Müll. Hal.) Broth	Mn, Mo, Zn, Cu, Pb	Jiang et al. (2018)
9	*Helicodontium capillare* (Hedw.) A. Jaeger	Zn, Cu, Pb, Cr	Mazzoni et al. (2012)
10	*Hylocomium splendens* (Hedw.) Schimp	Fe, Mn, Zn, Cu, As, Cd, Cr, Hg, V, Co, Mo, Ni, Pb	Zechmeister (1998); Galsomies et al. (1999), Parzych (2014)

(continued)

(continued)

S.n	Species name	Type of accumulated heavy metal	References
11	*Hypnum cupressiforme* Hedw	As, Cd, Cr, Cu, Fe, Hg, V, Zn	Galsomies et al. (1999)
12	*Marchantia polymorpha* L	Cd, Cu, Pb, Zn	Ares et al. (2018)
13	*Mielichhoferia japonica* Besch	Cu, Fe, Pb	Aikawa et al. (1999)
14	*Physcomitrium cyathicarpum* Mitt	Fe, Cu, Pb	Vats et al. (2010)
15	*Physcomitrella patens* (Hedw.) Bruch & Schimp.,	Cu	Antreich et al. (2016)
16	*Pleurochaete squarrosa* (Brid.) Lindb	Ni, Pb, Cu, Cr	Aydogan et al. (2017)
17	*Polytrichum commune Hedw*	F, Cr, Pb, Sr, Cu, Zn, Co, Mn, Fe, Cu, Cd, Zn	Wenstein and Davison (2003), Soltes and Gregusková (2013)
18	*Ricciocarpos natans* (L.) Corda	Zn, Cu, Cd, Hg	Boudet et al. (2011)
19	*Thuidium tomentosum* Schimp	Zn, Cu, Pb, Cr	Mazzoni et al. (2012)
20	*Timmiella barbuloides* (Brid.) Mönk	Ni, Pb, Cu, Cr	Aydogan et al. (2017)

16.4.4 Bryophytes in Soil Erosion

In the absence of bryophytes, the bare soil surface is prone to erosion by rain splash. On the other hand, when bryophytes cover the soil surface, it reduces the impact of rain splash as well as hold soil particles by its rhizoids, thus reducing soil erosion. They can absorb metals easily due to thin cuticle layer they have prominent ion-exchange properties (Little and Martin 1974); (they have large surface-to-weight ratio (Jiang et al. 2018); they are perennial plants, which can be collected throughout the year (Mahapatra et al. 2019). These factors allow them a faster response to mirror the differences in the level of heavy metals than most vascular plants (Zvereva and Kozlov 2011). In fact, some of the species of bryophytes are very sensitive to heavy metals and show visible injuries even in the presence of minutest amounts of metals, acting as "warning giver" (Sahu et al. 2007). On the other hand, some species can accumulate metals above their physiological needs and indicate the degree of metal pollution (Miranda et al. 2016) Mosses like *Atrichum, Nardia, Pogonatum, Pohlia* and *Trematodon* are inhibitors of soil erosion because of their trample resistant structure and their regenerative capacity Bahuguna et al. 2013.

16.5 Conclusion

Bryophytes have a broad range of uses for humans and provide important ecosystem services that keep communities healthy. As interest in a more sustainable future grows, bryophytes may help humans have a more sustainable life style particularly in urban areas. When used a substrate for green roofs, bryophyte will also help reduce problem such as the urban heat island effect while simultaneously being a better substrate choice in areas that experience times of arid weather due to their desiccation tolerance. In urban areas experiencing high levels of pollution, bryophyte can be used a bioindicators. Not only will this be a cheaper alternative to traditional air quality monitoring methods, combining bryophytes with green roofs may be more effective at monitoring pollution long term compared to just using epiphytic bryophytes. For individuals in right climate, mosses may be a good alternative to a traditional lawn as they provide habitat for local fauna while decreasing need for harmful upkeep like the use of pesticides. Although they have long been used by indigenous peoples as traditional medicines, interest in use bryophytes' secondary compounds in modern medicine is just beginning.

References

Alam A, Sharma V (2015) Horticultural importance of bryophytes-a review. Int J Hortic 5(19):1–9. https://doi.org/10.5376/ijh.2015.05.0019

Ares A, Itouga M, Kato Y, Sakakibara H (2018) Differential Metal Tolerance and Accumulation Patterns of Cd, Cu, Pb and Zn in the Liverwort Marchantia polymorpha L. Bull Environ Contam Toxicol 100(3):444–450

Aydogan S, Erdag B, Aktaş L (2017) Bioaccumulation and oxidative stress impact of Pb, Ni, Cu, and Cr heavy metals in two bryophyte species, Pleurochaete squarrosa and Timmiella barbuloides. Turk J Bot 41(5):464–475

Antreich S, Sassmann S, Lang I (2016) Limited accumulation of copper in heavy metal adapted mosses. Plant Physiol Biochem 101:141–148

Boudet LC, Escalante A, Von Haeften G, Moreno V, Gerpe M (2011) Assessment of heavy metal accumulation in two aquatic macrophytes: a field study. Ecotoxicology Environ Contamination 6(1):57–64

Bahuguna YM, Gairola S, Semwal DP, Uniyal PL and Bhatt AB (2013) Bryophytes and Ecosystem. In: Gupta RK, Kumar M (Ed), Biodiversity of lower plants. IK International Publishing House Pvt. Ltd., pp 279–296

Barkan VS, Lyanguzova IV (2018) Concentration of heavy metals in dominant moss species as an indicator of Aerial Technogenic Load 1. Russ J Ecol 49:128–134. https://doi.org/10.1134/S1067413618020030

Bates JW (2009) Mineral nutrition and substratum ecology. In: Goffinet B, Shaw AJ (eds) Bryophyte biology. Cambridge University Press. Pp, Cambridge, pp 299–356

Benscoter BW, Thompson DK, Waddington JM, Flannigan MD, Wotton BM, De Groot WJ, Bowman WD, Hacker SD, Cain ML (2017) Ecology. 4th Edition. Sinauer Associates, Inc

Cao W, Xiong Y, Zhao D, Tan H, Qu J (2020) Bryophytes and the symbiotic microorganisms, the pioneers of vegetation restoration in karst rocky desertification areas in southwestern China. Appl Microbiol Biotechnol 104:873–891. https://doi.org/10.1007/s00253-019-10235-0

Cardinale BJ, Duffy JE, Gonzalez A, Hooper DU, Perrings C, Venail P, Narwani A, Mace GM, Tilman D, Wardle DA, Kinzig AP, Daily GC, Loreau M, Grace JB, Larigauderie A, Srivastava DS (2012) Naeem S (2012)Biodiversity loss and its impact on humanity. Nature 486(7401):59–67. https://doi.org/10.1038/nature11148.Erratum.In:Nature.2012Sep13;489(7415):326. PMID: 22678280

Chen S, Yang Z, Liu X, Sun J, Xu C, Xiong D, Yang Y (2019) Moss regulates soil evaporation leading to decoupling of soil and near-surface air temperatures. J Soils Sediments 19:2903–2912

Chmielewski MW, Eppley SM (2019) Forest passerines as a novel dispersal vector of viable bryophyte propagules. Proc Royal Soc B Biol Sci 286(1897):20182253–20182253. https://doi.org/10.1098/rspb.2018.2253

Cornelissen JHC, Lang SI, Soudzilovskaia NA, During HJ (2007) Comparative cryptogam ecology: a review of bryophyte and lichen traits that drive biogeochemistry. Ann Bot 99(5):987–1001

Combertia C, Thorntona TF, de Echeverria VW, Patterson T (2015) Ecosystem services or services to ecosystems? Valuing cultivation and reciprocal relationships between humans and ecosystems. Glob Environ Chang 34:247–262. https://doi.org/10.1016/j.gloenvcha.2015.07.007

Crowley KF, Bedford BL (2011) Mosses influence phosphorus cycling in rich fens by driving redox conditions in shallow soils. Oecologia 167(1):253–265. https://doi.org/10.1007/s00442-011-1970-8

Costanza R and Daly HE (1992) Natural capital and sustainable development. Conserv Biol 6(1):37–46. https://www.jstor.org/stable/2385849

Costanza R, d'Arge R, de Groot R, Farber S, Grasso M, Hannon B et al (1997) The value of the world's ecosystem services and natural capital. Nature 387(6630):253–260. https://doi.org/10.1038/387253a0

Daily GC (1997a) Introduction: what are ecosystem services. In: Daily GC (ed) Nature's Services. Island Press, Washington DC, pp 1–10

Daily GC (1997b) Nature's services: societal dependence on natural ecosystems. Island Press, Washington, DC, pp 15–392

Dawson TP, Jackson ST, House JI, Prentice IC, Mace GM (2011) Beyond predictions: Biodiversity conservation in a changing climate. Science 332(6025):53–58

Decker EL, Reski R (2020) Mosses in biotechnology. Curr Opin Biotechnol 61:21–27. https://doi.org/10.1016/j.copbio.2019.09.021

Fernandez-Llamazares A, Lopez-Baucells A, Velazco PM, Gyawali A, Rocha R, Terraube J, Cabeza M (2021) The importance of Indigenous Territories for conserving bat diversity across the Amazon biome. Perspect Ecol Conserv 19(1):10–20. https://doi.org/10.1016/j.pecon.2020.11.001

Gavini SS, Suarez GM, Ezcurra C, Aizen MA (2019) Facilitation of vascular plants by cushion mosses in high-Andean communities. Alp Bot 129(2):137–148. https://doi.org/10.1007/s00035-019-00222-6

Gignac LD (2011) Bryophytes as predictors of climate change. In: Tuba Z, Slack NG, Stark LR (eds) Bryophyte ecology and climate change. Cambridge University Press. Pp, New York, pp 461–482

Glime JM (2017b) Gardening: Moss Garden development and maintenance (Chapter 7–4). In: Glime JM, Houghton MI, Glime JM (Eds). Bryophyte ecology Vol. 5. Michigan Technological University and the International Association of Bryologists, pp 1–22

Glime JM (2017d) Birds and bryophytic food sources (Chapter 16–2). In: Glime JM (Ed), Bryophyte Ecology Vol. 2. Michigan technological university and the international association of bryologists, pp 1–32

Glime JM (2017e) Invertebrate: Introduction (Chapter 4–1). In: Glime JM (ed), Bryophyte ecology Vol. 2. Michigan Technological University and the International Association of Bryologists, pp 1–22

Glime JM (2017) Chapter 7–4: gardening: moss garden development and maintenance. In: Glime JM (ed). Bryophyte ecology subchapters (Vol. 5, pp. 1–22). Houghton, MI: Janice M. Glime. Gornall JL, Jonsdottir IS, Woodin SJ, Vanderwal R (2007) Arctic mosses govern below-ground environment and ecosystem processes. Oecologia, 153(4), 931–941. https://doi.org/10.1007/s00442-007-0785-0

Govindapyari H, Leleekam M, Nivedita M, Uniyal PL (2010) Bryophytes: indicators and monitoring agents of pollution. NeBIO 1(1):35–41

Galsomies L, Letrouit MA, Deschamps C, Savanne D, Avnaim M (1999) Atmospheric metal deposition in France: initial results on moss calibration from the 1996 biomonitoring. Sci Total Environ 232(1–2):39–47

Ghatge MM, Shaikh SD, Dongare M (2011) Study on the metal absorption by two bryophytes from Koyana wildlife sanctuary (India). J Exp Sci 2(3):49–50

Gomez-Baggethun E, de Groot RS, Lomas PL, Montes (2010) The history of ecosystem services in economic theory and practice: from early notions to markets and payment schemes. Ecol Econ 69(6):1209–1218.https://doi.org/10.1016/j.ecolecon.2009.11.007

He X, He KS, Hyvonen J (2016) Will bryophytes survive in a warming world? Perspect Plant Ecol Evolut Systematics 19:49–60. https://doi.org/10.1016/j.ppees.2016.02.005

Huemmrich KF, Gamon JA, Tweedie CE, Oberbauer SF, Kinoshita G, Houston S, Oechel WC (2010). Remote sensing of tundra gross ecosystem productivity and light use efficiency under varying temperature and moisture conditions. Remote Sensing Environ 114(3):481–489.https://doi.org/10.1016/j.rse.2009.10.003

Hylander K, Frisk CA, Nemomissa S, Johansson MU (2021) Rapid post fire re-assembly of species rich bryophyte communities in Afroalpine heathlands. J Veg Sci 32(13033):1–11. https://doi.org/10.1016/j.rse.2009.10.003

Ito A (2011) A historical meta-analysis of global terrestrial net primary productivity: are estimates converging? Glob Change Biol 17(10):3161–3175. https://doi.org/10.1111/j.1365-2486.2011.02450.x

Itouga M, Hayatsu M, Sato M, Tsuboi Y, Kato Y, Toyooka K, Suzuki S, Nakatsuka S, Kawakami S, Kikuchi J, Sakakibara H (2017) Protonema of the moss *Funaria hygrometrica* can function as a lead (Pb) adsorbent. PLoS ONE 12(12):e0189726

Jassey VE, Chiapusio G, Binet P, Buttler A, Laggoun-Defarge F, Delarue F, Bernard N, Mitchell EA, Toussaint ML, Francez AJ, Gilbert D (2013) Above- and belowground linkages in Sphagnum peatland: Climate warming affects plant-microbial interactions. Glob Change Biol 19(3):811–823. https://doi.org/10.1111/gcb.12075

Jiang Y, Fan M, Hu R, Zhao J and Wu Y (2018) Mosses are better than leaves of vascular plants in monitoring atmospheric heavy metal pollution in urban areas. Int J Environ Res Public Health 15(6):1105. https://doi.org/10.3390/ijerph15061105

Jongmans AG, van Breemen N, Gradstein SR, Vanoort F (2001) How liverworts build hanging gardens from volcanic ash in Costa Rica. Catena 44:13–22. https://doi.org/10.1016/S0341-8162(00)00151-X

Kłos A, Ziembik Z, Rajfur M, Dołhańczuk-Śródka A, Bochenek Z, Bjerke JW, Świsłowski P (2018) Using moss and lichens in biomonitoring of heavy-metal contamination of forest areas in southern and north-eastern Poland. Sci Total Environ 627:438–449. https://doi.org/10.1016/j.scitotenv.2018.01.211

Koroleva Y, Napreenko M, Baymuratov R, Schefer R (2020) Bryophytes as a bioindicator for atmospheric deposition in different coastal habitats (a case study in the Russian sector of the Curonian Spit, South-Eastern Baltic). Int J Environ Stud 77(1):152–162. https://doi.org/10.1080/00207233.2019.1594301

Lett S, Jónsdóttir IS, Becker-Scarpitta A, Christiansen CT, During H, Ekelund F, Zuijlen KV (2021) Can bryophyte groups increase functional resolution in tundra ecosystems? Arctic Science 8(3):609–637. https://doi.org/10.1139/AS-2020-0057

Lindo Z, Whiteley JA (2011) Old trees contribute bio-available nitrogen through canopy bryophytes. Plant Soil 342(1):141–148. https://doi.org/10.1007/s11104-010-0678-6

Little P (1970) Martin MH (1974) Biological monitoring of heavy metal pollution. Environ Pollut 6(1):1–19

Mahapatra B, Dhal NK, Dash AK, Panda BP, Panigrahi KCS, Pradhan A (2019) Perspective of mitigating atmospheric heavy metal pollution: using mosses as biomonitoring and indicator organisms. Environ Sci Pollut Res 26:29620–29638. https://doi.org/10.1007/s11356-019-06270-z

Majumdar S, Dey M (2020) A handbook on bryophytes with special reference to type specimens of liverworts and hornworts in Indian Herbaria. Botanical Survey of India

Markham J, Fernández Otarola M (2021) Bryophyte and lichen biomass and nitrogen fixation in a high elevation cloud forest in Cerro de La Muerte. Costa Rica. Oecologia 195(2):489–497. https://doi.org/10.1007/s00442-020-04840-4

Martin A (2015) The magical world of moss gardening. Timber Press Inc., Portland, USA, pp 1–265

Miranda GM, Perez PA, Vargas PG, Zarazúa G, Sanchez-Meza JC, Gomez CZ, Tejeda S (2016) Accumulation of heavy metals in mosses: a biomonitoring study. Springerplus 5(1):715. https://doi.org/10.1186/s40064-016-2524-7

Mazzoni AC, Lanzer R, Bordin J, Schafer A, Wasum R (2012) Mosses as indicators of atmospheric metal deposition in an industrial area of southern Brazil. Acta Botanica Brasilica 26(3):553–558

Mori AS, Lertzman KP Gustafsson L (2017) Review: forest biodiversity and ecosystem services. Biodiversity and ecosystem services in forest ecosystems: a research agenda for applied forest ecology. J Appl Ecol 54:12–27.https://doi.org/10.1111/1365-2664.12669

Nonya C, Pherkop M, Vinay S, Heikham E, Uniyal3 PL (2022) Bryophytes in the Ecosystem Services: a review. J Bioresources 9(1):25–34, https://doi.org/10.5281/zenodo.8131443

Newman TR, Wright N, Wright B, Sjogersten S (2018) Interacting effects of elevated atmospheric CO_2 and hydrology on the growth and carbon sequestration of Sphagnum moss. Wetlands Ecol Manage 26:763–774. https://doi.org/10.1007/s11273-018-9607-x

O'Connell KB, Gower S and Norman J (2003) Comparison of net primary production and light-use dynamics of two boreal black spruce forest communities. Ecosystems 6(3):236–247. https://doi.org/10.1007/PL00021510

Oishi Y, Hiura T (2017) Bryophytes as bioindicators of the atmospheric environment in urban-forest landscapes. Landsc Urban Plan 167:348–355. https://doi.org/10.1016/j.landurbplan.2017.07.01

Oda T (1995) Characterization of Copper, lead and Zinc in metal tolerant mosses, *Atrichum undulatum* (Hedw.) P. Beauv., *Scopelophila cataracta* (Mitt.) Broth *Pohlia bulbifera* (Warnst.) Warnst. In heavy metals pollution areas. J Phytogeogr Taxon 43:91–102

Oluwole OO (2019) Nutrient uptake of Archidium acanthophyllum and Cyanotis lanata from Savanna Microsites on Baasi Inselberg in Southwestern Nigeria. African J Sci Nat 8:10–18. https://doi.org/10.46881/ajsn.v8i0.161

Parmar TK, Rawtani D, Agrawal YK (2016) Bioindicators: the natural indicator of environmental pollution. Front Life Sci 9(2):110–118. https://doi.org/10.1080/21553769.2016.1162753

Porada P, Lenton TM, Pohl A, Weber B, Mander L, Donnadieu Y, Beer C, Poschl U, Kleidon A (2016) High potential for weathering and climate effects of non-vascular vegetation in the Late Ordovician. Nat Commun 7(1):12113. https://doi.org/10.1038/ncomms12113

Porada P, Weber B, Elbert W, Poschl U, Kleidon A (2013) Estimating global carbon uptake by lichens and bryophytes with a process-based model. Biogeosciences 10(11):6989–7033. https://doi.org/10.5194/bg-10-6989-2013

Parzych A (2014) Accumulation of heavy metals in moss species Pleurozium schreberi (Brid.) Mitt. and Hylocomium splendens (Hedw.) BSG in Slowinski National Park. J Elementol 19(2):471–482

Rehm EM, Thomas MK, Yelenikc SG, Bouckc DL, D'Antonio CM (2019) Bryophyte abundance, composition and importance to woody plant recruitment in natural and restoration forests. For Ecol Manage 444:405–413. https://doi.org/10.1016/j.foreco.2019.04.055

Ryoma R, Lindberg SL (2005) Bryophyte recolonization on burnt soil and logs. Scand J for Res 20(6):5–16

Sahu V, Asthana AK, Nath V, Yunus M (2007) Bryophytes: a useful tool in heavy metal monitoring. EnviroNews 3(4):1–2

Salimi S, Berggren M, Scholz M (2021) Response of the peatland carbon dioxide sink function to future climate change scenarios and water level management. Glob Change Biol 27:5154–5168. https://doi.org/10.1111/gcb.15753

Schwartz MW (2012) Using niche models with climate projections to inform conservation management decisions. Biol Cons 155:149–156. https://doi.org/10.1016/j.biocon.2012.06.011

Shi XM, Song L, Liu WY, Lu HZ, Qi JH, Li S, Chen X, Wu JF, Liu S, Wu CS (2017) Epiphytic bryophytes as bio-indicators of atmospheric nitrogen deposition in a subtropical montane cloud forest: Response patterns, mechanism, and critical load. Environ Pollut 229:932–941. https://doi.org/10.1016/j.envpol.2017.07.077

Shaw AJ, Szovenyi P, Shaw B (2011) Bryophyte diversity and evolution: windows into the early evolution of land plants. Am J Bot 98(3):352–369. https://doi.org/10.3732/ajb.1000316

Singh MK (2006) Some aspects of accumulated carbon in few bryophyte dominated ecosystems: a brief mechanistic overview. Cereal Res Commun 34(1):37–40. http://www.jstor.org/stable/23788888

Smith RJ, Jovan S, Gray AN, McCun B (2017) Sensitivity of carbon stores in boreal forest moss mats—effects of vegetation, topography and climate. Plant Soil 421:31–42. https://doi.org/10.1007/s11104-017-3411-x

Soltes R, Greguskova E (2013) Accumulation characteristics of some elements in the moss Polytrichum commune (bryophytes) based on XRF spectrometry. J Environ Prot 4(6):522–528

Sun SQ, Liu T, Wu YH, Wang GX, Zhu B, DeLuca TH, Wang YQ (2017) Ground bryophytes regulate net soil carbon efflux: evidence from two subalpine ecosystems on the east edge of the Tibet Plateau. Plant Soil 417(1–2):363–376. https://doi.org/10.1007/s11104-017-3264-3

Tavili A, Jafari M, Chahouki MAZ, Sohrabi M (2017) How do cryptogams affect vascular plant establishment? Cryptogam Bryol 38:313–323. https://doi.org/10.7872/cryb/v38.iss3.2017.313

Turetsky MR (2003) The role of bryophytes in carbon and nitrogen cycling. The Bryologist 106(3):395–409

Vanderpoorteen A, Goffinet B (2009) Introduction to bryophytes. In: Vanderpoorten A, Goffinet B (ed), Introduction to bryophytes. Cambridge University Press, pp 34–39. https://doi.org/10.1017/CBO9780511626838

Vats SK, Singh A, Koul M, Uniyal PL (2010) Study on the metal absorption by two mosses in Delhi Region (India). In: Poster presentation in International Conference IAB

Vitt DH, House M (2021) Bryophytes as key indicators of ecosystem function and structure of northern peatlands. Bryophyte Diversity Evolut 043(1):253–264. https://doi.org/10.11646/BDE.43.1.18

Wielgolaski FE (1971) Vegetation types and primary production in tundra. In: Wiel FE, Roswall T (ed), Proceedings of the IV International Meeting on the Biological Productivity of Tundra. The Tundra Steering Committee, Leningrad USSR, pp 9–34

Weinstein LH, Davison AW (2003) Native plant species suitable as bioindicators and biomonitors for airborne fluoride. Environ Pollut 125(1):3–11

Xiao B, Hu K, Ren T, Li B (2016) Moss-dominated biological soil crusts significantly influence soil moisture and temperature regimes in semiarid ecosystems. Geoderma 263:35–46. https://doi.org/10.1016/j.geoderma.2015.09.012

Yang R, Dong J, Li C, Wang L, Quan Q and Liu J (2021) The decomposition process and nutrient release of invasive plant litter regulated by nutrient enrichment and water level change. PLoS ONE 16(5). https://doi.org/10.1371/journal.pone.0250880

Zanatta F, Engler R, Collart F, Broennimann O, Mateo RG, Papp B, Munoz J, Baurain D, Guisan A, Vanderpoorten A (2020) Bryophytes are predicted to lag behind future climate change despite their high dispersal capacities. Nat Commun 11(1):5601. https://doi.org/10.1038/s41467-020-19410-8

Zvereva EL, Kozlov MV (2011) Impacts of industrial polluters on bryophytes: a metaanalysis of observational studies. Water Air Soil Pollut 218:573–586. https://doi.org/10.1007/s11270-010-0669-5

Zechmeister HG (1998) Annual growth of four pleurocarpous moss species and their applicability for biomonitoring heavy metals. Environ Monit Assess 52(3):441–451

Chapter 17
Exploring Lichen Species Association and Their Indicative Significance Across Various Forest Types of Western Himalayas (Uttarakhand), India

Balwant Kumar, Pooja Bisht, and Ritika Tamta

Abstract The forests of Himalaya are rich in lichen diversity. On the basis of altitudinal changes lichen flora of the region has been divided into three major forest types viz. tropical, temperate and alpine. The composition of different lichen species in three major forest types of Western Himalaya has been discussed. The region's forests represent an abundance of lichens along with other vegetation. Lichens are ecologically crucial for any forest. Besides ecology, lichens are also used for preparing medicines, perfumes, spices, and other ingredients. Lichens reflect the nature of different types of forests and the woody nature found within them. Therefore, this article assesses the indicator value of lichens in different forest types of the region.

Keywords Lichens · Indicator value · Forest types · Western Himalayas

17.1 Introduction

Lichen is a group of plants that consists of two unrelated organisms, an alga, and a fungus, growing together in a close symbiotic association. They play a crucial role in forest nutrient cycling and create microclimates that facilitate the colonization of other vegetation. Lichens also help in forest succession as pioneer groups, and their presence increases the diversity in the region. These non-vascular, non-flowering, slow-growing thallophytes (0.5 mm per year) also show their indicator value in different forest types in the Himalayas. Besides, lichens are also a good source of spices, perfumes, secondary metabolites, dyes, cosmetics, fodder, etc. About 2000 species of lichens are reported from India, about 10% of the world's lichen diversity. In India, the temperate to alpine regions of 12 Himalayan states mainly contribute to the maximum diversity and biomass of lichens (Kumar 2009). Champion and Seth

B. Kumar (✉) · P. Bisht · R. Tamta
Department of Botany, Soban Singh Jeena University Campus, Almora, Uttarakhand 263601, India
e-mail: drbalwantkumararya@gmail.com

D. Arya et al. (eds.), *Climate Change Impact on Himalayan Biodiversity*,
Environmental Science and Engineering, https://doi.org/10.1007/978-3-031-77149-1_17

(1968) categorized several forest types in India, including Uttarakhand Himalaya, based on annual temperature, rainfall, and dry periods among the forests. In this context, lichen taxa were neglected and not considered to determine various forest types. However, various lichen species are also reflecting the differentiation in forest types. Based on previous studies on lichens in the Himalayas, most research is confined to taxonomic, molecular characterization, and lichenometry techniques. Until now, there has been scanty information on the ecological aspects of lichens. So, this chapter attempts to elucidate the indicator value of lichen flora in different forest types of the Western Himalayas.

Based on the countenance, growth forms, and distribution of different lichen species, the forests of Western Himalaya can be categorized into three types, namely: tropical forests (500–1000 m), temperate forests (1001–2000 m), and alpine forests (above 2001 m altitude).

17.2 Lichens of Tropical Forests

This region includes the foothills and the valley of the rivers of the Himalayan region. The lichens here are chiefly and bark inhabiting (corticolous) such as *Pertusaria, Lecanora, Lecidea,* and *Bocidia* growing on *Shorea robusta* trees commonly distributed in dry places of foothills (Upreti, 1997); however, in moist areas foliose and fruticose forms of lichens belonging to genera *Collema, Leptogium, Pyxine, Physcia* and *Dirinaira* found in good number. In more moist and humid places in this region, foliicolous (on leaves) and Graphidaceaeous genera commonly occur on perennial shrubs' leaves. The Pyrenocarpous genera *Pyrenuda, Parmentaria,* and *Porina* also occur abundantly in tropical regions (Upreti, 1997).

17.3 Lichens of Temperate Forests

Chir-pine forests are found in the lower elevation zone of the temperate forest. The trunk portion of *Pinus roxburghii* bears few species of *Parmelia* in dry regions. However, *Pinus* trees at altitudes around 1500 m exhibit luxuriant growth of lichen genera *Cetraria, Usnea,* and *Ramalina.* In higher altitude forests (above 2000 m), *Pinus wallichiana* is also preferred by some Usneoid lichens (Upreti, 1997). *Pinus* trees have dry bark, unsuitable for the growth of several lichens. The moist temperate region also supports the occurrence of some common lichen taxa that prefer to grow on the bark of *Quercus, Rhododendron, Cedrus, Aesculus,* etc. These lichens are *Chrysothrix*, Parmelioid, Usnioid, and Graphidiaceous lichens, while *Acarospora, Aspicilia, Umblicaria, Dermatocarpon, Porpidia, Buellia,* and *Caloplaca* mostly prefer to grow on a rock surface. Similarly, lichen genera *Peltigera, Lobaria,* and *Sticta* prefer to grow in soil (Kumar, 2009).

17.4 Lichens of Alpine Forests

Quercus semecarpifolia is one of the most common oaks in the Chopta region of Garhwal Himalaya, which supports 42 species of lichens. However, *Q. floribunda* and *Q. leucotrichophora* also support the occurrence of 16 and 12 species of lichens, respectively (Kumar & Upreti, 2008). Most of the lichen species are found on the twigs of all these trees, mainly twigs of *Q. semecarpifolia.* Because this oak is dome-shaped, its twigs retain more water, and hold high moisture. Therefore, it supports the growth of luxuriant lichen. All three *Quercus* species exhibit luxuriant growth of colonized lichens and members of the lichen family Physciaceae together with fruticose lichen genera *Ramalina* and *Usnea.*

It has also been observed that the thinned-out evergreen trees of silver fir in the Chopta- Tunganath landscape of Garhwal Himalaya provide a suitable habitat for Pyrenocarpous and Graphidaceaeous lichens, which are corticolous types. As in thinned forests, the wind and light pass more efficiently; therefore, these trees possess good lichen growth. Besides trees, 18 corticolous lichens were also recorded on different shrubs. These shrub species are *Cotoneaster mycrophyllus, Berberis petiolaris, B. asiatica, Prinsepia utili,* and *Drepanostachyum falcatum,* which act as a favourable host for the lichens in temperate and alpine forests (Tamta and Kumar 2023).

Lichens growing on bark are greatly influenced by the physical characteristics of the bark, i.e., smooth or rough surface. The undulated character of bark and its cracks and crevices help retain the collected moisture. More xerophytic species flourish on smooth surfaces, such as *Caloplaca, Pertusaria, Lecanora, Bacidia,* and *Lecidea.* The resinous bark of pine trees bears a few species of *Parmelia* or *Physcia.* Lichen genera *Evernia, Heterodermia, Anaptychia,* and *Usnea* abundantly grow on *Cedrus* trees, while species of *Perusaria, Ochrolecin, Leptogium,* and *Loharta* grow on *Betula* and *Rhododendron* trees in moist, wet, shaded, woody habitats between 2800–3500 m altitudes. In Alpine regions, *Cladonia* species primarily prefer dead wood and humus-rich soil near the trunks of trees. Only a few crustose lichens-*Pertusaria, Ochrolechia,* and *Lecidea* can grow on the *Quercus* species with rough, dry-bark in dry, cold, and open areas.

Besides, large trees in temperate regions with smooth bark, such as *Pyrus pasia* (a under canopy tree), provide an excellent substrate for fruticose lichen species such as *Ramalina* and *Usnea* and a few species of foliose lichens such as *Parmelia, Everniastrum, Pyxine,* and *Heterodermia.* When lichens grow on rock surfaces (saxicolous), rocks' physical and chemical characteristics are essential. Saxicolous species of lichens growing on dry, exposed rocks are *Acerospora, Pertusaria, Caloplaca, Diplorchistes, Lecidea, Lecanora, Rhizocarpon, Umbilicaria* and *Dermatocarpon.* Saxicolous lichens in areas inside dense forests with plenty of moisture and shade are *Leptogium, Collema,* members *of Verrucariaceae.* Near the stream, luxuriant growth of *Dermatocarpon* species occurs on the vertical faces of the boulders and stones.

The temperate Himalayan region's most common terricolous lichens are the *Cladonia* and *Thamnolia* species (Upreti 1997). Species of *Caloplaca, Rinodina,*

Diploschistes and *Bacidia* are muscicolous or humicolous in moist and shady places from temperate to alpine regions. The *Peltigera* species grow luxuriantly in moist areas in association with mosses. In vertical slopes in moist places, the *Peltigera* and *Lobaria* species grow luxuriantly.

1. **Lichen species association on different tree species**: On the basis of habitat preference a large number of foliose and fruticose lichens are reported on bark of *Quercus* trees. However, crustose and squamulose lichens are also observed on bark of coniferous trees at high altitude regions. In tropical forests *Pinus roxburghii* was observed as an excellent host tree for occurring foliose lichens on trunk portions; these lichens are *Cetraroipsis, Parmotrema, Everniastrum* etc. Similarly, on the canopy region of pine trees Ramalinoid and Usneioid (fruticose lichens) were recorded from Katarmal and Kasar forest of Almora (Uttarakhand). At a high-altitude region around 2800 m altitude *Q. semicarpifolia* tree was found a suitable phorophyte for abundance of lichen species followed by *Q. floribunda* and *Q. leucotrichophora* as co-dominant host trees for lichen growth. Besides these, *Pyrus pasia, Rhododendron arboreum* and *Abies pindrow* are also found suitable hosts particularly preferred by foliose forms to grow. In both temperate and alpine forests *Rhododendron* trees were recorded as an excellent phorophyte for the luxuriant growth of some crustose lichens on its smooth bark.

The thinner and smoother dry twigs were found an excellent medium to grow fruticose lichen taxa such as *Usnea* and *Ramalina* species which are hanging lichens and attached on the substratum with a single point. Thus, due to their point attachment with the bark of thinner twigs, they possess a best association of several lichen species. It was also observed that lichen mostly prefer to grow on dead or dry parts of trees. In higher elevation forest towards the alpine region the highest number of lichen species (42 species) were recorded in the forest sites having high density and canopy cover of forest trees along with higher forest cover. These canopy rich dense forests are also found rich in terms of ground vegetation and it provides high humus, moist, and precipitation. Dense, shady and closed canopied forests support luxuriant growth of terricolous, saxicolous, muscicolous and foliicolous types of lichens. Sometimes lichens also prefer to grow on age old rocks, in high altitudes with ample moisture lichen thallus sometimes attached loosely or tightly to the substratum.

It's crucial to examine into the study of lichens that are specific to particular hosts because the pH level of the host organism significantly impacts the accumulation of alkaline substances within the lichen structure. In essence, the acidity or alkalinity of the host environment can profoundly influence the chemical composition and metabolic processes of the lichen symbionts, ultimately affecting their ability to accumulate alkaline compounds. By understanding this relationship between host pH and alkaline substance accumulation, we gain valuable insights into the ecological dynamics and functional roles of lichens within their respective habitats.

2. **Lichen species association on different shrub species**: Study of Tamta and Kumar (2023) shows that maximum 18 species of macrolichens (foliose and fruticose) also prefer to grow on different shrub phorophytes. These 18 species of

lichens were recorded from a temperate forest of Morunaula, Kumaun Himalaya. In case of lichens species association on different shrubs a best association of nine species was found on *Cotoneaster microphyllus* followed by *Berberis petiolaris* with five species of lichens. Therefore, several shrubs also provide a good substratum to flourish the lichen species indifferent forest types of Himalayas.

3. **Role of lichen flora in relation to ecological behaviour of fauna in different forest types**: Musk deer is the state animal of Uttarakhand which is found in high altitude forests and alpine areas. Its famous food is *Usnea* lichen. Thus, the population of musk deer in these areas, its alternative food and availability of *Usnea* lichen should also be studied. In this sequence, in the higher Himalayan forests, the *Alectoria* and *Usnea* species of lichens are also used by some birds to build their nest. There is also a need to find out whether the concerned bird depends only on lichen for building its nest or uses other plants for this purpose and what is the level of population of such birds in that area? At some places in high altitude forests, the Runda species, a tailless rat, lives in burrows in the old walls along the roadside and uses the *Stereocaulon* lichen (saxicolous) growing on these walls as food. It will also be interesting to see what is the current number of Runda in these places and what is the status of this species of lichen. Thus, there is a great need to know this relationship between wildlife and lichens and their present status so that the importance of lichens in different types of forests can be understood in more depth.
4. **Nitrifying potential of lichens**: Nitrifier lichens are play a role in the nitrogen cycle by converting ammonia or ammonium ions into nitrates through the process of nitrification. This conversion is facilitated by certain species of lichen-forming fungi that contain specialized enzymes capable of oxidizing ammonia. Nitrifier lichens contribute to nitrogen cycling in ecosystems, making nitrogen available to other organisms and influencing nutrient dynamics in the environment. Their ability to participate in nitrification processes highlights their importance in ecosystem functioning and nutrient cycling. A sum of 14 fallen lichen species were recorded as nitrifier lichens which fix atmospheric nitrogen in different forest types of Western Himalayan region. These lichen species are- *Hypotrachyna cirrhata, Cetrariopsis wallichiana, Parmotrema tinctorum, P. praesorediosum, Usnea orientalis, Ramalina conduplicans, R. sinensis, Heterodermia diademata, Canoparmelia ecaperata, C. texana, Parmelia sp., Leptogium sp., Sticta sp.* and *Sulcaria sulcata.* Out of 14, maximum nine species of lichen taxa were recorded as fallen lichen (lichen litter) from three types of *Quercus* forests of Himalaya. These lichen species are regularly fallen from trees in the forest and after decomposition they help in nutrient cycling of the forest. In the decomposition process Usneoid lichens were found to be the fastest decomposer species as compared to foliose and Ramalinoid lichens. Hence, studying the presence or absence of these key lichen species (lichen litter) in the area can also provide valuable insights into mechanism of nutrient cycling of the forest and should be monitored across various forest types.
5. **Acidifying potential of lichens**: Some species have a good capacity of tolerant of pollutants and possess secondary metabolites. *Usnea* species forming usnic

acid. The acidifying potential of lichens refers to their capability to influence soil or substrate acidity through various processes, such as the release of organic acids during metabolism or the accumulation of acidic byproducts (alkaline, CO_2 etc.). Lichens contribute to soil acidification, particularly in areas with high lichen biomass or where lichens dominate the vegetation cover. This acidifying effect can impact soil chemistry, nutrient availability, and plant growth in ecosystems where lichens are abundant. Understanding the acidifying potential of lichens is essential for assessing their ecological role and their effects on ecosystem dynamics. Hence, it is essential to study key lichen species to assess their acidifying status.

6. **Lichen recovery in different forests**: In recent times, there has been a notable migration of households from the hilly areas, accompanied by a reduction in animal husbandry, resulting in the abandonment of animals on the streets. As a result, it appears that the exploitation of forests is diminishing. Additionally, the nature of people's occupations may have shifted due to the COVID-19 pandemic, with cooking gas replacing wood as a fuel source in the village. Given these circumstances, has the rate of lichen exploitation also declined? Urgent research is necessary to determine whether this has led to an enhanced recovery of lichens in forests.
7. **Conclusion**: In the last few decades, some lichen species have been in danger due to the deforestation of trees found in lower regions. These areas construct roads, parking, buildings, and other domestic uses. Every year, forest fires in pine forests cause immense damage to terricolous lichens and the lichen species found on the trunks of pine trees. Similarly, in temperate and alpine forest areas, lichen species, particularly those that grow as terricolous (soil inhabiting) and on shrubs and bushy trees (corticolous), are also severely harmed due to walking, picnicking, horse riding by the tourists and grazing of sheep and goats. It is also often seen that due to increasing population in the lower reaches of the tropical and temperate regions, encroachment on forests, increasing vehicle pollution, and climate change, the movement of lichen species is happening from these places to the upper areas, which is a matter of concern. Therefore, there is a need to understand these indicator values of lichen seriously.

References

Champion HG (1968) A revised survey of the forest type of India

Kumar B (2009) Assessment of lichen species in a temperate region of Garhwal Himalaya India. J Am Sci 5(4):107–112

Upreti DB (2008) An account of lichens on fallen twigs of three Quercus species in Chopta forest of Garhwal Himalayas, India Annals of Forestry. Ann For 15(1):92–98

Upreti D (1997) Recent researches in ecology, environment and pollution (vol XI). In: SC Sati (ed) Delhi, Delhi, India: Today and Tomorrow's

Chapter 18
Impact of Climate Change on Himalayan Ecosystem and Traditional Crops

Amit Kumar and Suresh Kumar

Abstract The Himalayan ecosystem, renowned for its biodiversity and cultural significance, is facing unprecedented challenges due to the impacts of climate change. These changes are analyzed in the context of their direct and indirect effects on the diverse species that inhabit the Himalayan ecosystems. Traditional crops including rice, wheat, barley, millets and numerous horticultural products are affected in terms of growth, production and geographic distribution by a number of climatic factors including rising temperatures, modified precipitation patterns and shifting weather extremes. The delicate balance of the Himalayan environment is seriously threatened by these changes, which also affect the availability of water, soil fertility, and pest dynamics, thereby challenging the traditional agricultural practices. The socio-economic ramifications of climate change on Himalayan communities heavily dependent on agriculture have several adverse effects. Shifts in traditional crop calendars, increased vulnerability to extreme weather events, and the potential loss of indigenous knowledge pose significant challenges to food security and livelihoods. This chapter will provide a comprehensive review of the effects of climate change on traditional crops cultivated in the Himalayan region, examining both the environmental and socio-economic dimensions. Additionally, it will examine the adaptation strategies employed by local communities, including the introduction of climate-resilient crop varieties, improved water management practices and community-based initiatives aimed at mitigating the adverse effects. It underscores the urgency of recognizing and addressing the multifaceted impacts of climate change on the Himalayan ecosystem and traditional crops. By fostering collaborative efforts among scientists, policymakers and local communities, we can develop adaptive strategies that not only preserve the rich cultural heritage of the region but also ensure the sustainability of agriculture in the face of a changing climate.

Keywords Biodiversity · Climate change · Traditional crops · Himalayas · Ecosystem

---

A. Kumar · S. Kumar (✉)
Department of Biosciences, Himachal Pradesh University, Shimla, H.P. 171005, India
e-mail: shahriask@gmail.com

D. Arya et al. (eds.), *Climate Change Impact on Himalayan Biodiversity*,
Environmental Science and Engineering, https://doi.org/10.1007/978-3-031-77149-1_18

18.1 Introduction

The Himalayan region is rich in its distinct environment, location and culture (Mani 1974). It is one of the hotspots of biodiversity. The Himalayan ranges exhibit a diverse climate, vegetation and land use pattern (Kumar and Gupta 2023). Because of the enormous changes in environment, ecoclimatic conditions and habitat types, the region has one of the most complex and diversified ecosystems (Singh 2006). The Indian Himalayan region encompasses approximately 4,19,873 km^2 and extends from Jammu and Kashmir in the north-west to Arunachal Pradesh in the east (Radha et al. 2019). Natural and cultural variety are considered to be abundantly stored in the Himalayan region. The International Panel on Climate Change has received scientific opinion indicating that climate change is occurring globally and affecting all forms of ecosystem services that humans depend on. The Himalayan mountain ecosystem is particularly vulnerable to climate change and its effects will increase the risks to livelihoods for those who directly depend on the products and services provided by these ecosystems for their survival and development (Singh et al. 2010; Houet et al. 2010; Schilling et al, 2012). In the Himalayan region, 70% of people are dependent on agriculture for their livelihoods which also significantly increases the household food security of local communities (Hussain et al. 2016). Discourses on agriculture in the Himalaya frequently brought attention to the region's ongoing agrarian distress, which is reflected in the region's declining yield in recent decades, a decline in food security and increasing poverty (Negi et al. 2012; Gautam and Andersen 2016; Ojha et al. 2017). Farmers in the area are suffering from enormous social, economic and environmental changes brought by climate change (Barua et al, 2014). The effects of climate change are expected to significantly alter social and ecological systems around the world and also the human behavior (Evans 2019). Additionally, Himalayan farmers' susceptibility is increased by their limited landholdings, high reliance on the rain and lack of irrigation facilities (Rasul 2014). Climate-related shift results in reduced crop diversity, increase in pest invasion and decline in crop yield (Kaul and Thornton 2014).

18.2 Climate Change Trends in the Himalayan Region

Climatic change is considered to be sensitive to the mountain biomes (Nogues-Bravo et al. 2008). According to the Third Assessment Report of Intergovernmental Panel on Climate Change, there will be an increase in precipitation (3% by 2020 and 11% by 2080) and temperature (1.4 °C by 2020 and 3.8 °C by 2080s) over the Indian subcontinent which would cause cyclones, flash floods, droughts and forest fires (IPCC2001; Kumar and Ashrit 2001). According to a thorough report developed by the International Centre for Integrated Mountain Development, the Eastern Himalayas are warming broadly by 0.01 to 0.04 ºC annually (Sharma et al. 2009a, b, c). Biodiversity is severely impacted by climate change which affects the phenology and distribution of species as well as biomes (Visser 2010). Due to rapid rate of warming, several species would not be able to adapt and results in extinction (Harley 2011). Several important aspects of the Himalayan region including ecosystems, agriculture, biodiversity, snow cover, glaciers, run-off processes and water supply are all impacted by climate change either directly or indirectly (Tewari et al. 2017; Bajracharya et al. 2018).

18.3 Glacier Change

Based on the recent estimates, the glacier coverage of the IHR ranges from approximately 25,000–30,000 km^2 (Pfeffer et al. 2014). The tongues of several of the Himalaya's glaciers are densely covered with debris. Across the Himalaya, debris covers percentages of the glacier area are approximately about 5% to 15% (Bajracharya et al. 2018). Thick covering of the debris make the glacier tongue more stable and significantly reduces melting of the glaciers (Bolch et al. 2008). The Himalayan glaciers have generally been decreasing at different rates since the nineteenth century, according to compilations of temporal changes; the Karakoram glaciers, on the other hand, have exhibited long-term unpredictable behavior (Hewitt 2011). According to regional estimates of changes in glacier length, the glaciers in the western Himalayas (Jammu and Kashmir, Himachal Pradesh) retreated more than those in the central (Uttarakhand) and eastern (Sikkim) Himalayas (Raina 2009). On the other hand, glacier mass balances also exhibit significant variability. The observed variability in glacier response to climate change can be explained by locally specific variations in topography, debris cover and meteorological reasons (Garg et al. 2017).

18.4 Change in Snow Cover

One of the most delicate natural resources, snow, requires a thorough temporal evaluation of numerous metrics at various scales (Wang and Li 2003). Reflective quality of snow has a significant global impact on energy exchange, surface hydrology and climatic changes. Seasonal snow cover provides significant short-term freshwater storage at the basin scale and is a crucial component of volumetric meltwater run-off modeling, predicting snow hazards and glacier mass balance (Haeberli et al. 2001). According to Gurung et al. 2011, there was an increase of 10–12% in snow cover of the eastern and western Himalaya and a decline of 12–14% in the central region of Hindu Kush Himalaya during 2000–2010. At the basin-scale, over the same period (2000–2011), the Indus basin has shown an increasing trend in snow cover, whereas the Ganga and Brahmaputra basins have shown decreasing trend (Singh et al. 2014). Rathore (2015) showed a statistically negligible increase in snow cover of approximately 1–2 percent in the Central Himalaya (Uttarakhand region) for the three sub-basins of the Ganga river.

18.5 Permafrost

Sediment or bedrock that has remained frozen for two years or more is known as permafrost. It primarily occurs in high mountain regions (Haeberli et al. 1993) and the Arctic zones (Westermann et al. 2015). It includes around 25% of the Northern Hemisphere's landmass (Bolch and Christiansen 2014). According to a number of studies, permafrost in the Northern Hemisphere has thawed over the last few decades (Nelson 2003). According to a preliminary regional modeling assessment, permafrost area in the Hindu Kush Himalayan region may encompass up to 1 million km^2, or about 14 times the area covered by glaciers (Gruber 2017). In the Kullu district of Himachal Pradesh, Allen et al. (2016) estimated a 420 km^2 permafrost spread. According to reports, permafrost and related landforms including ice-cored moraines and sorted patterned ground can occasionally be found in the cold, arid eastern regions of Jammu and Kashmir (Gruber 2017).

18.6 Landslides and Floods

The Indian Himalayan region is prone to frequent floods and landslides, which are distinguished by a remarkable capacity for sediment transport and fluvial erosion (Staubli 2018). The Himalayan relief's lifted humid monsoon air masses that are responsible for severe rainfall can be triggered by cloudburst and snowmelt processes (Gardner 2002). In populated downstream valleys, flood and landslide events annually result in significant economic losses as well as mortality. Severe occurrences cause damage to the infrastructure and may increase poverty in nearby vulnerable communities (Singh 2006). Severe floods can also cause epidemics to spring out, disrupt crops and ecosystems (Bhattachaiyya and Bora 1997). Massive amounts of sediment from landslides have an impact on fluvial geomorphology and can obstruct river corridors, creating transient lakes with a risk of dam failure and catastrophic downstream flooding (Ruiz-Villanueva et al. 2016).

18.7 Traditional Crops of the Himalaya

It had been previously determined that Indian Himalayan region was identified as center of crop diversity (Arora 1980). It was the result of geographical isolation and different environments that encouraged farmers to become more subsistence-minded which led to the emergence of such diversity. The early residents brought many crop varieties to the area.

Arora and Nayar (1984) noted that the indigenous crop varieties that are adapted to a certain place exhibit tremendous variance due to a variety of edaphic, topographic and climatic conditions combined with the varying selection pressure of centuries of farming. Globally significant agricultural diversification is found in high alpine regions where traditional crops include buckwheat, naked barley and many varieties of millet, amaranth, bean and highland rice (Table 18.1) with its special qualities of resistance to cold and drought, suited to hard, risky marginal situations (UNEP GEF 2013). Farmers have been growing these crops for millennia and as a result, they have assisted in ensuring the food security of underprivileged populations regardless climate change (Gauchan and Khanal 2011). With its extremely high ability to withstand cold, cold-tolerant rice is farmed in Nepal (Chhumchaur, Jumla) at the highest altitude in the world, 3030 masl. In Nepal, where poverty rates are highest, naked barley and wheat are also farmed at high altitudes close to the snow line, and provides a stable source of food and livelihood (CBS 2012; NPC 2016). In many of the isolated hills and hilly areas, local mountain communities heavily depend on these traditional underutilized crops which greatly aid in their site-specific food security, nutrition and adaptation (Joshi and Shrestha 2018). These crops are resilient to climate change and rich in nutrients. Small-scale farmers can rely on them for food, fodder, nutrition, livelihood and ecological security. They also present opportunities for value chain growth and income production. Amaranth, finger millet, proso millet and foxtail millet are a few of these crops that are gluten-free, high in dietary fiber, uncommon amino acids, antioxidants, vitamins, and higher in protein than major food staples like rice and wheat. They are also rich in micronutrients like calcium and iron. For this reason, they are regarded as both future crops and Himalayan superfoods. Additionally, majority of these crops are resilient and suitable for climate change adaptation because they can be cultivated in arid marginal lands with little water or inputs (Gauchan et al. 2018). These crops have a short production period which allows them to survive both biotic and abiotic challenges. They tolerate drought and cold temperatures and assure food availability in shorter periods of time during lean seasons. These traditional crops are regarded as "future smart foods" in the changing climate for site-specific food supply and adaptation because of their high nutritional content, resilience to climate change and risk diversification (Joshi et al. 2019).

18.8 Current Status of Himalayan Biodiversity

The biological and cultural richness of the Himalaya is well-known. According to Myers et al. (2000), Himalayan ecosystems are significant not just as repositories of biodiversity but also as sources of raw materials and for the sociocultural significance (Stepp et al. 2005). More than 10,000 different species of vascular plants are thought to be found in the Himalaya, with over a third (31.6%) being endemic (Mittermeier et al. 2004). However, the exact number of species is still up for debate and many more are still being discovered in the area. The existence of many species in the Himalaya is still threatened by rapid climate change, which has been exacerbated worst by the increase in anthropogenic activities like urbanization and deforestation, which have substantially altered the region's land cover (Cadotte et al. 2018). In fact, recent research and syntheses has shown that changes in biodiversity, both at

Table 18.1 Traditional crop diversity in the Himalayan region

Crop species	Vernacular name	Family	Ecological elevation (m)
Allium cepa L.	Pyaz	Amaryllidaceae	500–1100
Allium sativm L.	Lahusan	Amaryllidaceae	500–1500
Amaranthus oleracea L.	Chulae	Amaranthaceae	500–2000
Amaranthus spinosus L.	Ramdana	Amaranthaceae	1500–2500
Brassica. campestris L.	Sarson	Brassicaceae	500–2200
Brassica rapa L.	Shalgam	Brassicaceae	1000–2200
Chenopodium album L.	Bathu	Amaranthaceae	500–2500
Cicer arietinum L.	Chana	Fabacaeae	500–2000
Colocasia esculenta (L.) Schott	Kachyalu	Araceae	500–1500
Cyamopsis tetragonoloba (L.) Taub	Goar Fhali	Fabacaeae	800–1000
Daucus carota L.	Gazar	Apiaceae	800–2500
Echinochloa frumentacea (Link) Kunth	Jhangora	Poaceae	500–1800
Eleusine coracana Gaertn	Koda	Poaceae	500–2000
Fagopyrum esculentum Moench	Oggal	Polygonaceae	1500–2500
Fagopyrum tataricum (L.) Gaerrtn	Paphar	Polygonaceae	1500–2500
Glycine max (L.) Merr	Soyabean	Fabaceae	500–1400
Hordeum vulgare L.	Jau	Poaceae	500–2000
Lens esculenta Moench	Masoor	Fabaceae	500–2000
Lycopersicum esculentum L.	Tamatar	Solanaceae	500–2000
Macrotyloma uniflorum (Lam.) Verdc	Kolath	Fabaeae	500–1500
Oryza sativa L.	Chawal	Poaceae	500–1800
Phaseolus vulgaris L.	Rajmah	Fabaceae	1500–2500
Pisum sativum L.	Matar	Fabaceae	500–1500
Sesamum indicum L.	Til	Pedaliaceae	500–1500
Solanum tuberosum L.	Aalu	Solanaceae	500–2500
Triticum aestivum L.	Gehun	Poaceae	500–2500
Vigna mungo (L.) *Hepper*	Urad	Fabaceae	500–1800
Vigna radiata (L.) R. Wilczek	Moong	Fabaceae	500–1000
Zea mays L.	Makka	Poaceae	500–1800

the gene and species levels, are primarily caused by human activity (Woodbridge et al. 2020). Over the past 40 years, the Himalaya's ecosystem services have been severely affected by the rapid loss of ice (Bolch et al. 2012; Maurer et al. 2019), especially in the drier western Himalaya (Lutz et al. 2014) and results in disturbance of the ecosystem services. Rising air temperatures, which have increased by 1 °C since 1975, much faster than the global mean, has increased by 0.84 °C since 1,880 (Maurer et al. 2019). The combined effect of these disturbance factors may increase the risk of extinction for Himalayan biodiversity (Pandit et al. 2014).

The Indian Himalayan Region (IHR) exhibits a high biodiversity, with over 10,000 plant species, 300 animals, 977 birds, 281 herpetofauna species, 269 fish and numerous species of invertebrates and microorganisms, many of which are important for worldwide conservation (Kumar et al. 2019). The delicate mountain ecosystems in the IHR are gravely threatened by growing anthropogenic pressures despite their important ecological, hydrological and biological importance. The IHR is home to a number of wildlife species that are under threat from illegal wildlife trade, poaching for meat, unfavorable human-wildlife conflicts, habitat loss, fragmentation and degradation brought on by human development and use of natural resources (Sathyakumar and Bashir 2010). The National Mission for Sustaining the Himalayan Ecosystem (NMSHE) was established in 2008 by the Ministry of Environment and Forests as part of the National Action Plan on Climate Change (NAPCC) to address concerns specific to the Himalayan areas (Pandey 2015). Additionally, its aim is to analyze and recognize inputs for Himalayan sustainable development while keeping in mind the protection or fortification of this fragile ecosystem (Tiwari et al. 2017). It will also be essential to share ideas and information with south Asian provinces and territories that share or contribute to the Himalayan resources in order to restore and attract awareness in the Himalayan biodiversity. In order to assess and validate the forest and hydro resources along the health of the ecological unit, it is also necessary to establish an experimental, practical and insightful monitoring set of connections for the Himalayan environment (Raina 2009).

18.9 Impact of Climate Change on Biodiversity and Crop Productivity

Since agriculture is ultimately responsible for 20% of greenhouse gas emissions, it is both a major contributor to and an aggressor of climate change (Chivangulula et al. 2023). Agro-environmental and agricultural output are significantly impacted by precipitation extremes, such as droughts, floods and forest fires. Due to the severe effects of climate change on food and water resources, the decline in agriculture poses a challenge to farmers' quality of life and is a major contributor to poverty (Zhang et al. 2023; Li 2023). Particularly in developing countries, agricultural systems are essential to national economies and to the financial stability of individual households (Chikafa et al. 2023). The average global temperature is also expected to rise by

1.1–3.7 °C by the end of the twenty-first century (Urban et al. 2023). Increased temperatures will have significant detrimental effects on agricultural growth, making global temperature shift patterns very vulnerable to changes in crop productivity (Ratnayake et al. 2023). Crop productivity will be greatly impacted over the next few decades by changes in critical abiotic parameters such as temperature, precipitation, solar radiation and CO_2 emissions (Raihan 2023). Lobell and Field (2007) state that from 1962 to 2002, agricultural yield of wheat was significantly decreased by rising temperatures. Climate change-related temperature increases in the future will make these issues severe (Rezvi et al. 2023). It is anticipated that rising global temperatures will drastically lower crop output (Ma et al. 2023). In Future, rice crop will also be impacted by warm overnight temperatures. Because of its erratic blooming patterns, rice responds to high temperatures by producing lighter grains and reduced seed set (Ayyenar et al. 2023; Raihan 2023). Heat directly affects flowers during the day, accelerating the start of peak flowering and shortening their anthesis duration (Liu et al. 2023). According to various researches, the decline in wheat output can be attributed to rising temperatures (Karatayev et al. 2022; Kuriachen et al. 2022; Farhad et al. 2023). Heat stress accelerates the growth of plants, slows down photosynthetic activity and significantly affects reproductive activities (Farhad et al. 2023). Climate change-related weather extremes can have a detrimental effect on crops. For example, excessive fog and bare cold can cause betel leaves to fall and discolor and result in compressing lemon leaves; decomposing pineapple roots, ultimately giving rise to a bright red color (Pautasso et al. 2012; Rahman and Alam 2016; Lim et al. 2023). Therefore, there is an urgent need to combine short- and long-term solutions for mitigating the disruptive implications of climate change. It has been found that adopting adaptable practices for increasing crop diversity can result in increased resilience to the effects of climate change (Vernooy 2022). The world's biodiversity is suffering greatly as a result of climate change, which is one of the main drivers of species extinction (Manes et al. 2021). The speed and intensity of climate change is causing a shift in the ranges of marine, freshwater and terrestrial organisms that can thrive in their respective environments (Raihan 2023). A variety of factors including species abundance and distribution, migration patterns, activity schedules and the use of microhabitats are all impacted by shifts in the average climate regimes (Allan et al. 2021). The range of a species is often determined by its tolerance of biological interactions, environmental stressors and dispersion restrictions. Thus, the only options available to local species are to accept, adapt, relocate or die (Hankin et al. 2023). Therefore, the species that succeed best are those that have a greater capacity for change and a reduced requirement for stability in their existing surroundings (Nehe et al. 2023). Vulnerability to extreme heat wave events and climate change is largely influenced by poor habitat connectivity and restricted access to microclimates. For example, fluctuations in carbon sequestration rates are being caused by the expansion of mangroves due to climate change (Buhaug et al. 2023).

Reduced water availability as a result of climate change is becoming an increasingly serious issue at higher elevations (Wester et al. 2019). The Himalayas frequently experience less snowfall, forward shifts in the pattern of rainfall, and more front-free days, even if snowmelt can temporarily boost the amount of water available at lower elevations (Kohler and Maselli 2009). Farmers in Upper Mustang, Nepal are having less access to water and must obtain it from distant locations. Due to water constraint, rice growers in India have had to adjust the period for harvesting and transplanting (Rymbai and Sheikh 2018). Water stress is affecting crop productivity and income in the agricultural sector. Climate change poses a hazard since it will result in less rainfall and more frequent and intense droughts. Precipitation is expected to rise in certain places, however cropping patterns may be impacted by forward shifts in rainfall patterns. Crop yield and farm income is declined and impacted by these developments. Moreover, runoff and soil loss are more common in agricultural fields and it is also noted that trees are also essential for preserving the quality of the water and their disappearance due to agricultural growth may cause streams to become more sedimented (Price 2003).

18.10 Traditional Crops Under Climate Stress

Numerous environmental conditions severely affect the crop plants of commercial importance. Drought, salt and low (cold) temperatures are few factors that significantly contribute to the global decline in agricultural output (Boyer 1982). The productivity of agriculture and the geographic dispersion of many species, particularly significant agricultural crops, are severely restricted by low temperatures (Ait Barka and Audran 1997). The cold acclimatization process entails modifications to expression of genes (Seki et al. 2003) and results in increase of proline and sugar content (Ait Barka et al. 2006). The chilling stress results in the alterations in metabolism and physiology that result in the synthesis of Osmo protectants such as proline, sucrose and sugar which enable osmotic adjustment for continuous water absorption (Kasuga et al. 2004). The growth of the stressed plants can be improved with the help of plant growth promoting rhizobacteria (Nabti et al. 2010). The Himalayan region's traditional crops are under stress due to climate change. Farmers are trying to increase farm income by producing more contemporary cash crop varieties in response to increasing concerns about food availability and income (Sharma et al. 2009a, b, c). It is reported that since 1970s, the area planted with traditional crops in the Sikkim Himalaya has dropped by 50–60%. These crops include horse gram, buckwheat, finger millet, soybean, sweet potato, barley, amaranths and chenopods (Sharma et al. 2009a, b, c). In a similar vein, potatoes which are prized in the market are replacing buckwheat, hog millet, foxtail millet and other traditional crops (Semwal et al. 2004). In the Trans-Himalayan region, contemporary cash crops are displacing traditional crops including black pea, indigenous potatoes and barely (Sharma and Chauhan 2013). Despite the potential short-term financial benefits, adopting commercial crops carries considerable environmental hazards (Semwal et al. 2004). First, because minor millet and beans are drought-tolerant and adapted to less fertile soils, these traditional crops are highly valuable both nutritionally and medicinally. They are even thought to neutralize the negative economic effects of globalization and environmental change (Sharma et al. 2009a, b, c). Therefore, food and financial security would suffer from a decline in conventional crops. Second, extensive cultivation of contemporary crop varieties, such as commercial potato farming, is linked to increasing soil erosion and runoff (Semwal et al. 2004) and it is crucial that recently implemented agricultural systems in the agricultural frontiers affected by climate change seek to reduce rather than increase carbon emissions. In the Himalayan region, barley is a main crop, especially in nations like Nepal and Bhutan. Its yield may be affected by variations in temperature and water availability which make it extremely sensitive (Yadav et al. 2018). In many parts of the world, barley is primarily used to make malt and animal feed. On a global scale, it is the fourth most significant cereal crop (Zhou 2010). In the Himalayas, potatoes are an essential food crop, although cultivation can be impacted by variations in temperature and precipitation patterns. Other challenges arise from altered dynamics of pests and disease (Mishra et al. 2015). A significant grain crop in the foothills of the Himalayas is wheat. Wheat growth phases can be impacted by variations in temperature and precipitation which could result in lower yields (Tiwari et al. 2017). Although rice is a traditional crop that thrives in the Himalayas, its cultivation may be impacted by changes in growing circumstances. Concerns include altered flowering patterns and increased susceptibility to illnesses and pests (Pandey and Hartmann 2020). The Himalayan region is home to many apple orchards, particularly in states like Jammu & Kashmir and Himachal Pradesh. Warmer winters can change the amount of chilling needed which can change the patterns of flowering and fruiting (Kumar et al. 2019).

18.11 Socio-Economic Impacts of the Climate Change

Climate has a significant effect on development and economic productivity. Global environmental officials are becoming more and more concerned about climate change due to its effects on economic growth (Lamperti et al. 2021). The economic conditions of Himalayan population are impacted by climate change in a number of areas including forestry, fisheries, agriculture, cattle, tourism and human health (Sharma et al. 2009a, b, c). As people in the region primarily rely on forest resources, those who are economically weak and whose only source of income is natural resources are likely to suffer the most from the loss of biodiversity caused by climate change. Changes in the ranges of disease vectors (such as mosquitoes), water-borne diseases, water quality, air quality and food availability as well as quality are now showing signs of the effects of climate change on human health (McMichael et al. 2003). The majority of mountain farming is dependent on rainfed systems. Therefore, it continues to be a subsistence occupation and does not offer farmers opportunity to generate revenue. The negative effects of climate change are now more likely to affect agriculture in northwestern Himalaya (INCCA 2010; HPSAPPC 2012; Talib 2007; UKSAPCC 2014). Due to rising temperatures and lack of water, traditional Kashmiri apples and summer types of rice (nick cheena) vanished from several parts of J&K.

In the several districts of J&K, some 11,909 kanals of paddy land have been transformed into rainfed dry land due to the irregular rainfall in recent years. There is now a 44% food grain shortfall, a 33% vegetable shortage, and a 69% oilseed shortage in the Kashmir Valley, increasing the risk to food security. The agricultural production systems and eventually, the food security of the people in Himachal Pradesh are beginning to suffer from the chilly season and unpredictable rains (Basannagari and Kala 2013). Despite the fact that Uttarakhand is home to multiple climate zones, socioeconomic changes have recently led to a drop in the interest of the local mountain farming people in traditional crop cultivation as well as a decline in yields. The cultivation of mountain-specific crops such as *Fagopyrum tataricum*, *Hordeum himalayense*, *Solanum tuberosum*, *Hordeum vulgare* and *Panicum miliaceum* has been decreased from 5 to 16% in a few locations in Uttarakhand state between the 1970s and 2005 while the habit of cultivating wheat completely disappeared from a few communities (Maikhuri et al. 2001). The most significant horticultural crop in the state of Himachal Pradesh, apple, is susceptible to long-term temperature increases. The apple belt has been moved 30 km to the north during the past 50 years. The apple productivity has decreased recently by 2–3% and the largest decrease has been observed in marginal farms (4%). The area planted to apples and other fruits has decreased more in marginal farms (33.3%) than in small and big farms (4.9% and 5.6%, respectively). Due to the impact of winter droughts and hot summers on the sowing and yield of rabi season crops especially wheat and lentil in rainfed farming, marginal and small farmers are more susceptible to food insecurity.

18.12 Adaptation Strategies Against Climate Change

The nomadic farmers have evolved an advanced system of land use and management in response to many threats and difficulties posed by severe and extreme environmental conditions. This includes seasonal animal movements, sharing of grazing areas, rotational grazing, producing fodder for the lean season and other practices. The traditional farmers in the Himalayan region have been designing and reconstructing the land use attributes for maintaining agro-diversity systems. The various conventional methods are adapted in the Himalayas for sustainable agriculture and to control weather and climate risks (Sharma and Rai 2012).

- **Trans-Himalayan agropastoralism**—It is the conventional method used for sustainable agriculture and to control weather and climate risks. It includes rotational grazing, animal sharing and usage of grazing areas and production of fodder for the lean season. Growing potatoes, cabbage, green vegetables, barley, medicinal and fragrant plants in the North Sikkim regions of Thangu, Yumesamdong, Dambuchey-Bamzey and Lhonak valley (Muguthang area).
- **Traditional agriculture systems**—It includes integration and production of high-value cash crops in various agro-ecological zones under integrated farming such as giant cardamom, mandarin orange, ginger and potato. Many landrace varieties

of ginger, yams, tubers, beans, legumes, maize, buckwheat, rice and other crops are grown in rotation.

- **Land source management**—It include terrace-based farming. In order to protect the water table and lessen the harmful effects of the drought, farmers are building shelter belts around the boundaries of their agricultural fields. Raised bunds at the terrace borders and conservation tillage help stop runoff and washouts after heavy rains.
- **Livestock management**—Farm animals are fed throughout the winter in fields to maintain agricultural nutrients and lower labor expenses when transporting manure from the sheds. Numerous farmers in the villages have started cultivating a wide variety of fodder species on terraces including understorey grasses and tree fodder.

As the weather and temperature changed, the locals developed adaptation plans based on their traditional knowledge and past experiences to deal with the changing climate. Mitigation refers to actions taken to lower greenhouse gas emissions, whereas adaptation refers to steps taken to improve people's and communities' capacity to adjust to the effects of climate change and related effects that will affect many facets of human existence (Smith and Olesen 2010). In response to climate fluctuation, people began adopting cash crops, which led to a drop in the area under traditional crops (Negi et al. 2017). Diverse adaptation measures are necessary across regions due to the varying consequences of climate change on agricultural activities (Malhi et al. 2021). Strategies at the agricultural level can be securing insurance or diversification. Adjusting compensation programs or reducing the number of farms are two examples of adaptation on a regional or national level. It could, on a worldwide scale, need modifications to global food market patterns (Smit et al. 2000). Additional essential methods for adaptation include (i) choosing and cultivating novel crop types that are resistant to heat, drought or both (Jenkins et al. 2023); (ii) creating gene bank (Ramirez Villegas et al. 2012); (iii) development of agroforestry system (Shivakumar et al. 2005); (iv) development and implementation of the more advanced methods of the irrigation (Muchuru and Nhamo 2019) and (v) use of renewable sources of energy in modern agriculture (Magesa et al. 2023). In addition to the present adaption tactics and plans, government assistance for seasonal migrations and other livelihood options should be acknowledged as a way to enhance the sustainability of agriculture and its resilience to climate change (Bryan et al. 2013). There is alteration in cropping patterns in middle altitude villages (1000–1800 m asl): *Phaseolus vulgaris* is grown instead of *Macrotyloma uniflorum*; *Vigna unguiculata* is substituted for *Cajanus cajan*; *Phaseolus vulgaris* is substituted with *Glycine max*. In lower altitude, there is cultivation of *Glycine max*, *Echinocloa frumentaceus* and *Elusine coracana* in place of *Oryza sativa*.

As per information received from the Himachal Pradesh State Action Plan for Climate Change the various combination of planned approaches has been implemented in HP state agriculture sector as shown in the Fig. 18.1 (HPSAPCC 2012).

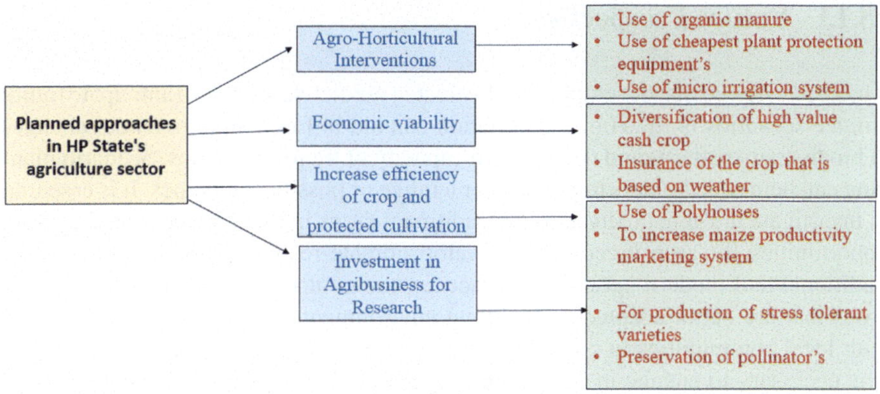

Fig. 18.1 Combination of planned approaches in HP State's agriculture sector

As per information received from the Jammu and Kashmir State Action Plan for Climate Change the various combination of planned approaches has been implemented in Jammu and Kashmir state agriculture sector as shown in the Fig. 18.2 (J&KSAPCC 2014).

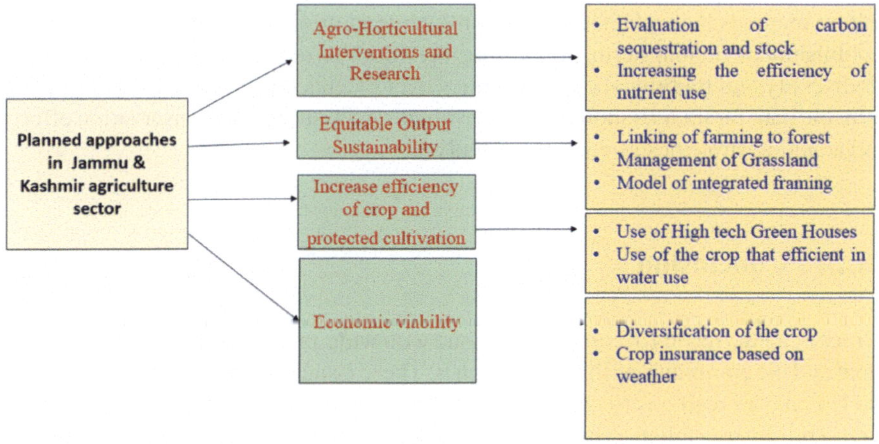

Fig. 18.2 Combination of planned approaches in Jammu and Kashmir state

18.13 Future Prospective

It has been viewed that developing sophisticated predictive models to anticipate future climate scenarios in the Himalayan region can aid in forecasting potential changes in biodiversity patterns and the adaptive capacity of traditional crops. Scenario planning can help stakeholders to prepare for a range of possible outcomes. It is essential to investigate the genetic diversity of traditional crops in the Himalayas and explore opportunities for crop breeding to develop climate-resilient varieties. Integrating traditional and modern agricultural practices can contribute to the development of crops that are better adapted to changing environmental conditions. Collaborating with local communities to co-create conservation strategies can be more effective. It is necessary to engage in participatory research that involves indigenous knowledge, incorporates community-based adaptation measures and respects traditional practices, ensuring the sustainability of both biodiversity and traditional agriculture. Research activities should involve implementation of ecosystem-based adaptation measures that leverage the natural functions of ecosystems to enhance resilience. This may include afforestation, habitat restoration and sustainable land management practices that contribute to biodiversity conservation while supporting traditional crop cultivation. There should be exploration and promotion of climate-smart agricultural practices that are tailored to the specific needs of the Himalayan region. This could include agroforestry, water-efficient irrigation techniques and precision agriculture methods that enhance productivity while minimizing environmental impact. Establishment of long-term monitoring networks is required to track changes in biodiversity and traditional crop dynamics over extended periods. This will provide valuable data for understanding trends, evaluating the success of conservation efforts and adapting strategies as conditions evolve.

18.14 Conclusion

Climate change results in complicated issue with wide-ranging effects on the environment and the communities that depend on it. These issues emphasize how susceptible the Himalayan region is to changing climatic circumstances such as rising temperatures, shifting precipitation patterns and glacier melting. The traditional crops that are the basis of the region's agriculture and means of subsistence are seriously threatened by these changes. Traditional crops at risk include barley, potatoes, apples, wheat and buckwheat. The production of these crops is closely related to the welfare of populations in the Himalayas. It is impossible to exaggerate the relationship between environmental changes and social as well as economic variables. Exacerbating pre-existing vulnerabilities, local people face disruptions in water resources, agriculture, livelihoods, health and infrastructure. The region's economy depends heavily on tourism which is in jeopardy and will affect the earnings of individuals who work in it. The adoption of climate-resilient measures, the advancement of sustainable agriculture practices and the preservation of traditional varieties of crops are critical going ahead. The growing seasonal changes and the rise in pests and illnesses highlight how urgently adaptation measures are needed. Developing comprehensive strategies to tackle the distinct difficulties presented by climate change in the Himalayan ecosystem necessitates cross-disciplinary collaborations between scientists, policymakers, local populations and non-governmental organizations. The Himalayan region and its people can only expect to lessen the negative effects and create a more resilient and sustainable future through such coordinated efforts.

Acknowledgements The financial support granted by University Grants Commission (UGC) to Mr. Amit Kumar (Award No NSFDC/E-71131) under JRF scheme is highly acknowledged.

References

Ait Barka E, Audran JC (1997) Response of champenoise grapevine to low temperature: changes of shoot and bud proline concentration in response to low temperatures and correlation with freezing tolerance. J Hortic Sci Biotechnol 72:577–582. https://doi.org/10.1080/14620316.1997.11515546

Ait Barka E, Nowk J, Clement C (2006) Enhancement of chilling resistance of inoculated grapevine plantlets with plant growth promoting rhizobacteria *Burkholderia phytofermans* strain. J Appl Environ Microbiol 72:7246–7252. https://doi.org/10.1128/AEM.01047-06

Allan JD, Castillo MM, Capps KA (2021) Stream ecology: structure and function of running waters. Springer Nature

Allen SK, Fiddes J, Linsbauer A, Randhawa SS, Saklani B, Salzmann N (2016) Permafrost studies in Kullu district, Himachal Pradesh. Curr Sci 11:257–260

Arora RK, Nayar ER (1984) Wild relatives of crop plants in India NBPGR. Sci Monograph 7:1–97

Arora RK (1980) Native food plants of the north-eastern tribal, Glimpses of Indian Ethnobotany Oxford IBH, New Delhi

Ayyenar B, Kambale R, Duraialagaraja S, Manickam S, Mohanavel V, Shanmugavel P, Muthurajan R (2023) Developing early morning flowering version of rice variety CO 51 to Mitigate the heat-induced yield loss. Agriculture 13(3):553

Bajracharya AR, Bajracharya SR, Shrestha AB, Maharjan SB (2018) Climate change impact assessment on the hydrological regime of the Kaligandaki Basin, Nepal. Sci Total Environ 625:837–848. https://doi.org/10.1016/j.scitotenv.2017.12.332

Barua A, Katyaini S, Mili B, Gooch P (2014) Climate change and poverty: building resilience of rural mountain communities in South Sikkim, Eastern Himalaya, India. Reg Env Change 14:267–280. https://doi.org/10.1007/s10113-013-0471-1

Basannagari B, Kala CP (2013) Climate change and apple farming in Indian Himalayas. A study of local perceptions and responses. PLoS ONE8 (10):e77976

Bhattachaiyya NN, Bora AK (1997) Floods of the Brahmaputra River in India. Water Int 22:222–229

Bolch T, Buchroithner MF, Pieczonka T, Kunert A (2008) Planimetric and volumetric glacier changes in Khumbu Himalaya since 1962 using Corona, Landsat TM and ASTER data. J Glaciol 54(187):592–600

Bolch T, Kulkarni A, Kaab A, Huggel C, Paul F, Cogley JG, Frey H, Kargel JS, Fujita K, Scheel M, Bajracharya S, Stoffel M (2012) The state and fate of Himalayan glaciers. Science 336:310–314

Bolch T, Christiansen HH (2014) Mountains, lowlands, and coasts: the physiography of cold landscapes. In: Snow and ice-related hazards, risks, and disasters. Elsevier Inc 201–217

Boyer JS (1982) Plant productivity and environment. Science 218:443–448
Bryan E, Ringler C, Okoba B, Roncoli C, Silvestri S, Herrero M (2013) Adapting agriculture to climate change in Kenya: household strategies and determinants. J Environ Manag 114:26–35
Buhaug H, Benjaminsen TA, Gilmore EA, Hendrix CS (2023) Climate-driven risks to peace over the 21st century. Clim Risk Manag 39:100471. https://doi.org/10.1016/j.crm.2022.100471
Cadotte MW, Wu ZY, Provan J, Zhu GF, Gao LM, Li DZ, Liu J, Milne RI (2018) Protect third pole's fragile ecosystem. Science 362:1368
CBS (2012) Nepal living standard survey (NLSS), Central Bureau of Statistics (CBS), National Planning Commission, Government of Nepal Kathmandu
Chikafa M, Nejadhashemi AP, Moller K, Razavi H, Bizimana JC (2023) Multidimensional evaluation of the impacts of agricultural interventions to achieve food security in Malawi. Food Energy Secur 12:e486. https://doi.org/10.1002/fes3.486
Chivangulula FM, Amraoui M, Pereira MG (2023) The drought regime in Southern Africa: a systematic review. Climate 11(7):147. https://doi.org/10.3390/cli11070147
Evans GW (2019) Projected behavioral impacts of global climate change. Annu Rev Psychol 70:449–474
Farhad M, Kumar U, Tomar V, Hossain A (2023) Heat stress in wheat: a global challenge to feed billions in the current era of the changing climate. Front Sustain Food Syst 7:1203721
Gardner JS (2002) Natural hazards risk in the Kullu District, Himachal Pradesh, India. Geogr Rev 92:282–306
Garg PK, Shukla A, Jasrotia AS (2017) Influence of topography on glacier changes in the central Himalaya, India. Global Planet Change 156:196–212
Gauchan D, Khanal U (2011) Socioeconomic dimension of conserving diversity, review of socioeconomic, institutional and market dynamics of agrobiodiversity in mountain ecosystems of Nepal. Annex 1-C pp 1–11; Project Document. In: Integrating Traditional Crop Genetic Diversity into Technology: Using a Biodiversity Portfolio Approach to Buffer against Unpredictable Environmental Change in the Nepal Himalayas
Gauchan D, Bhandari B, Gurung R, Joshi BK, Jarvis D (2018) Value chain development of underutilized food crops for nutrition sensitive agriculture in the mountains of Nepal. In: Proceeding 5th international agricultural marketing conference, Kathmandu, June 4–5, 2018. Nepal Agricultural Economic Society (NAES), Ministry of Agriculture and Livestock Development (MoALD), FAO, KISSAN-2, USAID, Nepal 2019
Gautam Y, Andersen P (2016) Rural livelihood diversification and household well-being: Insights from Humla, Nepal. J Rural Stud 44:239–249
Gruber S (2017) Inferring permafrost and permafrost thaw in the mountains of the Hindu Kush Himalaya region. Cryosphere 11(1):81–99
Gurung DR, Giriraj A, Anug KS, Shreshtha B, Kulkarni AV (2011) Snow-cover mapping and monitoring in the Hindu- Kush-Himalayas. ICIMOD, Kathmandu, Nepal
Haeberli W, Guodong C, Gorbunov AP, Harris SA (1993) Mountain permafrost and climatic change. Permafrost Periglacial Process 4(2):165–174
Haeberli W, Kääb A, VonderMühll D, Teysseire P (2001) Prevention of outburst floods from periglacial lakes at Grubengletscher, Valais, Swiss Alps. J Glaciol 47(156):111–122. https://doi.org/10.3189/172756501781832575
Hankin LE, Leger EA, Bisbing SM (2023) Reforestation of high elevation pines: direct seeding success depends on seed source and sowing environment. Eco Appl e2897
Harley C (2011) Climate change keystone predation and biodiversity loss. Science 334:1124–1127
Hewitt K (2011) Glacier change, concentration, and elevation effects in the Karakoram Himalaya, Upper Indus Basin. Mount Res Dev 33(3):188–200
Houet T, Lovel TR, HubertMoy L (2010) Exploring subtle land use and land cover changes: a framework for future landscape studies. Landscape Ecol 25:249–266
Hussain A, Rasul G, Mahapatra B, Tuladhar S (2016) Household food security in the face of climate change in the Hindu-Kush Himalayan region. Food Secur 8:921–937

INCCA (2010) Climate change and India-A 4x4 assessment. INCCA (Indian Network on Climate Change Assessment): a sectorial and regional analysis for 2030s:1–164
IPCC (2001) Climate change impacts, adaptation and vulnerability. Intergovernmental Panel on climate change (IPCC), Cambridge University Press, Cambridge
Jenkins B, Avis K, Willcocks J, Martin G, Wiltshire J, Peters E (2023) Adapting Scottish agriculture to a changing climate-assessing options for action. Ricardo Energy and Environment, London, UK
Joshi BK, Shrestha R, Gauchan D, Shrestha A (2019) Neglected, underutilized, and future smart crop species in Nepal. J Crop Improv 34(3):291–313. https://doi.org/10.1080/15427528.2019.1703230
Joshi BK, Shrestha R (2018) In future smart food- rediscovering hidden treasures of neglected and underutilized species for zero hunger in Asia, pp 161–178
Karatayev M, Clarke M, Salnikov V, Bekseitova R, Nizamova M (2022) Monitoring climate change, drought conditions and wheat production in Eurasia: the case study of Kazakhstan. Heliyon 8(1):e08660. https://doi.org/10.1016/j.heliyon.2021.e08660
Kasuga M, Miural S, Shinozaki K, Yamaguchi-Shinozaki K (2004) A combination of the Arabidopsis DREB1A gene and stress inducible rd29A promoter improved drought- and low-temperature stress tolerance in tobacco by gene transfer. Plant Cell Physiol 45:346–350
Kaul V, Thornton TF (2014) Resilience and adaptation to extremes in a changing Himalayan environment. Reg Env Change 14:683–698
Kohler T, Maselli D (2009) Mountains and climate change—from understanding to action. Published by Geographica Bernensia with the support of the Swiss Agency for Development and Cooperation (SDC), and an international team of contributors. Bern Switzerland
Kumar S, Gupta D (2023) Forest biodiversity and livelihood of tribal people in North Western Himalaya with special reference to Himachal Pradesh, India. Int J Plant and Soil Sci 35(19):888–907
Kuriachen P, Devi A, Sam AS, Kumar S, Kumari J, Suresh A, Jha GK (2022) Wheat yield responses to rising temperature: insights from northern plains of India. Theor Appl Climatol 150(3–4):1157–1172
Kuumar RT, Shrit RG (2001) Regional aspects of global climate change simulations validations and assessment of climate response of the Indian monsoon region to increase of greenhouse gases and sulphate aerosols. Museum 52:229–244
Lamperti F, Bosetti V, Roventini A, Tavoni M, Treibich T (2021) Three green financial policies to address climate risks. J Financ Stabil 54:100875. https://doi.org/10.1016/j.jfs.2021.100875
Li B (2023) Research on the path of promoting the common prosperity of farmers and countryside in the New Era. Acad J Humanit Soc Sci 6(13):105–111. https://doi.org/10.25236/AJHSS.2023.061317
Lim JA, Yaacob JS, Mohd Rasli SR, Eyahmalay JE, Enshasy HA, Zakaria MS (2023) Mitigating the repercussions of climate change on diseases affecting important crop commodities in Southeast Asia, for food security and environmental sustainability. A review. Front Sustain Food Syst 6:1030540
Liu P, Yin B, Gu L, Zhang S, Ren J, Wang Y, Zhen W (2023) Heat stress affects tassel development and reduces the kernel number of summer maize. Front in Plant Sci 14:1186921. https://doi.org/10.3389/fpls.2023.1186921
Lobell DB, Field CB (2007) Global scale climate–crop yield relationships and the impacts of recent warming. Environ Res Lett 2(1):014002
Lutz AF, Immerzeel WW, Shrestha AB, Bierkens MF (2014) Consistent increase in High Asia's runoff due to increasing glacier melt and precipitation. Nat Clim Change 4:587–592
Ma M, Huang D, Hossain SS (2023) Opportunities or risks: economic impacts of climate change on crop structure adjustment in ecologically vulnerable regions in China. Sustainability 15(7):6211
Magesa BA, Mohan G, Matsuda H, Melts I, Kefi M, Fukushi K (2023) Understanding the farmers' choices and adoption of adaptation strategies, and plans to climate change impact in Africa: a systematic review. Clim Serv 30:100362

Maikhuri RK, Rao KS, Semwal RL (2001) Changing scenario of Himalayan agroecosystems: loss of agrobiodiversity, an indicator of environmental change in Central Himalaya, India. The Environmentalist 21(1):23–39
Malhi GS, Kaur M, Kaushik P (2021) Impact of climate change on agriculture and its mitigation strategies: a review. Sustainability 13:1318
Mani MS (1974) Ecology and biogeography in India. Junk Publications The Hague, Dr. B.V
Maurer J, Schaefer J, Rupper S, Corley A (2019) Acceleration of ice loss across the Himalayas over the past 40 years. Sci Adv 5(6):eaav7266. https://doi.org/10.1126/sciadv.aav7266
Manes S, Costello MJ, Beckett H, Debnath A, Devenish-Nelson E, Grey KA, Vale MM (2021) Endemism increases species climate change risk in areas of global biodiversity importance. Bio Cons 257:109070. https://doi.org/10.1016/j.biocon.2021.109070
McMichael AJ, Campbell-Lendrum DH, Corvalan CF, Ebi KL, Githeko AK, Scheraga JD, Woodward A (2003) Climate change and human health risks and responses. WHO, Geneva
Mishra A, Salokhe VM, Tripathi KP (2015) Impact of climate change on potato production in India. J Agro 17(1):1–7
Mittermeier RA, Gil PR, Hoffmann M, Pilgrim J, Brooks T, Mittermeier CG, Da Fonseca GA (2004) Hotspots revisited: earth's biologically richest and most endangered eco regions. CEMEX
Muchuru S, Nhamo G (2019) A review of climate change adaptation measures in the African crop sector. Clim Dev 11:873–885
Myers N, Muttermeier RA, Muttermeier CA, Da-Fonseca AB, Kent J (2000) Biodiversity hotspots for conservation priorities. Nature 403:853–858
Nabti E, Sahnoune M, Ghoul M, Fischer D, Hofmann A, Rothballer M, Schmid M, Hartmann A (2010) Restoration of growth of durum wheat (*Triticum durum* var. *waha*) under saline conditions due to inoculation with the rhizosphere bacterium *Azospirillum brasilense* NH and extracts of the marine alga *Ulva lactuca*. J Plant Growth Regul 29:6–22
Negi GCS, Samal PK, Kuniyal JC, Kothyari BP, Sharma RK, Dhyani PP (2012) Impact of climate change on the Western Himalayan mountain ecosystems: an overview. Trop Ecol 53(3):345–356
Negi VS, Maikhuri RK, Pharswan D (2017) Climate change impact in Himalaya: people perception and adaptive strategies. J Mt Sci 14:403–416. https://doi.org/10.1007/s11629-015-3814-1
Nehe A, Martinsson UD, Johansson E, Chawade A (2023) Genotype and environment interaction study shows fungal diseases and heat stress are detrimental to spring wheat production in Sweden. PLoS ONE 18(5):e0285565
Nelson FE (2003) Frozen in time. Science 299(5613):1673–1675
Nogues-Bravo D, Araujo MB, Romdal T (2008) Scale effects and human impact on the elevational species richness gradients. Nature 453(7192):216–219
NPC (2016) Nepal and millennium development goals: final status report. National Planning Commission (NPC), Kathmandu, Nepal 2000–2015
Ojha HR, Shrestha KK, Subedi YR, Shah R, Nuberg I, Heyojoo B, Cedamon E, Rigg J, Tamang S, Paudel KP, Malla MP (2017) Agricultural land underutilisation in the hills of Nepal: investigating socio-environmental pathways of change. J Rural Stud 53:156–172
Pandey V, Hartmann TP (2020) Climate change and rice production in the Indian Himalayan region: a review. Sustainability 12(3):1127
Pandey K (2015) PM's climate mission plan for Himalaya yet to take off. Down to Earth News. P41431
Pandit MK, Manish K, Koh LP (2014) Dancing on the roof of the world: ecological transformation of the Himalayan landscape. Bio Sci 64:980–992
Pautasso M, Döring TF, Garbelotto M, Pellis L, Jeger MJ (2012) Impacts of climate change on plant diseases—opinions and trends. Eur J Plant Pathol 133:295–313
Pfeffer W (2014) The Randolph Glacier inventory: a globally complete inventory of glaciers. J Glaciol 60(221):537–552
Price MF (2003) Why mountain forests are important. For Chron 79(2):219–222

Radha PS, Chandel K, Pundir A, Singh M, Chauhan B, Simer K, Dhiman N, Shivani SY, Kumar S (2019) Diversity of ethnomedicinal plants in Churdhar Wildlife Sanctuary of district Sirmour of Himachal Pradesh, India. J App Pharm Sci 9(11):048–053

Rahman MH, Alam K (2016) Forest dependent indigenous communities' perception and adaptation to climate change through local knowledge in the protected area A Bangladesh case study. Climate 4(1):12

Raihan A (2023) A review of the global climate change impacts, adaptation strategies, and mitigation options in the socio-economic and environmental sectors. J Environ Sci Econ. https://doi.org/10.56556/jescae.v2i3.587

Raina VK (2009) Himalayan glaciers: a state of art review of glacial studies, glacial retreat and climate change. Ministry of Environment and Forest, New Delhi

Ramirez Villegas J, Salazar M, Jarvis A, Navarro-Racines CE (2012) A way forward on adaptation to climate change in Colombian agriculture: perspectives towards 2050. Clim Change 115:611–628

Rasul G (2014) Food, water and energy security in South Asia: a nexus perspective from the Hindu Kush Himalayan region. Env Sci Pol 39:35–48

Rathore BP (2015) Spatio-temporal variability of snow cover in Alaknanda, Bhagirathi and Yamuna sub-basins, Uttarakhand Himalaya. Curr Sci 108(7):1375–1380

Ratnayake SS, Reid M, Larder N, Kadupitiya HK, Hunter D, Dharmasena PB, Kariyawasam CS (2023) Impact of climate change on paddy farming in the village tank cascade systems of Sri Lanka. Sustainability 15(12):9271

Rezvi HUA, Tahjib-Ul-Arif M, Azim MA, Tumpa TA, Tipu MM, Najnine F, Brestič M (2023) Rice and food security: climate change implications and the future prospects for nutritional security. Food Energy Secur 12(1):e430

Ruiz-Villanueva V, Allen S, Arora M, Goel NK, Stoffel M (2016) Recent catastrophic landslide lake outburst floods in the Himalayan mountain range. Prog Phys Geogr 41(1):3–28

Rymbai D, Sheikh FM (2018) The insight of agricultural adaptation to climate change: a case of rice growers in Eastern Himalaya, India. Int J Biometeorol 62(10):1833–1845

Sathyakumar S, Bashir T (2010) Wildlife of the Himalaya: Conservation issues and the way forward. In: Arora S, Negi BS, Bhan S, Bali JS, Bharti VK (eds) Mountain ecosystem and man. Soil Conservation Society of India, New Delhi

Schilling J, FreierKorbinian P, HertigeElke SJ (2012) Climate change, vulnerability and adaptation in North Africa with focus on Morocco. Agric, Ecosyst Environ 156:12–26. https://doi.org/10.1016/j.agee.2012.04.021

Seki M, Kamei A, Yamguchi-Shinozaki K, Schinozaki K (2003) Molecular responses to drought salinity and frost: common and different plants for plant protection. Curr Opin Biotechnol 14:194–199

Semwal RL, Nautiyal S, Sen KK, Rana U, Maikhuri RK, Rao KS, Saxena KG (2004) Patterns and ecological implications of agricultural land-use changes: a case study from central Himalaya India. Agric Ecosyst Environ 102(1):81–92

Sharma HR, Chauhan SK (2013) Agricultural transformation in Trans Himalayan region of Himachal Pradesh: cropping pattern, technology adoption and emerging challenges. Agric Econ Res Rev 36:173–179

Sharma E, Chettri N, Tse-Ring K, Shrestha AB, Jing F, Mool P, Eriksson M (2009a) Climate change impacts and vulnerability in the eastern Himalayas. International Centre for integrated mountain development, Kathmandu

Sharma E, Sharma G, Liang L, Subba JR, Tanaka K (2009b) Sikkim Himalayan-agriculture: improving and scaling up of the traditionally managed agricultural systems of global significance. Resour Sci 31(1):21–30

Sharma E, Chettri N, Tsering K, Shrestha AB, Fan J, Mool P, Eriksson M (2009) Climate change impacts and vulnerability in the Eastern Himalayas. International Centre for Integrated Mountain Development (ICIMOD), Nepal

Sharma G, Rai, LK (2012) Climate change and sustainability of agrodiversity in traditional farming of the Sikkim Himalaya. Climate change in Sikkim. Patterns, impacts initiatives, 193–218

Singh JS (2006) Sustainable development of the Indian Himalayan Region: Linking ecological and economic concerns. Curr Sci 90(6):784–788
Singh SP, Singh V, Skutsch M (2010) Rapid warming in the Himalayas: ecosystem response and developmental options. Clim Dev 2:1–13
Singh SK, Rathore BP, Bahuguna IM, Ajai (2014) Snow cover variability in the Himalayan–Tibetan region. Int J Climatol 34(2):446–452
Sivakumar MVK, Das HP, Brunini O (2005) Impacts of present and future climate variability and change on agriculture and forestry in the arid and semi-arid tropics. Clim Change 70:31–72
Smit B, Burton I, Klein RJT, Wandel J (2000) An anatomy of adaptation to climate change and variability. In: societal adaptation to climate variability and change. Springer, Dordrecht, Netherlands
Smith P, Olesen J (2010) Synergies between the mitigation of, and adaptation to, climate change in agriculture. J Agric Sci 148:543–552
State Action Plan for Climate Change (HPSAPCC) (2012) Himachal Pradesh 1–172
State Action Plan for Climate Change (J&KSAPCC) (2014) Jammu and Kashmir 1–320
State Action Plan for Climate Change (UKSAPCC) (2014) Uttarakhand 1–227
Staubli A (2018) Analysis of weather and climate-related disasters in mountain regions using different disaster databases. In: Climate change, extreme events and disaster risk reduction. Springer, pp 17–41
Stepp JR, Castaneda H, Cervone S (2005) Mountains and bicultural diversity. Mt Res Dev 25:223–227
Talib AH (2007). On the brink-a report on climate change and its impact in Kashmir, action aid India, and report. Srinagar, Jammu and Kashmir
Tewari VP, Verma RK, Von Gadow K (2017) Climate change effects in the Western Himalayan ecosystems of India: evidence and strategies. For Ecosyst 4(1):13
Tiwari P, Joshi AK, Jagadish KSV (2017) Impacts of high temperature on productivity and quality of wheat: a review. Theo Appl Clim 128(3–4):1–15
UNEP GEF (2013) Integrating traditional crop genetic diversity into technology: using a biodiversity portfolio approach to buffer against unpredictable environmental change in Nepal Himalayas. Project Document, United Nation Environment Program (UNEP), Global Environment Facility (GEF) and Bioversity International, Kathmandu, Nepal
Urban MC, Nadeau CP, Giery ST (2023) Using mechanistic insights to predict the climate-induced expansion of a key aquatic predator. Ecol Monographs e1575
Vernooy R (2022) Crop diversification lead to climate-related resilience. Improving the theory through insights on practice. Agroecol Sustain Food Syst 46(6):877–901
Visser Marcel E (2010) Climate change: fatter marmots on the rise. Nature 466:445–447. https://doi.org/10.1038/466445a
Wang J, Li W (2003) Comparison of methods of snow cover mapping by analyzing the solar spectrum of satellite remote sensing data in China. Int J Remote Sens 24(21):4129–4136
Wester P, Mishra A, Mukherji A, Shrestha AB (2019) The Hindu Kush Himalaya assessment: mountains, climate change, sustainability and people. Springer
Westermann S, Qstby TI, Gisnås K, Schuler TV, Etzelmüller B (2015) A ground temperature map of the North Atlantic permafrost region based on remote sensing and reanalysis data. Cryosphere 9(3):1303–1319
Woodbridge J, Fyfe R, Smith D, Pelling R, De Vareilles A, Batchelor CR, Davies A (2020) What drives biodiversity patterns? Using long-term multi-disciplinary data to discern centennial-scale change. J Ecol 109:1396–1410
Yadav N, Sharma KD, Poonia TC (2018) Climate change and impact on barley yield in Western Himalayas. J Agromt 20(2):174–177

Zhang Y, Liu H, Qi J, Feng P, Zhang X, Li Liu D, Chen Y (2023) Assessing impacts of global climate change on water and food security in the black soil region of Northeast China using an improved SWAT-CO_2 model. Sci Total Environ 857:159482

Zhou MX (2010) In genetics and improvement of barley malt quality. Springer-Verlag, Berlin Heidelberg, Berlin, Heidelberg, p 117

Chapter 19
Influence of Climate Change on Himalayan Ecosystem: A Review

Neha Rawat, Neha Pandey, Aditya Joshi, and Deepa Pandey

Abstract The greatest challenges to economic development in the twenty-first century will come from climate change, which endangers humankind. Without a question, the most important environmental issues facing the world and humanity are those related to climate change and its uncertainties. Due to its distinctive biodiversity, delicate ecosystems, and rich cultural legacy, the Himalayan area is becoming more and more susceptible to the effects of climate change. With the goal of offering a thorough grasp of the issues and possible solutions, this research explores the complex effects of climate change on the Himalayan ecology, traditional agriculture, and renewable energy sources. With its varied flora and fauna, the Himalayan environment is undergoing changes in temperature, precipitation patterns, and glacier retreat. These alterations result in changed habitats, disturbed migration patterns, and a higher risk of species extinction, endangering the delicate balance of the ecosystem. Communities that depend on the region's biodiversity for their livelihoods are directly impacted by such ecological changes. The change in climate tends to change the growing seasons, precipitation levels, and insect dynamics, posing new problems for traditional crops that are essential to the livelihood of the local community. The challenges faced by subsistence farmers and indigenous groups include reduced yields, shifting crop appropriateness, and heightened susceptibility to harsh weather. To lessen the impact on food security, the study looks at adaptation tactics and the possible introduction of crop types resistant to climate change. This study emphasizes how urgent it is to solve the interconnected problems that climate change is posing in the Himalayan area. Sustaining the distinct biodiversity and guaranteeing food security for nearby residents need the use of efficient mitigation and adaptation tactics. Policymakers, academics, and local communities working to protect the

N. Rawat · D. Pandey (✉)
Department of Zoology, SSJWSSS Govt. PG College Ranikhet, Almora 263645, India
e-mail: pandeydeepa1506@gmail.com

N. Pandey
Department of Biotechnology, Graphic Era Deemed to be University, Dehradun 248002, India

A. Joshi
Department of Computer Science and Engineering, Graphic Era Deemed to be University, Dehradun 248002, India

D. Arya et al. (eds.), *Climate Change Impact on Himalayan Biodiversity*,
Environmental Science and Engineering, https://doi.org/10.1007/978-3-031-77149-1_19

Himalayan environment in the face of climate change may learn a great deal from the findings.

19.1 Introduction

The Himalayan area has a highly diverse range of climates and landscapes, which contribute significantly to its exceptionally rich biodiversity (Das et al. 2020; Wani et al. 2022b). Because of the region's widespread topographic gradients, the ecosystem is primarily susceptible to climate variations. The region consisting Himalaya is warming three times faster than the global average, according to Xu et al. (2009). The area of Himalaya gets warmed by 0.06 °C year between 1982 and 2006 (Shrestha et al. 2012). The area of Himalaya has seen significant increases in temperature, with wintertime changes often being larger than other seasons, exceeding the world mean trend of 0.85 °C between 1880 to 2012 (Schickhoff et al. 2015). Without a question, the biggest environmental issues facing the world and humanity are climate change and its unpredictability (Dad et al. 2021). According to estimates, the effects of climate change on the natural ecosystem will make it more likely that there will be more frequent forest fires, less frequent rainfall, and unexpected temperature changes. Long-term environmental changes would also disrupt seasonal precipitation patterns, have an impact on the mountain ecology, water supply, and food security (Misra 2014). The Himalayan region, frequently termed the "Third Pole," is a vital global ecosystem harboring unparalleled biodiversity and serving as a source of crucial freshwater for millions of people in Asia. However, the profound impact of climate change on this fragile ecosystem poses a significant threat to its ecological integrity and the sustenance of traditional agrarian practices. The exploration of renewable energy sources in the region introduces both challenges and opportunities, adding a layer of complexity to the dynamic interplay between environmental change and human activities. The Himalaya, and especially the Central Himalaya, are home to a wide variety of traditional crops. The Central Himalayan region's traditional agroecosystems, tended to for ages by the indigenous agricultural populations, are home to more than forty types of food grains (Maikhuri et al. 1996). The Himalayan region's agricultural output and food supply are significantly impacted by climatic variations, as mountain agriculture relies heavily on weather patterns and seasonal precipitation (Rao et al. 2012).

A variety of research articles, reports, and published documents provided secondary data for the study. With a special focus on traditional crops and changing Himalayan climate, a thorough literature search was done to find significant papers, research, and articles on the impact of climate change on the Himalayan environment.

19.2 Study Area

From tropical in the south to temperate and alpine in the north of the Himalayas, where high spots see constant winter snowfall, India has a great range of climatic zones (Attri 2010). The Indian Himalayan region is divided into eight biogeographic provinces: (Sikkim Plateau), (Ladakh Mountains), (Tibetan Plateau), (Cold arid regions of Eastern Himachal Pradesh and Uttarakhand), North-West Himalaya, West Himalaya, Central Himalaya, and Eastern Himalaya. These biogeographical zones are Trans-Himalaya, Himalaya, and Central Himalaya (Rodger and Panwar 1988). As per Negi et al. (2021), the region encompasses approximately 18% of India's entire geographical area, making up over 50% of the country. 40% of species that are indigenous to the Indian subcontinent and 40% of the country's forest cover (Sharma et al. 2010; Anthwal et al. 2010). Given that it offers ecosystem services, this area is important for India's ecological security. and advantages to millions of individuals living in or visiting the area (Negi et al. 2021).

19.3 Climate Change in the Himalaya

19.3.1 The Changing Landscape

Numerous observable occurrences in the Himalayas indicate a changing climate. The Intergovernmental Panel on Climate Change (IPCC) notes that extreme weather events are occurring more frequently and that precipitation patterns are changing, and the alarming rate of glacial retreat (IPCC 2019; Immerzeel et al. 2020). The area's traditional rural communities are using their traditional knowledge base and past experiences to help them deal with the consequences of climate change (Burnham et al. 2016; Negi et al. 2021). Rising temperatures have led to the melting of glaciers, affecting the region's major river systems, including the Ganges, Brahmaputra, and Indus (Bolch et al. 2019). This has significant implications for water resources downstream, impacting millions of people who depend on these rivers for agriculture, drinking water, and hydropower.

The Himalayas are well known for their abundant biodiversity, supporting a wide variety of animal and plant species. However, changing climatic conditions, such as altered temperature and precipitation patterns, are disrupting the ecosystems, and threatening the survival of many species (Paudel et al. 2020) by influencing the biodiversity of Himalaya (Fig. 19.1). Conservation efforts and sustainable land-use practices are becoming increasingly critical to safeguard the unique flora and fauna of the region. Communities in the Himalayas are at the forefront of adapting to the changing landscape. Indigenous knowledge, passed down through generations, is proving invaluable in developing adaptive strategies. To mitigate the effects of climate change, local efforts include building resilient infrastructure, establishing community-based early warning systems, and diversifying sources of income (Shah

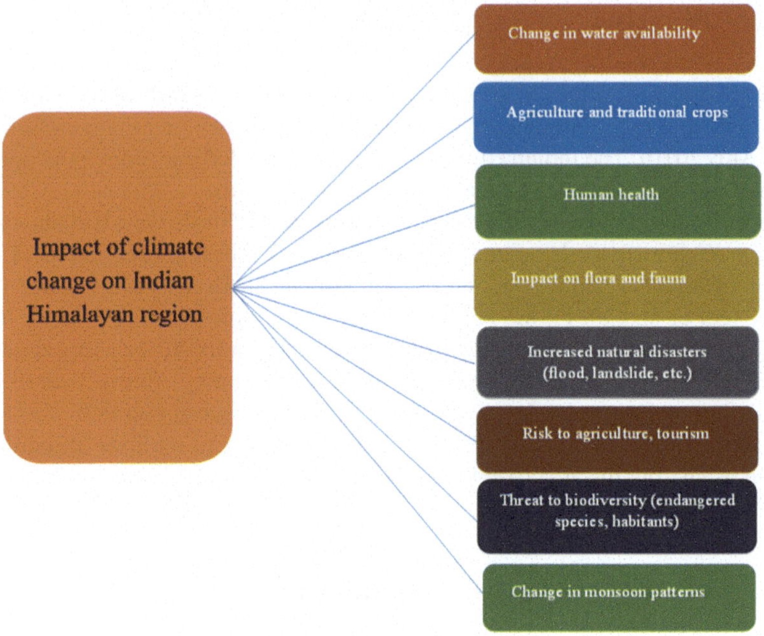

Fig. 19.1 Impact of climate change on various factors of Himalayan biodiversity

et al. 2021). The Himalayan landscape is undergoing significant transformations because of climate change. Glacial retreat, biodiversity loss, and changing agricultural patterns are challenges that require coordinated efforts at local, national, and international levels. Sustainable development, conservation, and community resilience are key elements in mitigating the impacts of climate change and preserving the unique and fragile ecosystems of the Himalayas.

19.3.2 Impact on Flora and Fauna

Changes in species composition, productivity, and biodiversity will influence most ecosystems (Leemans and Eickhout 2004). Most of the recent century has seen a 20% or greater decline in the average abundance of native plant and animal species (Shivanna 2020). phenological pattern prediction for the future. Due to global warming, late flowering plants are blossoming more often due to an elongation of their development cycle (Sahoo et al. 2019). These crops have supported the food and nutritional security of the hilly people for many centuries since they are suited to the regional climate and can survive environmental dangers as well as other natural calamities. However, the area cultivated with conventional crops has significantly reduced (>60%) in recent decades, and several crop species several species—*Panicum miliaceum, Glysine spp., Setaria italica, Hibiscus sabdarffia, Vigna* spp., and *Perilla frutescens*, to mention a few—are in danger of going extinct (Negi and Joshi 1996). Himalayan forests act as significant carbon sinks, absorbing and storing large amounts of carbon dioxide from the atmosphere. Deforestation or changes in vegetation patterns can affect this crucial role. The presence of diverse wildlife contributes to the overall health and resilience of ecosystems. Changes in climate can disrupt these ecosystems, affecting the distribution and behavior of species. Many Himalayan species, including migratory birds and animals, rely on specific climate conditions. Climate change can alter these conditions, affecting the timing and routes of migrations. Because of the amount and originality of biodiversity components at the gene, species, and ecosystem levels, the Indian Himalayan region is a well-known biodiversity hotspot. Locals believe that there is a greater chance of poor regeneration since many tree species have altered their phenology and displayed indicators of either early or late flowering. They understand how climate change has impacted the goods and services offered by the forest. There would be significant effects on the forest ecosystems due to the expected migration of butterflies up to greater altitude, the proliferation of mill bugs and other illnesses, and the spread of beetle species. The impoverished who depend on the forest would also face significant social and economic consequences.

19.3.3 Traditional Crop Under Stress

Mountain systems are among the most significant natural regions, offering vital ecosystem services to humankind (ICIMOD 2007). The inhabitants of the Himalayas have long engaged in integrated agriculture, which involves balancing the production of forestry, animal husbandry, and agroforestry. Mountain geography and inaccessibility have helped to maintain agro-biodiversity; however commercial agriculture is not as high yielding and profitable as plains (Nandy et al. 2006). Traditional crops cultivated by Himalayan communities are facing unique challenges. Shifts in temperature and precipitation patterns disrupt planting and harvesting cycles, leading to decreased yields and increased vulnerability to pests and diseases. Native knowledge, passed down through generations, is being tested as farmers adapt to these changing conditions.

The impacts of rising temperatures on traditional crops are extensively recorded. Studies by Lobell et al. (2014) and Challinor et al. (2014) highlight the vulnerability of crops to heat stress and the subsequent reduction in yields. Studies like Gregory et al. (2005) examine how the dynamics of pests and diseases in agriculture are affected by climate change.

Due to climate change, traditional farming methods—which are intricately linked to cultural legacy and community identity—are facing previously unheard-of difficulties. A complex knowledge of the difficulties traditional agricultural practices face is necessary given the complex interaction between traditional crops and climate change. This chapter intends to add to the continuing conversation about climate-related sustainable agriculture by utilizing a wide range of studies. Communities can negotiate the complicated terrain of climate change by identifying the stresses on traditional crops and investigating adaptation measures, protecting agricultural traditions as well as the livelihoods they support. Pollinator populations will be impacted by variations in floral diversity brought about by changes in land use and cover as well as the extinction of regional cultivars. Because they produce nutritious and therapeutic honey, honeybees have played a significant role in the Himalayan region's agricultural industry. According to recent reports, the region's native bee diversity is declining because of changing land use, growing monoculture, and the harmful effects of pesticides and herbicides (Pratap 1999).

19.3.4 Challenges to Agriculture

Agriculture is one of the industries most affected by climate change since it depends on temperatures and amounts of precipitation for its outside production activities. Because of the rising temperatures and decreasing amount of water available for irrigation, most crops would yield less. Climate change influences mountain agriculture, which is restricted to terraces, powered by biomass energy from nearby forests, and mostly rainfed (around 85% of it) area. According to Dhanai et al. (2014), apple trees' chilling hours are decreased due to variations in snowfall, which impacts when the buds burst. The increasing temperatures in Uttarakhand's Lower Hills have a major effect on the chilling need for apple fruit, which is a necessity for apple fruit to blossom (Nautiyal et al. 2020). Because of the unfavorable weather, crops like pomegranates and kiwis are replacing apples in terms of yield. Since shifting agriculture and subsistence farming are frequently coupled, marginal farmers will be most affected by climate change (Roy et al. 2018). According to Gupta et al. (2019), communities in the Himalayas are often dependent only on natural resources for their sustenance, rendering them susceptible to alterations in the surroundings caused by climate change. Mountain ecosystems and mountain inhabitants are particularly threatened by the current processes of global environmental change, population dynamics, economic globalization, and the consequent exploitation of natural resources (Joshi 2018). A significant supply of minerals, energy, water, and forest products comes from mountains. They also boast some of the world's most fragile ecosystems. Mountain tribes still have limited opportunities for a living because of their remoteness, the vulnerability of their alpine environment and the absence of motivation for them to maintain the equilibrium of their ecosystems (Panday et al. 2017).

According to Bhutiyani et al. (2007), long-term trends in the north-western Himalaya's mean, minimum, and maximum temperatures throughout the twentieth century point to a notable increase in air temperature there, with winter warming happening more quickly. This vulnerability extends beyond the local population.

Unpredictable, erratic, and sudden changes in climate pose a threat to rural people's agricultural output and stability of livelihood. Food quality, accessibility, and the food supply chain can all be impacted by climate change and unpredictability. Individuals are experimenting with new methods, which is making them more adaptable to climate change and variability; however, service providers may not be recording these innovations and disseminating them to others who may find value in them. Weather patterns have a significant impact on agricultural productivity and the availability of food, as agriculture is heavily Dependent on them (Negi et al. 2012).

19.3.5 Adaption Strategies

Scientists, governmental and non-governmental organizations, and other stakeholders should use technical, legislative, and financial strategies to assist farmers in adapting to climate change and enhance their livelihoods and food security (Dey et al. 2018). Enhancing farmers' skills and improving the socioeconomic viability of the farming system may be achieved by giving conventional agricultural goods more value. It is difficult to find detailed information regarding the well-being of humans in the Himalayan region but among other things, people's livelihoods, health, and access to natural resources would be impacted by climate change (Sharma et al. 2009). Numerous factors, including climate change, can affect the Himalaya's socioeconomic environment. It has the potential to impact the economy (agricultural, both human health and other sectors (livestock, forestry, tourism, fishing, etc.). The poor and marginalized people who depend on biodiversity are likely to suffer the most from the effects of biodiversity loss due to climate change primarily from natural resources Climate change has a substantial influence on human health in two ways: directly, through the consequences of heat stress, fatalities and injuries from floods and storms, etc.) and indirectly through modifications to the ranges of disease vectors, waterborne pathogens, water quality, air quality, food quality and availability (McMichael et al. 2003), respiratory disorders and cardiovascular mortality, infectious disease transmission, and malnourishment resulting from crop failures (Patz et al. 2005).

In the Himalayan Mountains, traditional agriculture has proven to be resilient to crop disease and a rich store of agrobiodiversity. According to Darwin's theory of "survival of the fittest," the surviving organisms undergo a variety of phenotypic and genetic adaptations. In the context of the phenomenon of acclimatization and adaptation to changing climate, living things adapt to their new surroundings by going through a variety of molecular and biochemical changes. It is implied that plant growth and productivity have been limited by the current climatic changes. However, to survive even in unfavorable climates, plants also develop completely different mechanisms and produce various biochemical molecules in counter defense crops. Because of unfavorable change in climate, several natural disasters, and stressors—like high temperatures, droughts, salinity, and rising CO_2 concentrations-have an increasing impact on plant growth and threaten sustainable agriculture.

19.4 Conclusion

In the Himalayan region the influence of climate change on community involvement is a complicated and ever-changing problem. Adaptive strategies are necessary, but maintaining cultural heritage while balancing both calls for a comprehensive approach. The Himalayan region can manage the uncertainties posed by climate change and create a sustainable future by empowering local populations, diversifying livelihoods, and sharing information to enhance community resilience. The Himalayan region witnesses an increase in the frequency and intensity of natural disasters, including landslides and floods. These events disrupt communities, damage infrastructure, and pose challenges to rebuilding efforts. Communities can lessen the negative effects of climate change on industries by encouraging them to diversify their sources of income. Initiatives that support handicrafts, ecotourism, and other sources of income improve stability in the economy. Empowering communities to actively participate in decision-making processes regarding their cultural and natural heritage is crucial. This involvement ensures that climate adaptation strategies respect and preserve local traditions (Pelling and High 2005). The impact of climate change on customary crops is extensive and diverse, presenting major obstacles to international agriculture. Traditional crops are affected in terms of growth, production, and general resilience by changes in temperature, precipitation patterns, and the frequency of extreme weather events. These crops become more susceptible to pests, illnesses, and limited water availability because of the delicate balance they have maintained with their native habitats being upset. Beyond just decreasing agricultural output, the effects of climate change on traditional crops also have an impact on food security and livelihoods of those who depend on agriculture. To lessen these effects and increase the agriculture sector's resilience, immediate action is needed. This entails putting climate-resilient crop types into cultivation, adopting sustainable agricultural methods, and making investments in infrastructure and technologies that can change with the environment.

To preserve the sustainability of the region's ecosystems, agriculture, and energy supplies, the study finishes by summarizing significant results, highlighting the necessity of tackling the Himalaya's changing climate, and suggesting directions for future research and cooperative efforts. It is necessary to gain competence via a deeper comprehension of farmers' responses to climate change and their continuing adaptation efforts. creation of plans and programs to increase the agriculture sector's output and yield.

References

Anthwal A, Gupta N, Sharma A, Anthwal S, Kim KH (2010) Conserving biodiversity through traditional beliefs in sacred groves in Uttarakhand Himalaya, India. Resour Conserv Recycl 54(11):962–971

Attri SD, Tyagi A (2010) Climate profile of India. Environment Monitoring and Research Center, India Meteorology Department, New Delhi, India

Bhutiyani MR, Kala VS, Power NJ (2007) Long-term trends in maximum, minimum and mean annual air temperatures across the northwest Himalaya during the twentieth century. Clim Change 85(1–2):159–177

Bolch T, Shea JM, Liu S, Azam FM, Gao Y, Gruber S, Zhang Y (2019) Status and change of the cryosphere in the extended Hindu Kush Himalaya region. The Hindu Kush Himalaya assessment: Mountains, climate change, sustainability and people, pp 209–255

Burnham M, Ma Z, Zhang B (2016) Making sense of climate change: hybrid epistemologies, socionatural assemblages and smallholder knowledge. Area 48(1):18–26

Challinor AJ, Watson J, Lobell DB, Howden SM, Smith DR, Chhetri N (2014) A meta-analysis of crop yield under climate change and adaptation. Nat Clim Change 4(4):287–291

Dad JM, Muslim M, Rashid I, Reshi ZA (2021) Time series analysis of climate variability and trends in Kashmir Himalaya. Eco Ind 126(10):76–90

Das DS, Rawat DS, Maity D, Dash SS, Sinha BK (2020) Species richness patterns of different life-forms along altitudinal gradients in the great Himalayan National Park, Western Himalaya, India. Taiwania 62(2):154–162

Dey T, Pala NA, Shukla G, Pal PK, Das G, Chakarvarty S (2018) Climate change perceptions and response strategies of forest fringe communities in Indian Eastern Himalaya. Env Devel Susta 20 (2)925:938.

Dhanai R, Negi RS, Singh S, Parmar MK (2014) The effects of climate change on natural resources and socio-economic condition of Himalayan communities of Uttarakhand, India. Int J Mod Commu Techno Res 2(11):21–25

Gregory PJ, Ingram JS, Brklacich M (2005) Climate change and food security. Philosophical transactions of the royal society. Bio Sci 360(1463):2139–2148

Gupta AK, Negi M, Nandy S, Alatalo JM, Singh V, Pandey R (2019) Assessing the vulnerability of socioenvironmental systems to climate change along an altitude gradient in the Indian Himalayas. Ecol Ind 106:105–512

ICIMOD (2007) Background note for regional meeting of experts on development of micro enterprises in mountain area 25–26 July, Unpublished Text

Immerzeel WW, Lutz AF, Andrade M, Bahl A, Biemans H, Bolch T, Baillie JE (2020) Importance and vulnerability of the world's water towers. Nature 577(7790):364–369

Joshi B (2018) Recent trends of rural out-migration and its socio-economic and environmental impacts in Uttarakhand Himalaya. J Urban Reg Stud Contemp India 4(2):114

Leemans R, Eickhout B (2004) Another reason for concern: regional and global impacts on ecosystems for different levels of climate change. Glo Envir Change 14(3):219–228. https://doi.org/10.1016/j.gloenvcha.2004.04.009

Lobell DB (2014) Climate change adaptation in crop production: beware of illusions. Glob Food Sec 3(2):72–76

Maikhuri RK, Rao KS, Saxena KG (1996) Traditional crop diversity for sustainable development of Central Himalayan agroecosystems. Int J of Sus Devel World Eco 3(3):8–31

McMichael AJ, Campbell-Lendrum CF, Ebi CKL, Githeko AK, JD, Scheraga & A. Woodward, (eds) (2003) Climate change and human health—risks and responses. WHO, Geneva

Misra AK (2014) Climate change and challenges of water and food security. Int J Sustain Built Environ 3:153–165

Nandy SN, Dhyani PP, Samal PK (2006) Resource information database of the Indian Himalaya. ENVISMonograph3, 108p. http://gbpihed.gov.in/envis/HTML/monograph3/Contents.html. Accessed 25 Dec 2009

Nautiyal P, Bhaskar R, Papnai G, Joshi N, Supyal V (2020) Impact of climate change on apple phenology and adaptability of Anna variety (low chilling cultivar) in lower hills of Uttarakhand. Int J Curr Microand App Sci 9(9):453–460

Neg GCS, Samal PK, Kuniyal JC, Kothyari BP, Sharma RK, Dhyani PP (2012) Impact of climate change on the western Himalayan Mountain ecosystems: an overview. Trop Ecol 53(3):345–356

Negi GCS, Joshi V (1996) Land use in a Himalayan catchment under stress: system responses. Ambio 25(2):126–128

Negi VS, Maletha A, Pathak R, Maikhuri RK (2021) Expansion of a native species and its impacts on alpine ecosystems, Indian Himalaya. Biologia 76(3):889–899

Pandey R, Jha SK, Alatalo JM, Archie KM, Gupta AK (2017) Sustainable livelihood framework-based indicators for assessing climate change vulnerability and adaptation for Himalayan communities. Eco Ind 79:338–346

Partap U (1999) Conservation of endangered Himalayan honeybee, *Apis cerana* for crop pollination. Asian Bee J 1:44–49

Patz JA, Campbell-Lendrum D, Holloway T, Foley JA (2005) Impact of regional climate change on human health. Nature 438:310–317

Paudel B, Zhang Y, Yan J, Rai R, Li L, Wu X, Khanal NR (2020) Farmers' understanding of climate change in Nepal Himalayas: important determinants and implications for developing adaptation strategies. Cli Change 158:485–502

Pelling M, High C (2005) Understanding adaptation: what can social capital offer assessments of adaptive capacity? Global Environ Change 15(4):308-19

Rao KS, Maikhuri RK, Semwal RL, Rawat LS, Saxena KG (2012) Climate change adaptation and mitigation: insights from Nanda Devi biosphere in Indian Himalaya. Land management in marginal mountain region: adaptation and vulnerability to global change. Bishen Singh mahendra Pal Singh, Dehradun, India

Rodgers WA, Panwar HS (1988) Planning a wildlife protected area network in India

Roy SS, Ansari MA, Sharma SK, Sailo B, Devi CB, Singh IM, Ngachan SV (2018) Climate resilient agriculture in Manipur. Curr Sci 115(7):1342–1350

Sahoo UK, Singh SL, Gogoi A, Kenye A, Sahoo SS (2019) Active and passive soil organic carbon pools as affected by different land use types in Mizoram, Northeast India. PLoS ONE 14(7):e0219969

Schickhoff U, Bobrowski M, Böhner J, Bürzle B, Chaudhary RP, Gerlitz L, Heyken H, Lange J, Müller M, Scholten T, Schwab N, Wedegärtner R (2015) Do Himalayan treelines respond to recent climate change? an evaluation of sensitivity indicators, Earth Syst. Dynam 6:245–265. https://doi.org/10.5194/esd-6-245

Sharma E, Chettri N, Tse-ring K, Shrestha AB, Jing F, Mool P, Eriksson M (2009) Climate change impacts and vulnerability in the Eastern Himalayas. International Centre for Integrated Mountain Development, Kathmandu

Sharma S, Pradhan G, Dahal DR (2010) Observations on vegetation richness over a gradient along the gangtokdeorali ridge in Sikkim. J Hill Res 23(2):28–30

Shah KK, Modi B, Pandey HP, Subedi A, Aryal G, Pandey M, Shrestha J (2021) Diversified crop rotation: an approach for sustainable agriculture production. Adv Agric 2021(1):8924087

Shivanna KR (2020) The sixth mass extinction crisis and its impact on biodiversity and human welfare. Resonance 25(1):93–109

Shrestha UB, Gautam S, Bawa KS (2012) Widespread climate change in the Himalayas and associated changes in local ecosystems. PLoS ONE 7(5):e36741

Shukla PR, Skea J, Calvo Buendia E, Masson-Delmotte V, Pörtner HO, Roberts DC, Malley J (2019) IPCC, 2019: climate change and land: an IPCC special report on climate change, desertification, land degradation, sustainable land management, food security, and greenhouse gas fluxes in terrestrial ecosystems

Wani ZA, Bhat JA, Negi VS, Satish KV, Siddiqui S, Pant S (2022b) Conservation Priority Index of species, communities and habitats for biodiversity conservation and management planning; a case of Gulmarg Wildlife Sanctuary, Kashmir Himalaya. Front For Glob Change 167

Xu J, Grumbine RE, Shrestha A, Eriksson M, Yang X, Wang YUN, Wilkes A (2009) The melting Himalayas: cascading effects of climate change on water, biodiversity, and livelihoods. Conserv Biol 23(3):520–530

Chapter 20
Forest Degradation and Carbon Sequestration: Assessing the Global Scenario

Stuti Rawat, Nitesh Kumar, Arti Jamwal Sharma, and Shreekar Pant

Abstract Forests are key role players in carbon sequestration storing around 25% of total global carbon. The trees and soil in the forests store carbon thus reducing carbon levels in the atmosphere. Different methods like remote sensing, allometric equations, and inventories are used to quantify carbon stocks. Studies used IPCC gain–loss approach, Landsat, SPOT, ASTER, and ALI imagery data in areas with selective logging. An extensive analysis of the research papers reveals that the world's forests contain around 861 gigatonnes of carbon. Out of this, 44% is stored in the soil up to a depth of one metre, 42% is found in live biomass both above and below the ground, 8% is present in dead wood, and 5% is located in litter, but this carbon stocks may be disturbed by land use pattern changes. A review of many findings suggest that soil carbon density reached its highest value in the silvopastoral system. SOC was affected by season and autumn season showed maximum soil carbon density. Forests get degraded due to various man-made factors as well as natural factors. This leads to loss of ecosystem services, habitat loss, and an increase in temperature. Sustainable forest management needs time to be followed as it guarantees the livelihood and well-being of people while maintaining the forests for future generations. The measures we propose for proper carbon sequestration is thatpolicy makers also take local people into considerations for management practices and formation of regulations and thus can be easily come into action.

Keywords Forests · Carbon store · Carbon sequestration · Forest management

S. Rawat · N. Kumar (✉)
Department of Biosciences, Himachal Pradesh University, Shimla 171005, Himachal Pradesh, India
e-mail: niteshchauhan7@gmail.com

A. J. Sharma (✉)
Department of Bio Sciences, Career Point University, Hamirpur 176041, Himachal Pradesh, India
e-mail: arti.botany@cpuh.edu.in

S. Pant
Centre for Biodiversity Studies, BGSB University, Rajouri 185234, Jammu Kashmir, India

D. Arya et al. (eds.), *Climate Change Impact on Himalayan Biodiversity*,
Environmental Science and Engineering, https://doi.org/10.1007/978-3-031-77149-1_20

20.1 Introduction

According to EU REDD Facility, "The slow process of a forest's biomass declining, species mix shifting, or soil quality deteriorating is known as "forest degradation". "Forest degradation" is the process of a forest area's ecological richness permanently declining for one or more reasons. The degradation of forests impacts carbon stores, ecological processes, structure, composition, and diversity (Bustamante et al. 2016). While the exact magnitude of these emissions remains uncertain, they are recognised as a substantial contributor to global carbon emissions. The Global Forest Resource Assessment of 2005 established a working definition while they were defining "forest." The source of this information is the Food and Agriculture Organisation of the United Nations in 2006. Therefore, any piece of land that is larger than 0.5 hectares (ha) and includes trees that are taller than 5 m and a canopy cover that exceeds 10%, or has trees that have the potential to meet these requirements naturally, is classified as a "forest".

20.2 Importance of Forests in Carbon Sequestration

The essential component of life on Earth is carbon. The carbon concentration of the atmosphere is regulated by life, but geological causes dominate over geological timeframes. The global carbon cycle encompasses the biological exchange of carbon (C) across the planet's atmosphere, ecosystem, water sphere, pedosphere, and geosphere. Differentiating between a long-lasting and a brief carbon cycle is feasible, as they take place over varying time spans ranging from hours to millions of years (Berner 2003). Forests affect climate change by either releasing or absorbing carbon, a potent greenhouse gas. The growth of forests can enhance their carbon reserves, while deforestation can contribute to increased carbon emissions (Berner 2003). Forest soils and tree biomass are two places where carbon is stored. Public policies and forest management have a big impact on the sequestration process (Lorenz et al. 2010; Sedjo and Sohngen 2012). Forest ecosystems have the largest area of non-icy land compared to other types of land ecosystems. Trees, which are essential elements of forest ecosystems, possess the highest amount or total volume of living forest biomass. The global forest biomass amounts to around 677 petagrams (Pg), with trees accounting for 80% of the total biomass. (Kindermann et al. 2008; Lorenz 2010). Figure 20.1 demonstrates the carbon cycle.

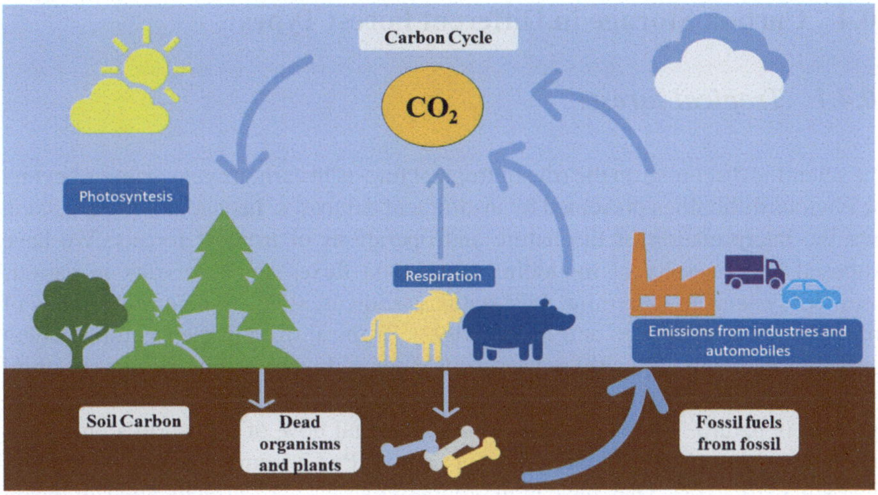

Fig. 20.1 Carbon cycle showing the cycling of carbon between animals, automobiles, forests and atmosphere

20.3 Role of Forests in Carbon Sequestration

Create carbon pools: Forests sequester carbon dioxide from the atmosphere and store it in various carbon sinks, such as living and dead trees, root systems, woods, and soils. At present, approximately 45% of the carbon found on land is contained within the living material and soils of existing forests. Live forests possess the greatest carbon density, closely followed by soils and the forest base. Harvested lumber and landfills serve as reservoirs for carbon.

Mitigate global warming: Carbon is emitted as CO_2 into the atmosphere and adds to greenhouse gas (GHG) emissions when a carbon reservoir decomposes or is combusted. Trees have absorbed around 25% of the carbon emissions generated by anthropogenic activities in the past forty years. This helps to reduce the rate of climate change. The U.S. Forest Service states that forests and forest products in the United States contribute to a reduction of around 16% in domestic carbon dioxide emissions. This is accomplished by sequestering 866 million metric tonnes of carbon dioxide per year, which is equal to the emissions generated by 50 million automobiles powered by gasoline or diesel fuel on an annual basis.

20.4 Carbon Storage in Different Forest Types

20.4.1 Tropical forests

The scientific discourse on the role of tropical forests in carbon cycle extends beyond the obvious difficulties presented by insufficient databases. Instead, it is a discussion over the interpretation of the nature and operations of tropical forests. We have proposed the necessity of modelling the carbon fluxes and reservoirs in various forest ecosystems to determine their status as sources, sinks, or neutrals in terms of atmospheric carbon (Lugo and Brown 1992). Tropical trees in their natural habitat store approximately 200–300 petagrams (Pg) of carbon, which is almost one-third of the amount of carbon present in the Earth's atmosphere. Tropical trees are highly dynamic in nature. They are responsible for around 60% of the world's photosynthesis, absorbing around 72 petagrams of carbon (Pg C) from the atmosphere annually (Mitchard 2018). They have high productivity rates, fast growth, and significant biomass buildup (Pirard et al. 2016). Here are some key aspects of how tropical forests contribute to carbon sequestration:

1. **High Net Primary Productivity (NPP)**:

They have a high level of Net Primary Productivity, which refers to the amount of carbon that plants fix through photosynthesis minus the amount lost through respiration (Pirard et al. 2016). The warm temperatures and abundant sunlight in tropical regions promote year-round growth, allowing for continuous carbon uptake by plants.

2. **Biomass Accumulation**:

They store vast amounts of carbon in their above-ground biomass, including the trunks, branches, leaves, and other plant structures. Old-growth tropical forests mayhave significant carbon stocks due to the accumulation of biomass over centuries (Holl and Zahawi 2014).

3. **Soil Carbon Storage**:

Tropical soils store substantial amounts of carbon, and the decomposition of plant material contributes to soil organic carbon. The complex interactions between vegetation, microorganisms, and soil in tropical ecosystems play a crucial role in carbon cycling (Dieleman et al. 2013).

4. **Biodiversity and Carbon Storage**:

Biodiversity is often linked to the resilience and carbon storage capacity of tropical forests. Diverse ecosystems may be more robust and capable of adapting to environmental changes, preserving their carbon sequestration functions (Daba and Dejene 2018).

5. **Deforestation and Degradation Impact**:

Tropical regions release great amounts of stored carbon into the air. Converting forests into agricultural land, logging, and other human activities contribute to carbon emissions.

6. **Conservation and Restoration**:

Conservation and restoration efforts in tropical forests are crucial for preserving their carbon sequestration capacity (Ota et al. 2020). Protecting intact forests and restoring degraded areas can enhance the ability of these ecosystems to sequester carbon (Chazdon and Guarigueta 2018).

7. **Indigenous and Local Knowledge**:

Indigenous and local groups possess ancestral wisdom regarding sustainable forest management techniques that can effectively preserve carbon reserves while simultaneously addressing local need.

Tropical forests are vital for global carbon sequestration, acting as significant carbon sinks. However, they are also highly vulnerable to deforestation and degradation, which can release stored carbon and contribute to climate change.

20.4.2 Temperate Forests

Temperate woods are predominantly situated in North America, the Mediterranean region Central and Western Europe, northeastern Asia, New Zealand and Southern Chile. The carbon stock of temperate woodlands and their soils is comparable to, and occasionally surpasses, that of tropical rainforest ecosystems. Change in the climate is predicted to result in a movement of the temperate woodland biome towards the poles, leading to changes in species composition and modifications to the carbon storage in the ecosystem. They have a high ability to absorb and store carbon due to their favourable climate, which is defined by four seasons, and their generally fertile soils. Temperate trees with rapid growth rates can amass approximately 20 Mg of wood ha^{-1} $year^{-1}$. Cool temperate moist woods store 625 tC/ha of above- and below-ground biomass, while warm temperate moist forests store 500 tC/ha. Cool temperate dry woods accumulated 280 tC/ha (Rajput et al. 2017).

20.5 Integrating Indigenous Knowledge in Forest Management

The knowledge employed by the local farmers in the Benue State community is considered a commendable illustration of the notion of performativity. Located in Nigeria, Benue State is a region within the country. Certain communities in Nigeria hold the belief that the forest is home to a divine deity that safeguards them from malevolence and ensures the safety of their families. Hence, it is considered abhorrent conduct to engage in deforestation or any illicit actions within the forest region (Tee et al. 2014). The indigenous population in the village has a comprehensive understanding of the significance of safeguarding their woodland for spiritual, cultural, and ecological reasons. They employ their native expertise to fiercely protect their forest resources, thereby impacting the long-term sustainability and preservation of these resources. Performativity can also be applied to the people of Ethiopia, providing another illustrative example. The indigenous population preserved forest plants, like *Croton macrostachycus*and *Schefflera abyssinica*, specifically for the purpose of beekeeping. In addition, they preserved *Abiziagummifera, Acacia abyssinica*, and *Militia ferruginea* as shade trees for the coffee farm. Indigenous knowledge has a substantial influence on the management of forest resources, effectively tackling climate control, safeguarding water bodies, and purifying the air (Hundera 2007).

This chapter focuses on the function of forests in the process of carbon sequestration and storage. It also examines the impact of forest degradation on both the ecosystem and carbon sequestration. Additionally, it explores the many techniques employed for quantifying the carbon storage of forests. In this context, we have also examined the obstacles that impede forest land restoration and the imperative for collaboration between individuals and policymakers to effectively attain their objectives (Emborg et al. 2012). We have also examined the methods for quantification and the latest developments in these techniques. Our planet's forests are invaluable ecosystems that provide us with clean air, biodiversity, and play a crucial role in combating climate change. However, unsustainable practices, deforestation, and illegal logging are threatening these vital resources. It is imperative that we unite in a concerted effort to promote and implement sustainable forest management practices. This call to action aims to inspire individuals, communities, businesses, and governments to collaborate in safeguarding our forests for future generations (Jenkins and Schaap 2018; Yuliani et al. 2018).

The Collaborative Partnership on Forests (CPF) held on 18 September, New York has urged for immediate worldwide measures to harness the full potential of forests for the benefit of humanity and the environment, as there are only seven years left to accomplish the 2030 Sustainable Development Goals (SDGs) and the Global Forest Goals. The CPF welcomed increased recognition of the importance of forests in achieving the Sustainable Development Goals (SDGs), which include helping to achieve sustainable and resilient agrifood systems, addressing climate change, biodiversity loss, degradation and desertification, and providing goods and services.

Partners also recognized the significant progress made in conserving, restoring and sustainably using the world's forests.

Our responsibility to protect and preserve the world's forests transcends borders and requires collective action. By heeding this call to action, we can ensure that future generations inherit a planet enriched by the vital benefits of healthy and thriving forests. Together, let us champion sustainable forest management as a cornerstone of our commitment to environmental stewardship and a sustainable future for all. The biomass enrichment of forest land with forest cover less than 70% results in an additional carbon storage of 1008.49 Tg C. This indicates the potential quantity of carbon that can be stored in the expanded biomass, litter, and forest product pool. To optimise the growth of Indian forests, it is recommended to implement selective thinning and protect areas with low biomass from grazing and fire. This will facilitate natural regeneration and promote the growth of new trees. It is worth noting that a significant proportion of stems per hectare, specifically 38.0% and 17.5%, are found in the 10–20 cm and 20–30 cm diameter classes, respectively. This potential can be realised through the current forest management trend of restoring damaged forest area, as indicated by the government's stated policy. The current National Forestry Action Programme (NFAP 1999) aims to enhance the density of forest cover in about 31 million hectares of forest areas with less than 40% canopy over a period of 20 years (1996–2016). This programme serves as a crucial means to obtain more carbon storage by enriching the biomass of existing forests. By implementing efficient strategies to halt the deterioration of forests and restore areas that have been damaged, the forests in India have the capacity to function as a net carbon sink in the future (Joshi Meenakshi 2003; Khare 2017). The storage of carbon in forests may account for 25% of all mitigation efforts and hence should not be disregarded (Mendelsohn et al. 2012).

20.6 Overview

This chapter highlights role of forests in carbon sequestration and storage, the impact of forest degradation in ecosystem and carbon sequestration, techniques used for quantification of carbon storage in forests. Here, we have also discussed the hurdles in the way of Forest Land Restoration and how the people and policy makers should work together hands-in-hand to achieve their goals. We have also discussed the techniques for quantification and recent advances in the techniques.

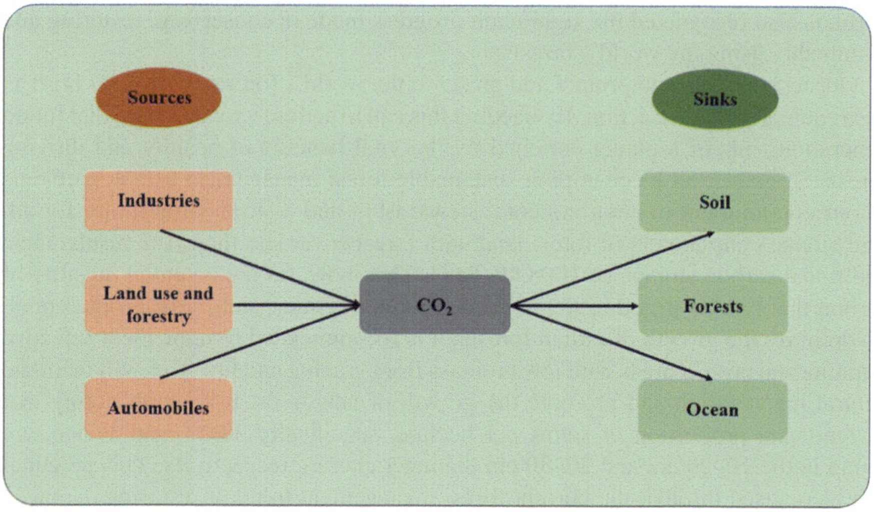

20.6.1 Study Area

The study reviews different research on carbon sequestration around the world on a global level. This article explores the various aspects of carbon sequestration on a global scale, highlighting the significance of understanding and implementing effective strategies to combat climate change. The world is home to a myriad of ecological systems, ranging from dense forests and expansive grasslands to vast oceans and intricate wetlands. Each of these ecosystems plays a unique role in the global carbon cycle while forests, for example, act as carbon sinks. Wetlands, on the other hand, are crucial in sequestering carbon in anaerobic conditions. Carbon sequestration studies acknowledge the regional disparities in carbon storage capacities. Tropical rainforests, with their high biodiversity and rapid biomass turnover, sequester substantial amounts of carbon. Arctic ecosystems, despite their seemingly sparse vegetation, store carbon in frozen soils, known as permafrost. Understanding these regional differences is crucial for developing targeted sequestration strategies that address the unique challenges and opportunities presented by each environment. Anthropogenic activities significantly contribute to the increase in atmospheric CO_2 levels. Carbon sequestration studies encompass the assessment of human-induced emissions and explore methods to capture and store carbon from various sources, such as power plants and industrial facilities. Implementing sustainable practices and technologies becomes imperative to mitigate the impact of these activities. Advancements in technology play a pivotal role in the global carbon sequestration efforts. Carbon capture and storage (CCS) technologies, for instance, aim to capture CO_2

emissions from industrial processes and power plants before they enter the atmosphere, subsequently storing the captured carbon in geological formations. Additionally, afforestation projects, ocean-based solutions, and soil carbon management strategies contribute to the global arsenal of tools for effective carbon sequestration. The discrepancies in the quantification of carbon stocks on a global level were an importantcriteria for upscaling the study to a global level.

20.7 Materials and Methods

To gather and analyze data on various types of forests, carbon stock, and sequestration pools in various forest ecosystems, a comprehensive literature search was conducted. This search utilized multiple sources, including the Web of Science, Google Scholar, Shod Ganga, ResearchGate, book chapters, journals. We also used IPCC Special Reports on Climate Change and Lands. The literature search methodology involves generating keywords related to biomass/carbon stock, sequestration, and pool estimation in various states of India. It also includes destructive and non-destructive methodologies, as well as the use of allometric equations for estimating biomass and carbon in different types of forests, such as plantation forests, natural forests, agroforestry, and shrub biomass. The focus is on understanding the biomass status of different forest types around the world. (a) Conduct a comprehensive literature search utilizing multiple online platforms. (b) Compiling the primary discoveries and emphasizing them in the review; (c) Analysis of the emphasized discovery.

20.7.1 Methods for Measuring Carbon Sequestration

Measuring carbon stocks is crucial for understanding and monitoring carbon sequestration and emissions in various ecosystems. Different methods are employed depending on the type of ecosystem (e.g., forests, grasslands, wetlands) and the specific goals of the assessment. Some common methods for measuring carbon stocks:-

20.7.2 Aboveground Biomass Estimation

Allometric Equations: Different equations exist for different tree species or vegetation types. These equations relate easily measurable parameters (such as diameter at breast height, height, or crown diameter) to biomass. Allometric equations are specific to tree species or ecosystems and are developed through field measurements and analysis (Kebede and Soromessa 2018).

20.7.3 Belowground Biomass Estimation

Excavation of soil pits or use soil cores to collect root samples, which are then dried and weighed to estimate belowground biomass. Similar to aboveground biomass, allometric equations can be used for estimating belowground biomass based on easily measurable parameters (Kuyah et al. 2012).

20.7.4 Soil Organic Carbon (SOC) Measurement

Soil Sampling and Analysis: Collect soil samples at different depths and analyze them for carbon content. This can be done using laboratory methods such as dry combustion or wet oxidation. Soil Carbon Maps can be utilized in remote sensing and geospatial technologies to create maps of soil carbon content over large areas (Ellert et al. 2007).

20.7.5 Remote Sensing

Remote sensing is essential for monitoring and evaluating carbon sequestration, which involves the collection and sequestration of atmospheric carbon dioxide over an extended period of time (CO_2) by plants, soils, and other ecosystems. Remote sensing helps monitor vegetation dynamics and health, providing data on the amount of biomass and changes in biomass over time. Sensor types like infrared and optical can detect variations in vegetation cover and health, helping researchers estimate the carbon content in plants. Remote sensing helps analyze changes like deforestation, afforestation, and changes in agricultural practices. These changes significantly impact carbon sequestration, and remote sensing data can provide insights into the spatial and temporal patterns of these changes. Satellite imagery and airborne sensors are used to map forest carbon stocks. By combining data on vegetation structure, biomass, and land cover, remote sensing can provide accurate estimates of carbon density in forests. This information is valuable for assessing the effectiveness of conservation efforts and sustainable forest management practices. Remote sensing data are integrated into ecosystem models and carbon accounting systems to improve the accuracy of carbon sequestration estimates. These models help researchers and policymakers assess the effect of anthropogenic activities on carbon stocks (Turner et al. 2004).

20.7.6 Ground-Based Inventories

Ground-based inventories are essential for accurately assessing and monitoring forest carbon stocks. These inventories involve on-site measurements and data collection to estimate the amount of carbon stored in forests. Ground-based inventories involve field measurements of tree attributes such as diameter at breast height (DBH), wood density, and height. These measurements are used to estimate aboveground biomass, which is a critical component of forest carbon stocks. Direct sampling methods, such as destructive sampling or non-destructive measurements, provide accurate data on tree biomass. Ground-based surveys help in identifying tree species within a forest. Precise identification of tree species is essential for assessing carbon stocks, as various species possess distinct capabilities for carbon storage. This information is often combined with biomass equations specific to each tree species (Espejo et al. 2020).

20.8 Results

The global forests contain a total of 861 gigatonnes of carbon. Out of this, 44% is stored in the soil up to a depth of 1 m, 42% is found in live biomass, 8% is present in dead wood, and 5% is found in litter (Pan et al. 2011). This equals nearly a century of yearly fossil fuel emissions. Only 30% of worldwide tree cover is tropical rainforest, but 50% of its carbon is stored in trees.2 Tropical forests store most carbon in biomass, while boreal forests store immense carbon in soils (Hengl et al. 2017). Using a conservative 1 m average soil depth, the 2012 worldwide mangrove carbon stock was calculated at 4.19 ± 0.62 Pg by IPCC 12 and other studies (Hamilton and Friess 2018). Nutrient cycle, productivity, and forest sustainability have been the focus of most tropical forest studies. Small, nonrandomly selected places with little global carbon study are used to obtain data (Brown et al. 1989). Large-girth tropical trees with commercial timber potential were in most inventories. Additionally, tropical forests are diverse in age and size. Omitting little or smaller girth class trees sometimes underestimates their contribution to forest biomass. In tropical forests, biomass estimation must first analyse the vegetation structure and species composition and include all or a major percentage of the vegetation component.

Bradshaw and Warkentin (2015) calculated that boreal forests store 367.3 to 1715.8 Pg (1015 g) of carbon, with the midpoint (1095 Pg) being 1.3 to 3.8 times more than prior mean estimates. As a whole, land use modifications have lowered soil C stores by 0.39 Mg ha^{-1} yr^{-1}) (Deng et al. 2016). The conversion of farmland to grassland led in an increase in soil carbon stocks at a rate of 0.30 Mg ha^{-1} yr^{-1}, while the conversion of forest to grassland resulted in a higher rise of 0.68 Mg ha^{-1} yr^{-1}. However, the conversion of grassland to farmland led to a loss in soil carbon stocks at a rate of 0.89 Mg ha^{-1} yr^{-1}, and the transformation of forest to agriculturalland resulted in an even greater drop of 1.74 Mg ha^{-1} yr^{-1}. Additionally, the converting forest to forest resulted in a decrease of 0.63 Mg ha^{-1} yr^{-1} in soil carbon stocks. India's carbon stock in 1984 and 1994 amounted to 1085.06 and 1083.69 Mt, respectively (Manhas et al. 2006; Richter et al. 2021).

Forest of Himachal Pradesh had 1158 t ha^{-1} as mean AGB (above ground biomass) (Sharma et al. 2008). In their study, they conducted research in 7 primary forest types within the temperate zone of the Garhwal Himalaya, ranging from 1500 to 3100 m above sea level. The objective was to examine the impact of slope aspects on density of carbon and provide suggestions for management of forest that prioritise carbon conservation and sequestration. Soil organic carbon (SOC) levels ranged from 40.3 CMg ha^{-1} in the *Pinus roxburghii* forest on the southwest aspect to 177.5 CMg ha^{-1} in the wet *Cedrus deodara* woodland on the northeast face. The total carbon density varied from 118.1 CMg ha^{-1} in the *Pinus roxburghii* forest on the southwest aspect to 469.1 CMg ha^{-1} in the wet *Cedrus deodara* woodland on the northeast face. The values of SOC and TCD were markedly greater on the northern slopes in comparison to southern slopes. For the aim of carbon sequestration, it is advisable to implement plantation silviculture on northern slopes. Furthermore, in order to promote the preservation of carbon, it is recommended to prioritise mature forest stands that are growing on northern slopes. Sheikh et al. (2021) examined the seasonal variations and fluctuations in soil at various depths. The findings indicated that the autumn season had the greatest influence on soil organic carbon (SOC), both in terms of seasonal dominance and depth. The soil had the largest concentration of organic carbon during the fall season, with spring and summer following in descending order. It was found by Sharma et al. (2011) Sharma et al. (2011) discovered that in the temperate zone of Garhwal Himalaya, the levels of soil organic carbon (SOC) and total carbon density (TCD) were much higher on the northern slopes compared to the southern slopes. The recommendation is to implement plantation silviculture on the northern aspects for carbon sequestration and to prioritise mature forest stands growing on the northern aspects for carbon conservation purposes. Sharma et al. (2023) found that soil carbon density reached its highest value (35.77 Mg ha^{-1}) in the silvopastoral system but decreased as the altitude increased.

The silvopastoral system had the highest recorded carbon storage potential (59.76 Mg ha^{-1}), followed by agrohortisilviculture, agrosilvohorticulture, agrisilviculture, pastoralsilviculture (Wani et al. 2023) found that subtropical soils experience greater carbon loss than temperate soils. Hence, the subtropical region necessitates a more rigorous application of C conservation farming techniques in contrast to the temperate climatic environment. Implementing C storage and conservation measures is crucial in all climatic regions to mitigate land deterioration. Encouraging horticultural land uses and applying effective soil management practices can aid in the restoration of soil carbon and improve the livelihood security of the hill population in the North Western Himalayas (Kumari et al. 2022). Kumar and Sharma (2015) evaluated the amount of carbon stored in the Balganga Reserved Forest (BRF) located in Tehri district, Uttarakhand, India. Site III has the highest total biomass density (TBD) and total carbon density (TCD) estimations at a depth of 1800–2600 m. Site II has the second highest TBD and TCD estimates at a depth of 1600–1800 m, with values of 83.92 and 41.96 Mg ha^{-1} respectively. Site I has the lowest TBD and TCD estimates at a depth of 1000–1400 m, with values of 57.22 and 28.61 Mg ha^{-1} respectively. On average, the TBD and TCD estimates are 83.13 and 41.56 Mg ha^{-1} respectively. The researchers employed volumetric equations derived from the Forest Survey of India (FSI) and Forest Research Institute (FRI) publications specific to each species.

Pearson et al. (2014) employed the IPCC gain–loss methodology to compute the specific reductions in live biomass resulting from the felling of trees, the unintended harm to other trees, and the associated logging infrastructure. Additionally, they accounted for the benefits derived from the regrowth occurring within and in the vicinity of the cleared areas. Asner (2009) found that Landsat, SPOT, ASTER, and ALI imaging data of medium spatial resolution detect selective logging forest clearings similarly. The classification of burned regions in Indonesia's 2015 fires and the estimation of emissions from them were conducted using high-spatial resolution Sentinel-1 A SAR images with a resolution of 10 m (Liesenberg and Gloaguen, 2013).

20.9 Discussion

Studies around the world showed that the carbon stocks decline with changes in land use pattern, anthropological activities, etc. Forest may get diminished due to various factors that are broadly categorised in two main categories: Human activities and Natural activities described as follows:

20.9.1 *Human Activities*

Forest degradation can result from slash-and-burn farming, commercial farmers removing forests for subsistence and cash crop production, and displaced farmers moving to new forest lands. Cattle ranchers deforest the area for pasture, often displacing slash-and-burn farmers who move to another forest. Increased herding and firewood collection may cause deforestation. Mining and petroleum can also deforest. These include relocation into woody areas and settlement projects that displace indigenous tribes and force them to seek refuge there. Construction of roads and highways gives infrastructure developers and other land users access to forested regions, while hydroelectric dams cause flooding. (Olagunju 2015).

20.9.2 *Logging and Deforestation*

Deforestation can cause temperature swings, desertification, reduced agricultural output, flood danger, greenhouse gas emissions, and many other problems for indigenous groups (Aju et al. 2015). Forests can be degraded by human activities, including slash-and-burn farming, commercial farmers cutting forests for subsistence and cash crop production, and displaced farmers moving to new forest lands. Cattle farmers also destroy trees for grazing, which displaces slash-and-burn farmers who shift to other forested areas. Intensifying herding may contribute to deforestation. Deforestation might also result from increased firewood collection. The mining and petroleum industries may also deforest. In other cases, people are moved to forested areas, and settlement initiatives cause indigenous groups to leave their homes and seek refuge in forests. Roads and highways give infrastructure developers and other land users access to forested areas, while hydroelectric dams cause flooding. (Broadbent et al. 2008).

20.9.3 Agricultural Expansion

Agricultural production results 80% of worldwide deforestation, making it the primary cause of habitat destruction (Bodo et al. 2021). Animal agriculture, including the rearing of livestock and production of animal feed, plays a substantial role in causing deforestation. Additionally, it is accountable for roughly 60% of direct worldwide greenhouse gas (GHG) emissions. Collectively, the food system, encompassing both production and consumption, accounts for approximately 37% of anthropogenic greenhouse gas emissions worldwide. The agricultural sector is increasingly expanding its land utilisation to grow animal feed in order to fulfil the growing need for animal products, milk, and biofuels. This form of expansion is just exacerbating the strain on forests and intensifying the devastation of crucial ecosystems (Tee et al. 2014).

20.9.4 Infrastructure Development

According to Nayak et al. (2020), the presence of linear infrastructure, such as roads and electricity transmission lines, has caused a significant decrease of 71.5% in the number of big forest patches in India. This reduction is in comparison to a hypothetical situation where all linear infrastructures were not present. It is advisable to redirect infrastructure projects in order to circumvent extensive forest areas. Furthermore, the scientist suggests that linear infrastructure can be consolidated to reduce fragmentation.

20.10 Natural Factors for Degradation

20.10.1 Climate Change

It is exacerbating the severity and frequency of droughts, tropical storms, heat waves, fires, and frosts. As a result, forest health, resilience, productivity, and composition are declining (Lloret and Batllori 2021). Internationally, drought is acknowledged as the predominant element that impacts forest operations and leads to tree death on a large scale. However, forests in many regions also encounter additional extreme climatic phenomena that can cause major stress and even destruction. As a result, their ability to mitigate climate changeis greatly diminished (Lloret and Batllori 2021).

20.10.2 Pest and Disease Outbreaks

Insect pests infesting trees provide a formidable biotic menace. The majority of the pest species that are frequently reported belong to the orders Coleoptera and Lepidoptera. Hemiptera species are also a terrible menace. Insects navigate the tree to forage for food and/or construct a dwelling. Insects have the ability to consume every component of a tree. Species that use the same type of resource can be categorised. This categorization highlights the ecological roles of insects, particularly their effects on the forest. Typically, every type of tree has a distinct range of insects that are commonly found with it (Nair 2007). The FRA 2015 report indicated that temperate forests were significantly impacted by insects and pests, resulting in a disturbance of approximately 69,582 hectares of forest land (Nair 2007). According to the FRA 2015 Report, insects and pests significantly impacted temperate forests, causing the disturbance of about 69,582 hectares of forest land (Sturrock et al. 2011).

Forest degradation can disrupt carbon sequestration process in several ways:

Reduced Biomass Accumulation: Forest degradation often involves the removal of trees or damage to their structure. This directly reduces the capacity of the forest to accumulate biomass. Trees are major carbon storage units, and their removal decreases the overall carbon sequestration potential of the forest.

Altered Carbon Allocation: Forest degradation changes the allocation of carbon within the ecosystem. For example, if certain tree species are selectively removed, the remaining trees may allocate more carbon to growth or defense mechanisms, altering the overall carbon balance (Berner 2003).

Soil Disturbance: Degradation activities such as logging, improper land use, or infrastructure development can disturb the soil. This can accelerate the decomposition of organic matter and release stored carbon into the atmosphere as CO_2. Healthy soils play a crucial role in carbon sequestration, and their disturbance can have long-lasting effects on the carbon balance (Ilintsev et al. 2020).

20.10.3 Conclusion

Forests, through the process of photosynthesis, capture carbon dioxide and convert it into oxygen, providing a breathable atmosphere for all living organisms (Wilson and Cagalanan 2016). Simultaneously, the carbon is stored in trees and soil, preventing it from contributing to the greenhouse effect. One such approach includes sustainable forest management discussed below.

20.11 Sustainable Forest Management

20.11.1 Selective Logging

In tropical forests globally, selective logging is prevalent. Production-used woods are cleared of commercially useful trees (Osazuwa-Peters et al. 2015). Tropical forests were selectively logged in India, Burma, Malaysia, and West Africa during colonial control (Puettmann et al. 2015). The 1950s saw selective tree harvesting, but the 1970s saw debates regarding its merits and cons. Selective logging affects 400 million hectares—10% of global forests (Martin et al. 2015). This approach intentionally eliminates economically useful trees, despite growing concerns about forest destruction. Successful tree removal logging requires selective tree removal. It requires strict forest management to meet aims. To maintain a consistent supply of forest commodities and services while preserving the forest's worth and development potential. Selective logging, a prevalent method for harvesting forest timber, can address numerous challenges associated to rising wood product demand. This approach reduces tropical forest carbon emissions and terrestrial biodiversity loss (Milne et al. 2015). Selective logging reduces animal habitat product availability, carbon storage, and biodiversity (Van Breugel et al. 2011).

20.11.2 Agroforestry

Agroforests have more biodiversity than plantations but less than native forests. Diverse species help ecosystems recover from shocks and maintain ecological processes. Agricultural "working forests" can replace natural forests. But their existence doesn't guarantee mature forest preservation (Sharma et al. 2023).

20.11.3 Reforestation

Reforestation and afforestation cool Earth. Landowners and commercial loggers need both approaches to produce trees sustainably and meet demand. Deforestation must be treated and prevented with reforestation. Deforestation harms animals, reduces CO_2 absorption, and increases flooding danger. Afforestation prevents desertification, which occurs when drought or excessive agriculture turn rich land into desert.

20.11.4 Emerging Technologies for Monitoring Carbon Sequestration

Advancements in technology have provided new and improved ways to monitor carbon sequestration in forests. These emerging technologies offer more accurate, efficient, and scalable methods for assessing carbon stocks and tracking changes over time. Continual advancements in technology and scientific understanding contribute to the development of more precise and efficient methods for carbon quantification, supporting efforts to mitigate and adapt to climate change Here are some notable emerging technologies for monitoring carbon sequestration:

1. Remote Sensing: LiDAR laser beams measure distances and create three-dimensional forest structure maps. Rodríguez-Veiga et al. (2017) found that the approach accurately assesses aboveground biomass and predicts carbon stocks. Large-scale forest cover monitoring is possible with Landsat and Sentinel high-resolution satellite photography. Remote sensing data on deforestation, degradation, and regeneration aids carbon accounting (Gao et al. 2020).
2. Drones:LiDAR or multispectral drones can survey forest structure and composition. They provide high-resolution data for monitoring small or inaccessible areas (Tang and Shao 2015). Drone LiDAR is more flexible and cost-effective than airborne. 3D forest canopy data allows reliable carbon stock measurement.
3. Machine Learning/Artificial Intelligence: Data Analysis Machine learning systems can automatically detect forest cover changes and estimate carbon reserves using satellite and LiDAR data. These algorithms improve data analysis speed and accuracy. Advanced machine learning-based carbon flux models recreate ecological fluxes. These models predict carbon dynamics using climate, land use, and forest structure (Sahith and Lal 2024)
4. SensorNetworks: Local carbon flows are monitored by ground-based sensors. Soil sensors can provide real-time soil carbon content and microbial activity data, enhancing our understanding of belowground carbon processes. Artificial intelligence-enabled camera traps track woodland animals. Understanding biodiversity's role in carbon cycling and ecosystem health requires this knowledge (Sarkar and Chapman 2021).

5. Global Forest Watch: Open Data Platforms provide real-time forest cover data to track deforestation and replanting worldwide. Data Sharing and Collaboration: Open data projects and collaborative platforms improve carbon accounting accuracy and transparency by sharing data and methodology (Talberth et al. 2015).
6. Quantum Sensing: Quantum sensors measure environmental parameters precisely. Quantum sensing enhances carbon stock estimates. Blockchain technology creates a secure, irreversible record of carbon offset transactions, improving carbon accounting transparency. Forest conservation carbon credit markets may benefit (Crawford et al. 2021).

New techniques can monitor and manage forest carbon sequestration, enhancing climate change mitigation.

References

Aju PC, Iwuchukwu JJ, Ibe CC (2015) Our forests, our environment, our sustainable livelihoods. Eur J Acad Essays 2(4):6–19

Asner GP (2009) Tropical forest carbon assessment: integrating satellite and airborne mapping approaches. Environ Res Lett 4(3):034009

Berner RA (2003) The long-term carbon cycle, fossil fuels and atmospheric composition. Nature 426(6964):323–326

Bradshaw CJA, Warkentin IG (2015) Global estimates of boreal forest carbon stocks and flux. Global and Planetary Change 128:24–30. https://doi.org/10.1016/j.gloplacha.2015.02.004

Broadbent EN, Asner GP, Keller M, Knapp DE, Oliveira PJ, Silva JN (2008) Forest fragmentation and edge effects from deforestation and selective logging in the Brazilian Amazon. Biol Cons 141(7):1745–1757

Bustamante MM, Roitman I, Aide TM, Alencar A, Anderson LO, Aragão L, Asner GP, Barlow J, Berenguer E, Chambers J (2016) Toward an integrated monitoring framework to assess the effects of tropical forest degradation and recovery on carbon stocks and biodiversity. Glob Change Biol 22(1):92–109

Chazdon RL, Guariguata M (2018) Decision support tools for forest landscape restoration. CIFOR, Bogor Barat, Indonesia 70

Crawford SE, Shugayev RA, Paudel HP, Lu P, Syamlal M, Ohodnicki PR, Chorpening B, Gentry R, Duan Y (2021) Quantum sensing for energy applications: review and perspective. Adv Quantum Technol 4(8):2100049

Daba MH, Dejene SW (2018) The role of biodiversity and ecosystem services in carbon sequestration and its implication for climate change mitigation. Environ Sci Nat Resour 11(2):1–10

Deng L, Zhu Gy, Tang Z-s, Shangguan Z-p (2016) Global patterns of the effects of land-use changes on soil carbon stocks. Global Ecol Conserv 5:127–138. https://doi.org/10.1016/j.gecco.2015.12.004

Dieleman WI, Venter M, Ramachandra A, Krockenberger AK, Bird MI (2013) Soil carbon stocks vary predictably with altitude in tropical forests: implications for soil carbon storage. Geoderma 204:59–67

Ellert BH, Janzen HH, VandenBygaart AJ, Bremer E (2007) Measuring change in soil organic carbon storage. Soil sampling and methods of analysis, 25–38

Emborg J, Walker G, Daniels S (2012) Forest landscape restoration decision-making and conflict management: applying discourse-based approaches. Forest landscape restoration: integrating natural and social sciences, 131–153

Espejo AB, Federici S, Green C, Olofsson P, Sánchez MJ, Waterworth R, Carter SL, Herold M (2020). Integrating remote-sensing and ground-based observations for estimation of emissions and removals of greenhouse gases in forests: methods and guidance from the global forest observations initiative. Edition 3.0

Gao Y, Skutsch M, Paneque-Gálvez J, Ghilardi A (2020) Remote sensing of forest degradation: a review. Environ Res Lett 15(10):103001

Hamilton SE, Friess DA (2018) Global carbon stocks and potential emissions due to mangrove deforestation from 2000 to 2012. Nat Clim Chang 8(3):240–244

Hengl T, Mendes de Jesus J, Heuvelink GBM, Ruiperez Gonzalez M, Kilibarda M, Blagotić A, Shangguan W, Wright MN, Geng X, Bauer-Marschallinger B, Guevara MA, Vargas R, MacMillan RA, Batjes NH, Leenaars JGB, Ribeiro E, Wheeler I, Mantel S, Kempen B (2017) SoilGrids250m: global gridded soil information based on machine learning. PLoS ONE 12(2):e0169748. https://doi.org/10.1371/journal.pone.0169748

Holl KD, Zahawi RA (2014) Factors explaining variability in woody above-ground biomass accumulation in restored tropical forest. For Ecol Manag 319:36–43

Hundera K (2007) Traditional forest management practices in Jimma zone, south west Ethiopia. Ethiopian J Educ Sci 2(2):1–10

Ilintsev A, Bogdanov A, Nakvasina E, Amosova I, Koptev S, Tretyakov S (2020) The natural recovery of disturbed soil, plant cover and trees after clear-cutting in the boreal forests, Russia. Iforest-Biogeosciences and Forestry, 13(6):531

Jenkins M, Schaap B (2018) Forest ecosystem services. Background analytical study, 1

Joshi Meenakshi PPS (2003) Carbon sequestration by rehabilitating degraded forests in India. XII World Forestry Congress 0405:(B2)

Kebede B, Soromessa T (2018) Allometric equations for aboveground biomass estimation of Olea europaea L. subsp. cuspidata in Mana Angetu Forest. Ecosyst Health Sustain 4(1):1–12. https://doi.org/10.1080/20964129.2018.1433951

Khare TMN (2017) Carbon sequestration potential of selected tree species in the campus of Shuats. IJSRD- Int J Scientific Res Dev 5(6):2321–613

Kindermann G, McCallum I, Fritz S, Obersteiner M (2008) A global forest growing stock, biomass and carbon map based on FAO statistics. Silva Fennica 42(3):387–396

Kumar A, Sharma M (2015) Estimation of carbon stocks of Balganga reserved forest, Uttarakhand, India. Forest Sci Technol 11(4):177–181

Kumari R, Kumar A, Saikia P, Khan ML (2022) Vulnerability assessment of the indian himalayan forests in terms of biomass production and carbon sequestration potential in changing climatic conditions. In: Lackner M, Sajjadi B, Chen W-Y (Eds), Handbook of climate change mitigation and adaptation (pp 147–161). Springer International Publishing. https://doi.org/10.1007/978-3-030-72579-2_101

Kuyah S, Dietz J, Muthuri C, Jamnadass R, Mwangi P, Coe R, Neufeldt H (2012) Allometric equations for estimating biomass in agricultural landscapes: II. Belowground biomass. Agric Ecosyst Environ 158:225–234

Liesenberg V, Gloaguen R (2013) Evaluating SAR polarization modes at L-band for forest classification purposes in Eastern Amazon, Brazil. Int J Appl Earth Obs Geoinf 21:122–135

Lloret F, Batllori E (2021) Climate-induced global forest shifts due to heatwave-drought. Ecosystem collapse and climate change, 155–186

Lorenz K (2010) Carbon sequestration in forest ecosystems. Springer

Lorenz K, Lal R, Lorenz K, Lal R (2010) The importance of carbon sequestration in forest ecosystems. Carbon Sequestration in Forest Ecosystems, 241–270

Lugo A E, Brown S (1992) Tropical forests as sinks of atmospheric carbon. Forest Ecol Manag 54(1):239–255. https://doi.org/10.1016/0378-1127(92):90016-3

Manhas R, Negi J, Kumar R, Chauhan P (2006) Temporal assessment of growing stock, biomass and carbon stock of Indian forests. Clim Change 74:191–221

Martin PA, Newton AC, Pfeifer M, Khoo M, Bullock JM (2015) Impacts of tropical selective logging on carbon storage and tree species richness: A meta-analysis. Forest Ecol Manag 356:224–233. https://doi.org/10.1016/j.foreco.2015.07.010
Mendelsohn R, Sedjo R, Sohngen B, Mooij RA, d Keen, M, Parry IWH (2012) Fiscal policy to mitigate climate change fiscal policy to mitigate climate change: a guide for policymakers. In Chapter 5. Forest Carbon Sequestration* (pp ch05). International Monetary Fund. https://doi.org/10.5089/9781616353933.071.ch00510.5089/9781616353933.071
Milne E, Banwart, SA, Noellemeyer E, Abson DJ, Ballabio C, Bampa F, Bationo A, Batjes N H, Bernoux M, Bhattacharyya T, Black H, Buschiazzo DE, Cai Z, Cerri CE, Cheng K, Compagnone C, Conant R, Coutinho HLC, de Brogniez D, Zheng J (2015) Soil carbon, multiple benefits. Environ Dev 13:33–38. https://doi.org/10.1016/j.envdev.2014.11.005
Mitchard ETA (2018) The tropical forest carbon cycle and climate change. Nature 559(7715):527–534. https://doi.org/10.1038/s41586-018-0300-2
Nair KS (2007) Tropical forest insect pests: ecology, impact, and management. Cambridge University Press
Nayak R, Karanth KK, Dutta T, Defries R, Karanth KU, Vaidyanathan S (2020) Bits and pieces: Forest fragmentation by linear intrusions in India. Land Use Policy 99:104619
Olagunju TE (2015) Impacts of human-induced deforestation, forest degradation and fragmentation on food security. New York Sci J 8(1):10
Osazuwa-Peters OL, Jiménez I, Oberle B, Chapman CA, Zanne AE (2015) Selective logging: do rates of forest turnover in stems, species composition and functional traits decrease with time since disturbance?—A 45year perspective. Forest Ecol Manag 357:10–21. https://doi.org/10.1016/j.foreco.2015.08.002
Ota L, Chazdon RL, Herbohn J, Gregorio N, Mukul SA, Wilson SJ (2020) Achieving quality forest and landscape restoration in the tropics. Forests 11(8):820
Pan Y, Birdsey A, Fang J, Houghton R, Kauppi PE, Kurz WA, Phillips OL, Shvidenko A, Lewis SL, Canadell JG, Ciais P, Jackson RB, Pacala SW, McGuire AD, Piao S, Rautiainen A, Sitch S, Hayes D (2011) A large and persistent carbon sink in the world's forests. Science 333(6045):988–993. https://doi.org/10.1126/science.1201609
Pearson TR, Brown S, Casarim FM (2014) Carbon emissions from tropical forest degradation caused by logging. Environ Res Lett 9(3):034017
Pirard R, Dal Secco L, Warman R (2016) Do timber plantations contribute to forest conservation? Environ Sci Policy 57:122–130
Puettmann KJ, Wilson SM, Baker SC, Donoso PJ, Drössler L, Amente G, Harvey BD, Knoke T, Lu Y, Nocentini S (2015) Silvicultural alternatives to conventional even-aged forest management-what limits global adoption? Forest Ecosyst 2(1):1–16
Rajput BS, Bhardwaj DR, Pala NA (2017) Factors influencing biomass and carbon storage potential of different land use systems along an elevational gradient in temperate northwestern Himalaya. Agrofor Syst 91(3):479–486. https://doi.org/10.1007/s10457-016-9948-5
Richter F, Jan P, El Benni N, Lüscher A, Buchmann N, Klaus VH (2021) A guide to assess and value ecosystem services of grasslands. Ecosyst Serv 52:101376
Rodríguez-Veiga P, Wheeler J, Louis V, Tansey K, Balzter H (2017) Quantifying forest biomass carbon stocks from space. Current Forestry Reports 3:1–18
Sarkar D, Chapman CA (2021) The smart forest conundrum: contextualizing pitfalls of sensors and AI in conservation science for tropical forests. Tropical Conserv Sci 14:19400829211014740
Sedjo R, Sohngen B (2012) Carbon sequestration in forests and soils. Annu Rev Resour Econ 4(1):127–144
Sharma RK, Sankhayan PL, Hofstad O (2008) Forest biomass density, utilization and production dynamics in a western Himalayan watershed. J Forestry Res 19(3):171–180
Sharma C, Gairola S, Baduni N, Ghildiyal S, Suyal S (2011) Variation in carbon stocks on different slope aspects in seven major forest types of temperate region of Garhwal Himalaya, India. J Biosci 36:701–708

Sharma H, Pant KS, Bishist R, Lal Gautam K, Ludarmani Dogra R, Kumar M, Kumar A (2023) Estimation of biomass and carbon storage potential in agroforestry systems of north western Himalayas, India. CATENA, 225, 107009. https://doi.org/10.1016/j.catena.2023.107009

Sheikh MA, Tiwari A, Anjum J, Sharma S (2021) Soil carbon pool and dynamics of different fractions in subalpine temperate coniferous forests of Western Himalaya, India. Environ Monit Assess 193:1–13

Sturrock R, Frankel S, Brown A, Hennon P, Kliejunas J, Lewis K, Worrall J, Woods A (2011) Climate change and forest diseases. Plant Pathol 60(1):133–149

Talberth J, DellaSala D, Fernandez E (2015) Global forest watch report. Wilderness 503:657–7336

Tang L, Shao G (2015) Drone remote sensing for forestry research and practices. J Forestry Res 26:791–797

Tee F, Agbidye F, Ogwuche J (2014) Indigenous forest conservation practices in Benue state, Nigeria. J Agric Forestry Soc Sci (JOAFSS) 12:(1)

Turner DP, Guzy M, Lefsky MA, Ritts WD, Van Tuyl S, Law BE (2004) Monitoring forest carbon sequestration with remote sensing and carbon cycle modeling. Environ Manag 33:457–466

Van Breugel M, Ransijn J, Craven D, Bongers F, Hall JS (2011) Estimating carbon stock in secondary forests: decisions and uncertainties associated with allometric biomass models. For Ecol Manage 262(8):1648–1657

Wani OA, Sharma V, Kumar SS, Babu S, Sharma KR, Rathore SS, Marwaha S, Ganai NA, Dar SR, Yeasin M, Singh R, Tomar J (2023) Climate plays a dominant role over land management in governing soil carbon dynamics in North Western Himalayas. J Enviro Manag 338:117740. https://doi.org/10.1016/j.jenvman.2023.117740

Wilson SJ, Cagalanan D (2016) Governing restoration: Strategies, adaptations and innovations for tomorrow's forest landscapes. World Dev Perspect 4:11–15

Yuliani EL, de Jong E B, Knippenberg L, Bakara DO, Salim MA, Sunderland T (2018) Keeping the land: Indigenous communities struggle over land use and sustainable forest management in Kalimantan, Indonesia

Chapter 21
Assimilating Biodiversity, Conservation and Sustainable Development in the Himalayas

Nitesh Kumar, Sanjeev Kumar, Arti Jamwal Sharma, Ruchika Devi, Harinder Singh Banyal, Pratibha, and Vishal Rana

Abstract The Himalayan region, known for its breathtaking landscapes and unique biodiversity, stands at the crossroads of conservation and sustainable development. The delicate balance between these two imperatives is critical, as the immediate natural resources of the area are indispensable to the livelihoods of the local communities. However, this fragile environment faces severe threats from the rapid expansion of infrastructure, urbanization, and the impacts of climate change. The need of the hour is an all-encompassing strategy that integrates community involvement, scientific research, and traditional knowledge to conserve the Himalayas' biodiversity effectively. In this context, it is imperative to recognize the interconnectedness of conserving diverse flora and fauna to pursue sustainable human development. Scientific research is crucial for understanding the complex ecosystems of the Himalayas and identifying sustainable practices. Community involvement ensures the preservation of indigenous wisdom and fosters a sense of ownership and responsibility among the people. The challenge is to strike a harmonious balance that considers the environment, the well-being of local communities, and economic development concurrently. Collaborative efforts involving stakeholders such as government bodies, local communities, and environmental organizations are essential. In essence, achieving a sustainable coexistence in the Himalayas requires a holistic perspective that blends environmental conservation with community well-being and economic progress.

N. Kumar · R. Devi (✉) · H. S. Banyal · Pratibha
Department of Biosciences, Himachal Pradesh University, Shimla 171005, Himachal Pradesh, India

S. Kumar
Department of Basic Sciences, College of Forestry, Dr. Y.S. Parmar University of Horticulture and Forestry, Nauni, Solan 173230, Himachal Pradesh, India

A. J. Sharma (✉)
Department of Bio Sciences, Career Point University, Hamirpur 176041, Himachal Pradesh, India
e-mail: arti.botany@cpuh.edu.in

V. Rana
Department of Chemistry, Thakur PG College of Education Dhaliara, Kangra, Himachal Pradesh, India

D. Arya et al. (eds.), *Climate Change Impact on Himalayan Biodiversity*,
Environmental Science and Engineering, https://doi.org/10.1007/978-3-031-77149-1_21

Keywords Himalayan · Sustainable development · Environment · Indigenous · Communities

21.1 Introduction

The highest mountain range on Earth is symbolized by the Himalayas. The Himalayas, which span more than 2,400 kms, have a vast range of climatic climates, from subtropical to northern (Dhital 2015; Rana et al. 2021). The diversity of the Himalayas is remarkable and has many different aspects. The western part, which includes the Indian regions of Ladakh and Zanskar, is mostly cold, dry, and semi-desert (Hussain et al. 2023). With an excess of 10 m of precipitation annually, the eastern end, or the Brahmaputra watershed, contains some of the wettest locations on earth (Petterson 2023). It creates the world's largest bioclimatic gradient, with vascular plants found between 60 and 6,400 m above sea level (Eckstein et al. 2022). The Himalayan region, which is sometimes referred to as the "land of snow," is a stunning region that crosses five nations in South Asia: China, Pakistan, India, Nepal, and Bhutan (Goswami 2022). The Himalayas are not just famous for having the world's highest peaks, like Mount Everest. They are also home to an amazing variety of life, known as Himalayan biodiversity (Sherpa and Puschiasis 2023). Spanning a wide range of elevations, the Himalayan ecosystem unfolds like a natural staircase, each step revealing a distinct vegetation zone. At the bottom of the hills, there are lots of green, tropical forests with many different kinds of plants and animals. As you go up the hills, the scenery changes to cooler forests, and if you keep going higher, you'll reach meadows with colourful flowers (Weaner and Christopher 2016). The unique topography of the Himalayas, coupled with the extreme variations in altitude and climate, has given rise to an unparalleled array of flora and fauna. This biological diversity is so exceptional that the Himalayas are recognized as one of the world's biodiversity hotspots, harbouring a treasure trove of both unique and endemic species (Koulelis et al. 2023). In the Himalayas, there are famous flowers like rhododendrons and orchids, as well as useful plants for medicine, making the area look and smell nice (Guo et al. 2022). The Himalayas are also known for a variety of valuable timber species. The animals there are interesting too, with special species that have learned to live in the tough conditions of the mountains (Zhao 2022). Steep inclines and deep forests are home to majestic species like the red panda, the endangered Bengal tiger, and the elusive snow leopard. Numerous rivers originating from the Himalayas are also home to a variety of fish species. The Himalayan area is not just a home to many different plants and animals; it's also very important to the culture and spirituality of the people (Bisht 2008). These mountains are considered sacred by various religions, and they've influenced stories, myths, and customs that have been part of local communities for a long time (Francaviglia 2016). The people who live near the Himalayan mountains depend a lot on the natural things they get from there, like farming and making traditional medicine (Gagné and Chostak 2023). But, these mountains are facing big problems from the modern world, like the changing

21.3 Materials and Methods

The methodology used for the thorough literature assessment on the conservation of Himalayan biodiversity comprised a careful examination of a variety of information from reliable sources. The primary platforms utilized for gathering information included the Web of Science, where a targeted search was conducted to identify peer-reviewed articles and scientific papers relevant to Himalayan biodiversity conservation. Another important resource used was Google Scholar, which used a thorough search algorithm to find academic journals and research papers about the preservation of Himalayan biodiversity. The approach adopted for Google Scholar aimed at incorporating a multidisciplinary perspective, gathering insights from various academic disciplines to provide a comprehensive understanding of the subject. The investigation also extended to Shod Ganga, the repository of Indian electronic thesis and dissertations, where the focus was on thesis and dissertations relevant to Himalayan biodiversity conservation. A comprehensive review of articles from reputable journals specializing in ecology, conservation biology, and environmental science was conducted. The combination of these diverse sources ensured a thorough and holistic exploration of the conservation strategies employed for Himalayan biodiversity.

21.4 Threat to Biodiversity

1. **Extinction**:

Extinction is the complete disappearance of a species from the entire planet. Throughout Earth's history, there have been five significant occasions known as mass extinctions, where a large number of species vanished simultaneously (Butchart et al. 2006). During these periods, many different kinds of organisms went extinct relatively quickly. Currently, we are facing the ongoing sixth mass extinction event, and human activities are the primary cause (Cowie et al. 2022; Pievani 2014). This means that due to what people are doing, many species are disappearing at an alarming rate (Lévêque 2022). Even when there isn't a mass extinction happening, some species still naturally disappear over time, but it occurs at a slower, regular pace known as the background extinction rate (Chakrabarty 2022). What's particularly worrying for conservation efforts is when a species is locally wiped out from a specific area, known as extirpation, even if it's not extinct globally (Löbl et al. 2023). This localized loss can disrupt ecosystems and reduce overall biodiversity in a particular region (Landmann et al. 2023). So, whether on a global or local scale, the disappearance of species is a critical concern for maintaining the balance and health of Earth's ecosystems.

2. **Habitat loss**: The term "habitat loss" describes two primary processes that have a substantial influence on different species' living spaces: habitat fragmentation and habitat destruction (Zhang et al. 2023).

- **Habitat Destruction**: When a natural environment that is vital to a species is destroyed or severely disrupted, habitat destruction takes place (Prakash and Verma 2022). Developments such as urbanization, industrialization, and deforestation might cause this. For instance, the original habitat is destroyed when forests are cleared for agriculture or lumber, depriving animals of their essential dwellings (Rahman 2023). By changing the circumstances necessary for some species to live, pollution and climate change can also lead to the destruction of habitat (Prakash 2021).
- **Habitat Fragmentation**: The process of dividing a continuous habitat into smaller, isolated sections is known as habitat fragmentation. This process occurs when human activities, such as building roads, urban areas, or agricultural fields, create barriers within a habitat (Fuller et al. 2015). These barriers can isolate different parts of the habitat from each other, preventing species from moving freely between them (Carvalho et al. 2016). This isolation can lead to smaller and more vulnerable populations, as well as difficulties in finding mates or suitable resources (Allendorf et al. 2012). Over time, the separated patches may become increasingly inhospitable for certain species, further contributing to declines in population.

The natural spaces that species rely on for survival are essentially disrupted by habitat loss, whether it be through destruction or fragmentation. It is a major cause of biodiversity loss and may cause certain species to become extinct or decrease because they cannot adapt to their changed or fragmented surroundings.

3. **Climate Change**: The release of greenhouse gases via processes like burning fossil fuels for electricity, such as carbon dioxide and methane, is a major driver of climate change (Ndubuisi et al. 2023). This phenomenon not only leads to a rise in the average global temperature but also causes unpredictable changes in weather patterns. The impact of climate change poses serious threats to biodiversity through various mechanisms, including:

- **Species Range Shifts**: As temperatures change and ecosystems are altered, many species may need to shift their geographic ranges to find suitable conditions. This can lead to challenges for species that are unable to move or adapt quickly, potentially resulting in population declines or local extinctions.
- **Mismatched Biotic Interactions**: Climate change has the potential to alter the timing of important biological processes, including flowering, migration, and reproduction. If different species within an ecosystem respond to climate change at different rates, it can create mismatches in their interactions. For example, if the timing of flowering plants and the emergence of pollinators becomes desynchronized, it can affect pollination and reproduction.
- **Sea Level Rise**: The melting of polar ice caps and glaciers, primarily due to global warming, contributes to rising sea levels. This poses a direct threat to coastal habitats, including wetlands and estuaries, which are crucial for many species (Santojanni et al. 2023). Sea level rise can result in habitat loss, affecting both terrestrial and marine biodiversity.

- **Ocean Acidification**: Ocean acidification is the result of the world's seas absorbing extra atmospheric carbon dioxide. Aquatic life may be adversely affected by acidification (Jewett and Romanou 2017). This threatens the health of marine ecosystems and the species that depend on them.

Therefore, a complex and linked threat to biodiversity is climate change. Changes in temperature and weather patterns upset the delicate ecosystem balance, which has an impact on species distribution and behavior. To guarantee the survival of a variety of species in the face of shifting environmental conditions, conservation efforts must incorporate climate change mitigation techniques, such as cutting greenhouse gas emissions, safeguarding sensitive habitats, and enhancing ecosystem resilience.

4. **Overexploitation**: Overexploitation occurs when organisms are harvested or collected at a pace that exceeds their ability to reproduce and replenish. This unsustainable practice includes activities like the illegal hunting of elephants, the unsustainable pursuit of bush meat through excessive hunting, overfishing in aquatic ecosystems, and the excessive collection of slow-growing plants and fungi. Such practices can lead to significant population declines, disrupt ecosystems, and threaten the long-term survival of targeted species. Conservation efforts are crucial to ensure responsible and sustainable management of resources to prevent the negative impacts of overexploitation on biodiversity (Rawat and Agarwal 2015).
5. **Invasive Species**: Biodiversity is seriously threatened by invasive species. This species is a non-native organism that, when introduced to a new environment, establishes itself and spreads rapidly, often outcompeting or displacing native species. They may have no natural predators in the new environment, allowing them to thrive and disrupt the balance of the ecosystem. Invasive species can alter the physical structure and composition of ecosystems. New diseases could be brought in by exotic species to native species that have little or no resistance. This can result in disease outbreaks that harm or wipe out native populations, particularly when the invasive species acts as a carrier for pathogens. These species can interbreed with native species, leading to genetic hybridization. This can dilute the genetic integrity of native populations, potentially reducing their adaptability and resilience to environmental changes. These are responsible for disrupting established ecological relationships, such as pollination or seed dispersal partnerships. The entire ecosystem could potentially be impacted by a domino effect due to this situation impacting multiple species and ecological processes. Controlling and managing invasive species is crucial for biodiversity conservation. Strategies include early detection and rapid response, habitat restoration, and the development of integrated management plans. International cooperation is often necessary, as invasive species can move across borders and impact ecosystems on a global scale (Keller et al. 2011).

21.5 Reason of Biodiversity Loss

Biodiversity loss happens when species are displaced or become extinct. Losing a specific species might not seem important to some people, especially if it's not a well-known one like the Bengal tiger or bottlenose dolphin. However, scientists say that species are disappearing much faster now than in the past, and this could mean losing tens of thousands of species during our lifetimes. This could seriously affect us because it might lead to ecosystems breaking down. The loss of biodiversity has a big impact on ecosystems because species are connected in complicated ways. For example, if one species disappears, it could lead to the extinction of another. To figure out how much biodiversity is being lost, scientists look at which species are in danger of disappearing and study how ecosystems are declining. The greatest concern to biodiversity is the combination of more people on Earth and the resources they use. As the human population grows, we need more resources to live and expand, but we're taking these resources from the environment in a way that can't be sustained. The five main dangers to biodiversity are losing habitats, pollution, using up resources too much, introducing foreign species, and temperature Variability. The increase in travel and trade has led to the spread of invasive species, while the other threats come from the fact that more people are using more resources.

21.6 Conservation Strategies of Biodiversity

- **In situ conservation**: The preservation of biodiversity within its native habitat is known as "in situ conservation.". This approach focuses on protecting and managing ecosystems, species, and genetic diversity in the places where they naturally occur. In situ conservation aims to maintain the ecological processes and interactions that sustain life, ensuring the long-term survival of diverse forms of life in their native environments. Because it preserves the complex linkages and ecological processes that characterize natural ecosystems in addition to protecting species, in situ conservation is regarded as a fundamental and all-encompassing approach to biodiversity conservation. It highlights how crucial it is to maintain biological communities' dynamic and ever-changing characteristics within their natural environments.

21.7 Protected Areas

National Parks: Designating areas where human activities are limited to protect natural habitats and wildlife. Examples include Yellowstone National Park in the United States and Serengeti National Park in Tanzania.

Wildlife Reserves: Establishing specific areas to conserve and protect wildlife, allowing them to thrive without significant human disturbance. An example is the Sundarbans Reserve Forest in Bangladesh.

Biosphere Reserves:

Designating large areas that integrate conservation, sustainable development, and scientific research. Biosphere reserves aim to reconcile the conservation of biodiversity with the needs of local communities. For instance, the Western Ghats in India has a designated biosphere reserve.

Community-Based Conservation:

Involving local communities in the conservation of biodiversity, recognizing their traditional knowledge and practices. This approach fosters a sense of ownership and responsibility among residents. Examples include community-managed reserves in the Amazon rainforest.

Habitat Restoration:

Undertaking activities to restore degraded habitats and ecosystems to their natural state. This may involve reforestation, wetland restoration, or other interventions to improve the health and functionality of ecosystems.

Conservation of Threatened Species:

Implementing specific conservation programs for endangered or threatened species within their natural habitats. For example, the efforts to protect the mountain gorillas in the Virunga Massif of Central Africa involve in situ conservation measures.

Conservation of Crop Wild Relatives:

Protecting the natural habitats of wild varieties of plants connected to cultivated crops. These wild relatives often contain valuable genetic traits that can be crucial for crop improvement and adaptation to changing environmental conditions.

Conservation of Genetic Diversity:

Protecting the genetic diversity of crops and domesticated animals within their centers of origin or diversity. This involves maintaining traditional farming practices that preserve diverse varieties and breeds.

- ***Ex-situ* conservation**: *Ex-situ* conservation is the preservation of biodiversity outside its natural habitat. In this approach, living organisms or genetic material are taken out of their native environment and placed in controlled settings such as botanical gardens, zoos, seed banks, or captive breeding programs. *Ex-situ* conservation is employed to complement *in-situ* conservation efforts and serves as a safeguard against the loss of species, particularly when their natural habitats are threatened. *Ex-situ* conservation plays a crucial role in biodiversity conservation, especially when species are facing imminent threats in their natural habitats. It provides a safety net, allowing for the preservation of genetic diversity, the potential reintroduction of species into the wild, and scientific research that contributes to overall conservation efforts. However, it is essential to recognize that *ex-situ* conservation is most effective when integrated with comprehensive *in-situ* conservation strategies to address the broader challenges facing ecosystems and their inhabitants. Examples of *ex-situ* conservation:

Botanical Gardens:

- **Living Collections**: Maintaining living collections of plants, including rare and endangered species, in botanical gardens. These gardens often serve as educational resources and contribute to research on plant biology and ecology.

Zoos and Aquariums: Captive Breeding Programs: Breeding and maintaining populations of endangered or threatened animal species in controlled environments to ensure their survival. Examples include breeding programs for giant pandas and California condors.

Seed Banks and Germplasm Repositories:

- **Storage of Seeds**: Collecting, storing, and managing seeds of diverse plant species in seed banks. These repositories act as a genetic reservoir for crop plants and endangered species.

Cryopreservation:

- **Preservation of Genetic Material**: Freezing and storing genetic material, such as sperm, eggs, or embryos, to maintain genetic diversity and facilitate future reintroduction efforts. This technique is commonly used for endangered species with limited population sizes.

In Vitro Conservation:

- **Tissue Culture**: Growing and maintaining plant cells, tissues, or organs in controlled laboratory conditions. In vitro conservation helps preserve plant species and allows for the propagation of rare or endangered plants.

DNA Banks.

- **Genetic Information Storage**: Storing genetic information in the form of DNA samples to preserve the genetic diversity of species. DNA banks contribute to research, genetic studies, and potential future cloning or reintroduction efforts.

Sustainable Development

Sustainable development is an approach to advancement that aims to fulfill current needs without jeopardizing the capacity of future generations to satisfy their requirements. To maintain the planet's ecosystems and natural resources while fostering harmony and equity within society, it stands for striking a balance between economic, social, and environmental factors. The idea of sustainable development became well-known in the latter half of the 20th century as worries about social injustice, environmental damage, and unstable economies increased. In its 1987 report "Our Common climate, destroying where animals and plants live, hunting, and building things in ways that hurt the environment (Ikechukwu and Shamijah 2022). It's really important to work hard to take care of these mountains by creating safe areas for nature, doing things that don't harm the environment, and getting the local communities involved in protecting their natural home (Uralovich et al. 2023). The variety of life in the Himalayas shows how living things can be strong and able to adapt to tough conditions (Kattel 2022). This amazing place, with its many plants, interesting animals, and cultural importance, needs to be taken care of well (Zerbe 2022). We should make sure to protect it and use practices that don't harm it, so it stays wonderful for many generations in the future (Crutchfield 2023).

21.2 Research Objectives

The study aims to investigate the intricate relationship between protecting the distinctive Himalayan flora and fauna and promoting sustainable development in the area, with a focus on the conservation of biodiversity against sustainable development. The study's focus areas are outlined in the following primary objectives:

1. Assessment of Biodiversity: Analyze the biodiversity of the Himalayan region in great detail, noting important species and habitats that support the ecological balance. This entails being aware of the distribution, quantity, and importance of different animal and plant species.
2. Impact of Development Activities: Investigate the direct and indirect effects of development activities such as infrastructure expansion and urbanization on the Himalayan biodiversity. Examine the ecological footprint of these activities and their potential consequences for the delicate balance of the ecosystem.
3. Community Involvement and Traditional Knowledge: Explore the role of local communities in biodiversity conservation. Assess the value of traditional knowledge held by indigenous populations and involve these communities in developing sustainable practices that align with both conservation and development goals.
4. Climate Change Impacts: Investigate the specific impacts of climate change on the Himalayan region and how these changes might affect biodiversity. Understand the vulnerabilities of different species and ecosystems to climate-related shifts.
5. Economic Valuation of Biodiversity: Assess the economic value of Himalayan biodiversity, considering ecosystem services, tourism, and other tangible and intangible benefits. This evaluation will contribute to understanding the economic implications of conservation efforts.

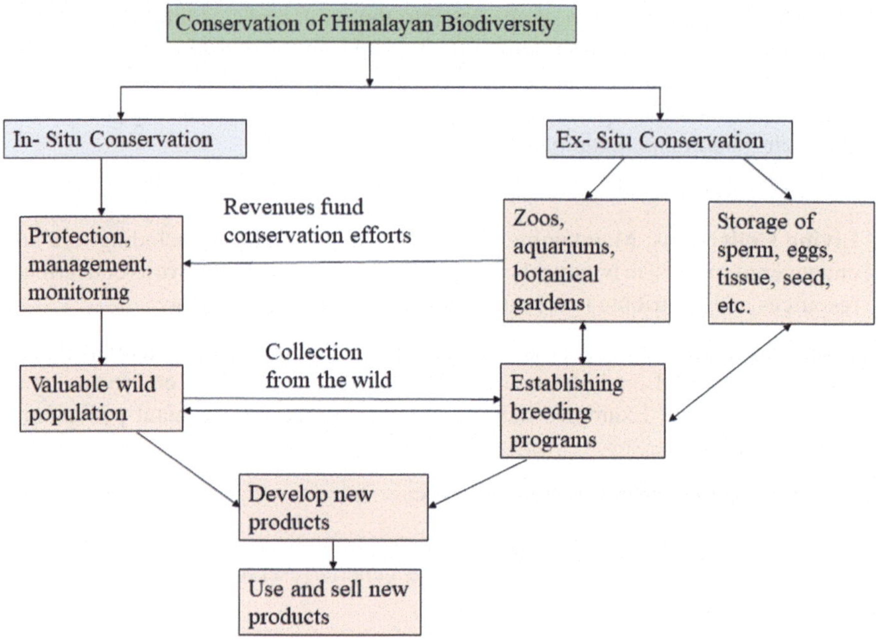

Fig. 21.1 Conservation of Himalayan Biodiversity

Future," the Brundtland Commission popularized the phrase "sustainable development" by defining it as "development that meets the needs of the present without compromising the ability of future generations to meet their own needs" (Fig. 21.1).

Balancing Economic, Social, and Environmental Considerations:

Sustainable development recognizes the interconnectedness of economic, social, and environmental factors. It aims to avoid the pitfalls of a one-dimensional focus on economic growth, acknowledging that true progress involves a delicate balance between these three dimensions. This balance ensures that economic prosperity does not come at the expense of social well-being or environmental health.

Harmonious and Equitable Society:

The goal of sustainable development is not just to attain growth in the economy but also to foster a harmonious and equitable society. This means addressing social issues such as poverty, inequality, and access to education and healthcare. By promoting inclusivity and social justice, sustainable development aims to create a society where the benefits of progress are shared by all.

Preserving Natural Resources and Ecosystems:

One of the central tenets of sustainable development is the responsible use and conservation of natural resources. This involves recognizing the finite nature of resources and working towards their sustainable management. Additionally, the preservation of ecosystems is crucial to maintaining biodiversity and the overall health of the planet.

Historical Context and Emergence:

The concept of sustainable development gained prominence in response to growing concerns in the late 20th century. Environmental degradation, social inequality, and economic instability became pressing issues, prompting the need for a more holistic and forward-thinking approach to development.

Inter-generational Responsibility:

Sustainable development emphasizes a sense of inter-generational responsibility. This means that the actions taken today should not hinder the capacity of coming generations to fulfill their requirements. It calls for a long-term perspective that considers the impacts of current decisions on the well-being of those who will come after us.

Global Challenges and Cooperation:

Many sustainability challenges are global, such as climate change and loss of biodiversity. Sustainable development recognizes the need for international cooperation to address these challenges effectively. This involves collaboration between governments, businesses, and communities on a global scale.

Sustainable development is a comprehensive and forward-looking approach to societal progress that seeks to address current needs without compromising the ability of future generations to meet their own needs. It emphasizes the importance of balance, inclusivity, environmental stewardship, and global cooperation in creating a better and more equitable world. Sustainable development includes:

Environmental Stewardship: Prioritizing the responsible use and conservation of natural resources, minimizing pollution and waste, and protecting biodiversity.

Social Equity: Ensuring that development benefits all members of society, addressing issues of poverty, inequality, and social injustice. This includes promoting access to education, healthcare, and other essential services for all.

Economic Viability: Fostering economic growth and development that is inclusive, resilient, and does not undermine long-term economic stability. This often involves promoting sustainable business practices and innovation.

Inter-generational Equity: Recognizing the interconnectedness of generations and the importance of leaving a positive legacy for future inhabitants of the planet.

Global Cooperation: Acknowledging that many sustainability challenges are global and require international collaboration to address issues such as climate change, biodiversity loss, and resource depletion.

Sustainable development is a holistic and integrated approach that considers the interdependence of economic, social, and environmental factors. It emphasizes the need for long-term thinking and the adoption of practices that promote resilience, inclusivity, and environmental responsibility. Governments, businesses, communities, and individuals all play crucial roles in advancing sustainable development to create a better and more equitable world for present and future generations.

21.8 General Discussion About Conservation of Himalayan Biodiversity V/s Sustainable Development

It can be challenging to strike a balance between protecting the distinctive ecosystems of the Himalayan region and advancing economic development and human well-being when it comes to the conservation of Himalayan biodiversity versus sustainable development (Khawas 2007). The Himalayas support local communities, offer ecosystem services, and are home to a wide variety of flora and wildlife that are essential to preserving ecological equilibrium. The Himalayan region, known for its magnificent vistas and abundant wildlife, is at a crossroads where preservation efforts and the necessity of sustainable development must be carefully balanced (Negi 2022). This geographical wonder, which spans multiple nations, must balance meeting the expanding demands of human populations with protecting its distinctive flora and animals. The Himalayas, commonly referred to as the "abode of snow," are home to an incredible diversity of species that are unrivaled in their ecological value (DeBuys 2021). Millions of people rely on this region for their life-giving resources, including food, water, and a wide range of ecosystem services (Brooks et al. 2006). But the insatiable drive for development has put the very things that make the Himalayas an ecological gem in jeopardy. The Himalayas have emerged as a focal point for conversations on how to find a balance between human growth and the preservation of this special and delicate ecosystem as the world struggles with the necessity of sustainable development (Thompson and Warburton 1985). There is an incredible diversity of plants and animals in the area, many of which are endemic and in danger of going extinct because of habitat loss, climate change, and irresponsible behavior. To shed insight into the challenges involved in balancing environmental preservation with socioeconomic advancement, this conversation delves into the dynamic interaction between the protection of Himalayan biodiversity and the goal of sustainable development (Prasad 2023). A complex strategy that takes into account both the immediate requirements of the surrounding communities and the long-term preservation of biodiversity is needed to achieve this delicate balancing act (Beatley 2000). The discourse delves into the multifaceted nature of this intricate predicament, scrutinizing the function of conservation tactics, sustainable development methodologies, and the necessity of cooperation between governmental bodies, local communities, and environmental advocacy groups (Santos et al. 2023). Here are several crucial aspects to consider in this conversation:

21.9 Biodiversity Importance

There are many different plant and animal species in the Himalayan region, many of which are native to the area, contributing to the region's high biodiversity. The importance of this biodiversity goes much beyond its aesthetic value because it is essential to preserving the region's overall ecological health (Rawal et al. 2013). These varied species support intricate ecological connections that maintain the ecosystem's resilience and equilibrium. But habitat loss, climate change, and human activity still pose dangers to the Himalayan region. To protect these species from the impending threat of extinction and to maintain the integrity of the Himalayan ecosystem, conservation activities are therefore essential (Pandit and Kumar 2013). By concentrating on conservation efforts, we sustain the interdependent web of life that characterizes the ecological richness of the Himalayan region in addition to protecting individual species. The recognition of the inherent importance of these endeavours underscores their urgency (Kothari et al. 2013).

Ecosystem Services:

The Himalayas provide vital ecosystem services, like water regulation, climate regulation, and soil conservation. The Himalayas are a primary source of vital ecosystem services, supporting many people living downstream as well as the region. The most important of these essential services is water regulation (Joshi and Joshi 2019). The Himalayas' complex system of rivers, lakes, and glaciers acts as a natural water regulator, affecting the flow and distribution of water across the area. This guarantees a steady and dependable supply of water for the many ecosystems found in the Himalayas, from dense forests to alpine meadows (Joshi and Negi 2011).

Local Communities:

Sustainable development should consider the needs of local organizations that depend on the resources of the Himalayas for their livelihoods. Initiatives that support local populations economically and protect the environment should be included in sustainable development in the Himalayas (De Graaf et al. 1996). This entails developing sustainable alternative livelihood options that are good for the environment and the economy. Promoting eco-friendly tourism, community-based natural resource management, and sustainable agriculture are a few examples of how to improve the economic well-being of the surrounding areas without sacrificing the integrity of the Himalayan ecosystems (Jansen 2003).

Climate Change:

The Himalayas are especially susceptible to changing weather patterns, with rising temperatures, glacial retreats, and changes in precipitation patterns. The Himalayan ecosystem makes a major contribution to climate regulation, among other things (Negi et al. 2012). The enormous tracts of forest that blanket the slopes serve as an essential carbon sink, drawing carbon dioxide from the atmosphere and lessening the effects of climate change. Glaciers' reflecting surfaces also help to chill the surrounding air, which affects regional and global temperature patterns. Thus, the Himalayas are essential to preserving climate stability, which has a significant impact on subsequent weather patterns and ecological processes (Chettri et al. 2020). In essence, the challenge lies in finding a harmonious equilibrium between safeguarding the biodiversity of the Himalayas and making sure that progress satisfies current demands without endangering the capacity of future generations to fulfill their own. This requires a holistic and thorough approach, involving collaboration among local communities, governments, academic institutions, and the business sector to discover and implement sustainable solutions.

Outcomes of Study:

The study on biodiversity, conservation, and sustainable development in the Himalayas aims to provide a comprehensive understanding of the region's ecological dynamics. Through a thorough Biodiversity Assessment, key species, ecosystems, and conservation status of endemic species are identified. This study catalyses raising awareness among individuals about the critical importance of biodiversity conservation in the Himalayas, fostering a deeper understanding of the ecological dynamics and promoting a sense of responsibility towards sustainable practices for the well-being of both the region and its inhabitants. Sustainable Development Integration explores the incorporation of goals into conservation, socio-economic aspects, and practices promoting both biodiversity and human well-being. Guiding Future Research Directions, the study recommends priorities, identifies emerging issues, and integrates findings into a broader research agenda for sustainable development. Overall, these aspects contribute to a concise yet comprehensive understanding of the intricate interplay between biodiversity, conservation, and sustainable development in the Himalayas.

21.10 Conclusion

The study investigating the assimilation of biodiversity, conservation, and sustainable development in the Himalayas has yielded a profound and holistic understanding of the region's ecological intricacies. The meticulous biodiversity assessment conducted throughout the research identified key species, evaluated biodiversity hotspots, and documented the conservation status of endemic species, laying a foundation for informed conservation efforts. The study underscored the intricate interplay between biodiversity conservation and sustainable development by exploring how sustainable development goals are integrated into conservation efforts. The scientific contributions of the research, coupled with its educational impact, emphasized the significance of disseminating knowledge and awareness about Himalayan biodiversity. The guidance provided for future research directions serves as a roadmap for sustained efforts, identifying emerging challenges and fostering a continual research agenda for the preservation of Himalayan biodiversity and the promotion of sustainable development. The study advocates for concerted efforts, recognizing that the preservation of biodiversity is inseparable from the sustainable development crucial for the region's enduring ecological health and the well-being of its inhabitants.

References

Allendorf FW, Luikart GH, Aitken SN (2012) Conservation and the genetics of populations. John Wiley and Sons. 1–484

Beatley T (2000) Preserving biodiversity: challenges for planners. J Am Plann Assoc 66(1):5–20

Bisht RC (2008) International encyclopaedia of Himalayas (5 Vols. Mittal Publications, Set). 1–1322

Brooks TM, Mittermeier RA, Da Fonseca GA, Gerlach J, Hoffmann M, Lamoreux JF, Mittermeier CG, Pilgrim JD, Rodrigues AS (2006) Global biodiversity conservation priorities. Science 313(5783):58–61

Butchart SH, Stattersfield AJ, Brooks TM (2006) Going or gone: defining possibly extinct species to give a truer picture of recent extinctions. Bulletin-British Ornithologists Club 126:7

Carvalho F, Carvalho R, Mira A, Beja P (2016) Assessing landscape functional connectivity in a forest carnivore using path selection functions. Landscape Ecol 31:1021–1036

Chakrabarty P (2022) Explaining life through evolution. Penguin Random House India Private Limited. 1–279

Chettri N, Shrestha AB, Sharma E (2020) Climate change trends and ecosystem resilience in the Hindu Kush Himalayas. Himalayan weather and climate and their impact on the environment, 525–552

Cowie RH, Bouchet P, Fontaine B (2022) The sixth mass extinction: fact, fiction, or speculation? Biol Rev 97(2):640–663

Crutchfield P (2023) Protecting future generations by enhancing current generations. The Routledge Handbook of the ethics of human enhancement, Roultage. 282–292

De Graaf H, Musters C, Ter Keurs W (1996) Sustainable development: looking for new strategies. Ecol Econ 16(3):205–216

DeBuys W (2021) The Trail to Kanjiroba: rediscovering earth in an age of loss. Seven Stories Press. 1–237

Dhital MR (2015) Geology of the Nepal Himalaya: regional perspective of the classic collided orogen, Springer. 1–498

Eckstein RL, Welk E, Klinger YP, Lennartsson T, Wissman J, Ludewig K, Hansen W, Ramula S (2022) Biological Flora of Central Europe–Lupinus polyphyllus Lindl. Perspectives in Plant Ecology, Evolution and Systematics, 125–715

Francaviglia RV (2016) Believing in place: a spiritual geography of the Great Basin, University of Nevada Press. 1–312

Fuller MR, Doyle MW, Strayer DL (2015) Causes and consequences of habitat fragmentation in river networks. Ann N Y Acad Sci 1355(1):31–51

Gagné K, Chostak S (2023) Climate change beyond technocracy: citizenship and drought practices in the Indian Himalayas. J Peasant Studies, 1–23

Goswami K (2022) Reimagining tibet: politics of literary representation. Taylor & Francis

Guo C-A, Ding X, Hu H, Zhang Y, Yang H, Wang Y (2022) An ethnobotanical study on wild plants used by Tibetan people in Gyirong Valley, Tibet. China. J Ethnobiol Ethnomedicine 18(1):1–20

Hussain S, Sharma S, Bhatti RC, Singh AN (2023) Floral diversity expedition in Ladakh: an insight into the exploration, distribution pattern, ethnobotanical, phytochemical studies, and conservation strategies. Ethnobot Res Appl 26:1–32

Ikechukwu AM, Shamijah BM (2022) Philosophy and the impact of climate change on the environment. Oracle Wisdom J Philos Public Affairs (OWIJOPPA) 6(2)

Jansen L (2003) The challenge of sustainable development. J Clean Prod 11(3):231–245

Joshi A, Joshi P (2019) Forest ecosystem services in the central Himalaya: local benefits and global relevance. Proc Nat Acad Sci India Section B Biol Sci 89:785–792
Joshi G, Negi GC (2011) Quantification and valuation of forest ecosystem services in the western Himalayan region of India. Int J Biodiversity Sci Ecosyst Serv Manag 7(1):2–11
Jewett L, Romanou A (2017) Ocean acidification and other ocean changes. Climate Sci Special Report Fourth Nat Climate Assess 1:364–392
Kattel GR (2022) Climate warming in the Himalayas threatens biodiversity, ecosystem functioning and ecosystem services in the 21st century: is there a better solution? Biodivers Conserv 31(8–9):2017–2044
Keller RP, Geist J, Jeschke JM, Kühn I (2011) Invasive species in Europe: ecology, status, and policy. Environ Sci Eur 23:1–17
Khawas V (2007) Environmental challenges and human security in the Himalaya. In: National seminar 'himalayas in the new millennium: environment, society, economy and polity, centre for himalayan studies, North Bengal University, Darjeeling, January, 11–12
Kothari A, Camill P, Brown J (2013) Conservation as if people also mattered: policy and practice of community-based conservation. Conserv Soc 11(1):1–15
Koulelis PP, Proutsos N, Solomou AD, Avramidou EV, Malliarou E, Athanasiou M, Xanthopoulos G, Petrakis PV (2023) Effects of climate change on greek forests: a review. Atmosphere 14(7):1155
Landmann T, Schmitt M, Ekim B, Villinger J, Ashiono F, Habel JC, Tonnang HE (2023) Insect diversity is a good indicator of biodiversity status in Africa. Commun Earth Environ 4(1):234
Lévêque C (2022) Biodiversity erosion: issues and questions. Wiley, 1–240
Löbl I, Klausnitzer B, Hartmann M, Krell F-T (2023) The silent extinction of species and taxonomists—an appeal to science policymakers and legislators. Diversity 15(10):1053
Ndubuisi OG, KolawoleAyotunde EDA, Lukeman EDLS (2023) Green house effect and global climate change: the african perspective. IPHO-J Adv Res Appl Sci 1(07):06–14
Negi G (2022) Trees, forests, and people: the Central Himalayan case of forest ecosystem services. Trees Forests People 8:100–222
Negi G, Samal P, Kuniyal J, Kothyari B, Sharma R, Dhyani P (2012a) Impact of climate change on the western Himalayan mountain ecosystems: an overview. Trop Ecol 53(3):345–356
Pandit MK, Kumar V (2013) Land-use change and conservation challenges in the Indian Himalayas: Past, present, and future. conservation biology: voices from the tropics, 121–133
Petterson MG (2023) Himalayan thick-skin basement deformation of the Ladakh Batholith, Leh-Ladakh Region, NW India. Springer Nature, 1–149
Pievani T (2014) The sixth mass extinction: Anthropocene and the human impact on biodiversity. Rendiconti Lincei 25:85–93
Prakash S (2021) Impact of climate change on aquatic ecosystem and its biodiversity: an overview. Int J Biol Innov 3(2)
Prakash S, Verma A (2022) Anthropogenic activities and Biodiversity threats. Int J Biol Innov IJBI 4(1):94–103
Prasad S (2023) Gender, environment and sustainable development: challenges and responses from India. Taylor & Francis. 1–276
Rahman MM (2023) Habitat loss and biodiversity loss go hand in hand in Bangladesh: causes and consequences. Available at SSRN 4443842
Rana SK, Rawal RS, Dangwal B, Bhatt ID, Price TD (2021) 200 years of research on Himalayan biodiversity: trends, gaps, and policy implications. Front Ecol Evol 8:603422
Rawat U, Agarwal N (2015) Biodiversity: concept, threats and conservation. Environ Conserv J 16(3):19–28
Rawal R, Bhatt I, Sekar KC, Nandi S (2013) The Himalayan biodiversity: richness, representativeness, uniqueness, and life-support values. GB Pant Institute of Himalayan Environment and Development (GBPIHED), Kosi-Katarmal, Almora, Uttarakhand, India, 1–84
Santos E, Carvalho M, Martins S (2023) Sustainable water management: understanding the socioeconomic and cultural dimensions. Sustainability 15(17):13074

Santojanni FB, Miner H, Hain H, Sutton G (2023) The impact of climate change on biodiversity in coastal ecosystems. Jurnal Ilmu Pendidikan Dan Humaniora 12(3):167–182
Sherpa PY, Puschiasis O (2023) A reflexive approach to climate change engagement with Sherpas from Khumbu and Pharak in Northeastern Nepal (Mount Everest Region). In: Anthropology and climate change. Routledge. 224–241
Thompson M, Warburton M (1985) Knowing where to hit it: a conceptual framework for the sustainable development of the Himalayas. Mountain Research and development, 203–220
Uralovich KS, Toshmamatovich TU, Kubayevich KF, Sapaev I, Saylaubaevna SS, Beknazarova Z, Khurramov A (2023) A primary factor in sustainable development and environmental sustainability is environmental education. Caspian J Environ Sci 21(4):965–975
Weaner L, Christopher T (2016) Garden revolution: how our landscapes can be a source of environmental change. Timber Press. 1–327
Zerbe S (2022) Restoration of multifunctional cultural landscapes: merging tradition and innovation for a sustainable future (Vol 30). Springer Nature. 1–716
Zhang H, Bearup D, Barabás G, Fagan WF, Nijs I, Chen D, Liao J (2023) Complex nonmonotonic responses of biodiversity to habitat destruction. Ecology 104(12):4177
Zhao W (2022) Beginning: China's national park system. Nat Sc Rev 9(10):nwac150

Chapter 22
Endemic Species of Himalayan Region in Sikkim

Pranjal Sharma, Priyanka Sharma, Arti Jamwal Sharma, Ankita, Monika Rathour, and Shreekar Pant

Abstract Himalayas covers about five countries named as: India, China, Nepal, Bhutan and Pakistan. Mount Everest is the highest peak located in the Himalaya (27°59′N 86°55′E). Himalayan range is bordered on the north-west by Karakoram and Hindu Kush ranges, TibetanPlateau and Indo-Gangetic Plain. Some rivers like the Indus, the Ganges and the Tsangpo-Brahmaputra emerged from the vicinity of the Himalayas. There are different and unique type of species found in the flora and fauna of Himalayan region. Some of these species are endemic and critically endangered. Snow leopard is the main predator found at high altitudes but fall in endangered category. There are many biodiversity hotspots, 353 new species that includes 242 plants, 16 amphibians, 16 reptiles, 14 fishes, 2 birds, 2 mammals and more than 61 invertebrates have been found in the eastern Himalaya in the year between 1998–2008.In Sikkim, the climate varies from sub-tropical in the south to tundra in the north and the average annual temperature in most of the area of Sikkim is about 18 °C. For the protection of the biodiversity in Sikkim the government made many protected areas such as Khangchendzonga National Park, Maenam wildlife sanctuary, Shingba Rhododendron sanctuary etc. The present study is based on the vulnerability of endemic species found in the Sikkim area covers the Himalayan region due to climate change. Sikkim contains different types of plant and animal species which are of great economic importance. The livelihood of people was mostly dependent upon it. There is total 48 plant species and 20 animal species present in the flora and fauna of Sikkim are endemic due to deforestation, illegal poaching, global warming and some climate factors etc. Sikkim had lost 123 ha of its natural forest which is equal to 66.8 kt of CO_2 emissions. There is an urgent need of conservation of the biodiversity through various methods such as creation of protected areas, captive breeding and reintroduction or by increasing awareness among public for sustainable use of species.

P. Sharma · P. Sharma (✉) · A. J. Sharma · Ankita · M. Rathour
Department of Bio Sciences, Career Point University, Hamirpur 176041, Himachal Pradesh, India
e-mail: p.priyanka144@gmail.com

S. Pant
Centre for Biodiversity Studies, BGSB University, Rajouri 185234, Jammu Kashmir, India

D. Arya et al. (eds.), *Climate Change Impact on Himalayan Biodiversity*,
Environmental Science and Engineering, https://doi.org/10.1007/978-3-031-77149-1_22

Keywords Biodiversity hotspot · Critically endangered · Himalaya · Endemic · Vulnerability

22.1 Introduction

The name "Himalayas" is mainly originated from Sanskrit in which "hima" means snow or frost and "a-laya" means dwelling. It is a mountain range located in asian continent and covers about five countries such as Nepal, China, Pakistan, Bhutan and India. Himalaya is mainly divided into three regions named as: Western Himalaya includes Zanskar range, the Pir Pranjal Range, the Dhauladhar Range and western parts of Shivalik range and the Great Himalayas of Himachal Pradesh. Eastern Himalaya is extended from Eastern Nepal across North-East India, Bhutan, The Tibet region to Yunnan in China and Northern Myanmar. Indian Himalayan Region consist of Western Himalaya which covers the hilly region of Jammu and Kashmir and Himachal Pradesh, central Himalaya which includes Uttarakhand, eastern Himalaya which includes Sikkim and Arunachal Pradesh and North-Eastern hill region covers Assam, Manipur, Mizoram, Nagaland and Tripura. The Indian Himalayan Region covers around 13 states in India that includes Jammu & Kashmir, Ladakh, Uttarakhand, Himachal Pradesh, Arunachal Pradesh, Manipur, Meghalaya, Mizoram, Nagaland, Sikkim, Tripura, Assam and West Bengal. The glaciers of the Himalayan region are continuously melting due to global warming which is of great concern because due to this the level of sea also increases. The ecology of Himalayas i.e. flora and fauna is dependent upon various factors such as climate, rainfall, altitude and soil. The common indicators responsible for climate change are erratic rainfall, increasing temperature, snowfall at low rate and increase in temperature. The temperature of Himalaya was increased by 1.5 °C as compared to the average increased temperature which is 0.74 °C globally in the last century. The temperature of higher altitude of Himalayan and trans-Himalayan zone are increasing at higher rate (Negi et al. 2022).

The change in climate is one of the major problems of our planet. Due to increase in temperature the glaciers of the Himalayan region is melting continuously mainly due to global warming. There are also other reasons which are responsible for the climate change such as increase rate of green house gases like carbon dioxide, chlorofluorocarbon, methane, hydrofluorocarbons, perfluorocarbons, nitrogen trifluoride etc. Changes in the climate also affect the vegetation of the Himalayan region and cause the endangerment of many species.

The vegetation of Himalayan region is mainly divided into four categories: (i) tropical (ii) subtropical (iii) temperate (iv) alpine. The most dominant species occur in the flora of Himalayan region is *Pinus roxburghii* commonly known as Chir pine. Apart from these trees, there are around 4000 flowering species of plants that occurs in the eastern Himalaya. The eastern Himalayan region covers the area from Sikkim eastwards and encircling Darjeeling, Kurseong and the temperate region consists of forests which includes oaks, maples, alder, laurels, rhododendrons etc. The Himalayan region contains 3,160 plant species which are endemic.

Endemic species are those species which are found rarely and only in a particular area. The factors which induced endemism are habitat change, degradation and destruction, global warming etc. The characteristics result in endemism generally involves species found only in limited area, population present in small amount or depopulation, low reproductive ability or species harvested by people (Isik et al. 2011). Many changes were occurred in the land of Himalaya due to deforestation. Deforestation is mostly caused due to increased population The wildlife of eastern Himalaya is also in danger due to deforestation such as Himalayan black bear, common leopard (*Uncia uncial*) etc. While in birds koklas pheasant (*Pucrasia macrolopha*), Himalayan snow cock (*Tetraogallus himalayensis*), Chestnut-breasted Partridge (*Arborophila* mandellii), White-naped Yuhina (*Yuhina* bakeri) etc. are highly vulnerable species. *Buxus wallichiana* is a tree species which is of great economic use is under threat due to its over exploitation (Wani et al. 2023). There are about 1,8440 plant species out of which 1,748 species are medicinal plants while 675 wild edibles occurred in the Indian Himalayan region. *Angelica glauca* which is commonly known as Choru is a medicinal plant species which is endemic and found in Uttarakhand, Jammu & Kashmir and Himachal Pradesh. This plant species is commonly used in the western Himalaya as a medicine (Singh et al. 2020).

Himachal Pradesh (30°22′40″-33°12′40″ N to 75°04·20″ E) that involves area of the Trans Himalaya and north-west Himalaya and spread about 55,673Km^2 area (Pandit et al. 2007). There are total 881 plant species out of which 108 (12.2%) are endemic and 27 i.e. 3% are threatened species in Himachal Pradesh according to IUCN (Chawla et al. 2012). Most of the land area in Sikkim is used for forestry. Sikkim is considered rich in biodiversity i.e. flora and fauna. Its flora includes 4500 flowering species of plant, 1,681 medicinal plant species, 8 tree ferns, 257 small grasses, 38 rhododendrons, 23 bamboos, 16 conifers, 11 oaks etc. and its wildlife includes 7000 moths, 5892 insects, 689 butterflies 568 birds, 144 mammals, 50 amphibians, 48 fish, 33 reptiles etc. There is total 1000 species of Rhododendron all over the world and some of these species are endemic and found in Sikkim.

There are many protected areas in the Sikkim out of which the Khangchendzonga Biosphere Reserve is considered as the most protected areas located in the eastern Himalaya. Sikkim possesses different kind of vegetation. The tree species which are dominant includes Acer, Michelia, Juglans, Rhododendron, Larix etc. These species are commonly or mainly found in North-Sikkim such as Lachen, Lachung, Yakthang and Zemu. In Sikkim bamboo forests is considered important for various resources for example house construction, ladders, mats, baskets, fodder, fishing, handicrafts etc. Sikkim contains unique type of plant and animal species and some of them are still unexplored. But today, Sikkim's biodiversity is under threat contains many endemic species because of habitat fragmentation, loss of food security illegal poaching, global warming etc. and there is an urgent need for the conservation of its biodiversity. Our objective is to review the endemic species of Sikkim's Himalayan zone that include both flora and fauna.

22.2 Study Area

Sikkim is in the eastern Himalayan zone of India (27°4′46″ to 28°7′48″N and 88°58″ to 88°55″25″E). It is the second smallest state and covers about 7,096 km^2 of area with 619,000 population is considered as biodiversity 'Hot Spot' (Fig 22.1). It is the second smallest state and covers about 7,096 km^2 of area with 619,000 population is considered as biodiversity 'Hot Spot'. Sikkim's two-third area is mainly covered with the mountains and snow. It has been found that there are many sanctuaries present for the conservation purpose in Sikkim. Khangchendzonga National Park covers most of the area i.e. 2,931.12 km^2 constitutes about 82% of the total protected area of Sikkim and includes 18 glaciers (Fig 22.2).

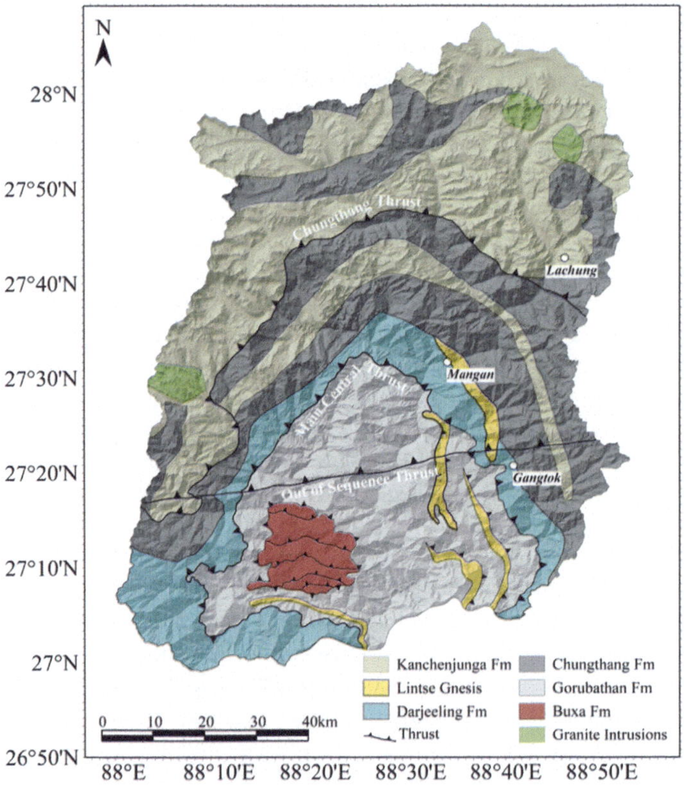

Fig. 22.1 Geological map of the Sikkim Himalayas overlaying the transparent 30-m shadedtopography (Bhasin et al. 2023)

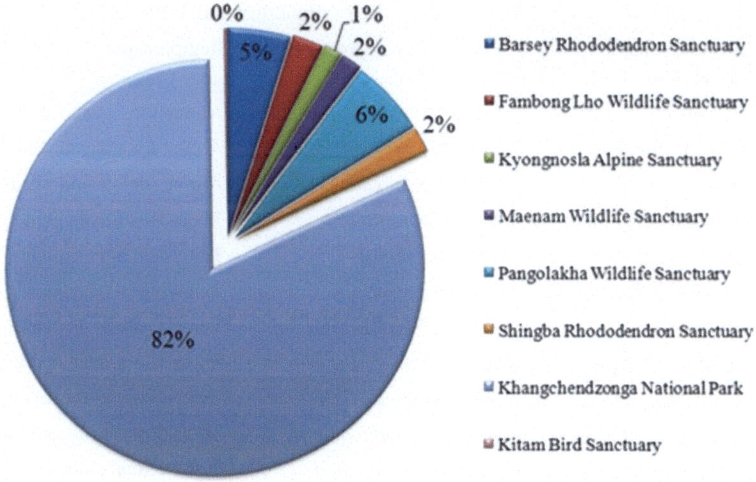

Fig. 22.2 Protected area networks of Sikkim, where Khangchendzonga National Park has highest coverage area and Kitam Bird Sanctuary is home to the lowest coverage area network (km^2) (Banerjee et al. 2019)

22.3 Materials and Methods

The present study has been focused to gather and analyse data on different types of endemic species present in the Himalayan region of Sikkim through literature. Different kind of sources were used in this review such as google scholar, journals, book chapters etc. The main aim of this search was to find out various endemic species present in the flora and fauna of Sikkim's Himalayan region: (i) Conduct a comprehensive literature search by using multimedia (ii) Organizing previous discoveries and emphasizing them in the review (iii) Analysis of the emphasized discovery.

22.4 Results and Discussion

In Table 22.1, there are total 18 species of animals which are endemic out of which 14 species was belonged to trans-himalayan ecoregion and 4 species was belonged to Alpine forests and scrubland.

Table 22.2 showed medicinal plant species which are under threat in the west Sikkim. There are total 39 species in which 7 plant species are nearly threatened, 16 species are vulnerable, 8 species are endangered and 4 species are fall in the least count category. *Dactyolorhiza hatagirea*plant species is critically endangered, *Chamaecostus cuspidatus* species is rare and there is not enough data about *Berberis aristata* species.

Table 22.1 Some endemic species of fauna in Sikkim (Banerjee et al. 2019)

S. no	Common name of species	Scientific name of species	Region name
1	Alpine musk deer	*Moschus chrysogaster*	Alpine forests and scrubland
2	Himalayan tahr	*Hemitragus jemlachicus*	Alpine forests and scrubland
3	Blue sheep	*Pseudios nayaur*	Alpine forests and scrubland
4	Pallas's cat	*Otocolobus manul*	Trans-Himalayan ecoregion
5	Snow leopard	*Panther uncial*	Trans-Himalayan ecoregion
6	Black-necked crane	*Grus nigricollis*	Trans-himalayan ecoregion
7	Tibetan fox	*Vulpes ferrilata*	Trans-Himalayan ecoregion
8	Tibetan wild ass	Equus kiang	Trans-Himalayan ecoregion
9	Tibetan argali	*Ovis ammon*	Trans-Himalayan ecoregion
10	Tibetan gazelle	*Procapra picticaudata*	Trans-Himalayan ecoregion
11	Tibetan wolf	*Canis lupus chanco*	Trans-Himalayan ecoregion
12	Eurasian lynx	*Lynx lynx*	Trans-Himalayan ecoregion
13	Brown bear	*Ursus arctos*	Trans-Himalayan ecoregion
14	Tibetan snow-cock	*Tetraogallus tibetanus*	Trans-Himalayan ecoregion
15	Lammergeier	*Gypaetus barbatus*	Trans-Himalayan ecoregion
16	Golden eagle	*Aquila chrysaetos*	Trans-Himalayan ecoregion
17	Shelduck	*Tadorna ferruginea*	Trans-Himalayan ecoregion
18	Caterpillar-fungus (half insect-half plant)	*Cordyceps sinensis*	Alpine forests and scrubland

There are total 48 flowering plant species which are endemic as showed in Table 22.3. Out of these species the maximum number of endemic flowering plant species belongs to family Ericaceae.2 species belongs to family Apiaceae, 10 species from Asteraceae family, 2 species from Ranunculaceae family, 1 species from Caryophyllaceae family, 2 species from Fabaceae family, 2 species from Berberidaceae family, 1 species from Cyperaceae family, 1 species from Campanulaceae family, 1 species from Orchidaceae family, 1 species from Juncaceae family, 1 species from Podophyllaceae family, 1 species from Annonaceae family, 12 species from Ericaceae family, 1 species from Dipsacaceae family, 3 species from Rosaceae family, 1 species from Lauraceae family, 3 species from Papaveraceae family, 1 species from Tamaricaceae family and 1 species from Primulaceae family

There are total 19 species of Rhododendron in Sikkim Himalayan region which are rarely found as showed in Table 22.4. Out of these species, *R. arboretum* and *R. thomsonii* are vulnerable. *R. grande*, *R. triflorum*, *R. ciliatum*, *R. setosum*, *R. anthopogan*, *R. decipiens*, *R. aeruginosum*, *R. nivale* are threatened species. *R. leptocarpum*, *R. maddeni*, *R. niveum*, *R. pumilum*, *R. sikkimense* are endangered and *R. Pendulum*, *R. wightii*, *R. flugens* are came under rare category.

Table 22.2 Some threatened medicinal plants in west Sikkim (Tamang et al. 2017)

S. no	Botanical name	Family	Uses	Status
1	*Abrus precatorius*	Fabaceae	Tonsil, pneumonia, skin disease, cough, cold, menstrual problems, abortion	Nearly Threatened
2	*Aconitum ferox* Wall Ex. Ser	Ranunculaceae	Asthma, blood clots, cough, diabetes, rheumatism, snake bite, stomach ache and toothache	Vulnerable
3	*Aconitum heterophyllum* Wall. Ex Royle		Antidote for snakebite, food poisoning, stomachache, antiperiodic, piles, body sore throat and antifertility	Vulnerable
4	*Acorus calamus* L	Arecaceae	Asthma, epilepsy, fever, pneumonia, bronchitis, gastritis, sore throat, body pain, cuts, wounds and skin problems	Nearly Threatened
5	*Allium wallichi* kunth	Alliaceae	Cholera, dysentery, cold and cough and reducing altitude sickness	Endangered
6	*Berberis aristata* DC	Berberidaceae	Fever, jaundice, malaria, diarrhea, swelling, eye problems and rabies	Data deficient
7	*Bergenia ciliate* (Haw.) Sternb	Saxifragaceae	Juice of whole plant is used in treating urinary problems, leaf juice is used in treating earaches, root is used in fever and diarrhea, root juice is used in cold, cough and asthma	Vulnerable
8	*Betula utilis* D. Don	Betulaceae	Decoction used for cleaning wounds as antiseptic	Nearly threatened
9	*Calamus erectus* Roxb	Arecaceae	Diabetes	Vulnerable

(continued)

Table 22.2 (continued)

S. no	Botanical name	Family	Uses	Status
10	*Cardiocrinum giganteum* (Wall.) Makino	Liliaceae	Leaves are used as cooling agent to reduce pains of wounds and bruises. Roots paste is used as poultice in bone dislocation	Vulnerable
11	*Dactyolorhiza hatagirea* (D.Don) Soo	Orchidaceae	Diarrhea, dysentery, fever, diabetes. Paste used in cuts and wounds, body ache, gastritis and bone fracture	Critically endangered
12	*Dioscorea deltoidea* Wall.ex Griseb	Dioscoreaceae	Oral contraceptives and lice problems	Vulnerable
13	*Ephedra gerardiana* Wall.ex Stapf	Ephedraceae	Low blood pressure, fever, asthma, cuts, indigestion and headache	Vulnerable
14	*Evodia fraxinifolia* (D.Don) Hook.f	Rutaceae	Dysentery and fever	Vulnerable
15	*Franxinus floribunda* Wall	Oleaceae	Gout and bone fractures	Endangered
16	*Hedgsonia heteroclite* Hook.f. & Thomson	Cucurbitaceae	Leaves smoke is inhaled to allay irritation from small insects, leaves boiled and liquid is taken for nose complaints and fever	Endangered
17	*Heracleum wallichii* DC	Apiaceae	Diarrhea and gastritis	Least count
18	*Horsfieldia kingie* (J.D. Hooker) Warburg	Myristiceae	Piles	Endangered
19	*Juniperus recurva* buch.-Ham. Ex D.Don	Cupressaceae	Smoke inhaled in vomiting and headache	Nearly threatened
20	*Chamaecostus cuspidatus* (Nees & Mart) C. Specht & D.W. Stev	Costaceae	Diabetes, skin problems, fever, asthma, bronchitis and intestinal worm disease	Rare
21	*Macropanax dispermus* (Wallich ex G.Don) Seemann	Araliaceae	Diabetes	Least count
22	*Mahonia napaulensis* DC	Berberidaceae	Dysentery, diarrhea and urinary disorders	Nearly threatened

(continued)

Table 22.2 (continued)

S. no	Botanical name	Family	Uses	Status
23	*Martynia annua* L	Martyniaceae	Root decoction is used in snakebite, leaf juice in leprosy, tuberculosis and sore throat	Endangered
24	*Meconopsis horridula* Hook.f. & Thomson	Papaveraceae	Bone fractures, fever, back pain	Nearly threatened
25	*Mensua ferrea* L	Clusiaceae	Skin diseases, menstrual problems and paste applied in hydrocoel and wound	Vulnerable
26	*Nardostachys grandiflora* DC	Valerianceae	Nervine tonic, antiseptic, stomachic, sedative, anti-bacterial	Vulnerable
27	*Nyctanthes arbortristis* L	Oleaceae	Leaves decoction is take in malarial fever, bark paste is used in dislocated joints and leaves and flower infusion is given to cause expulsion of placenta in women	Least count
28	*Panax sokpayensis* Shiva K. Sharma & Pandit	Araliaceae	Fever, weakness, liver disorders, Menstrual disorders indigestion and vomiting	Endangered
29	*Panax bipinnatifidum* Seem	Araliaceae	Fever, weakness, liver disorders, menstrual disorders indigestion	Endangered
30	*Paris polyphylla* Smith	Liliaceae	Stomachache, fever, diarrhea and dysentery	Vulnerable
31	*Phlogocanthus thyrsiflora* (Hardwick.) Mabberley	Acanthaceae	Diabetes, cold and cough and fever	Least count
32	*Picrorhiza kurrooa* Royole ex Benth	Plantaginaceae	Root paste applied in cuts, wounds and injuries	Vulnerable
33	*Rheum austral* D.Don	Polygonaceae	Constipation, skin eruption, liver disorders, internal injuries, body pain, diarrhea, dysentery, ulcers, cuts and wounds	Vulnerable

(continued)

Table 22.2 (continued)

S. no	Botanical name	Family	Uses	Status
34	*Rheum nobile* Hook.f. & Thomas	Polygoneaceae	Ulcers, bronchitis, fever, fractures, rheumatism arthritis, heart complaints, tonic after delivery of baby and swelling	Nearly threatened
35	*Rhodiola himalensis* (d.Don) S.H. Fu	Crassulaceae	Kidney problem, urinary disorders, asthma, lung infection, poisoning, arthritis, fever, oral infection and to the women after child birth	Nearly threatened
36	*Rhododendron anthopogon* D.Don	Ericaceae	Indigestion, vomiting, respiratory disorders, stomachache and dysentery	Vulnerable
37	*Rhododendron setosum* D.Don	Ericaceae	Young leaves are used in wounds	Vulnerable
38	*Sapindus mukorossi* Gaertn	Sapindaceae	Burns, epilepsy, tonic, anthelmintic, purgative, asthma and in removing dandruff and lice	Endangered
39	*Sassurea gossypiphora*.D.Don	Asteraceae	Wool is used in cuts and other parts are used in wounds	Vulnerable

22.5 Conclusion

Throughout the world, the tribal groups had more traditional knowledge about the biodiversity. Many species are threatened or endemic in the nature and are on the verge of extinction. In this study, there are 20 species in the fauna in which trans-himalayan ecoregion contains more endemic species i.e. 14 species. Apart from that, there are 48 plant species from which Ericaceae family contains most of the endemic species i.e. 12 species found in Sikkim. The government had made some laws for their conservation but we should gather and help in the conservation of these species. There are also many sanctuaries and one national park in Sikkim for the protection of these species and prevent them from being threatened or endangered.

Table 22.3 Some endemic flowering plant species of Sikkim (Banerjee et al. 2019)

S. no	Common name	Botanical name	Family
1	Falsely tender Acronema	*Acronema pseudotenera*	Apiaceae
2	Cave's Anaphalis	*Anaphalis cavei*	Asteraceae
3	Hooker,s Pearly everlasting	*Anaphalis hookeri*	Asteraceae
4	Floppy Himalayan Anemone	*Anemone demissa var. monantha*	Ranunculaceae
5	–	*Angelica nubigena*	Apiaceae
6	–	*Arenaria thangoensis*	Caryophyllaceae
7	–	*Astragalus zemuensis*	Fabaceae
8	Sikkim barberry	*Berberis sikkimensis*	Berberidaceae
9	-	*Cacalia chola*	Asteraceae
10	Sikkim calamus palm	*Calamus inermis*	Asteraceae
11	–	*Caragana spinifera*	Fabaceae
12	–	*Carex kingiana*	Cyperaceae
13	–	*Codonopsis affinis*	Campanulaceae
14	–	*Coelogyne treutleri*	Orchidaceae
15	Palm-leaf coneflower	*Cremanthodium palmatum subsp. benthamii*	Asteraceae
16	Atripappa False Dandelion	*Crepis atropappa*	Asteraceae
17	Mung-chyit-mungyal	*Juncus sikkimensis*	Juncaceae
18	–	*Jurinea coopri*	Asteraceae
19	King's Ligularia	*Ligularia kingiana*	Asteraceae
20	Big-Fruited Ligularia	*Ligularia pachycarpa*	Asteraceae
21	–	*Ligularia yakla*	Asteraceae
22	–	*Mahonia sikkimensis*	Berberidaceae
23	–	*Podophyllum sikkimensis*	Podophyllaceae
24	–	*Ranunculus sikkimensis*	Ranunculaceae
25	–	*Uvaria lurida var. sikkimensis*	Annonaceae
16	–	*Agapetes incurvata*	Ericaceae
27	–	*Agapetes siikimensis*	Ericaceae
28	Atratus Dipsacus	*Dipsacus atratus*	Dipsacaceae
29	Saro kat	*Eriobotrya hookeriana*	Rosaceae
30	–	*Geum macrosepalum*	Rosaceae
31	Lekh pipli	*Lindera heterophylla*	Lauraceae
32	–	*Lloydia flavonutans*	Liliaceae
33	Maddenia Himalayica	*Maddenia himalaica*	Rosaceae
34	Himalayan blue poppy	*Meconopsis grandis*	Papaveraceae
35	Blue poppy	*Meconopsis superba*	Papaveraceae
36	Himalayan woodland-poppy	*Meconopsis villosa*	Papaveraceae

(continued)

Table 22.3 (continued)

S. no	Common name	Botanical name	Family
37	–	*Myricaria albiflora*	Tamaricaceae
38	Primrose	*Primula whitei*	Primulaceae
39	Bailey's Rhododendron	*Rhododendron baileyi*	Ericaceae
40	Chia-phule Gurans	*Rhododendron camelliaeflorum*	Ericaceae
41	Fringed Leaf Rhododendron	*Rhododendron ciliatum*	Ericaceae
42	Rosy Rhododendron	*Rhododendron glaucophyllum*	Ericaceae
43	Patle Korlinga	*Rhododendron grande*	Ericaceae
44	Bhutle Gurans	*Rhododendron lanatum*	Ericaceae
45	Sano Lahare Chimal	*Rhododendron lindleyi*	Ericaceae
46	Wallich Ko Chimal	*Rhododendron wallichii*	Ericaceae
47	Dr. Wight Ko Gurans	*Rhododendron wightii*	Ericaceae
48	–	*Saussurea conica*	Asteraceae

Table 22.4 Rarity status of Sikkim Himalayan rhododendron (Singh et al. 2009)

S. no	Species	Altitudinal (m)	Maximum height	Status
1	*R. arboretum* Smith Scarlet Arborescent Rhododendron	1700–3400 m	15 m	Vulnerable
2	*R. grande* Wt	2000–3000 m	10–15 m	Threatened
3	*R. triflorum*	2300–4000 m	2–3 m	Threatened
4	*R. leptocarpum*	3000–3500 m	1–2 m	Endangered
5	*R. maddeni*	2500–4000 m	1.5–2.5 m	Endangered
6	*R. falconeri*	3000–3500 m	10–15 m	Threatened
7	*R. niveum*	3100 m	2.5–5 m	Endangered
8	*R. ciliatum*	3000–3800 m	1–1.5 m	Threatened
9	*R. setosum*	3000–5500 m	0.5 m	Threatened
10	*R. anthopogan*	3000–5000 m	45 cm	Threatened
11	*R. decipiens*	3300–3800 m	2.5–3 m	Threatened
12	*R. thomsonii*	3300–4500 m	3–5 m	Vulnerable
13	*R. pendulum*	3300–4000 m	1–1.5 m	Rare
14	*R. pumilum*	3500–4500	50–200 mm	Endangered
15	*R. wightii*	3500–4500 m	3–5 m	Rare
16	*R. sikkimense*	3700 m	3–5 m	Endangered
17	*R. flugens*	4000–5000 m	2–5 m	Rare
18	*R. aeruginosum*	4500–5000 m	1.5 m	Threatened
19	*R. nivale*	4500–6000 m	1–1.5 m	Threatened

References

Ahmed JM, Murtaza G, Shaheen H, Habib T (2020) Distribution pattern and associated flora of *Jurinea dolomiaea* in the western Himalayan highlands of Kashmir: an indicator endemic plant of alpine phytodiversity. Elsevier. 116.https://doi.org/10.1016/j.ecolind.2020.106461

Banerjee A, Chakraborty P Bandopadhyay R (2019) Urgent conservation needs in the Sikkim Himalaya biodiversity hotspot. Biodiversity 20(2–3):88-97.https://doi.org/10.1080/14888386.2019.1656547

Bhasin R, Aslan G Dehls J (2023) Ground investigations and detection and monitoring of landslides using SAR Interferometery in Gangtok, Sikkim Himalaya. GeoHazards 4(1):25–39.https://doi.org/10.3390/geoharzards4010003

Burlakova EL, Karatayev YA, Karatayev AV, May EM, Benett LD, Cook JM (2011) Endemic species: contribution to community uniqueness, effect of habitat alternation, and Conservation priorities. Elsevier 144(1):155–165. https://doi.org/10.1016/j.biocon.2010.08.010

Chawla A, Parkash O, Sharma V, Rajkumar S, Lal B, Gopichand Singh DR, Thukral KA (2012) Vascular plants, Kinnaur, Himachal Pradesh. India. Check List 8(3):321–348

Chhetri B, Badola KH Barat S (2021) Electronic Supplement of the paper, How people perceive resilience of Himalayan pheasents, Phasianide, in relation to climate warming in Eastern Himalaya. Nat Conserv Res 6(3):87–96

Dhakal P, Sharma KR, Rajak NP Laxman K (2023) Population and distribution of Himalayan Tahr (*Hemitragus jemlahicus*) in Lamtang National Park of the Nepal Himalaya. J Zool 4(1):2–10. https://doi.org/10.54393/mjz.v4i01.60

Isik K (2011) Rare and endemic species: why are they prone to extinction? Turk J Bot 35(4):411–417. https://doi.org/10.3906/bot-1012-90

Lucksom SZ (2008) 10.0 Endemic and threatened orchids of sikkim and their conservation. Special Habitats Threatened Plants India 11(1):63

Maiti GG, Maity D (2009) Diversity of vascular plants of Kanchenjunga Biosphere Reserve, Sikkim and its Conservation. Indian Forester, 1416–1436

Negi SCG, Samal KP, Kuniyal CJ, Kothyari PB, Sharma KR Dhyani, PP (2012) Impact of climate change on the western Himalaya mountain ecosystems: an overview. Int Soc Tropical Ecol 53(3):345–356

Negi SV, Tiwari CD, Singh L, Thakur S Bhatt I (2022) Review and synthesis of climate studies in the Himalayan region. Environ Dev Sustain 24(9):10471–10502

Pandit KM, Sodhi SN, Koh PL, Bhaskar A Brook WB (2007) Unreported yet massive deforestation driving loss of endemic biodiversity in Indian Himalaya. Biodivers Conserv 16:153–163.https://doi.org/10.1007/s10531-006-9038-5

Samant SS, Joshi CH (2005) Plant diversity and conservation status of Nanda Devi National Park and comparision with highland National Parks of the Indian Himalayan Region. Int J Biodiv Sci Manag 1(1):65–73. https://doi.org/10.1080/17451590509618081

Samant SS, Pant S, Singh M, Lal M, Singh A, Sharma A, Bhandari S (2007) Medicinal plants in Himachal Pradesh, north western Himalaya, India. Int J Biodivers Sci Manag 3(4):234–251. https://doi.org/10.1080/17451590709618177

Singh BH, Sundriyal CR (2005) Composition, economic use and nutrient contents of alpine vegetation in the Khangchendzonga Biosphere Reserve, Sikkim Himalaya, India. Arct Antarct Alp Res 37(4):591–601. https://doi.org/10.1657/1523-0430(2005)037[059:CEUANC]2.0.CO;2

Singh G, Chandra N, Lingwal S, Bisht SPM, Tiwari ML (2020) Distribution and threat assessment of an endemic and endangered species *Angelica glauca* in high ranges of Western Himalaya. J Herbs Species Med Plants 26(4):394–404. https://doi.org/10.1080/10496475.2020.1748783

Singh KK, Rai KL, Gurung B (2009) Conservation of Rhododendron in Sikkim Himalaya: an overview. World J Agric Sci 5(3):284–296

Singh P, Sanjappa M (2011) Flowering plants of Sikkim-an analysis, 65–88

Singh P, Subramani PS (2015) Wild Edible Plant Genetic Resources For Sustainable Food Security and Livelihood of Kinnaur District, Himachal Pradesh. India Int J Conserv Sci 6(4):657–668

Singh PS, Sharma S, Dhyani PP (2019) Himalayan arc and treeline: distribution, climate change responses and ecosystem properties. Biodivers Conserv 28:1997–2016. https://doi.org/10.1007/s10531-019-01777-w

Subba S, Lachung R Subba S, Nepal S (2015) Analysis of vegetation in a Representative Temperate Plant Community in Lachung range of the Sikkim Himalaya. Int J Environ Biodiversity 6(3):18–24

Sundriyal M, Sundriyal CR, Sharma E, Purohit NA (1998) Wild edibles and other useful plants from the Sikkim Himalaya. India. Oecologia Montana 7(1–2):43–54

Tamang KD, Dhakal D, Gurung S, Sharma PN Sreshtha GD (2013) Bamboo diversity, distribution Patttern and its uses in Sikkim (India) Himalaya. Int J Scientific Res Publ 3(2):1–6

Tamang M, Pal K, Rai KS, Kalam R, Ahmad RS (2017) Ethnobotanical survey of threatened medicinal plants in West Sikkim. Int J Botany Studies 2(6):116–125

Thapamagar T, Bhusal RD, Bhandari S, Ghimire CK (2020) Report: Alpine Musk Deer Moschus chrysogaster conservation outreach program in Khaptad National Park. Nepal. Zoos' Print J 35(7):26–27. https://doi.org/10.11609/zp.v35i7

Wani AZ, Satish KV, Islam T, Dhyani S, Pant S (2023) Habitat suitability modelling of *Buxus wallichiana* Bail.: an endemic tree species of Himalaya. Vegetos 36(2):583–590

Chapter 23
Role of Bamboo-Based Agroforestry in Climate Change Mitigation and Livelihood: Indian Himalayan Region Perspective

Ashutosh Kumar, Himanshu Bargali, Kavita Bisht, and Kanchan Pant

Abstract According to the United Nations Framework Convention on Climate Change (UNFCCC), carbon sinks need to be improved to gradually reduce the impacts of global warming by achieving and maintaining carbon neutrality. Studies indicated that the Himalayan subalpine, alpine and temperate forests are vulnerable to climate change. In the Indian Himalayan Region (IHR), mountainous communities rely on forests and forest-based resources for their daily needs. However, increasing anthropogenic activities are changing the land use pattern of the region. In this connection, development of sustainable land-use practices that ensure several ecological services and reduce pressure on the forest are requisite. The present study advocates the role of bamboo and agroforestry in carbon sequestration and adaptability of bamboo-based agroforestry as sustainable land use options which can play a significant role in climate change mitigation. Through the Scopus databases, we conducted a systematic review of papers using the PRISMA method. A total of 73 research documents of Scopus database consisted of 55 (81%) research articles were analysed. It was found that bamboo-based agroforestry increases the growth of bamboo, yield of crops, physiochemical properties, and above-ground biomass in degraded agricultural lands. However, in some studies, the growth rate was not significant due to different factors like spacing, canopy of bamboo, etc. Considering the carbon sequestration potential of the bamboo species and its adaptability as an agroforestry element, the bamboo-based agroforestry system needs to be investigated scientifically so that it can be taken into consideration as a viable nature-based solution for livelihood and climate change mitigation strategies for IHR.

A. Kumar · H. Bargali · K. Bisht
Centre for Biodiversity Conservation and Management, G.B. Pant National Institute of Himalayan Environment (NIHE), Kosi-KatarmalAlmora, Uttarakhand 263643, India

K. Pant (✉)
Hemwati Nandan Bahuguna, Swami Ram Teerth (SRT) Campus, Badshahi Thaul, Tehri, Uttarakhand 249145, India
e-mail: dr.kanchanfrm@gmail.com

D. Arya et al. (eds.), *Climate Change Impact on Himalayan Biodiversity*, Environmental Science and Engineering, https://doi.org/10.1007/978-3-031-77149-1_23

Keywords Carbon sequestration · Bamboo based agroforestry · Indian Himalayan region

23.1 Introduction

Global climate change is the key environmental challenge of the twenty-first century. International agreements are made to achieve goals for reducing Greenhouse gases emissions. Recently Intergovernmental Panel on Climate Change (IPCC) submitted its sixth assessment report on March 2023, according to the report, climate and weather extremes are increasingly driving displacement in Asia if global warming exceeds 1.5 °C, the world will experience permanent negative effects on the ecosystem with low resilience in mountainous region affected by ice-sheet, glacier, melt, or by accelerating and increasing sea level rise (IPCC 2023). According to the United Nations Framework Convention on Climate Change (UNFCCC), carbon source emissions should be decreased and carbon sinks should be improved. Climate change and the rise in land-use issues demand the development of sustainable land-use methods that balance agricultural production with the supply of several ecological services. According to worldwide estimates, changes in the utilization of land are crucial for carbon reduction since they help reduce emissions by roughly 30% through the process of carbon sequestration (Aynekulu et al. 2020).

The Indian Himalayan Region (IHR), which covers 16.2% of the country's total land area and covers a region of approximately 5.3 lakh km^2, occupies India's northern boundary. It is a significant biodiversity hotspot and carbon sink and it is a sensitive landscape that is particularly vulnerable to natural disasters, constant concern exists regarding the effects of climate change due to global warming, which could include unusual floods, hazards to food security, ecological degradation, and landslides and loss of biodiversity (Xu et al. 2009; Tewari et al. 2017). Studies indicated that changes in vegetation types on a moderate to large scale are noticed (Tripathi et al. 2022). Mountain-dwelling plant species have already begun to move to higher elevations due to an increase in average temperature in the IHR (Singh and Gupta 2016). the highest temperature increase is reported in Ladakh, Jammu-Kashmir, Himachal Pradesh, and Uttarakhand. North-Eastern states like Arunachal Pradesh and Sikkim, as well as Ladakh, Jammu & Kashmir, and Himachal Pradesh, are experiencing reduced rainfall (FSI 2021). The vulnerable ecosystems of mountains are probably in more danger of being impacted by climate change than other ecosystems (Tewari et al. 2017). Accelerated forest degradation and a shift in land use patterns raise questions regarding a decrease in carbon storage capacity and have a big impact on carbon pools (Sahoo et al. 2019). The ideas of REDD (Reduction Emissions from Deforestation and Forest Degradation) and AFOLU (Agriculture, Forestry, and Other Land Uses) have identified agroforestry as a major land-use system in landscape-scale mitigation plans (Verchot et al. 2007). To achieve India's responsibilities to Nationally determined contributions (NDC), IHR must be protected (Ahirwal et al. 2021). Amongst the range of options suggested to mitigate climate change, manipulation of the global carbon cycle to sequester atmospheric CO_2 into long-term storage has been extensively promoted (Stockmann et al. 2013). Future losses and damages from climate change for people and ecosystems would be decreased by Land use change, forestry (via reforestation and reduced deforestation), and the energy supply sector all reach net zero CO_2 emissions in the majority of global modelled paths (IPCC 2023). However, improved land use can increase carbon sequestration in vegetation and soil (Nath et al. 2021a, b).

Agroforestry models are needed to reduce climate change, sustain local biodiversity, improve livelihood, and have the capacity to mitigate and adapt to climate change. In IHR the agroforestry systems show promising potential for carbon storage and would be valuable for climate change mitigation and adaptation (Nath et al. 2021a, b). Behari (2001) created seven effective agroforestry models with three bamboos (*B. bamboos, B. nutans, and D. strictus*) to recover damaged agricultural lands in central India. Soybean (Glycine max), Niger (*Guizotia abyssinica*), Moong (*Phaseolus aureus*), Wheat (*Triticum aestivum*), Urad (*Phaseolus mungo*), Pigeon pea (*Cajanas cajan*), and Mustard (*Brassica campestris*) were the intercrops. In northeast India's traditional agroforestry system, bamboo plays a significant role. They are mostly handled for domestic use in home gardens and for commercial usage in bamboo groves (Nath and Das 2008).

Additionally, the bamboo-based agroforestry system helps in restoring degraded lands due to the system's association with continuous agriculture and forestry operations (Nath et al. 2009). Thus, attempts should be taken to grow bamboo on degraded and uncultivated land (Devi and Singh 2021; Nigatu et al. 2020). The bamboo-based agroforestry system can offer additional environmental services like enhanced land and water quality and, a source of carbon dioxide sink (Nath and Das 2009). The major goals of cultivating crops, such as market accessibility and early maturity, as well as the function of bamboo as a fodder plant, are reduced with the adoption of bamboo-based agroforestry as an alternate land use option (Akoto et al. 2018). The bamboo-based agroforestry system can offer additional environmental services like enhanced land and water quality and, a source of carbon dioxide sink (Nath and Das 2009).

Since India is the second largest bamboo producer after China. More than half of the nation's bamboo resources are found in the North-Eastern States but the nation's overall bamboo-bearing area has decreased by 1.06 million ha since 2019 (FSI 2021). The area of bamboo was reported to have decreased by 19.4% in Uttarakhand while increasing by 58% in Himachal Pradesh (ISFR 2021). In thirteen Indian Himalayan states, 79 species of bamboo from 22 genera have been recorded (Sundriyal and Joshi 2015). Expanding bamboo production has a positive impact on the environment and economic benefit (Kumar et al. 2023). Bamboo sequesters carbon quickly, and the warming effect has been mitigated by the large carbon reserves in bamboo (Gu et al. 2019; Manandhar et al. 2019) which continue to increase more quickly each year (Zhang et al. 2022). Bamboo can be a key component of "carbon trading" and contribute to worldwide efforts to combat climate change (Seethalakshmi et al. 2009). They have a far greater ability to accumulate biomass than the majority of highland tree species that grow quickly (Nigatu et al. 2020). It is one of the greatest non-timber forest products regarded as a good complement to the use of timber in things like building and construction, pulpwood, flooring, panel products, and furniture. Currently, bamboo is being promoted for the establishment of forest plantations to reduce the demand on natural forests for lumber and wood fuel (Opoku et al. 2016). Nearly 90 species of bamboo exist in the northeastern hill states of India, 41 of which are indigenous to the area (Devi and Singh 2021) and 14 species of bamboo are in the western Himalayas (Kumar et al. 2023). In the state of Uttarakhand, 8 bamboo species belonging to five genera have been identified as naturally occurring. These species include *Bambusa bambos, Ampelocalamus patellaris, Dendrocalamus somdevae, Dendrocalamus strictus, Arundinaria falcata, Chimonobambusa jaunsarensis, Thamnocalamus spathiflorus* and *Thamnocalamus falconeri* (Sundriyal and Joshi 2015). In Himachal Pradesh, eight bamboo species that belong to five genera in forest and non-forest areas i.e., *Bambusa nutan and Dendrocalamus strictus* in the lower Himalayas or the plains, *Drepanostachyum falcatum and hamnocalamus spathiflorus* in the middle or outer Himalaya, and *Dendrocalamus parishii, Munro, Phyllostachys aurea and Phyllostachys bambusoides* used in cultivation (Richa and Sharma 2012). Only two bamboo species, *Bambusa nutans* and *Drepanostachyum falcatum* are found in the Jammu region (Kumar 2021). This article aims to explore the land use pattern of bamboo-based agroforestry model in IHR, to understand the current status of the existing knowledge domain related to bamboo-based agroforestry and how bamboo-based agroforestry respond to carbon sequestration and growth and yield of bamboo and crop species and to investigate existing research gaps and suggest prospects of bamboo-based agroforestry in IHR.

23.2 Material and Methods

The research methodology adopted for this study involves completing a bibliometric analysis of international literature related to agroforestry based on bamboo. A comprehensive review of the Scopus database (www.scopus.com) was carried out to gather research papers related to bamboo-based agroforestry. Referred to as the "science of science" (Li et al. 2021), bibliometric research provides a quantitative method for examining large amounts of data from the literature (Kumar et al. 2019; Rodrigo-Comino et al. 2023). More than 82 million papers from more than 25,100 publications published by 5000 international publishers with SCI indexes are available in Scopus. Given that, 99% of the journals catalogued in the Web of Science database are also included in Scopus (Kumar and Pandey 2024; Bargali et al. 2024), we directed our subsequent analysis toward the Scopus database. Our search was restricted to a single database, which might not have included all the important literature on agroforestry based on bamboo. We applied the same methodology, employing the keywords 'Bamboo-based agroforestry,' to search the Web of Science database to address this limitation. A total of 68 papers were retrieved. The Preferred Reporting Items for Systematic Reviews and Meta-Analyses (PRISMA) guidelines (Moher et al. 2009) were adhered to for this study (Fig. 23.1).

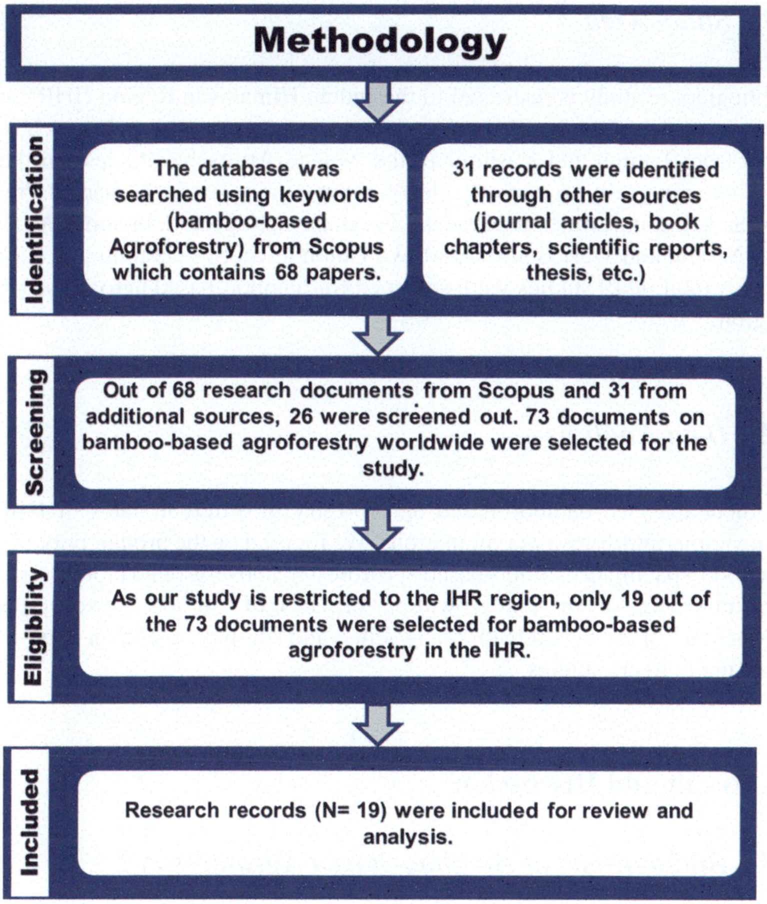

Fig. 23.1 Flow diagram of different phases of a systematic literature review

We gathered details from each selected publication, including subject area, journal/publisher name, publication year, country contribution, affiliations, funding sources, and keywords. we measured the yield of crops, growth of bamboo, aboveground biomass, and physio-chemical property-related variables of bamboo-based agroforestry. Review the synthesis taken by calculated data and different observational types. The selected papers included qualitative studies, which were assessments of the area distribution of bamboo-based agroforestry all over the world and in the Indian Himalayan region.

23.2.1 Study Area

Our bibliographic study is restricted to the Indian Himalayan Region (IHR), which occupies from the West to the northeast region of the country and stretches over 2500 km from Jammu and Kashmir in the west to Arunachal Pradesh in the east region, containing fully or partially eleven mountainous states (Himachal Pradesh, Uttarakhand, Sikkim, Arunachal Pradesh, Meghalaya, Nagaland, Manipur, Mizoram, Tripura, Assam, and West Bengal) and two Union territories (Jammu and Kashmir, Ladakh). A total of 19 studies were screened for bamboo-based agroforestry in the IHR region.

23.2.2 Data Analysis

In this meta-analysis, bamboo-based agroforestry in different states of IHR was found in various agroforestry combinations. We focused on the productivity of crops with bamboo, spacing in bamboo-based agroforestry, above-ground biomass, physio-chemical properties of soil, and growth performance of bamboo. However, data on the productivity of crops and bamboo, spacing, and soil physio-chemical properties were lacking in many studies.

23.3 Result and Discussion

23.3.1 Bibliography of Bamboo-Based Agroforestry

The first publication record on bamboo-based agroforestry was recorded in the year 1985 and only a few publication records were observed up to 2014. Thereafter an increasing trend was seen with maximum publication records in the year 2020 (Fig. 23.2). India with maximum publication records (38) is the major contributor to bamboo-based agroforestry research, followed by China (12), Indonesia (5), Ethiopia (3), etc. (Fig. 23.3). Of the total 68 research documents in the Scopus database, 81% (55) were research articles, 6% (4) were conference papers, 4% (3) were book chapters, 4% (3) were review papers, 3% (2) were conference papers, and 2% (1) was a book (Fig. 23.4). The keyword "bamboo" followed by "agroforestry" was the most commonly utilized term in the publications. The relative size of each keyword is illustrated in (Fig. 23.5). The predominant research areas were Agricultural and Biological Sciences (35%), followed by Environmental Science (26%), Earth and Planetary Sciences (7%), Engineering (6%) Social sciences (5%), and others. Most of the studies were affiliated with Assam University (4) and the Indian Council of Agricultural Research (4) from India followed by the National

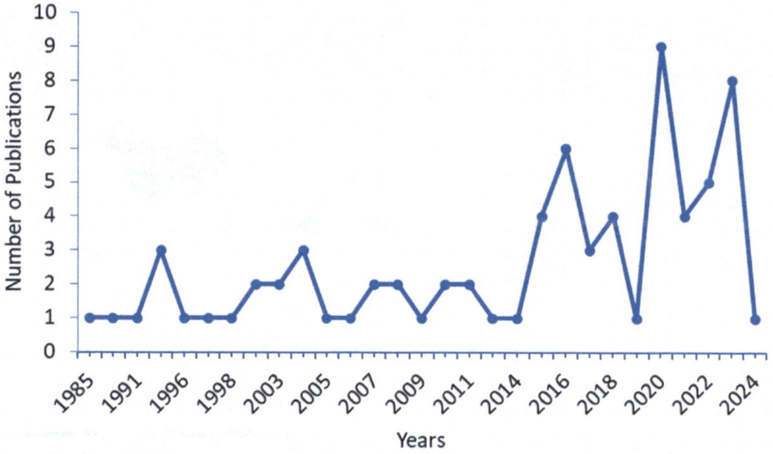

Fig. 23.2 The temporal trend of studies conducted on bamboo-based agroforestry region, along with the corresponding number of studies for each year

Long-term Observation and Research Station for Forest Ecosystem in Hangzhou-Jiaxing-Huzhou Plain (3), Dr. Yashwant Singh Parmar University of Horticulture and Forestry (3), Nanjing University of Information Science & Technology (3), and several others. A significant number of research projects on bamboo-based agroforestry have received funding from 58 funding agencies across various countries and international organizations. The Department of Biotechnology, the government of West Bengal, India (3), the Japan Society for the Promotion of Science of Japan (3), Bundesministerium für Bildung und Forschung (3), the National Key Research and Development Program of China (3), have supported or been acknowledged in the maximum number of studies followed by others. The Fig. 23.6 illustrates the top journals utilized in bamboo-based agroforestry. Although India emerged as a leading research contributor in the field of bamboo-based agroforestry the Indian Himalayan Region was far behind with limited publications recorded.

The tabulation of the global scenario of bamboo-based agroforestry is depicted in Table 23.1. A total of 73 research documents on bamboo-based agroforestry were found globally, and the majority of studies on bamboo-based agroforestry were found in India, i.e., 38 followed by China (12) etc. The numbers of bamboo species globally shown in (Fig. 23.2) which were part of agroforestry systems.

The Table 23.2 shows the tabulation of crops with different bamboo species in IHR. A total of 19 studies on bamboo-based agroforestry were found in 8 states of the IHR. Out of which, the most studies were found in eastern Himalayan states, which were 8 in number, and only 2 states from western Himalayas. The major bamboo genus was *Bambusa* and *Dendrocalamus,* which were found with different agroforestry crops in IHR. The major species were rice, cassava, maize, colocasia, capsicum, ginger, potatoes, pineapple, and bananas. Ginger, soyabean, groundnut,

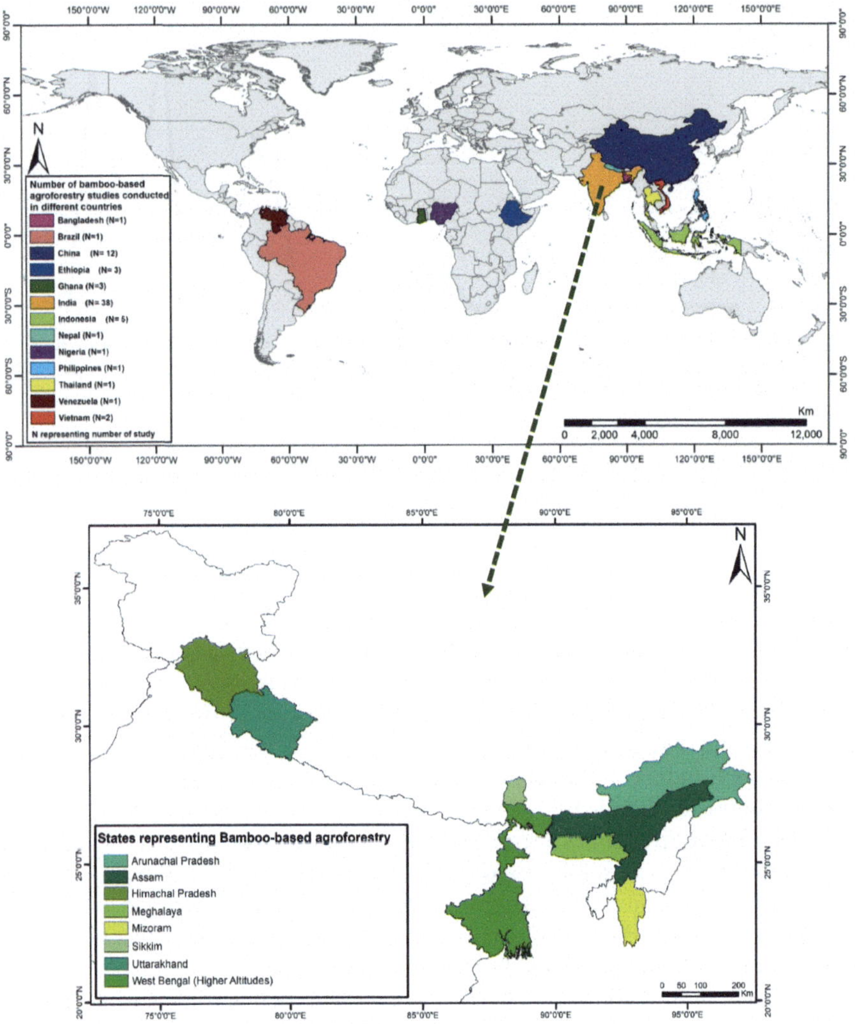

Fig. 23.3 The top country contributing to the number of studies carried out in the field of bamboo-based agroforestry and contribution of Indian Himalayan Region

pigeonpea, cowpea, lady's finger, bottle gourd, turmeric, elephant foot yam, bael, citrus, kathal, moringa, neem, semul, siris, aonla, bakain, banana, betelnut, etc.

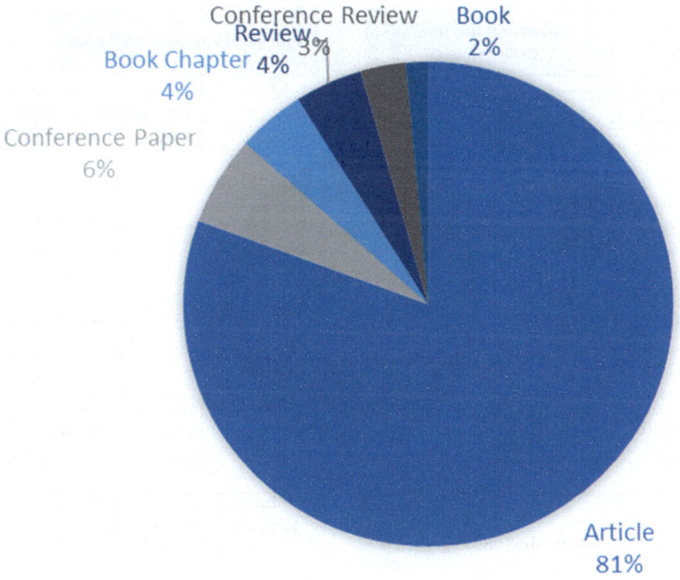

Fig. 23.4 Publication status over the years

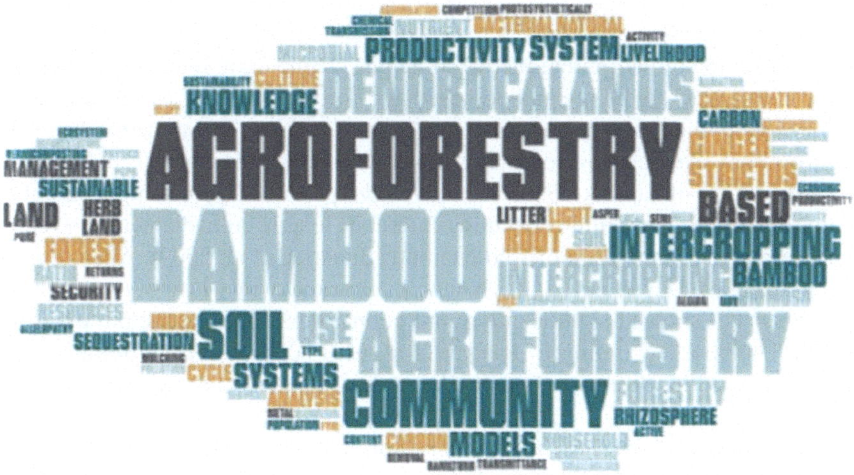

Fig. 23.5 Word cloud of keywords used in the studies of bamboo-based agroforestry

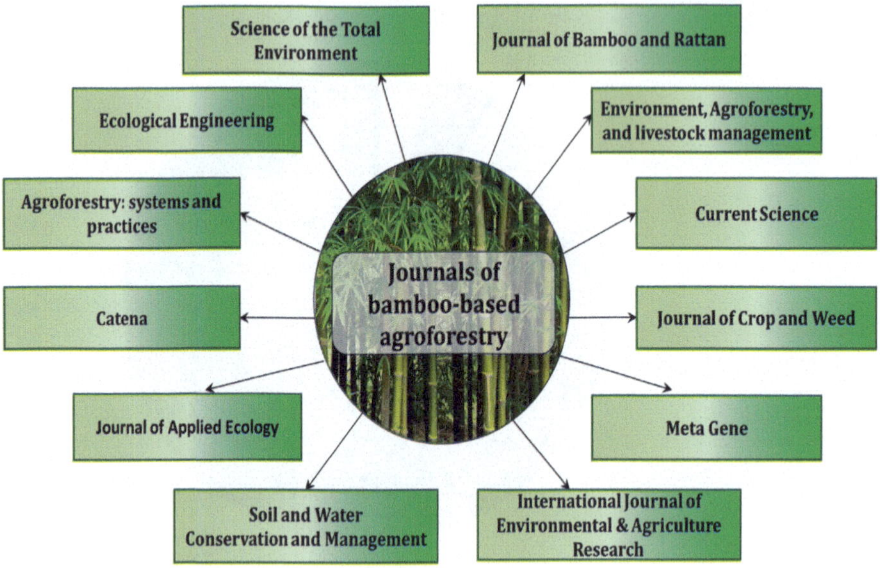

Fig. 23.6 Top publishing journals on bamboo-based agroforestry in Indian Himalayan Region

23.3.2 Growth Response in Bamboo-Based Agroforestry

In comparison to solo cropping, increased height, diameter, length, and number of bamboo culms were found in bamboo-based agroforestry. In the Eastern Himalayas region, it was found that bamboo cultivated in intercropping grows better than solo cropping. Crops like rice, cowpea, lady's finger, turmeric, and Colocasia that grew in between two rows of bamboo (*B. balcooa, B. tulda*) produced a much higher number of culms or clusters. The number of culms, internode length, culm diameter, and culm height were higher than those of solo bamboo (Banerjee et al. 2009). According to a related study by Dev et al. (2020), bamboo cultivated using the sesame-chickpea cropping system showed significantly more culms, internodes, and culm diameter than solo bamboo. According to Panday et al. (2014) the number of clumps and clums, as well as the height of the culm, increase as a result of the growth of ringal alongside farm crops.

In IHR, the growth of agricultural crops with bamboo yielded better than solo cropping. A study reported in Uttarakhand by Panday et al. (2014) that the local farmers' use of the ringal improved crop output and, on the other hand, household income and welfare. A study indicates that the juvenile stage of bamboo showed no significant crop yield in a bamboo-based agroforestry system (Banerjee et al. 2009). A related study in Ghana (Akoto et al. 2020) reported that bamboo (*B. balcooa*) agroforestry considerably increased soil moisture, pH, and crop yield levels in the third year, which showed that the yield of crops was not significant in juvenile bamboo. A study reported in Central India that the higher productivity observed

Table 23.1 Global status of bamboo-based agroforestry

Country	Number of studies	Number of species	Bamboo species	References
India	38	10	*Bambusa bambos, Dendrocalamus asper, Dendrocalamus Hamiltomi, Bambosa bamboos, Bambusa nutans and Dendrocalamus strictus, Dendrocalamus brandisii, Bambusa cacharensis, Bambusa tulda, Bambusa balcooa, Borassus flabellifer, Dendrocalamus giganteus, and Melocanna baccifera, Phyllostachys bambusoides, Bambusa vulgaris, Dendrocalamus longispathus, Dendrocalamus sikkimensis, Bambusa arundinacea, Sinarundinaria facuta, Thamnoclamus falconeri, Thamnoclamus spathiflora, and Thamnoclamus jaunsarensis*	Dev et al. (2020), Deb and Arunachalam (2005), Sharma et al. (2022), Seshadri (1985), Behari (2001), Kittur et al. (2016)
Ghana	3	6	*Bambusa vulgaris, Dendrocalamus, Guaduella and Oxytenanthera, Bambusa balcooa and Oxytenanthera abyssinica*	Garima et al. (2021), Akoto et al. (2018, 2020)

(continued)

Table 23.1 (continued)

Country	Number of studies	Number of species	Bamboo species	References
China	12	3	*Phyllostachys praecox, Phyllostachys edulis Phyllostachus Pubescens*	Bian et al. (2023), Shi et al. (2022), Zhang et al. (2020), Wang et al. (2023)
Ethiopia	3	1	*Yushania alpina*	Nigatu et al. (2020)
Vietnam	2	1	*Bambusa Bluemeana*	Nguyen (2004), Mailly et al. (1997)
Nepal	1	3	*Bambusa balcooa, Bambusa nutans and Bambusa bambos*	Banik et al. (2008)
Nigeria	1	3	*Bambusa vulgaris, Dendrocalamus, Guaduella and Oxytenanthera*	Igbokwe et al. (2016)
Indonesia	5	1	*Bambusa vulgaris*	Fauziyah et al. (2021), Purba et al. (2020)
Thailand	1	1	*Thyrsostachys siamensis*	Jha and Lalnunmawia; (2003)
Bangladesh	1	3	*Bambusa balcooa, Bambusa bambos, Bambusa nutans, Bambusa tulda, Bambusa vulgaris*	Banerjee et al. (2009), Riyadh et al. (2020)
Brazil	1	3	*Merostachys neesii bamboo Dendrocalamus asper and Dendrocalamus strictus*	Mesquita et al. (2023)
Philippines	1		Specific species of bamboo is not mentioned	Floresca (2015)
Venezuela	1	3	*Bambusa vulgaris, Guadua amplexifolia, and Guadua angustifolia*	Escalante et al. (2011)

Table 23.2 Status of bamboo-based agroforestry across Indian Himalayan Region

IHR states	Crops	Bamboo species	References
Assam	Oryza *sativa* (Rice)	*Bambusa cacharensis, Bambusa vulgaris* and *Bambusa balcooa*	Jyoti et al. (2015)
Arunachal Pradesh	*Manihot esculenta* (Cassava), *Colocasia esculenta* (Colocasia), *Capsicum annuum* (Capsicum), *Zingiber officinale* (Ginger), *Solanum tuberosum* (Potato), *Ananas comosus* (Pineapple), *Musa acuminata* (Banana)	*Bambusa tulda* and *Dendrocalamus hamiltonii*	Deb and Arunachalam (2005), Tangjang and Nair (2016)
Himachal Pradesh	*Zingiber officinale* (Ginger)	*Dendrocalamus asper, Bambusa balcooa*	Sharma et al. (2022)
Mizoram	*Zingiber officinale* (Ginger), *Glycine max* (Soya bean)	*Dendrocalamus hamiltonii, Dendrocalamus longispathus* and *Bambusa tulda, Melocanna baccifera* and *Dendrocalamus longispathu*	Jha and Lalnunmawia; (2003)
West Bengal	*Oryza sativa* (Rice), *Arachis hypogaea* (Groundnut), *Cajanus cajan* (Pigeon pea), *Vigna unguiculata* (Cowpea), *Abelmoschus esculentus* (Lady's finger), *Lagenaria siceraria* (Bottle gourd), *Curcuma longa* (Turmeric), *Colocasia esculenta* (Colocasia), and *Amorphophallus paeoniifolius* (Elephant foot yam)	*Bambusa tulda and Bambusa balcooa*	Bhattacharjya et al. (2017)
Uttarakhand	*Aegle marmelos* (Bael), *Citrus limon* (Citrus spp.), *Moringa oleifera* (Moringa), *Azadirachta indica* (Neem) and *Albizia lebbeck* (Siris), *Phyllanthus emblica* (Aonla), *Melia azedarach* (Bakain), *Musa acuminata* (Banana), *Areca catechu* (Betel nut)	*Bambusa balcooa, Bambusa nutans, Bambusa tulda, and Dendrocalamus hamiltonii, Sinarundinaria facuta, Thamnoclamus falconeri, Thamnoclamus spathiflora, and Thamnoclamus jaunsarensis*	Tewari et al. (2015)
Sikkim	Oryza *sativa* (Rice), *Zea mays* (Maize)	*Dendrocalamus hamiltonii and Dendrocalamus sikkimensis*	Banik et al. (2008)

in the bamboo-based agroforestry system was primarily due to the utilization of the intercropping space for sesame and chickpea production (Dev et al. 2020). Integration of bamboo with croplands was confirmed to be a suitable technique to increase the productivity of food crops and non-food biomass (Mailly et al. 1997).

23.3.3 Effect of Canopy

It was found that the bamboo canopy increased the yield and showed better growth of shade-loving species. According to the study, *D. asper* showed better results for ginger yield traits than *B. balcooa*. Additionally, the soil moisture content under the canopies of bamboo species was higher than sole cropping (Sharma et al. 2022). It may be because *D. asper* canopy provided shade and reduced the photoactive radiation for the ginger crop, and ginger being a shade-loving species, its yield is much higher than solo cropping. When bamboo canopies were present, the ginger crop yielded better than monocropping (Garima et al. 2021), which showed that intercropping ginger with bamboo species promotes a good microclimate for its healthy growth and the crop's preference for shade. Therefore, we can say that *D. asper* provides a better yield for shade-loving crops.

23.3.4 Effect of Soil Property

Soil plays a major role in sequestering carbon, as the highest carbon stock is stored in the soil organic carbon pool (FSI 2021). The perusal of the data shows that soil nutrients in bamboo-based agroforestry were found to be higher than those in solo cropping, and available N, K, and P also increased significantly in bamboo-based agroforestry. However, SOC density was greatly improved by increased nutrients. In comparison to a solo crop, bamboo-based agroforestry had more nutrients (N, P, and K) due to leaf litter and the fine root decomposition of bamboo (Singh et al. 2014; Rao and Ramakrishan 1989). The research conducted by Jagadish et al. (2015) on bamboo litter decomposition and nutrient release has also validated the increased soil nutrients in the order of N > P > K. According to related studies, decomposing bamboo leaf litter releases nutrients that are advantageous to intercrops and bamboo (Baruah and Borah 2019). A study recommended that bamboo litter can be employed in agroforestry systems to regulate soil nutrients (Deb et al. 2005). As reported by Deb et al. (2005), *B. tulda* soil moisture content was higher than that of *D. hamiltonii*. Due to its lower moisture content, *B. tulda* was suitable for nutrient enrichment in traditional agroforestry. The available N, P, and K were filled in the following order: *D. asper* > *D. hamiltoni* > solo cropping. The relative increased leaf litter production and its deposition may be responsible for the higher status of the nutrients under *D. asper*.

Bamboo-based agroforestry shows higher soil organic carbon content as compared to solo cropping. A related study by Sharma et al. (2021) reported that ginger output increased by 10.5–15.6% when intercropped with *D. asper and D. hamiltonii. D. asper* with ginger showed greater organic carbon than *D. hamiltonii, as D. asper* showed greater litter fall, due to which the organic carbon increased in the soil. But a similar study in Solan reported that *D. asper* with ginger showed less organic carbon than *B. balcooa* because the bulk density of *B. balcooa* was lower than *D. aper* (Sharma et al. 2022). In terms of soil type, north-east farmers consider black soil to be the most advantageous since it has a higher organic carbon content and is more fertile (Jyoti et al. 2015). The significance of soil moisture and nutrient availability (Wang et al. 2023) and the potential of bamboo under agroforestry management techniques highlighted that a dense network of fine roots, a high rate of nitrogen fixation, the absence of hazardous materials in the litter or root wastes, and a potential for profuse mycorrhizal association are all characteristics of bamboo (Igbokwe et al. 2016).

23.3.5 Effect of Above-Ground Biomass

Bamboo-based agroforestry showed a temporal change in above-ground biomass. It was found that above-ground biomass carbon increases over time, and the increased rate was recorded more in fast-growing species. We can explain this by the fact that fast-growing bamboo species show more litterfall than slow-growing bamboo. According to a study in the eastern Himalayan region from 2004 to 2007, there was a clear trend in the rise of C estimates for the leaves, branches, and culm components. As new culm increased continuously throughout the years, the increased culm density of *B. cacharensis* was attributed to the increased aboveground carbon stock. As reported by Nath et al. (2011) the three species, i.e., *B. cacharensis* contributed 36–38%, *B. vulgaris* 34–36%, and *B. balcooa* 28–30%, to the total carbon stock of bamboo in home gardens. The potential of bamboo-based Agroforestry to restore fallow land was higher than that of tree species. A study reported by Rao and Ramakrishan (1989) in the Eastern Himalayas of Garo, Khashi, and Naga Hills was an experiment performed in fallow land where bamboo-based agroforestry shows greater above-ground biomass than trees, shrubs, and herbs. Above-ground biomass carbon increases continuously at 10 years of fallow age; 15 years follow. As reported by the increase in biomass due to the annual accumulation of N, P, and K in bamboo-based agroforestry, nutrients increased sharply in follow land between 5 and 10-years-old. Enrichment ratios of nitrogen in the total vegetation and in the bamboo, component increased sharply in follow land between 5 and 10. The potassium concentration was lower due to quick collection by the more dominant *D. hamiltonii* and slow release by litterfall.

23.3.6 Effect of Spacing in Bamboo Growth and Crop Yield

The yield of the crop and the growth of bamboo in terms of height, number of culms, diameter, and length of culms were both influenced by spacing. The most common spacing in bamboo-based agroforestry that has gained popularity is 8 m × 8 m, 10 m × 10 m, and 12 m × 12 m. In IHR, when bamboo was intercropped with closer spacing, the yield of the crop was less as compared to wider spacing, whereas the growth of bamboo showed the highest growth in close spacing. This result corroborates many related studies. A study in Jhashi by Dev et al. (2020) found that bamboo cultivated with the sesame-chickpea cropping system showed a considerable increase in longer culms, internodal length, culm diameter, and quantity of culms than solo cropping. A similar result was reported by Kittur et al. (2016): smaller spacing results in a higher height of bamboo species, whereas larger planting spacing was recorded to be slower height growth of bamboo. This is explained by the fact that intercrops use sunshine, space, moisture, and nutrients more efficiently when there is less competition between them, ultimately increasing the yield of crops. According to Kittur et al. (2016) bamboo roots compete with understory crops at different spacings ranging from 4 m × 4 m to 8 m × 8 m. Wider spacing was beneficial for the growth characteristics, especially at 8 m by 8 m, 10 m × 10 m, and 12 m × 12 m. Due to less root competition, understorey turmeric absorbed more N, P, and K. In the north-east, crops such as rice, groundnut, pigeon peas, and cowpea were interplanted at 10 m × 12 m and 10 m × 10 m with *Bambusa tulda and B. balcooa* (Banerjee et al. 2009). Additionally, Dev et al. (2020) reported that higher productivity was observed in the bamboo-based agroforestry system, primarily due to the utilization of the intercropping space for sesame and chickpea production. Another study carried out in Ghana found that *Bambusa balcooa* intercropped with maize, cassava, or cowpea at a 5 m × 5 m spacing produced more than monocropping. (Akoto et al. 2020; Tangjang et al. 2016) reported that *Phyllostachys bambusoides* with *P. wallichiana* at 4 m × 4 m spacing provided ecosystem services to maximize carbon sequestration. The growth of *Dendrocalamus hamiltonii, D. longispathus, and Bambusa tulda* was not significant in the initial years, as the bamboo clumps were established only after the second year, after which the effect of spacing became significant (Jha and Lalnunmawia 2003). However, it revealed that crop yield is almost unaffected by the juvenile bamboo.

23.3.7 Land Use with Bamboo

According to the review, 53% of bamboo-based agroforestry in IHR was used on agricultural land, followed by degraded land (26%), home hardening (10%), and tree species (11%). In agricultural land, bamboo was used mostly for intercropping. According to a survey by Tangjang et al. (2016), *Phyllostachys bambusoides* with *P. wallichiana* agroforestry in Arunachal Pradesh can offer ecosystem services to maximize carbon sequestration and produce renewable biomass as carbon–neutral energy. A similar study reported in Vietnam by Nguyen et al. (2004) found that incorporating Acacia *mangium* or *Tephrosia candida* with *Bambusa blumeana* decreased erosion rates in comparison to other agroforestry systems. Bamboo in home gardens is practiced in Assam (Nath et al. 2015). In Nepal, bamboos were cultivated around the margins of agricultural fields to protect crops from high-speed wind and soil erosion (Banik et al. 2008).

Bamboo-based agroforestry was used to restore degraded land. Many studies have reported the ecological significance of bamboo-based agroforestry in the restoration of degraded areas, particularly on hilly slopes (Pande et al. 2012). Jha and Lalnunmawia (2003) reported that north-east India uses an agroforestry model with bamboo and ginger to restore degraded lands. Even in an environment of competition, *B. vulgaris* and *D. strictus* may grow quickly and do well in a system of agroforestry on poor soils (Igbokwe et al. 2016), but they show resiliency in water-stressed circumstances (Nirala et al. 2018). Findings in Africa suggest that we can maximize bamboo's potential for recovering degraded land and maximizing the carbon sink by introducing it into mixed-use agroforestry complexes (Partey et al. 2017). Farmers must grow bamboo along the boundaries of land to reduce erosion, as recommended by Fauziyah et al. (2021). A study was conducted in China to look at the impact of Lei bamboo (*Phyllostachys praecox*) on soil that had been contaminated with heavy metals (Cd, Zn, and Cu). The results showed that bamboo roots removed heavy metals (Cd, Zn, and Cu) from the rhizosphere soil and transferred them in significant amounts to the stems, and amounts of both Cu and Cd were removed from agroforestry, which were 1.4 times greater than those in pure forest (Bain et al. 2023). The bamboo-based agroforestry system had the highest cumulative CH_4 uptake in the spring. According to the study, bamboo-based agroforestry showed potential for methane reduction. Additionally, it shed light on the microbiological processes that explained the observed variations in GHG fluxes between the different agroforestry systems (Shi et al. 2022). In Ethiopia, agricultural sectors and institutions are experimenting with bamboo plantation lands for gully restoration and rehabilitation of degraded areas (Nigatu et al. 2020). In India, 96.3 million ha of land have been degraded, which is 29.32% of the nation's entire geographic area (Negi et al. 2022). Since bamboo species shows adaptable potential in degraded land, a considerable role in soil conservation, and a fast growth rate, it should be used in degraded agricultural land in hilly areas in IHR.

23.3.8 Issues and Challenges in Bamboo Based Agroforestry

Bamboo is not prioritized for cultivation in the agriculture and forestry subsectors of the economy since it is categorized as a minor forest product, lacking the necessary technologies to process bamboo. Studies related to bamboo-based agroforestry are limited, especially in the North-Western Himalayan region and surrounding areas, and installing bamboo-based agroforestry models in hilly areas is a challenging part.

23.4 Conclusion and Future Prospects

In this study, we reviewed 68 research documents found on Scopus database and other sources like Google Scholar, Research gate, etc. We documented the distribution of bamboo-based agroforestry globally and in the IHR. Additionally, bamboo species paired with agriculture crops across IHR were documented. These studies encompass various states such as Assam, Arunachal Pradesh, Himachal Pradesh, Mizoram, West Bengal, Uttarakhand, Sikkim, and Meghalaya.

We found that the yield of crops and the growth of bamboo both was higher in mix cropping than solo cropping. Integrating bamboo with crops significantly enhanced the nutrients, soil organic carbon, above-ground biomass, physiochemical properties, and yield of crops. Various attributes like spacing, litterfall, canopy, and suitable species selection for agroforestry combinations played a major role in bamboo-based agroforestry systems. It was found that bamboo-based agroforestry played a major role in the restoration of degraded land and secondary succession. This shows that bamboo-based agroforestry has the potential to increase the productivity and carbon sequestration, and can be used to meet the need for timber requirement. Based on the study, following are the key recommendations for bamboo based agroforestry in the Indian Himalayan region especially in the North-Western states.

- It is important to identify the combination of crop species that would be suitable to grow with bamboo species, e.g., shade-loving species like ginger and turmeric and shade-tolerant species like cowpea, which should be brought under high canopy bamboo species like *Dendrocalamus asper* and *D. hamiltonii,* as they performed better under shade conditions. For light-demanding species like pigonpea, rice should be brought under a low canopy or draft bamboo like Ringal (*Sinarundinaria falcata*, etc.).
- It is important to incorporate a suitable spacing pattern that is suitable for both crop and bamboo growth. The better bamboo growth was observed in a large spacing i.e. 12 m x 10 m. In hilly districts of IHR, bamboo should be preferred in bio fencing rather than intercropping because of small land holdings and terrace farming lands, for better crop productivity.
- An extensive survey of bamboo-based agroforestry is needed to identify it in the western Himalayan region, and the sequestration potential of bamboo-based agroforestry in the IHR, which is still underutilized due to a lack of documentation.
- The current documentation of bamboo-based agroforestry in IHR could help policymakers to enhance their recognition for of land degradation in REDD+. It can also be taken into consideration for CDM projects under the Kyoto Protocol and ensuring proper utilization, management, and innovative strategies, which can become an effective carbon sink. In addition to supporting the various needs of livelihood in IHR, primarily in the Noth-Western Himalayas.

Acknowledgements The authors express their gratitude to the Director of the G.B. Pant National Institute of Himalayan Environment (GBPNIHE) for their encouragement and support. They also acknowledge Dr. I.D. Bhatt and Dr. Aseesh Pandey for their invaluable assistance during the study, the inputs received from them greatly contributed to improving the content of the manuscript. Colleagues at GBPNIHE are thanked for their cooperation and assistance during the study.

References

Ahirwal J, Nath A, Brahma B, Deb S, Sahoo UK, Nath AJ (2021) Patterns and driving factors of biomass carbon and soil organic carbon stock in the Indian Himalayan region. Sci Total Environ 770:145292

Akoto DS, Denich M, Partey ST, Frith O, Kwaku M, Mensah AA, Borgemeister C (2018) Socioeconomic indicators of bamboo use for agroforestry development in the dry semi-deciduous forest zone of Ghana. Sustainability 10(7):2324

Akoto DS, Partey ST, Denich M, Kwaku M, Borgemeister C, Schmitt CB (2020) Towards bamboo agroforestry development in Ghana: evaluation of crop performance, soil properties and economic benefit. Agrofor Syst 94:1759–1780

Aynekulu E, Suber M, Van Noordwijk M, Arango J, Roshetko JM, Rosenstock TS (2020) Carbon storage potential of silvopastoral systems of Colombia. Land 9(9):309

Banik RL, Tewari S, Kaushal R (2008) Bamboo in homestead and agroforestry system of India. In: Kundu SS, Dagar JC, Chaturvedi OP, Sirohi SK (eds) Environment, agroforestry, and livestock management. International Book Distributing, Lucknow, New Delhi, pp 351–365

Banerjee H, Dhara PK, Mazumdar D (2009) Bamboo (Bambusa spp.) based agroforestry systems under rainfed upland ecosystem. J Crop Weed 5(1):286–290

Baruah A, Borah IP (2019). Bamboo-based agroforestry for sustainable utilization in land under shifting cultivation of Assam. Indian Forest 145(1):15–22

Bargali H, Pandey A, Bhatt D, Sundriyal RC, Uniyal VP (2024) Forest fire management, funding dynamics, and research in the burning frontier: a comprehensive review. Trees For People 100526

Behari B (2001) Agroforestry model of bamboo cultivation on degraded agricultural land. Doctoral dissertation, Ph. D. thesis. Guru Ghasidas University, Bilaspur

Bhattacharjya S, Bhaduri D, Chauhan S, Chandra R, Raverkar KP, Pareek N (2017) Comparative evaluation of three contrasting land use systems for soil carbon, microbial and biochemical indicators in North-Western Himalaya. Ecol Eng 103:21–30

Bian F, Zhang X, Zhong Z, Wen X, Xiu C, Li Q, Huang Z (2023) Introducing sedum affects root-soil interface phytoremediation of heavy metals in lei bamboo forest and potential risks from edible bamboo shoots. Land Degrad Dev 34(6):1820–1829

Deb S, Arunachalam A, Arunachalam K (2005) Cell-wall degradation and nutrient release pattern in decomposing leaf litter of Bambusa tulda Roxb. and Dendrocalamus hamiltonii Nees. in a bamboo-based agroforestry system in north-east India. J Bamboo Rattan 4(3):257–277

Dev I, Ram A, Ahlawat SP, Palsaniya DR, Singh R, Dhyani SK, Prasad J (2020) Bamboo-based agroforestry system (Dendrocalamus strictus + sesame–chickpea) for enhancing productivity in semi-arid tropics of central India. Agrofor Syst 94:1725–1739

Devi AS, Singh KS (2021) Carbon storage and sequestration potential in aboveground biomass of bamboos in North East India. Sci Rep 11(1):837

Escalante E, Guerra A, Martínez R, Piñuela A (2011) The multispecies agroforestry system of the DANAC Foundation in tropical dry forest landscapes of Yaracuy, Venezuela (a case study). Agroforestry as a tool for landscape restoration. Nova Science Publishers Inc., New York, USA, pp 69–81

Fauziyah E, Widyaningsih TS, Kuswantoro DP (2021) Management of village treasury land based on bamboo agroforestry. In: IOP conference series: earth and environmental science, vol 940, no 1. IOP Publishing, p 012007

Floresca JP (2015) GIS-based crop suitability and climate change ulnerability of farming systems in Cagayan Valley. In: Proceedings of the 36th Asian conference on remote sensing: fostering resilient growth in Asia (ACRS 2015)
Forest Survey of India (2021) India state of forest report 2021. Ministry of Environment, Forest & Climate Change, Government of India. https://fsi.nic.in/forest-report-2021
Garima B, D. R., Thakur, C. L., Kaushal, R., Sharma, P., Kumar, D., & Kumari, Y. (2021) Bamboo-based agroforestry system effects on soil fertility: ginger performance in the bamboo subcanopy in the Himalayas (India). Agron J 113(3):2832–2845
Gu L, Zhou Y, Mei T, Zhou G, Xu L (2019) Carbon footprint analysis of bamboo scrimber flooring—implications for carbon sequestration of bamboo forests and its products. Forests 10(1):51
Igbokwe GO, Bello AG (2016) Application of agroforestry techniques in the management of bamboo. Int J Sustain Agricul Res 3(1):29–34
Intergovernmental Panel on Climate Change (IPCC) (2023) Sixth assessment report cycle. https://www.ipcc.ch/report/sixth-assessment-report-cycle/
ISFR (India State of Forest Report) 2021 by Forest Survey of India. Ministry of Environment and Forests & Climate Change, Government of India. Dehradun, India. Retrieved December 14, 2024, from, Available online: https://fsi.nic.in/forest-report-2021
Jagadish MR, Biradar PI, Nethravathi B, Parvathy S, Anil kumar KS, Vishvanath S (2015) Nutrient dynamics through litter fall and decomposition in bamboo agroforestry systems in humid tropics of Karnataka, India. Ecoscan (Special issue) 7:497–501
Jha LK, Lalnunmawia F (2003) Agroforestry with bamboo and ginger to rehabilitate degraded areas in North East India. J Bamboo Rattan 2(2):103–109
Jyoti NA, Lal R, Das AK (2015) Ethnopedology and soil quality of bamboo (Bambusa sp.) based agroforestry system. Sci Total Environ 521:372–379
Kittur B, Sudhakara K, Mohan Kumar B, Tk K, Sureshkumar P (2016) Bamboo based agroforestry systems in Kerala, India: performance of turmeric (Curcuma longa L.) in the subcanopy of differentially spaced seven year-old bamboo stand. Agrofor Syst 90:237–250
Kumar V, Parihar RD, Sharma A, Bakshi P, Bali AS, Rodrigo-Comino J (2019) Global evaluation of heavy metal content in surface water bodies: a meta-analysis using heavy metal pollution indices and multivariate statistical analyses. Chemosphere 236:124364–124364
Kumar S, Rawat D, Singh B, Khanduri VP (2023) Utilization of bamboo resources and their market value in the Western Himalayan Region of India. Adv Bamboo Sci 100019
Kumar M (2021) Bamboo species of district Bilaspur of Himachal Pradesh, India. J Plant Sci Res 8
Kumar A, Pandey A (2024) A bibliometric analysis of 100 years of research on Himalayan cedar: research trends, gaps, and future implications. Front For Glob Chang 7:1485083
Li T, Cui L, Xu Z, Hu R, Joshi PK, Song X, Tang L, Xia A, Wang Y, Guo D, Zhu J, Hao Y, Song L, Cui X (2021) Quantitative analysis of the research trends and areas in grassland remote sensing: a scientometrics analysis of Web of Science from 1980 to 2020. Remote Sens 13:1279
Mailly D, Christanty L, Kimmins JP (1997) Without bamboo, the land dies: nutrient cycling and biogeochemistry of a Javanese bamboo talun-kebun System. For Ecol Manag 91:155–173
Manandhar R, Kim JH, Kim JT (2019) Environmental, social and economic sustainability of bamboo and bamboo-based construction materials in buildings. J Asian Archit Build Eng 18(2):49–59
Mesquita M, Battisti R, de Araújo DS, de Moraes DHM, de Araújo Almeida R, Flores RA, Salvador PRI (2023) Bamboo species, size, and soil water define the dynamics of available photosynthetic active solar radiation for intercrops in the Brazilian savanna biome. Adv Bamboo Sci 3:100025
Moher D, Liberati A, Tetzlaff J, Altman DG, PRISMA Group T (2009) Preferred reporting items for systematic reviews and meta-analyses: the PRISMA statement. Ann Intern Med 151(4):264–269
Nath AJ, Das AK (2008) Bamboo resources in the homegardens of Assam: a case study from Barak Valley. J Trop Agric 46:58–61
Nath A, Das A (2011) Carbon storage and sequestration in bamboo-based smallholder homegardens of Barak Valley, Assam. Curr Sci 100:229–233
Nath AJ, Das G, Das AK (2009) Above ground standing biomass and carbon storage in village bamboos in North East India. Biomass Bioenerg 33(9):1188–1196

Nath AJ, Lal R, Das AK (2015) Ethnopedology and soil properties in bamboo (Bambusa sp.) based agroforestry system in North East India. CATENA 135:92–99

Nath AJ, Kumar R, Devi NB, Rocky P, Giri K, Sahoo UK, Pandey R (2021a) Agroforestry land suitability analysis in the Eastern Indian Himalayan region. Environ Chall 4:100199

Nath AJ, Sileshi GW, Laskar SY, Pathak K, Reang D, Nath A, Das AK (2021b) Quantifying carbon stocks and sequestration potential in agroforestry systems under divergent management scenarios relevant to India's nationally determined contribution. J Clean Prod 281:124831

Nath AJ, Das AK (2009) Carbon farming through village bamboos in rural landscape of Northeast India as affected by traditional harvest regimes. Vol 7 Bamboo Commun Econ Dev 13

Negi VS, Pathak R, Dhyani V, Durgapal M, Joshi RK, Bhatt ID (2022) Land restoration in the Himalayan Region: Steps towards biosphere integrity. Land Use Policy 121:106317

Nguyen L (2004) Bamboo, its filter effect in different agroforestry systems and its role in the household economy in Northern Vietnam. Master of Science Thesis. Sveriges Lantbruks, Universitet (SLU), Uppasala, Swedia

Nigatu A, Wondie M, Alemu A, Gebeyehu D, Workagegnehu H (2020) Productivity of highland bamboo (Yushania alpina) across different plantation niches in West Amhara, Ethiopia. For Sci Technol 16(3):116–122

Nirala DP, Kumar J, Ahmad MS, Kumari P (2018) Bamboo based agroforestry system for livelihood and ecological security in North Chhotanagpur division of Jharkhand. J Pharmacogn Phytochem 7(1S):1996–1999

Opoku D, Ayarkwa J, Agyekum K (2016) Factors inhibiting the use of bamboo in building construction in Ghana: perceptions of construction professionals. Mater Sci Appl 7(2):83–88

Pandey R, Dasgupta S, Kumar B, Todaria NP (2014) Bamboo in the courtyard: case study of Adoption of Hill Bamboo in Rudraprayag, Kedarnath valley, Uttarakhand

Pandey R (2012) Base paper for the committee to study development in hill states arising from management of forest lands

Partey ST, Sarfo DA, Frith O, Kwaku M, Thevathasan NV (2017) Potentials of bamboo-based agroforestry for sustainable development in Sub-Saharan Africa: a review. Agricul Res 6. https://doi.org/10.1007/s40003-017-0244-z

Purba JH, Manik IWY, Sasmita N, Komara LL (2020). Telajakan and mixed gardens landscape as household based agroforestry supports environmental aesthetics and religious ceremonies in Bali. In IOP Conference Series: Earth and Environmental Science 449(1):012041

Rao K, Ramakrishnan P (1989) Role of bamboos in nutrient conservation during secondary succession following slash and burn agriculture (Jhum) in North-East India. J Appl Ecol 26:625. https://doi.org/10.2307/2404087

Richa S, Sharma ML, N. K. (2012) Bamboos of Himachal Pradesh. Ind For 138(12):1141–1145

Riyadh ZA, Rahman MA, Miah MG, Saha SR, Hoque MA, Rahman MM, Miyajima I (2020) Performance of spices as lower-storey crop in jackfruit-papaya multistorey agroforestry system in Bangladesh. J Fac Agri Kyushu Univ 65(2):223–231

Rodrigo-Comino J, Seeling S, Seeger MK, Ries JB (2023) Light pollution: a review of the scientific literature. Anthrop Rev 10(2):367–392. https://doi.org/10.1177/20530196211051209

Sahoo UK, Singh SL, Gogoi A, Kenye A, Sahoo SS (2019) Active and passive soil organic carbon pools as affected by different land use types in Mizoram, Northeast India. Plos One 14(7):e0219969

Seethalakshmi KK, Jijeesh CM, Balagopalan M (2009) Bamboo plantations: an approach to carbon sequestration. In: National workshop on Global warming and its implications for Kerala, Department of Forests and Wildlife, Kerala Forest Research institute and Bamboo Development Agency, Thiruvananthapuram, Kerala. January, pp 19–21

Seshadri P (1985) Intercropping of bamboo (*D. strictus*) with Soybean—an agroforestry study. Ph. D. thesis, Tamil Nadu Agricultural University, Coimbatore 480 p

Sharma U, Bhardwaj DR, Sharma S et al (2022) Assessment of the efficacy of various mulch materials on improving the growth and yield of ginger (Zingiber officinale) under bamboo-based agroforestry system in NW-Himalaya. Agroforest Syst 96:925–940. https://doi.org/10.1007/s10457-022-00753-8

Sharma G, Bhardwaj DR, Thakur C, Kaushal R, Sharma P, Kumar D, Kumari Y (2021) Bamboo-based agroforestry system effects on soil fertility: Ginger (Zingiber officinale Rosc.) performance in the bamboo sub-canopy in the Himalayas (India). Agron J 113. https://doi.org/10.1002/agj2.20684

Shi M, Li Q, Zhang H, Sun J, Zhang J, Song X (2022) Agroforestry alters the fluxes of greenhouse gases of Moso bamboo plantation soil. Environ Res Lett 17(11):115003

Singh A, Gupta A (2016) Changing climate and glacio-hydrology in Indian Himalayan Region: a review. Wiley Interdisc Rev Clim Change 7(3):393–410

Singh R, Garg KK, Wani SP, Tewari RK, Dhyani SK (2014) Impact of water management interventions on hydrology and ecosystem services in Garhkundar-Dabar watershed of Bundelkhand region, Central India. J Hydrol 509:132–149

Stockmann U, Adams MA, Crawford JW, Field DJ, Henakaarchchi N, Jenkins M et al (2013) The knowns, known unknowns and unknowns of sequestration of soil organic carbon. Agric Ecosyst Environ 164:80–99

Sundriyal M, Joshi K (2015) Bamboo status and trade vulnerability: a central Himalayan case study. In: 10th World bamboo congress, Korea, pp 1–9

Tangjang S, Nair PKR (2016) Integrated bamboo + pine home-gardens: a unique agroforestry system in Ziro valley of Arunachal Pradesh, India. Int J Environ Agric Res 2(2):25–34

Tewari VP, Verma RK, Von Gadow K (2017) Climate change effects in the Western Himalayan ecosystems of India: evidence and strategies. For Ecosyst 4(1):1–9

Tewari S, Banik RL, Kaushal R, Bhardwaj DR, Chaturvedi OP, Gupta A (2015) Bamboo based agroforestry systems. ENVIS. Centre on Agroforestry Systems. http://www.frienvis.nic.in

Tripathi AK, Pandey PC, Sharma JK, Triantakonstantis D, Srivastava PK (2022) Climate change and its impact on forest of Indian Himalayan Region: a review. Clim Change Impacts, Responses Sustain Ind Himalaya 207–222

Verchot LV, Van Noordwijk M, Kandji S, Tomich T, Ong C, Albrecht A, Palm C (2007) Climate change: linking adaptation and mitigation through agroforestry. Mitig Adapt Strat Glob Change 12:901–918

Wang H, Cao X, Fan W, Deng P, Xu X (2023) Effects of intercropping systems of Phyllostachys edulis and Bletilla striata on soil bacterial community composition and function. Agrofor Syst 97(4):617–630

Xu J, Grumbine RE, Shrestha A, Eriksson M, Yang X, Wang YUN, Wilkes A (2009) The melting Himalayas: cascading effects of climate change on water, biodiversity, and livelihoods. Conserv Biol 23(3):520–530

Zhang X, Gao G, Wu Z, Wen X, Zhong H, Zhong Z et al (2020) Agroforestry alters the rhizosphere soil bacterial and fungal communities of moso bamboo plantations in subtropical China. Appl Soil Ecol 143:192–200

Zhang X, Lu J, Zhang X (2022) Spatiotemporal trend of carbon storage in China's bamboo industry, 1993–2018. J Environ Manag 314:114989

Correction to: Population Status of High-Value Medicinal Plants and Decadal Climate Variability in Niti Valley, Nanda Devi Biosphere Reserve, Western Himalaya

Arun Pratap Mishra, Himanshu Sahu, and Amit Kumar

Correction to: Chapter 4 in: D. Arya et al. (eds.), *Climate Change Impact on Himalayan Biodiversity*, Environmental Science and Engineering, https://doi.org/10.1007/978-3-031-77149-1_4

The original version of the book has been updated with the title "Population Status of High-Value Medicinal and Decadal Climate Variability in Niti Valley, Nanda Devi Biosphere Reserve, Western Himalaya" changed to "Population Status of High-Value Medicinal Plants and Decadal Climate Variability in Niti Valley, Nanda Devi Biosphere Reserve, Western Himalaya" in Chapter 4. The book has been updated with the change.

The updated version of this chapter can be found at
https://doi.org/10.1007/978-3-031-77149-1_4

D. Arya et al. (eds.), *Climate Change Impact on Himalayan Biodiversity*,
Environmental Science and Engineering, https://doi.org/10.1007/978-3-031-77149-1_24

Conclusion

In the diverse landscape of the Himalayan region, the impact of climate change is increasingly evident across various sectors, from forests to agriculture, biodiversity, and historical heritage. Forest degradation poses a significant threat to carbon sequestration, exacerbating the effects of greenhouse gas emissions on global climate patterns. Sustainable forest management practices emerge as a critical solution, offering a balance between conservation and resource utilization. Assimilating biodiversity conservation with sustainable development is imperative for the region's ecological health and socio-economic well-being. Efforts to conserve endemic species in Sikkim and catalog threatened plants in Uttarakhand highlight the importance of preserving unique ecosystems and traditional knowledge systems. Avifauna in the Ganges floodplain faces numerous challenges, including habitat loss and bird hunting, necessitating urgent conservation interventions to protect vital ecosystems and migratory bird species. The exploration of prehistoric forts not only presents an opportunity for historical tourism but also highlights the need for cultural preservation and economic diversification in the region. The vulnerability of lichen species to deforestation and forest fires emphasizes the interconnectedness of biodiversity conservation and ecosystem resilience. Research on medicinal plants reveals their significant therapeutic value and underscores the importance of indigenous knowledge in biodiversity conservation and healthcare practices. Climate change's far-reaching impacts on Himalayan ecosystems call for adaptive strategies and collaborative efforts to mitigate its effects and enhance resilience. Millet crops emerge as resilient alternatives amidst changing climate patterns, offering food security and sustainable livelihood options for local communities. Indigenous knowledge of medicinal flora in Arunachal Pradesh highlights the importance of traditional medicine in healthcare systems and the need for conservation efforts to preserve medicinal plant diversity. Population studies of medicinal plants in Niti Valley provide valuable insights into climate variability's effects on high-value species and the importance of data-driven conservation measures. Observations on crop cultivation using geo-informatics techniques underscore the importance of

D. Arya et al. (eds.), *Climate Change Impact on Himalayan Biodiversity*, Environmental Science and Engineering, https://doi.org/10.1007/978-3-031-77149-1

technology in monitoring and mitigating climate change impacts on agricultural productivity. Bryophytes' ecosystem services in urban environments demonstrate their potential in sustainable development and environmental monitoring, offering innovative solutions to urban environmental challenges. Scoping forest resources in the Western Himalayas emphasizes the importance of community-based conservation and market interventions to ensure the sustainable management of natural resources. Transitioning from traditional cropping patterns to alternative economic activities is essential for livelihood security amidst climate uncertainties, requiring innovative approaches and policy support to facilitate economic diversification and resilience-building initiatives. In conclusion, the book "Impact of Climate Change on Himalayan Biodiversity" provides a comprehensive understanding of climate change's effects on the region's ecosystems, highlighting the urgent need for adaptive strategies, conservation efforts, and community engagement to safeguard the Himalayas' rich biodiversity and cultural heritage for future generations.

Editor(S)

The manufacturer's authorised representative in the EU is Springer
Nature Customer Service Centre GmbH, Europaplatz 3, 69115 Heidelberg,
Germany. If you have any concerns regarding our products, please
contact ProductSafety@springernature.com

Printed and bound by CPI Group (UK) Ltd, Croydon, CR0 4YY
01/10/2026
02225932-0019